INTRODUCTION TO HAZARDOUS WASTE INCINERATION

INTRODUCTION TO HAZARDOUS WASTE INCINERATION

SECOND EDITION

Joseph J. Santoleri
Consultant

Joseph Reynolds
Louis Theodore
Manhattan College

WILEY-INTERSCIENCE

A JOHN WILEY & SONS, INC., PUBLICATION

New York • Chichester • Weinheim • Brisbane • Singapore • Toronto

This book is printed on acid-free paper. ♾

Published simultaneously in Canada.

For ordering and customer service, call 1-800-CALL-WILEY.

Library of Congress Cataloging-in-Publication Data:

Santoleri, Joseph J.
Introduction to hazardous waste incineration / Joseph Santoleri, Joseph Reynolds, Louis Theodore.—2nd ed.
p. cm.
Includes bibliographical references and index.
ISBN 0-471-01790-6 (cloth : alk. paper)
1. Hazardous wastes—Incineration. I. Reynolds, Joseph. II. Theodore, Louis.
III. Title.
TD796 .S26 2000
628.4′457—dc21 00-035917

Printed in the United States of America.

10 9 8 7 6 5 4 3 2 1

To Betty,
for your many years of faith and unselfish help
to me and our family while I have been deeply involved in this
and other projects

To Barbara,
My principal reviewer and
soulmate of 27 years

To the Lakes (formerly Apostolakos)
Harry (ole moneybags)
Anne (always reliable)
Margaret (did Charley ever get lucky)
Nora (a born-again RN)
Helen (my favorite cheerleader and CIA agent)

PREFACE

The engineering profession has expanded its responsibilities to include the management of not only general wastes but also hazardous wastes. One of the hazardous waste management options available is by incineration. Increasing numbers of engineers, technicians, and maintenance personnel are being confronted with problems in this most important area. The problem of hazardous waste has become a major concern in the 1990s. The environmental engineer of today and tomorrow must develop a proficiency and an improved understanding of the incineration of hazardous wastes in order to cope with these problems. Although this is not the first professional book to treat this particular subject, it is one of few books dealing with the technical and engineering aspects of waste incineration that may be used as a text. Although much of the material is concerned with the incineration of hazardous wastes, the application to other combustion/incineration systems is also considered.

This text–reference book is intended primarily for practicing engineers and engineering students. It is assumed that the reader has already taken basic courses in physics and chemistry and has a minimum background in mathematics through calculus. The authors' aim is to offer the reader the fundamentals of incineration with appropriate practical application to the incineration of wastes and to provide an introduction to the specialized literature in this and related areas. The readers are encouraged, through the references, to continue their own development beyond the scope of this book.

As is usually the case in preparing a book, the problem of what to include and what to omit has been particularly difficult. However, every attempt has been made to offer engineering course material to individuals with a technical background at a level that should enable them to better cope with some of the complex problems encountered in waste incineration today.

The book is divided into four parts: an overview of the hazardous waste field, incineration principles, equipment, and facility design. In Part I, the general subject of waste incineration is examined. A separate chapter is devoted to other options for controlling all wastes and in particular hazardous wastes. The recent history of standards and regulations are also reviewed including the recently enacted MACT standards for hazardous waste combustion. Part II covers the broad subject of incineration principles. This includes chapters on basic concepts, stoichiometric calculations, and thermochemical considerations.

Essentially, all of the material can be applied directly to the incineration of other wastes, including municipal solids, medical and infectious, sewage sludge, vent furnaces, and so forth. Much of this material is original and appeared in print for the first time in the first edition. Part III is highlighted by individual chapters devoted to equipment that may be found in a waste incineration facility. These include the incinerator, waste heat boiler, quench system, and air pollution control equipment. The last chapter examines some of the storage, materials handling, and ancillary equipment. Part IV contains three chapters devoted to design principles and their application to a hazardous waste incineration facility. The book concludes with two extensive illustrative examples detailing the design of waste facilities. A hazardous waste incineration software package (including disk) is described and documented in Appendix A. A short write-up on the International System of units (SI/metric) is also provided in Appendix B.

No engineering book is complete without information on how to use it. By the same token, no engineering text is complete without its illustrative examples, which serve the important purpose of demonstrating the use of the procedures, equations, tables, graphs, and so on, presented in the text. There are many such examples as well as practice problems at the end of each of the technical chapters.

During the preparation of this book, the authors were ably assisted in many ways by a number of practicing engineers with expertise in the fields of combustion, incineration, and air pollution control. We gratefully acknowledge their invaluable assistance; the names of these engineers are listed under the titles of the chapters to which they contributed (*contributing author* for those who played a major role in writing a part or several parts of the chapter and *partial contributor* for those who made a finite—though less major—contribution to the chapter).

The authors also wish to acknowledge the contributors to the first edition. The following were listed as contributors for these chapters. Chapter 1, Kent Murphy; Chapter 2, Elizabeth Girardi Schoen and Catherine Frega; Chapter 3, Madho Ramnarine Singh; Chapter 7, Raymond B. Bartone; Chapter 11, Abdool Jabar; Chapter 13, Benedetto Conetta; and Chapter 14, Rocco Grassi. The following were partial contributors: Chapter 4, Maritza Montesinos; and Chapter 5, Mitchell Summerfield.

Our sincere gratitude is due to Nancy L. Becker of RMT/Four Nines for her dedicated and patient efforts in assisting, revising, reformatting, and updating this manuscript.

Somehow the editor usually escapes acknowledgment. We were particularly fortunate to have Jim Smith of John Wiley & Sons serve as the editor for the first edition. He had the vision early on to realize the present need and timeliness for a project of this nature. He tolerated our mistakes and ignored our idiosyncrasies. His patient handling of the numerous problems that arose was particularly appreciated. It also would have been difficult to complete the second edition without the able direction of Jackie Kroschwitz (Publisher), Robert Esposito (Editor), and Danielle Lacourciere (Assistant Managing Editor).

JOSEPH J. SANTOLERI
JOSEPH REYNOLDS
LOUIS THEODORE

Berwyn, Pennsylvania

CONTENTS

I

HAZARDOUS WASTE PROBLEM

Part I serves the purpose of placing the subject of hazardous waste incineration in historical perspective and in perspective with respect to the field of hazardous waste management. Three chapters are included in this part.

Chapter 1 presents some of the historical background of the hazardous waste problem, which has been compounded over the last half-century and, as a result, now requires enormous efforts to solve. This chapter also covers the topic of where the hazardous waste problem stands today. Hazardous waste management is a rather broad field with many different solutions and techniques for the disposal of hazardous waste material, of which incineration is only one option.

Chapter 2, entitled *Other Options*, describes many of these methods. Some of them are alternatives to incineration and are sometimes used instead of incineration to handle a particular waste stream. Others are techniques that are used to process wastes that cannot, because of the physical or chemical properties of the waste, be treated by incineration.

The legal aspects of the hazardous waste problem (i.e., the standards and regulations imposed on industry by federal, state, and local governments), which have a significant impact on the methods industry uses to handle particular wastes, are covered in Chapter 3. The possible future of incineration is also discussed—a future determined just as much by legal standards as by social, economic, and technological considerations.

1

HAZARDOUS WASTE INCINERATION OVERVIEW

1.1 INTRODUCTION

Hazardous wastes are legally defined as those wastes that may cause adverse or chronic effects on human health or the environment when not properly controlled. Hazardous wastes are generated either because processes have converted harmless materials into hazardous substances or because natural materials that are hazardous to begin with have been concentrated and released into the environment. These substances may be ignitable, reactive, corrosive, radioactive, infectious, or toxic. They may exist as solids, liquids, sludges, powders, or slurries. About 90% of them are liquid or semiliquid. Some of these wastes are nondegradable and may persist in nature indefinitely.[1]

The hazardous waste issue is a by-product of this nation's economic growth over the past 40–50 years. As industry has expanded, society has developed a greater reliance on a broader range of chemicals and chemical by-products in plastics, automobiles, and other major manufacturing sectors. When chemicals are used to develop new products, wastes are almost always produced. As the economy has grown and as industries have diversified, the result has been the advent of many new waste streams. However, with waste minimization and recycling processes used in many industries today, a more concentrated waste stream residue and solids are often left for final disposal.

Table 1.1.1 shows the type of wastes generated by various industries as well as the usual methods of disposing of these wastes.[2] Some industries produce waste streams that are typical of the industry, for example, textile, pulp and paper, leather tanning, and plating wastes. For a given industry, the characteristics of these waste streams do not vary much from facility to facility. Other industries have highly variable waste streams; an example is the chemical

Table 1.1.1 Hazardous material use, production, and waste by industry

	Properties of Products Used[a]					Properties of Products Made[a]					Toxic Materials Used[b]					Toxic Materials Produced[b]					Disposal Methods[c]				Toxic Wastes Produced[b]				
Industry	C	E	F	R	P	C	E	F	R	P	S	H	G	O	I	S	H	G	O	I	DW	SEA	CON	AEC	S	H	G	O	I
Mining	×	×		×										×	×											×	×		×
Food			×							×				×															
Textiles	×		×		×						×	×		×												×		×	
Paper and allied products	×		×																			×					×		×
Alkalis and chlorine						×						×						×		×		×							×
Cyclic intermediates	×	×	×			×	×	×			×	×	×	×	×	×		×	×		×	×	×		×	×	×	×	×
Organic chemicals	×	×	×			×	×	×			×	×	×	×	×	×		×	×		×	×	×		×	×	×	×	×
Inorganic chemicals	×		×			×	×	×			×	×		×	×		×			×	×	×	×			×	×	×	×
Plastic materials	×	×	×				×	×			×		×	×	×				×		×	×	×		×		×	×	
Drugs	×		×	×	×				×		×	×		×	×		×		×	×			×	×			×	×	×
Soaps and cleaners	×		×			×					×			×	×				×									×	
Paints and allied products			×					×			×				×	×			×	×		×			×			×	×
Agricultural chemicals	×	×	×				×	×			×	×		×	×			×	×	×	×	×	×		×			×	×

Explosives	×					×	×		×			×	×			×	×			×					×	×
Petroleum and coal products	×	×	×			×	×		×	×	×	×	×	×	×	×	×	×	×	×		×	×	×	×	×
Leather tanning	×		×						×	×		×	×										×		×	×
Asbestos products													×				×							×		×
Blast furnaces and steel	×											×	×							×			×	×	×	×
Nonferrous metals	×											×	×				×							×		×
Metal services	×		×						×	×		×								×		×	×			×
Department of Defense	×	×	×	×		×	×	×	×	×		×	×		×			×	×		×	×	×	×	×	×
Atomic Energy Commission	×			×				×				×	×					×		×			×	×	×	×
National Aeronoutics and Space Administration	×	×	×			×	×		×		×	×						×		×		×		×	×	
Consumers	×	×	×	×	×				×	×		×	×							×		×	×		×	×

[a]Hazard code: C = corrosive, E = explosive, F = flammable, R = radioactive, P = pathogenic.
[b]Toxicity code: S = toxic solvents, H = toxic metals, G = toxic gases, O = toxic organics, I = toxic inorganics.
[c]Disposal code: DW = deep well, SEA = deep sea dump, CON = contract disposal, AEC = radioactive disposal.

industry where over a thousand different products may be produced at a single facility.

It is estimated that 264 million metric tons of hazardous waste were generated in 1981, a quantity equal to more than 70 billion gallons. These are enormous numbers, although large portions of the total were actually mixtures of hazardous and nonhazardous wastes, such as wastes mixed with industrial process liquids. Since 1981, the amount of hazardous waste generated annually has increased. About 14,000 installations generate regulated quantities of hazardous waste, while 4800 companies manage these wastes using various treatment, storage, and disposal (TSD) techniques. The overwhelming fraction of hazardous waste generated in this country, 96%, is managed on-site. Of this, about 14.7 billion gallons are disposed of in or on the land each year and about 500 million gallons are incinerated.[3]

Hazardous waste generators and management facilities are concentrated in the manufacturing industries. An estimated 85% of all generators and 72% of all treatment, storage, and disposal facilities are associated with industrial manufacturing operations. Of the total quantity of wastes generated, manufacturers account for 92%; the chemical industry alone generates about 68% of the total.[2]

1.2 HISTORICAL BACKGROUND

Wastes that are now identified as *hazardous* have been around for a long time. The present-day problems with hazardous wastes seem to be fairly recent, however. Some reasons for this follow:

- *Expansion* Urban centers are more concentrated than they were years ago; the population has increased; industry in general, and the chemical industry in particular, have greatly expanded.
- *Reaction of the Environment* It takes time for the land to react to abuse, and there are fewer places to leave wastes around without causing problems.
- *Public Awareness* Above all, society has become much more aware of the health and environmental dangers associated with hazardous wastes.

To fully understand the reasons why there is a hazardous waste problem today, it is necessary to examine the past. The U.S. government has generally been inclined to leave industry alone and avoid overregulation. This national attitude of freedom has been clearly reflected in a reluctance to regulate and enforce health and environmental issues, either on a state or federal level. Prior to World War II, there were no environmental regulatory programs. What federal and state activities there were can best be described as *advisory* and mildly *persuasive*. No authority existed to ensure that good waste management was practiced. Rivers handled the liquid wastes, and the open burning dump took care of the solids. With the end of World War II and the economic prosperity that followed, the amounts and types of wastes produced by society expanded exponentially. At first, waste streams were handled in the same manner as before the war when lesser amounts of waste were produced. The same technologies were used to treat the waste and the same institutional structures were employed to oversee their management. For the most part, these wastes ended up in the streams and skies; the balance went into the land.

Substantive water and air pollution control legislation was not passed until as late as the 1950s and 1960s. (The history of solid and hazardous waste legislation is covered in detail in Chapter 3.) These landmark efforts by the federal government led to the development of federal and state regulatory programs to require adequate treatment of wastewater and waste airstreams before they were discharged into the ambient environment. However, as wastewater treatment plants and air pollution control systems were built, the residual wastes produced by such efforts were diverted to the land. At that time, there was no federal legislation to protect the land because few people considered it necessary. Some protection was provided by a myriad of controls emanating from local governments. Local control, however, was normally directed at regulating the use of land and not preserving its quality. Consequently, there were almost no restrictions on diverting concentrated waste streams resulting from air and water cleanup to the land.

The technology for land disposal of wastes was not adequate for the types of wastes received. The use of a sanitary landfill was not established until after World War II, and even then the ability of the landfill to handle liquids, sludges, and persistent chemicals in hazardous wastes was not adequate. Pits, ponds, and lagoons utilized for these wastes were also inadequate. More sophisticated technologies, such as incineration, were not considered because the less expensive option of using the land remained.[4]

Technology is available today to render hazardous wastes sufficiently innocuous so that they can be disposed of in ways that will not create unreasonable risks to public health and the environment. There are many methods for handling process waste streams. With the exception of incineration, these are covered in Chapter 2. Of these, seven are quite commonly used, although not all are suitable for the disposal of hazardous wastes. Brief descriptions are given in the following list:

1. *Incineration* This involves oxidative conversion of combustible materials to harmless gases suitable for atmospheric release. Undesired gaseous products such as HCl, SO_x, NO_x, and Cl_2 must be removed prior to release. (*Note*: the symbol NO_x represents various oxides of nitrogen.) Heats of combustion may sometimes be recovered, and occasionally additional combustible material must be added to ensure adequate combustion. Residual solids, such as ash, are landfilled. Any scrubber liquids are usually treated in a wastewater treatment plant.
2. *Dispersal into Contiguous Water Bodies* Historically this procedure has been widely employed, but the Environmental Protection Agency (EPA)

action has now curtailed it in many industries. Usually the waste solid is temporarily stored in a lagoon, wherein it is treated to minimize environmental impact. Overflow from lagoons is treated in a wastewater treatment plant prior to discharge to a body of water, usually a river. For example, acids and bases are neutralized and suspended solids are permitted to settle. In the case of river discharge, attempts are made to release waste during periods of high flow.

3. *Ocean Dumping* This practice involves transporting the material out to sea and then releasing it. Usually the material is conveyed by barge. It may be released directly or within containers. Costs depend on distance of transport, and whether the material can be loaded in-plant or must be intermediately transported. Materials disposed in this manner have been restricted and, as a result, mileage to disposal points have increased.
4. *Lagooning* Lagoons may be employed for either temporary or permanent storage. Permanent storage occurs when the solid material tends not to settle. For example, phosphate slimes have undesirable properties if water cover is not maintained. Other examples include red mud (*blows*) from alumina and phossy water (*burns*) from phosphorus. *Phossy water*, also known as "red water," is water that contains particles of phosphorus metal. The metal bursts into flames when exposed to oxygen.
5. *Sanitary Landfill* This pertains to the burial of nonhazardous solid wastes under controls sufficient to preclude degradation of the surrounding environment.
6. *Chemical Landfill* This is the extension of sanitary landfill to enable the acceptance of hazardous waste.
7. *Subsurface Injection* In this case, the solid material is slurried and then pumped into underground cavities. The EPA opposes such disposal practice for hazardous waste.

The EPA is currently in the process of further examining alternative technologies for treating hazardous wastes.[5] Included in the list are biological treatment processes, dechlorination, carbon adsorption, recovery–reuse–recycle processes, and solidification and stabilization processes.

Today, in spite of its acceptance by EPA as the most reliable process for destruction of hazardous waste, incineration is no longer a popular method for hazardous waste disposal even though it has a proven track record. The following is a brief history of incineration systems in the United States.

Incineration of chemicals now considered hazardous is a technology that is less than 40 years old. After World War II, the chemical industries in the United States began to grow rapidly and with that growth came a commensurate growth in chemical wastes that needed to be disposed. Solid materials, whether municipal wastes or by-products of process operations, were often burned in open dumps. In the early 1950s, the widespread installation of municipal incinerators for the treating of these solid wastes occurred, but most liquid process wastes were released directly to sewers or nearby streams and rivers.

In specific industries, there were some early attempts at waste incineration before 1960, but such attempts were fairly crude. By banning waste discharge to sewers and streams, the *Federal Water Quality Act* promulgated in 1966 provided a driving force for the incineration of combustible waste materials. During the late 1960s and early 1970s, industrial incineration became an important disposal method for processing wastes from the chemical and allied industries. Most of the early systems, however, were not designed to meet current air pollution control regulations and would not be permitted to operate today.

Dow Chemical Company began a large on-site incineration facility at Midland, Michigan, in the late 1960s, utilizing a rotary kiln as the basis of the incineration process. (Various types of incinerators are discussed in Chapter 7.) Liquid injection incinerators for liquid process wastes were first introduced by companies such as John Zink in Tulsa, Oklahoma, Thermal Research and Engineering Corporation (now Selas T-Thermal Company) in Blue Bell, Pennsylvania, and other combustion equipment manufacturers who anticipated liquid process incineration as a new market. Liquid injection incinerators of this period were often vertically fired and discharged hot gases from the combustion chamber directly to the atmosphere. In most cases, the environmental criterion was opacity, which sometimes permitted high particulate discharge and often concentrations of acid gases such as hydrogen chloride and sulfur dioxide.

After the Clean Air Amendments of 1970, certain limitations were placed on the types of emissions that could be discharged to the atmosphere. Incinerator manufacturers had to add particulate removal devices such as absorbers, high-energy scrubbers, electrostatic precipitators, or baghouses. The capital costs required for a complete system thereby increased substantially over that for the incinerator itself.

During the 1970s more sophisticated liquid injection systems were developed and refined to handle a greater variety of process wastes. For example, new incinerator designs were needed to handle some wastes containing inorganic salts that attacked refractories and for the destruction of waste gases from vinyl chloride monomer plants. During the same period, rotary kiln technology was the major system employed for the destruction of process wastes in solid or sludge form and that could not be atomized in a liquid injection system. The rotary kiln was equipped with the same type of air pollution control devices that were needed in the liquid injection system. In 1974, about the time of the oil crisis, the first heat recovery system was employed with a liquid injection incinerator. In 1975, additional waste heat boilers were installed, some of these even on flue gases with corrosive atmospheres. (Refer to Chapter 8 for a discussion of waste heat boilers.)

In 1979 and 1980, the *Resource Conservation and Recovery Act* (RCRA), passed in 1976, began to exert its influence through regulatory promulgation and enforcement. Temporary Part A permits were issued for incineration facilities under RCRA authorization and EPA began changing the rules for incineration. First, detailed design requirements were proposed, followed by destruction and removal efficiency (DRE) standards. These were later changed to the current performance standards. The new design criteria stressed the well-known combustion parameters of retention time, adequate operating temperature, turbulence, and sufficient oxygen.

Federal regulation of chemical waste incineration under RCRA quite naturally brought about significant progress in source testing. Under RCRA, EPA proposed the first regulation for hazardous waste incinerators in May, 1980, and promulgated an interim final regulation in July, 1982. The regulation set performance standards for three process parameters: particulate emissions [0.08 gr/dscf (grains per dry standard cubic foot) or 180 mg/dscm (milligrams per dry standard cubic meter) of exhaust, corrected to 7% oxygen dry volume], HCl emissions (<4 lb/h or, if a scrubber is employed, better than 99% removal efficiency), and destruction and removal efficiency for organic materials (a minimum of 99.99%).* The *trial burn* for compliance demonstration was established. Over a 3-year period following proposal of the regulation, earlier analytical methods were upgraded and refined. In 1983, a national program sponsored by EPA for source testing of hazardous waste incinerators spurred the development of more sensitive and reliable field sampling methods such as the volatile organic sampling train (VOST). In 1984, the Hazardous and Solid Waste Act (HSWA) modified the original RCRA.

In the mid-1980s, issues regarding toxic metal and organic emissions from combustion of hazardous wastes in boilers and other process furnaces as well as incinerators became of concern to the EPA. The boiler and industrial furnace (BIF) regulations were promulgated in 1991 covering compliance plans and emission standards for these systems. The emissions of products of incomplete combustion (PICs) and especially the polychlorinated-dibenzodioxins (PCDD) and polychlorinated-dibenzofurans (PCDF) from all hazardous waste combustors (HWCs) caused EPA to issue a moratorium in 1993 on all new permits for incinerators. The Maximum Achievable Control Technology (MACT) rules were initially drafted and released on April 19, 1996, for HWCs: hazardous waste incinerators (HWIs), cement kilns (CKs), and lightweight aggregate kilns (LWAKs). Boilers were not included at that time. The final rule was issued on September 30, 1999. All of these changes are covered in more detail in Chapter 3.

Today, permitting of hazardous waste incinerators is a lengthy and complex process, using testing methods that are continually being refined. After all general facility standards have been met and a detailed trial burn plan has been approved by the appropriate state or EPA region, performance, and operating limitations for the incinerator are established through the trial burn. More ex-

*Additional details on these standards are provided later in the text.

tensive details on the current permitting process for hazardous waste incinerators are given in Chapter 3.

1.3 TODAY'S PROBLEM

Two decades have passed since hazardous waste control became a public concern. The key starting point for this concern was the passing of the RCRA in 1976. In the ensuing years, EPA gradually issued regulations based on the RCRA; states became actively involved in managing hazardous waste programs, and another law, the *Comprehensive Environmental Response, Compensation and Liability Act* (CERCLA), better known as *Superfund* came into being in 1980. During the same period the Chemical Process Industries (CPI) began tightening up production operations to generate fewer or less harmful wastes. Also, an emerging waste-management service industry became a big time business; current estimates are that this represents a $5 to 6 billion/year market of its own.

From many perspectives, however, this does not mean that the hazardous wastes problem is under control. Despite the efforts of many companies, and sometimes state or local governments, to build new waste management facilities, local resistance to their siting is still extremely intense. The lack of available sites has caused many problems in disposing of wastes and, especially where wastes must be transported long distances, has created other risks, for example, the possibility of truck or rail accidents. The NIMBY (not in my back yard) syndrome remains the overriding response of most citizens to the siting of treatment facilities. Similar problems abound with most alternative waste treatment technologies, such as landfills. Regulatory, industrial, or environmental structures have prevented many such technologies from being widely applied, even though these methods may be safer or cheaper in the long term. Legal, economic, and societal hurdles have also hampered the Superfund program, the intent of which has been the cleaning up of derelict dump sites by using a fund generated from taxes on chemical feedstocks.[7]

The difficulty of identifying hazardous wastes and assessing the resulting health and environmental risks is another current problem. Some questions that need answers are:

1. How do hazardous wastes move through the environment?
2. To how much of the wastes are human populations actually exposed?
3. Based on the answer to question 2, what degree of health hazard and risk do these amounts represent?
4. After their final treatment and disposal, what is the ultimate fate of pollutants?

Finding the answers to these and more questions like these will require time and money. The present list of hazardous wastes defined by RCRA is primarily based upon short-term studies. Though the list is quite extensive, it is certain to become much longer.

Sampling and analytical methods for identifying hazardous wastes are generally costly, complex, time consuming and were developed for specific media. The proper identification of hazardous wastes present at a site and their concentration is the first step in assessing the potential risks posed. Problems arise, however, when dealing with the complex and heterogeneous waste mixtures and conditions normally found at a hazardous waste dump site. As a result, it is difficult to obtain representative samples. Standardized and validated sampling methods are therefore needed to ensure reproducible results. If representative samples are not obtained, the true extent of the risks posed to human health and the environment by a particular site cannot be measured.

In addition to high costs for chemical analysis, there are other analytical and quality assurance/quality control (QA/QC) problems, particularly with nonaqueous samples. Some limitations include: difficulties in preserving samples at the concentration levels found on-site; problems in preparing samples for analysis; and complexities in identifying interference effects, where the presence of some compounds may be hidden from analysis by the presence of others. Biological tests may provide quick screening tools to reduce chemical analysis costs. Biological tests use animals, plants, microorganisms, and cells to evaluate the toxicity of sample materials. Biological analysis may also overcome some of the difficulties posed by mixtures and the synergistic-antagonistic reactions of co-disposed wastes, since these tests study reactions to the whole waste stream and, unlike chemical tests, do not identify individual compounds in that stream. A synergistic-antagonistic reaction involves two or more compounds reacting with one another to create a greater or lesser hazardous waste stream.[8]

Adequate treatment and disposal capacity is critical to the management of hazardous wastes. The development of environmentally sound disposal facilities is essential to the successful implementation of the hazardous waste regulatory program currently mandated by law. This capacity does not exist. There is a shortage of suitable disposal sites, and this problem will become even more acute as additional wastes are determined to be hazardous, existing sites are closed because they do not meet environmental criteria, and wastes that are being disposed of on private property are taken to off-site facilities. Although not all states lack appropriate capacity, other problems prevail. In New York, for example, where adequate capacity does exist, the disposal facility is often not within a reasonable distance of the waste generator. This increases the cost and energy associated with long transportation hauls. There are also legislative barriers to aggravate the capacity problem. The state of Connecticut, for example, passed a law allowing local governments to prohibit, through zoning, land usage for hazardous waste disposal. No disposal site can be built, established, or altered without the approval of both the local zoning planning board and the Connecticut Department of Environmental Protection. This law almost guarantees that the department will not get approval from local governing bodies for a site to be used for regional purposes. Advocates of the law contend that the smaller towns should not be forced to serve as dumping grounds for wastes from other communities.

There are techniques to reduce waste volumes, but these are not expected to alleviate the capacity problem at the present time. Such techniques include the restriction of hazardous chemicals used in operations, substitution of less hazardous materials, and better quality control to reduce production spoilage. The more hazardous or toxic waste streams can be isolated from mixtures in which they occur, and wastes can be concentrated by dewatering, resulting in reduced waste volumes and disposal costs. Many wastes also contain valuable basic materials, which makes material recovery logical from both resource conservation and environmental viewpoints. However, these techniques have not gained general acceptance or wide use because at present they are usually more expensive than land disposal.

1.4 SOCIOECONOMIC CONCERNS

Socioeconomic and environmental considerations, together with public acceptance, dictate the location of a proposed site for a hazardous waste incinerator. Unfortunately, the controlling factors seem to be the socioenvironmental concerns rather than economic or technical factors. *Economic* concerns include installation and operating costs of the incinerator, costs of transporting and storing the hazardous wastes, and disposal costs of the residue. The costs include adaptation of the site, which involves access roads, water and sewage problems, topography modifications, soil problems, aesthetic treatment, and compensation for possible losses by neighboring taxpayers from air pollution damage to and devaluation of their property. Social and environmental concerns mainly involve potential effects upon public health. The adverse consequences of living within an environment that receives low levels of various air contaminants on a daily basis is not known. However, the inhalation of high levels of many specific metallic or salt particulates, polynuclear organic compounds, and acid gases is known both to have serious respiratory effects and to contribute to development of chronic illnesses such as cancer. These contaminants can be emitted continuously from a hazardous waste incinerator if not properly removed from the stack gas.

Siting the Incinerator

The siting of a proposed incinerator has been one of the main social problems. A large segment of the public is apparently unconvinced that hazardous wastes can be disposed of safely. Furthermore, the siting process is not always perceived by the public as being fair, open, and equitable. Those who live near the selected site feel that they carry too much of the cost, including health concerns, lower property values, and the stigma of living near a site. The challenge to the industries involved and the federal and state governments is to convince the public that the site is needed, that reasonable steps have been taken to protect public health, that the solution is fair, and that appropriate incentives have been developed for community acceptance. This is known today as *environmental justice*.

Despite the ever-increasing quantities of hazardous wastes, there are at present approximately 28 commercially available hazardous waste incinerators and industrial furnaces in the United States.[9] During the past several years, disposal of oil and hazardous substances resulting from cleanup of accidental spills and abandoned waste sites has become increasingly difficult. As described previously, greater public awareness of dangers associated with hazardous waste disposal has led to local opposition to proposed land disposal sites. The justifications for new incineration facilities are obvious and critical but finding a site is extremely difficult.

Under today's regulations, the private sector would not under normal circumstances be able to obtain a site for a hazardous waste incineration facility. For this reason, some public role in siting has become necessary. This does not guarantee a site, however. To meet the need for a commercial hazardous waste site, New York State proposed to site and finance a high-technology hazardous waste disposal facility in 1980 with an approved budget of $60 million.

The facility would have consisted of the necessary equipment and structures for transporting, receiving, and storing wastes; a laboratory for analytical and testing services; various processing units; an incinerator; secure landfills; waste lagoons and ponds; and storage tanks and wastes piles. After almost 2 aggressive years of searching, a rural site of 600 acres near Sterling in central New York State was finally chosen. The selected site met with such strong public opposition that the plan for the high-technology hazardous waste disposal facility was suspended.

In a more positive example, Dow Chemical Company in Midland, Michigan, has treated and disposed of virtually all of its wastes on-site for the past 40 years in an effort to avoid involving other parties and to control its liability. In addition to an intensive program of waste prevention and recovery, Dow Chemical has stressed thermal destruction rather than land disposal. Less than 1% by volume of Dow's chemical wastes is deposited in the ground. This approach has proven to be cost effective and has not invoked credible local criticisms. On-site treatment, therefore, appears to be a viable approach for the larger corporations with adequate land resources.

Economic Factors

There is a lack of consistent information on unit costs of incineration, as well as that of other current hazardous waste management practices. Unit costs vary widely depending on such factors as: (1) the physical state, the British thermal unit (Btu) content, ash content, and the toxicity of the particular waste, (2) the form of the waste (i.e., drum versus bulk), and (3) the competition at the regional level among alternative waste management facilities. For all hazardous waste management practices, costs are also strongly affected by changes in the federal regulatory program.

Recently, commercial incineration has become increasingly cost competitive with land disposal. The price of landfilling hazardous waste has increased substantially because of increased landfill design requirements, operating costs and the requirements for pretreating and special handling of many waste streams prior to landfilling. For example, the prices charged by waste management firms generally increased about 25–75% per year from 1978 to 1996, depending on the type and form of waste and the geographical location; at some facilities, the price increase exceeded 80%. This cost increase for landfilling of hazardous waste was the highest reported for any treatment or disposal option, a trend tending to make incineration a more attractive disposal alternative. Recent changes in the RCRA program and the potential long-term liability costs associated with landfills provide additional incentives for incineration.

Nevertheless, the cost of incineration, which may be reduced as technology advances, is generally higher than many other disposal alternatives. Although this cost may be considerable, it is the NIMBY syndrome or the NIMEY (not in my election year) syndrome, and not economic factors, that present the main barrier to incineration as a more widely used hazardous waste management option.

1.5 WASTE CHARACTERIZATION

Waste characterization is a major factor in assessing the feasibility of converting a hazardous waste material by process incineration. It effects the design of the process incinerator and its emission control system and helps determine the compatibility of a waste with a proposed or available facility. It also plays a part in determining process incineration operating conditions for complete conversion of a specific waste.

Hazardous wastes can be classified under five categories: (1) waste oils and chlorinated oils, (2) flammable wastes and synthetic organics, (3) toxic metals, etchants, pickling, and plating wastes, (4) explosive, reactive metals, and compounds, and (5) salts, acids, and bases. A particular waste may overlap into any number of these five categories. Flammable wastes are comprised mainly of contaminated solvents; this category also includes many oils, pesticides, plasticizers, complex organic sludges, and off-specification chemicals. Synthetic organic compounds include halogenated hydrocarbon pesticides, polychlorinated biphenyls, and phenols. Only combustible wastes are candidates for incineration; if a waste contains a significant amount of heavy metals, the metals should be concentrated and removed prior to incineration.

A number of wastes present special problems when incinerated. Metals, for example, become extremely fine metal oxide and metal salt particles that may not be collectable by conventional air pollution control equipment. Resins may polymerize, coating incinerator surfaces and plugging nozzles designed to atomize liquid wastes. Polyolefins and nitrocellulose may detonate rather than burn. When the wastes being incinerated contain a significant concentration of halogenated compounds, the formation of undesirable combustion products such as hydrogen chloride, chlorine, hydrogen fluoride, fluorine, hydrogen bromide, and bromine results. For design purposes, it is necessary to know (a) whether the waste is a gas, liquid, or solid; (b) the fraction of the waste that is organic and whether the inorganic material includes water; and (c) why the waste is hazardous and to what degree.

Characterization of waste is usually accomplished using available analytical techniques, although problems do arise because almost all of the traditional analytical methods are applicable for the analysis of either pure chemicals or nearly homogeneous materials.[10] Chemical wastes are typically highly heterogeneous and the sample matrix is almost always complex. Hence, many of the usual methods or instruments do not work well on waste samples. This problem is further complicated because almost all of the state-of-the-art instruments are designed to analyze microsized specimens. It is very unlikely that the few milligrams of a heterogeneous waste specimen are representative of the whole sample.

Proper sample collecting and handling are obviously critical steps in determining waste characteristics. Because sampling situations vary widely, no universal sampling procedure is recommended. The EPA published a checklist of suggested steps to be followed to help maximize the safety of sampling personnel, minimize sampling time and cost, reduce errors in sampling, and protect the integrity of samples after sampling.[11]

To properly characterize the waste for incineration, the following data must be determined during the analysis.[11]

The Percentages of Carbon, Hydrogen, Oxygen, Nitrogen, Sulfur, Halogens, Phosphorus, Ash, Metals, and Salts in the Waste, as Well as Its Moisture Content The amounts of these components need to be known in order to calculate stoichiometric combustion air requirements and predict combustion gas flow and composition. (Part II covers these stoichiometric calculations in detail.) The presence of halogenated and sulfur-bearing wastes can result in the formation of HCl, Cl_2, HF, Fl_2, H_2, HBr, Br_2, and SO_x in the incinerator gases. These must be removed with suitable scrubbing equipment before discharge into the atmosphere. (Chapter 10 covers the topic of scrubbing equipment in detail.) Also, in the incineration of organic waste containing chlorine, sufficient hydrogen should be provided by either the waste or the auxiliary fuel so that hydrogen chloride, rather than chlorine gas, is formed. The nitrogen content of waste material is generally low, but the presence of nitrogen-containing materials (nitrates, ammonium compounds, etc.) can greatly increase the NO_x emissions.

Trace Metals Trace metals are a potential cause for concern in incineration emissions, and analysis for them should generally be performed unless it is known that they are not present in the waste. As mentioned earlier, wastes containing significant amounts of metals will generally be poor candidates for incineration. Such wastes will require postcombustion emission control of a special type. The effluent or solid waste from the emission control device must in turn be treated as a hazardous waste, even though it will be considerably reduced in volume and weight from the original hazardous waste.

Ash Content The ash content of the waste should be determined in order to evaluate the potential for excessive slag formation as well as potential particulate emissions from the incinerator. Kinematic viscosity and the size and concentration of solids in a liquid waste are the most important physical properties to consider in evaluating a liquid waste incineration design. The physical handling system and burner atomization techniques are dependent on viscosity and solid content of the waste. Chemically complex sludges may contain such elements as sodium, potassium, magnesium, phosphorus, sulfur, iron, aluminum, calcium, silicon, oxygen, nitrogen, carbon, and hydrogen. Several chemical reactions can be expected to take place in the high-temperature oxidizing atmosphere of an incinerator operation treating such sludges. Resulting ash may contain Na_2SO4, Na_2CO_3, NaCl, and other salts. Sludges containing substantial amounts of sodium can cause defluidization of a fluidized bed by forming low-melting eutectic mixtures. (The fluidized-bed incinerator is described in Chapter 7.) Furthermore, if the particles of the fluidized bed are components of silica sand, Na_2SO_4 will react with the silica to form a viscous sodium silicate glass that will cause rapid defluidization.

Heating Value The heating value of a waste corresponds to the quantity of heat released when the waste is burned and is commonly expressed in units of British thermal units per pound (Btu/lb). This quantity must be considered in establishing an energy balance for the combustion chamber and in assessing the need for auxiliary fuel firing. As a rule of thumb, a minimum heating value of about 5000 Btu/lb is required to sustain combustion in most incinerators. Fluid bed units using preheated air will burn materials with heating values as low as 3000 Btu/lb.

Special Characteristics Special characteristics of the waste such as extreme toxicity, mutagenicity or carcinogenicity, corrosiveness, fuming, odor, pyrophoric properties, thermal instability, shock sensitivity, and chemical instability should be considered in incinerator facility design. Thermal or shock instability are of particular concern from a combustion standpoint, since wastes with these properties pose an explosion hazard. Other special properties relate more directly to the selection of waste-handling procedures and air pollution control requirements, topics to be considered in later chapters.

In Table 1.5.1 a scheme for grouping combustible hazardous wastes according to elemental content is shown.[12] The four groups represent varying degrees of ease of incineration. For example, classes 3 and 4 are more difficult to burn

Table 1.5.1 Chemical waste classification

Waste Class	Elemental Composition	Example
1	C, H and/or C, H, O	Tars from production of styrene Off-specification phenol
2	C, H, N and/or C, H, N, O	Solid residue from manufacture of aromatic amines TDI manufacture reactor tar bottoms[a]
3	C, H, Cl and/or C, H, Cl, O	Vinyl chloride monomer manufacturing wastes Phenolic tar from 2,4-D manufacture[b]
4	C, H, N, Cl and/or C, H, Cl, N, O	Nitrochlorobenzene manufacturing wastes
	C, H, S and/or C, H, S, O	Petroleum refining sour waste
	C, H, F and/or C, H, F, O	Fluorinated herbicide wastes
	C, H, Br and/or C, H, Br, O	Ethylene bromide manufacturing wastes
	C, H, P and/or C, H, P, O	Malathion
	C, H, Si and/or C, H, Si, O	Tetraethyl orthosilicate wastes
	C, H, Na and/or C, H, Na, O	Refinery spent caustic

[a]TDI is toluene diisocyanate.
[b]2,4-D is 2,4-dichlorophenoxyacetic acid.

by virtue of their content of chlorine, bromine, fluorine, sulfur, phosphorus, silicon, and sodium. Combustion of these elements generates acids and oxides and, hence, special emission control techniques are required.

It may not be necessary to follow the complete, elaborate analysis protocol for each shipment of waste from the same source, unless the material is entirely different from earlier shipments. Regulations have been established by USEPA on how often the shipments should be sampled and for what parameters samples should be analyzed.

1.6 THE NEED TO INCINERATE

From a regulatory point of view, incineration now appears to be the most viable means of waste treatment for organics. The next several decades should increase the demand for incineration because of limitations placed on other hazardous waste management options as a result of the 1984 RCRA amendments. Incineration of wastes offers the following advantages or potential advantages:

- *Volume reduction*, especially for bulky solids with a high combustible content

- *Detoxification*, especially for combustible carcinogens, pathologically contaminated material, toxic organic compounds, or biologically active materials that would affect sewage treatment plants
- *Regulatory compliance*, especially for fumes containing odorous compounds, photoreactive organics, carbon monoxide, or other combustible materials subject to regulatory emission limitations
- *Environmental impact mitigation*, especially for organic materials that would leach from landfills or create odor nuisances
- *Energy recovery*, especially when large quantities of waste are available and reliable markets for by-product fuel or steam are nearby

These advantages have justified development of a variety of incineration systems, of widely different complexities and functions to meet the needs of municipalities, commercial and industrial firms, and institutions. The most developed and commonly used incinerators are the liquid injection and rotary kiln. Although liquid injection incinerators have been the most popular in Europe and the United States, rotary kiln incinerators are better suited to handle all physical forms of hazardous wastes. High-temperature incinerators are capable of destroying highly toxic organic wastes [e.g., organochlorine, herbicide orange, and polychlorinated biphenyls (PCBs)] with high efficiency. Fluidized-bed and other technologies such as cement kilns and lightweight aggregate kilns have proven they are economically, technically, and environmentally acceptable. The various types of incinerators are further discussed in Chapter 7.

Incineration is not a new technology and has been used for treating organic hazardous waste for many years. The major benefits of incineration are that the process actually destroys most of the waste rather than just disposing of or storing it.[13] Incineration can be used for a variety of specific wastes and is becoming competitive in cost compared to other disposal methods.

Although incineration has become more attractive in hazardous waste management, it is not without its problems.[14] Among these are:

- *Cost* In most instances, incineration is a costly processing step, both in initial investment and in operation.
- *Operational Problems* Variability in waste composition and the severity of the incinerator environment result in many practical waste-handling problems, high maintenance requirements, and—at some installations—equipment unreliability.
- *Secondary Environmental Impacts* Many waste combustion systems result in the emission of odors, sulfur dioxide, hydrogen chloride, carbon monoxide, carcinogenic polynuclear hydrocarbons, nitrogen oxides, fly ash and particulate fumes, and other toxic or noxious materials into the atmosphere. Wastewaters from residue quenching or scrubber-type air pollution control are often highly acidic and contain high levels of dissolved solids, abrasive suspended particles, biological and chemical oxygen de-

mand, heavy metals, and pathogenic organisms. Finally, residue disposal can present a variety of aesthetic, water pollution, and health-related problems.

- *Public Sector Reaction* Few incinerators are installed without arousing concern, close scrutiny, and, at times, hostility from the public at large and/or regulatory agencies.[15]

With all these disadvantages, incineration has persisted as an important concept in waste management. The American Society of Mechanical Engineers (ASME) Research Committee on Industrial & Municipal Wastes in a recent publication describes "Why Engineering Experts Say We Need Waste Incineration." While the ASME strongly supports the reduction and/or recycle of hazardous wastes, it also realistically recognizes that some wastes are inevitably generated. Therefore, proper treatment and disposal are required. Incineration is a full-scale, proven, effective technology for treatment of a broad range of organic wastes.[16]

A comprehensive analysis of some of the above disadvantages has been performed by two of the authors of this book plus others as a result of the public's concern with incineration.[17,18]

REFERENCES

1. The Comptroller General, *How to Dispose of Hazardous Wastes—A Serious Question that Needs to Be Resolved*, U.S. General Accounting Office, Washington, DC, CED 79–13, December 1978.
2. P. W. Powers, *How to Dispose of Toxic Substances and Industrial Wastes*, Noyes Data Corp., Park Ridge, NJ, 1976.
3. L. M. Thomas, "EPA Fights Hazardous Waste," *EPA Journal*, **10**, 4–7, October (1984).
4. H. L. Hickman, "Why We Have a Hazardous Waste Problem," *EPA Journal*, **10**, 10–11, October (1984).
5. D. White and B. Burke, "Choices in Disposal of Hazardous Waste," *EPA Journal*, **10**, 20–21, October (1984).
6. W. E. Sweet, R. D. Ross, and G. V. Velde, "Hazardous Waste Incineration: A Progress Report," *J. Air Pollution Control Assoc.*, **35**, 138–143, February (1985).
7. N. Basta, R. V. Hughson, and C. F. Mascone, "What Are Your Views on Hazardous Wastes?" *Chem. Eng.*, **58**, March 4, 1985.
8. The Comptroller General, *Hazardous Waste Sites Pose Investigation, Evaluation, Scientific, and Legal Problems*, U.S. Government Accounting Office, Washington, DC, CED 81–57, April 1981.
9. Map provided by Environmental Information, Ltd., Minneapolis, MN 55439, www.envirobiz.com, 1998.
10. K-C. Lee, J. L. Hasen, and G. M. Whipple, "Characterizing Petrochemical Wastes for Combustion," Union Carbide Corporation, So. Charleston, WV; ASTMs Haz-

ardous and Industrial Solid Waste Testing: Fourth Symposium, Arlington, VA, May 2–4, 1984.

11. E. R. DeVera, B. P. Simmons, R. D. Stephens, and D. L. Storm, *Samplers and Sampling Procedures for Hazardous Waste Streams*, Municipal Environmental Research Lab, Cincinnati, OH, EPA-600/2-80/018, January, 1980.
12. T. Shen, U. Choi, and L. Theodore, *Hazardous Waste Incineration Manual*, U.S. EPA Air Pollution Training Institute, Research Triangle Park, NC, 1985.
13. E. T. Oppelt, "Incineration of Hazardous Waste: A Critical Review," *Journal Air Pollution Control Association* (JAPCA), **37**(5), (1987).
14. W. R. Niessen, *Combustion and Incineration Processes*, 2nd ed., Dekker, New York, 1995.
15. P. Costner and J. Thornton, "Playing With Fire: Hazardous Waste Incineration," a Greenpeace Report, Washington, DC, 1990.
16. The American Society of Mechanical Engineers, "Hazardous Waste Incineration: What Engineering Experts Say," *ASME CRTD*, **32**, (1984).
17. J. J. Santoleri, J. D. Lauber, and L. Theodore, "Facts or Myths: The Burning Issue of Incineration," AWMA—86th Annual Meeting, Denver, CO, June 13–18, 1993.
18. P. Chrostowski and S. Foster, "Scientific Peer Review of Greenpeace's Position on Hazardous Waste Incinerator Impacts," Report by Clement International Corporation for Laidlaw Environmental Services, Ltd., March 17, 1992.

2

OTHER OPTIONS

2.1 INTRODUCTION

Although incineration is presently considered the method of choice to manage organic wastes, there are other options. Incineration provides essentially complete destruction and produces relatively small volumes of waste, most of which is nonhazardous. However, capital and operating costs for incineration are high, and the permitting process is lengthy, expensive, and complicated. Pretreatment methods such as concentration, volume reduction, and removal of solids or inorganics may be required for incineration. Modification of the process generating the waste may also be necessary in order to reduce the quantity of waste.

This chapter discusses other options to incineration as well as pretreatment and process modification techniques associated with incineration and pollution prevention. These options are:

- Chemical treatment
- Biological treatment
- Physical treatment
- Ultimate disposal method
- Process modification(s)/pollution prevention

The chapter concludes with a discussion of the selection process involved in choosing a form of hazardous waste treatment.

Typically, when options other than incineration are chosen, more than one unit process (or step) is required. For example, a typical process sequence that could serve as an alternative to incineration might be sedimentation with neutralization, biodegradation, followed by sludge separation, and finally landfilling the waste. These alternative options are often less costly and are possibly easier to permit and start up than incineration.

2.2 CHEMICAL TREATMENT

There are various chemical processes used to treat hazardous waste. Relatively few are used as the primary or the only treatment step. Most of the chemical processes prepare the waste for further processing and are but one step in a scheme of waste processing. The following chemical processes will be discussed: calcination or sintering, catalysis, electrolysis, hydrolysis, microwave discharge, neutralization, oxidation, photolysis, precipitation, and reduction. Some of the processes such as neutralization and precipitation are commonly used; others, microwave discharge, for example, are still in the development stage. For more detail on these processes the reader is referred to the literature.[1,2]

Calcination

Calcination is a thermal decomposition process without any interaction with a gaseous phase. Operating temperatures of about 1800°F and atmospheric pressure can produce dry powder from slurries, sludges, and tars as well as aqueous solutions, by driving off volatiles. Typical calciners are open hearth, rotary kiln, and fluidized bed. Calcination is a well-established process and recommended as a one-step process for the treatment of complex wastes containing organic and inorganic components. It is also one of the few processes that can satisfactorily handle sludges. A real advantage of calcination is that it can concentrate, destroy, and detoxify in a single step. Organic components are usually destroyed while inorganics are reduced in volume and leachability, thus making them suitable for landfill.

Treatment applications include recalcination of lime sludges from water treatment plants, coking of heavy residues and tars from petroleum-refining operations, concentration and volume reduction of liquid radioactive wastes, and treatment of refinery sludges containing hydrocarbons, phosphates and compounds of calcium, magnesium, potassium, sodium, iron, and aluminum. The use of calcination is likely to expand in the future since it handles tars, sludges, and residues that are difficult to handle by other methods.

The energy requirements for calcination are high; depending upon the amount of water and organics in the waste stream, the energy needed can be as much as 20 times greater than that used in other chemical methods. If the waste stream contains a combustible organic, the energy requirement would be reduced. Operating costs are variable and dependent upon the content of the waste stream. Investment costs for calcination must also include air pollution control equipment such as particulate removal devices and gas absorbers.

Catalysis

Catalysis is a modification of the mechanism and rate of a chemical reaction. If a waste stream can be modified by a chemical reaction, the successful use of a catalyst will reduce operating cost and energy requirements. Catalysts can be *selective* such as in the detoxification of chlorinated pesticides by dechlorination or *versatile* as in the complete destruction of cyanides by air oxidation. While catalytic oxidation is commonly used as an alternative to incineration in the decomposition of organic wastes, it is not widely used in the treatment of hazardous waste. The technology is not well developed, and because of catalyst specificity, progress in the development of one application does not necessarily mean progress in the development of another. In addition, the many components in a waste stream, in varying concentrations, can inhibit the effectiveness of a catalyst on the target reaction. Some laboratory research has been done on the following catalytic processes: oxidation of cyanides, sulfides and phenols, and decomposition of sodium hydrochlorite solutions. A catalyst will normally reduce operating temperature and therefore conserve energy. Catalytic processes may be higher in capital costs, but lower in operating costs, and thus result in lower overall expenditure.

Electrolysis

Electrolysis is the reaction of either oxidation (loss of electrons) or reduction (gain of electrons) taking place at the surface of conductive electrodes immersed in an electrolyte under the influence of an applied potential. Electrolytic processes can be used for reclaiming heavy metals, including toxic metals from concentrated aqueous solutions. Electrolysis is not useful for organic waste streams or viscous tarry liquids. Application to waste treatment has been limited because of cost factors. A frequent application is the recovery for recycle or reuse of metals, such as copper, from waste streams. Pilot applications include oxidation of cyanide waste and separation of oil–water mixtures. Electrolysis has energy costs that can be 10–35% of operating costs. Gaseous emissions may occur and, if these are hazardous and cannot be vented to the atmosphere, further treatment such as scrubbing is required. Wastewater from the process may also require further treatment. Costs are dependent upon concentration of the feed and desired output.

The most common waste treatment application of electrolysis is the partial removal of heavy metals from spent copper pickling solutions. When the typical concentration of the spent pickling solution is 2–7% copper, the system design is similar to that of a conventional electroplating bath. When recovering from more dilute streams, mixing and stirring are necessary to increase the rate of diffusion; it may also be necessary to use a large electrode surface area and a short distance between electrodes.

Another consideration with this method is the removal of collected ions from the electrodes. This removal may or may not be difficult, but it must be addressed. The material collected must be ultimately disposed of if it is not suitable for reuse.

Hydrolysis

Hydrolysis is the reaction of a salt with water to produce both an acid and a base:

$$XY + H_2O \rightarrow HY + XOH$$

Reactions generally require high temperature and pressure, acid or alkali, and sometimes catalysts. Feed streams can be aqueous, slurries, sludges, or tars. The process could be useful for a variety of wastes, but few applications have been used. The petroleum industry uses hydrolysis to recover sulfuric acid from the sludge of acid treatment of light oils. Hydrolysis has also been used to detoxify waste streams of carbamates, organophosphorus compounds, and other pesticides.

Energy requirements vary with application but are generally high. Some products of hydrolysis are toxic and so require further treatment. An additional drawback is that, depending on the waste stream, the products may not be predictable and the mass of *toxic* substance discharged may be greater than that of the waste originally inputted for treatment. Operating costs are completely dependent upon application and can be extremely high.

Microwave Discharge

In this technique, high-frequency microwave power is used to establish a plasma or gaseous discharge in which neutral molecules are partially decomposed into metastable, free radical, and ionic species. This technique is not yet used for waste treatment. One problem is that, when the decomposition products recombine to form stable molecules, the products may be of higher molecular weight or greater toxicity than the original reactants. Some research has been done on organic vapors.

Power requirements and capital cost appear to be high. A possible future use could be for small quantities of highly toxic materials.

Neutralization

Neutralization is a reaction that adjusts either acid (low pH) or base (high pH) by addition of the other to achieve a pH of 7 (neutrality). (For a definition and discussion of the pH, see Chapter 4.) The treated stream undergoes essentially no change in physical form, except possible precipitation or gas evolution. The process has extremely wide applications to aqueous and nonaqueous liquids, slurries, and sludges. It is very widely used in waste treatment. Some applications include pickle liquors, plating wastes, mine drainage, and oil emulsion breakage. It is commonly used in industries such as batteries, aluminum, coal, inorganic and organic chemicals, photography, explosives, metals, pharmaceuticals, power plants, and textiles.

The reaction can take place in batch or continuous reactors (see Chapter 6 for details on reactors). The flow can be cocurrent or countercurrent depending upon desired results and waste stream characteristics. The reaction can be in the liquid or gaseous phase or both. Limitations include temperature dependence and residual formation. The reaction may be exothermic and cause a temperature increase that may be undesirable; for example, higher temperature may cause the formation of other toxins or hazardous wastes, or the high-temperature stream may require special handling. Residuals such as precipitates or gases may also form and then must be treated by other types of processing such as clarifying or absorbing. Neutralization may not make the waste less hazardous; it may only be a treatment for the next step of waste handling. For example, changing the pH may help precipitate heavy metals, prevent metal corrosion or damage of process equipment, affect operation of biological treatment, or provide a neutral recycle material.

Energy requirements are fairly low and are related to pumping and agitation. Capital and operating costs are highly variable depending upon specific process requirements. Neutralization is a very common process step but rarely comprises the sole hazardous waste treatment.

Some common methods of neutralizing wastes are mixing acidic and alkaline streams together, passing acid wastes through packed beds of limestone, mixing acid waste with lime slurries, adding solutions of concentrated bases such as caustic (NaOH) or soda ash (Na_2CO_3) to acid streams, blowing waste flue gas from a boiler through alkaline waste liquids, adding compressed CO_2 to basic wastes, or adding acid such as H_2SO_4 or HCl to basic (alkaline) waste streams. The choice of the acid or alkali used is often based on process requirements, but cost is the primary consideration. Lime and sulfuric acid are relatively inexpensive. Lime will, in the process of neutralizing sulfate-bearing wastes, form calcium sulfate, which will precipitate and may be undesirable. Caustic and soda ash are more expensive, but may be the better choice of base for wastes containing sulfates. If sulfides or cyanides are present in the waste stream, toxic gas may be evolved during neutralization; these gases require special handling by gas treatment devices such as scrubbers. A schematic of a neutralization system is given in Fig. 2.2.1.[2]

Oxidation

Oxidation is a process in which one or more electrons are transferred from the chemical being oxidized to the chemical initiating the transfer (*oxidizing agent*). The main use in treating hazardous wastes is detoxification. For example, oxidation is used to change cyanide to less toxic cyanate or completely oxidize cyanide to carbon dioxide and nitrogen. Oxidation can also aid in the precip-

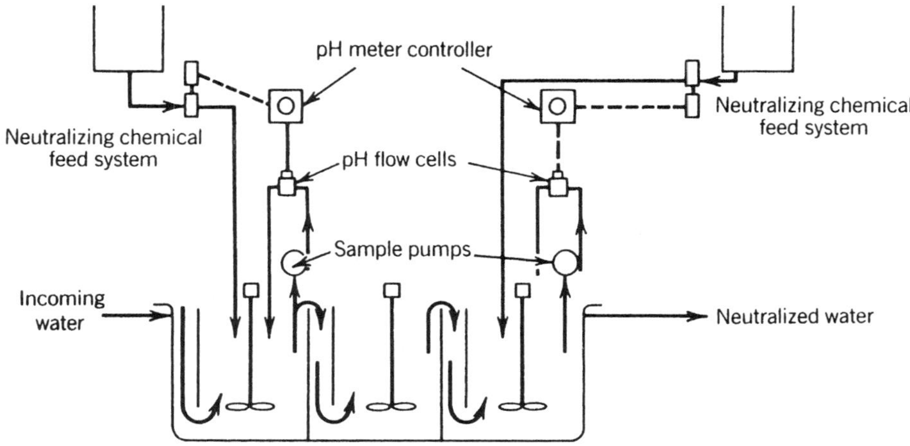

Figure 2.2.1 Schematic of a neutralization system.

itation of certain ions in cases where the more oxidized ion has a lower solubility.

Some common oxidizing agents and their uses are:

- *Chlorine* as a gas or in hypochlorite salt is often used to oxidize cyanide. It can also be used on phenol-based chemicals, but this application is somewhat limited because of the formation of toxic chlorophenols when the process is not adequately controlled.
- *Ozone* is used in the oxidation of cyanide to cyanate and to oxidize phenols to nontoxic compounds. Because ozone is an unstable molecule, it is highly reactive as an oxidizing agent. There are no inherent limitations to oxidation with ozone, but components of the wastewater should be examined to determine that enough ozone is being added to also oxidize other components that could compete with the toxic components for ozone.
- *Ozone with ultraviolet (UV) light* can oxidize some compounds. Halogenated organic compounds are resistant to ozone but in the presence of UV light can be oxidized.
- *Hydrogen peroxide* (H_2O_2) is a powerful oxidizing agent and is successful with phenols, cyanides, sulfur compounds, and metal ions. The process is sensitive to pH, which must be controlled. The optimum pH is ~3 or 4, with process efficiency decreasing at higher or lower pH.
- *Potassium permanganate* ($KMnO_4$) is an excellent oxidizing agent and reacts with aldehydes, mercaptans, phenols, and unsaturated acids. It has been used to destroy organics in wastewater and potable water. The reduced form is manganese dioxide, which is mostly insoluble and can be

easily removed by filtration. The process is pH sensitive and is faster at high pH (up to 9.5).

Oxidation is used in the treatment of waste, the most common applications being dilute waste streams. The energy consumption is relatively low since only pumps and mixing equipment are needed. One disadvantage is that some oxidizing agents add other metal ions that may have to be removed. The costs are lower when *continuous* rather than *batch* processes are used. The oxidation processes are not too successful with sludges; because of the difficulty in achieving good contact mixing, incomplete reaction occurs.

An additional limitation to the use of ozone oxidation is a result of its high instability—it must be generated on-site, which can add considerably to its expense. On the positive side, ozone possesses antibacterial and antiviral properties; this disinfecting characteristic has been widely used to scrub gases to remove odor. Oxidation with ozone is very effective for a wide variety of components and as a result is used extensively, despite the expense.

Photolysis

Photolysis is the breakage of chemical bonds under the influence of UV or visible light. The extent of the degradation varies according to the compound and the light used. Complete conversion of an organic to CO_2 and H_2O by this method is highly improbable, and partly degraded components are likely to be hazardous. Large-scale applications of photolysis have not been developed and are not likely to be developed in the future. There are some UV-assisted ozonation and chlorination applications that have been successful. The process has not been studied in great detail but would surely be highly energy intensive. Specialized application to detoxification of pesticide-contaminated solvents is a possible future use.

Precipitation

Precipitation involves the alteration of the ionic equilibrium to produce insoluble precipitates. Chemical precipitation is allied with solids separation processes such as filtration in order to remove the sediment. Undesirable metal ions and anions are commonly removed from waste streams by converting them to an insoluble form. The process is sometimes preceded by chemical reduction of the metal ions to a form that can more easily be precipitated.

Chemical equilibrium can be affected by a variety of means in order to change the solubility of certain compounds. For example, precipitation can be induced by:

- *Alkaline Agents* An alkaline agent such as lime or caustic soda, when added to waste streams, raises the pH. The higher pH decreases the solubility of the metal ions that precipitate out of solution as hydroxides. The optimum pH for precipitation varies depending on the metal ion to be removed. The hydroxide precipitation can also be affected by organic radicals that form chelates. Chelates mask the typical precipitation.

- *Sulfides* Soluble sulfides, such as hydrogen or sodium sulfide, and insoluble sulfides, such as ferrous sulfide, are used to precipitate heavy-metal ions as insoluble metal sulfides. Sodium sulfide and sodium bisulfide are most commonly used for this purpose. The sulfides have lower solubilities than hydroxides in the high pH range and also have low solubility around the pH 7 range or below. Both the sulfide ion and hydrogen sulfide are toxic so that sulfide precipitation usually requires pretreatment and close control. Pretreatment is normally an adjustment to a pH of 7 or 8. Posttreatment usually involves aeration or chemical oxidation.
- *Sulfates* Zinc sulfate or ferrous sulfate are used to precipitate cyanide complexes. Precipitation of cyanide complexes does not destroy the cyanide molecules, and the collected sludge must therefore be appropriately handled. Reports have shown that this type of sludge in the presence of sunlight can break down to free cyanide.
- *Carbonates* Carbonate precipitation is successful in removing metals. Direct precipitation of metal ions with calcium carbonate, or the conversion of hydroxides into carbonates with carbon dioxide, is common since carbonates are easily filtered out. The solubilities of carbonates are usually between those of hydroxides and sulfides.

Precipitation with chemicals is a common waste stream treatment process and is effective and reliable. Energy requirements are fairly low as compared to other processes. The disposal of the resulting sludge can propose a serious problem, however. Some examples of industries using precipitation for waste removal include: inorganic chemicals, metal finishing, copper forming, foundries, explosive manufacturing, pharmaceutical manufacturing, electronic component manufacturing, and textile mills.

Reduction

In chemical *reduction*, one or more electrons are transferred from the reducing agent to the chemical being reduced. The lowering of the valence state may reduce toxicity or encourage a particular chemical reaction. The first step in a reduction process is normally pH adjustment. The reducing agent can be a gas, solution, or finely divided powder. Mixing is critical in order to assure contact between the reducing agent and the waste for complete reaction. The reduction reaction is normally followed by a separation step, such as precipitation, to remove the reduced compound.

Some common reducing agents are sulfur dioxide, sodium metabisulfite, sodium bisulfite, and ferrous salts for reducing chromium; sodium borohydride to reduce mercury; and alkali metal hydride to reduce lead. Sulfur dioxide is normally used as a gas. The reduction of chromium with sodium metabisulfite and sodium bisulfite is highly dependent on the pH and temperature. Sodium hydroxide must be added to control the pH during the reaction. Since the reduction of chromium with ferrous sulfate works best when the pH is below 3, acid must be added during the reaction.

The major application of chemical reduction is for the treatment of chromium waste. Chromium is very toxic in its hexavalent state (Cr^{6+}); trivalent chromium (Cr^{3+}) is much less toxic and will precipitate in an alkaline solution by forming a hydroxide. Chromium reduction is used for waste treatment in industries such as: metal finishing, inorganic chemical manufacture, coil coating, battery manufacture, iron and steel manufacture, aluminum forming, electronic manufacture, porcelain enameling, and pharmaceutical manufacturing.

Since energy is required only for pumping and mixing, energy requirements for reduction are low. Continuous processes are more economical than batch processes. Treatment by reduction is effective and reliable. The main disadvantage is the potential formation of compounds as toxic or hazardous as the compounds originally in the waste. In the treatment of mixed wastes, interference of oxidizing agents can decrease the efficiency of the reducing agent. The use of sulfur dioxide as the reducing agent has the added disadvantage of special handling and storage requirements, since sulfur dioxide is somewhat hazardous. An illustration of hexavalent chromium reduction with sulfur dioxide is shown in Fig. 2.2.2.

2.3 BIOLOGICAL TREATMENT

Biological processes also involve chemical reactions but are differentiated from the chemical category in that these reactions take place in or around microorganisms. The most common use of biological processes in waste treatment is for the decomposition of organic compounds.

The different biological processes described here are activated sludge, aerated lagoons, anaerobic digestion, composting, enzyme treatment, trickling filter, and waste stabilization ponds. All of these, except enzyme treatment, use microorganisms to decompose the waste. Enzyme treatment generally involves extracting the enzyme from the microorganism and using it to catalyze a particular reaction. With proper control, these processes are reliable and environmentally sound; additive chemicals are usually not needed and operational expenses are relatively low.

Activated Sludge

The *activated sludge* process uses microorganisms to decompose organics in aqueous waste streams. The microorganisms take the organics into the cell, through the cell wall, and into the cytoplasm where the organics are broken down by enzyme oxidation and hydrolysis to produce energy and other cellular

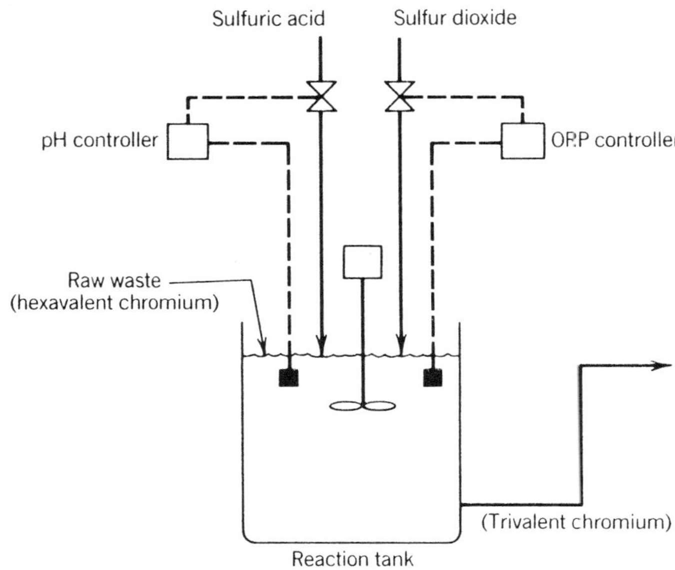

Figure 2.2.2 Hexavalent chromium reduction with sulfur dioxide.

material. Besides taking in the organics as food, the biomass acts as a filter to collect colloidal matter and suspended solids. Volatile organics can be driven off somewhat by the aeration process. Some metals are collected in the organisms and in the sludge.

The controlling factors in the process are the extent of mixing to ensure contact of the microorganisms with the waste, the amount of dissolved oxygen for the respiration of the organisms, the concentration of metals that are toxic to the microorganisms, and the types and concentration of the organics to be decomposed. Organics that have been successfully decomposed by microorganisms include: polysaccharides, proteins, fats, alcohols, aldehydes, fatty acids, alkanes, alkenes, cycloalkanes, and some aromatics. Requiring more careful process control, isoalkanes, halogenated hydrocarbons, and lignin have also been decomposed by microorganisms.

The activated sludge process must usually be preceded by neutralization, some metal removal, and possible solids removal. Most activated sludge processes work with <1% suspended solids. The process is normally followed up by a separation step, usually sedimentation, to remove the biological sludge from the waste liquid stream. Much of the biological sludge is recycled, although some portions are disposed of in landfills.

Activated sludge is the most widely used of the biological waste treatment processes. The energy requirements are a large part of operating cost. The pumping, mixing, and aeration require power. The system typically requires no chemical additions and degradation is natural. The main process disadvantage is that it does not handle slurries, tars, or streams with a high concentration of suspended solids. The reason it is best suited for liquid streams is the critical need for good mixing to assure contact of the waste with the microorganisms. The activated sludge process has been used to treat the waste streams from iron and steel manufacture, canneries, pulp and paper mills, petroleum refining, organic chemical manufacture, and pharmaceutical manufacture.

Aerated Lagoon

The *aerated lagoon* is an earthen basin that is artificially aerated; the basin is generally lined to make it impermeable. The microbial reactions are the same as those that take place in the activated sludge process except the biological sludge is not recycled. In this process microorganisms are used to decompose organics in waste streams containing <1% solids; solid contents >1% usually result in the settling of some solid matter because of limited mixing and aeration. These solids often undergo anaerobic microbial decomposition.

The aerated lagoon process has been used for petrochemical, textile, pulp and paper mill, cannery, leather tanning, gum and wood processing, and some other industrial waste streams. The aerated lagoons usually require less energy and are therefore less expensive to operate since recycle systems are not needed; more land area is a requirement, however. The removal efficiencies are not as high as those for the activated sludge process since mixing and aeration are less effective. The retention time in an aerated lagoon is typically longer than that for the activated sludge process. One other disadvantage is that aerated lagoons do not handle streams with variable organic and metal constituents as well as activated sludge.

Anaerobic Digestion

In *anaerobic digestion*, microorganisms that do not require oxygen for respiration are employed. This process is useful for the degradation of simple organics. The microbiology is not well understood, but essentially the cells use part of the organic compounds for cell growth and part is converted to methane and carbon dioxide. A delicate equilibrium is required, and this makes the process less suitable for industrial waste streams.

Anaerobic digestion has been used for sewage treatment, but few applications for hazardous waste have been implemented. There has been some study done for the treatment of brewery, alcohol distillery, and cotton kiering waste streams.

The anaerobic digestion takes place in a closed vessel with no agitation mechanisms. The gases rise to the top and are collected for possible use as a heating fuel. The digested sludge settles to the bottom, and once it is stable and inert can be disposed of by landfilling.

The power requirements are low and consequently operating costs are fairly low. There are disadvantages because of the need for control of pH, temperature, and waste concentration and constituents. There is potential for application to the degradation of specialized wastes from pharmaceutical or other processes.

Composting

Composting is aerobic digestion by microorganisms in the soil. The organisms decompose organics and multiply. This process, unlike other biological processes, can tolerate some toxicants and metals. This is accomplished essentially by piling waste in the ground and aerating it occasionally by turning and moving the soil. The collection of leachate and runoff is normally required to protect the groundwater.

Composting is commonly used for organic wastes, and complete digestion requires about 3 or 4 months. Since there is a market for the soil and the decomposed organics (humus) in Europe, it is frequently used there. Most of the attempts to use this technique in the United States have failed because of a lack of this market. Composting has been successful with municipal refuse, high concentration organic sludges, and some petroleum refineries.

Energy requirements are somewhat higher than for other biological processes. Equipment for landmoving and possibly pumping is required. Composting occasionally requires the addition of some nutrients as well as alkali for pH control. The few possible emissions are carbon dioxide, steam, liquid effluent of partly oxidized organics, and some unpleasant odors, all of which are fairly simple to control. The main disadvantages are the large land requirements and the necessity of a market for the humus.

Enzyme Treatment

Enzyme treatment involves the application of specific proteins (simple or combined), which act as catalysts in degrading the waste. Enzymes work on specific types of compounds, specific molecules, or a specific bond in a particular compound. Enzymes are inhibited by the presence of insoluble inorganics, are sensitive to pH and temperature fluctuations, and do not adapt to variable concentrations.

Enzyme treatment may prove useful for specific industrial applications but at present is not commonly used in waste treatment. The specificity of enzymes and their fragile nature make them fairly useless in most waste applications. Addition of enzyme in an attempt to catalyze activated sludge processes have failed. Since it is not used on a large scale for waste treatment, little data are available on the economics of this process.

Trickling Filters

Trickling filters are also used for the decomposition of organic waste streams. The microbiological reactions are similar to those that occur with activated sludge and aerated lagoons. The microorganisms are held in a support media and wastewater is trickled over them. Usually waste streams are sprayed to absorb oxygen before passing through the support media. Microorganisms and filter design vary depending upon the waste and the efficiency required.

Some industries that use trickling filters are refineries (with oil, phenol, and sulfide wastes), canneries, pharmaceutical manufacture, and petrochemical manufacture. The energy requirements for this process are low, although recycled pumps are required for recirculation of the filter effluent. Efficiency of removal is not high, about 50–85%; because of this, trickling filters are normally used in combination with other methods, such as activated sludge.

Waste Stabilization Ponds

Waste stabilization ponds are shallow basins into which wastes are fed for biological decomposition. The chemical reactions involved are the same as those that occur in the other biological processes. Aeration is provided by the wind, and anaerobic digestion may also occur near the bottom of deeper ponds. This method can handle only low concentration of organics with <0.1% suspended solids.

The ponds are very commonly used for sewage treatment and dilute industrial wastes. Some industries using them include: steel, textiles, oil refineries, paper and pulp, and canneries. Waste stabilization ponds are normally used as a final treatment step for effluent because they are not efficient enough to be used on their own. Some advantages are low-energy needs and, assuming appropriate feeds, few effluent handling problems. Disadvantages include a high sensitivity to inorganics and suspended solids and large land acreage requirements. Occasionally the basin must be lined for leaching control.

2.4 PHYSICAL TREATMENT

There are more than 20 known types of physical treatment processes used in the handling of hazardous wastes; however, very few of these are fully developed or commonly used in industry. Some treatments have been found to have little potential use so that further research in these areas is unlikely. Zone refining, freeze drying, electrophoresis, and dialysis all fall into this category. The difficulty of the operation and/or the high cost of these processes overshadows any future use they may have. The most common processes used today are sedimentation, filtration, flocculation, and solar evaporation. Most other processes fall in between these two extremes, that is, they show some potential for future use but are not presently used to any great extent.

Physical treatments may be separated into two categories: *phase separation* and *component separation* processes; in the latter, a particular species is separated from a single-phase, multicomponent system. The various physical treatments may fall into one or both of these categories. Sedimentation and centrifugation are used in phase separation, liquid ion exchange and freeze crystallization are used for component separation, and distillation and ultrafiltration are used in both.

Phase separation processes are employed to reduce waste volume and to concentrate the hazardous waste into one phase before further treatment and material recovery are performed. Such waste streams as slurries, sludges, and emulsions, which contain more than one phase, are the usual candidates for this category.

Filtration, centrifugation, and flotation may be used on slurries that contain larger particles. If the slurry is colloidal, flocculation and ultrafiltration are generally used. If the slurry or sludge is known to contain any volatile components, evaporation or distillation is used to remove them from the waste stream. Because emulsions are so difficult to separate, the type of physical treatment required is usually decided on a case-by-case basis.

Component separation processes remove particular ionic or molecular species without the use of chemicals. Most of these are used in wastewater treatment and include such techniques as: liquid ion exchange, reverse osmosis, ultrafiltration, air stripping, and carbon adsorption. The first three are used to remove ionic and inorganic components; the last two techniques are used to remove volatile components and gases.

Many factors need to be considered when selecting a particular type of physical treatment. These include: the characteristics of the waste stream and the desired characteristics of the output stream, the technical feasibility of the different physical treatments when applied to a particular case, regulatory constraints, and economic, environmental, and energy considerations. The selection of waste treatment methods is discussed in Section 2.7.

In this section, the more established types of physical treatment used today are examined. These include: carbon adsorption, resin adsorption, centrifugation, distillation, electrodialysis, evaporation, liquid–liquid extraction, filtration, flocculation, flotation, freeze crystallization, high-gradient magnetic separation, ion exchange, liquid solidification, air stripping, steam stripping, and ultrafiltration.

Carbon Adsorption

Carbon adsorption is a recommended and established process used in separating organic and certain inorganic species from aqueous waste streams. It is suggested that the concentration of the waste species be 1% or less so that carbon regeneration is less frequent. A good number of carbon adsorption systems are being used today in municipal and industrial wastewater treatment facilities.

In carbon adsorption, a filter bed of activated carbon is placed in a vessel and used to adsorb certain components. A large adsorptive surface area, about 2.5–7.5 million ft^2/lb, is needed to accomplish the process. Best results are achieved when the component is slightly soluble in water, has a high molecular weight, high polarity, and low ionization capability, and when the concentration of any suspended solids is <50 parts per million (ppm).[2] Once the carbon has been fully used, regeneration of the carbon may be done simultaneously with either the destruction (thermal regeneration) or recovery (nonthermal regeneration) of the *adsorbate* (i.e., the material adsorbed by the carbon).

Energy costs for this process range from 5 to 25% of the total operating cost depending upon whether thermal regeneration is used. If the carbon is thermally regenerated, equipment costs increase since a furnace, afterburner, and scrubber are needed. If the carbon is not regenerated, a means of carbon disposal is needed.

Resin Adsorption

Resin adsorption is used in the removal of organic solutes from aqueous waste streams. Solute concentration in the wastewater may be as high as 8%. It is the preferred method when recovery of the adsorbate is desired since thermal regeneration of carbon destroys the organic material. It is also useful when there is a high concentration of dissolved inorganic salts in the waste stream.

Synthetic resins may be used to remove hydrophobic or hydrophylic solutes, which may be recovered by chemical means. Their applications include phenol recovery, fat removal, and color removal from wastewaters. Possible future uses include the removal of pesticides, carcinogens, and chlorinated hydrocarbons.

A schematic of a typical resin adsorption system is shown in Fig. 2.4.1.[2] Resin adsorption is similar to carbon adsorption except that two filter beds are used—one bed is used for adsorption while the other is being regenerated. The waste stream flows downward in the system at a rate of 1–10 gall/min$\cdot$ft^2 of cross section, and adsorption stops when the bed becomes saturated or the effluent concentration reaches a certain level.

Energy costs for this system depend on whether the resin is regenerated or not. If the resin is not regenerated, it must be disposed. Because the high cost of resins results in a high capital cost, regeneration is usually desirable to keep overall costs low.

Centrifugation

Centrifugation is a well-established, popular liquid–solid separation process used in commercial and municipal waste treatment facilities. It is usually used to reduce slurry and sludge volumes (*sludge dewatering*) and to increase the solids concentration in these waste streams. It is a technically and economically competitive process and is commonly used on waste sludges produced from water pollution control systems and on biological sludges produced in industry and municipal treatment facilities. Possible future uses include the removal of soluble metals from wastewaters and the dewatering of sludges produced in sulfur dioxide air pollution control systems.

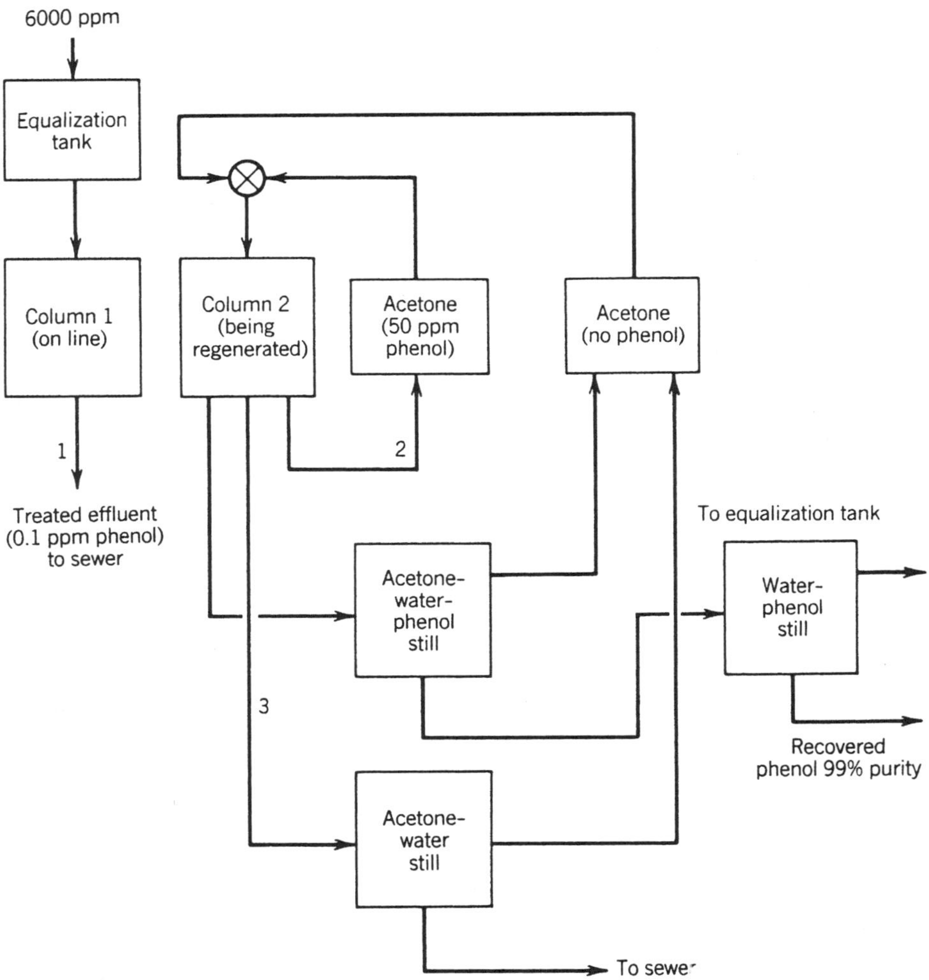

Figure 2.4.1 Schematic of a resin adsorption system. *Note*: The three wash streams resulting from regeneration are sequential cuts, taken in the order listed. The last cut (No. 3) is a postrinse with water.

Centrifugation is performed in a closed system and is therefore an excellent choice when treating volatile or toxic fluids. The liquid and solid are mechanically separated by centrifugal force. The removal of most of the liquid increases the solid concentration in, and reduces the volume of, the waste stream. The collected solid waste may then be treated and disposed or recovered. Three types of units are available for centrifugation: These are solid bowl, disk type, and basket type. The first two are used in large plants, the third, in smaller plants. A diagram of the solid-bowl centrifuge in shown in Fig. 2.4.2.[2]

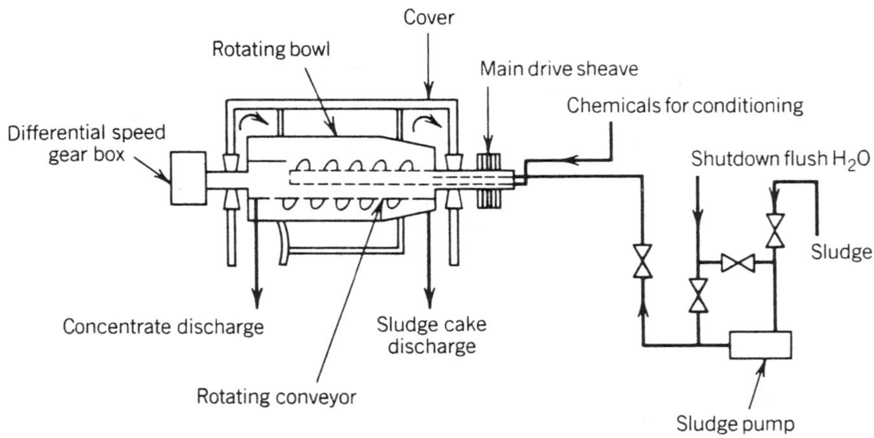

Figure 2.4.2 Solid-bowl centrifuge.

Distillation

Distillation is a liquid-phase separation process used to recover organic components from hazardous waste product streams. Because it is a nondestructive process with no effluent problems, rising chemical prices and stricter emissions regulations are making distillation a more competitive and popular recovery process. Some of its uses include recovering resins from resin adsorption systems after regeneration and recovering solvents from carbon adsorption systems and paint wastes. Distillation cannot be used to separate thick wastes such as sludges or slurries.

There are five types of distillation processes: batch, continuous fractional, azeotropic, extractive, and molecular. In general, the basis for separation is the difference in vapor pressures of the components in the mixture. The waste mixture fed to the vessel is heated to produce a liquid phase and a vapor phase. Components with relatively high volatilities rise to the top and those with lower volatilities flow to the bottom. After the components are separated, the vapor is usually condensed. The condensed vapor (*overhead product*) and the *bottoms product* stream may be recovered or treated and disposed. This process is illustrated in Fig. 2.4.3.[2] Capital, operating, and utility costs for distillation systems depend upon the nature of the feed and product streams.

Electrodialysis

Electrodialysis is a fully developed process used in separating ionic components and is commonly used in the desalination of brackish water. Its use for hazardous waste treatment, however, is still in the pilot plant stage. The potential uses of electrodialysis are in acid mine drainage treatment, the desalting of sewage plant effluents, and in sulfite-liquor recovery. One disadvantage of this

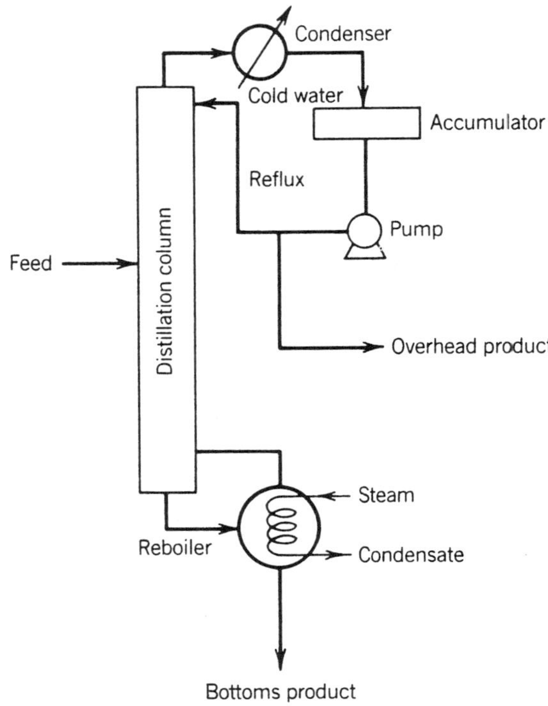

Figure 2.4.3 Continuous fractional distillation column.

process is that it may produce low-level toxic or flammable gases during operation. A schematic of an electrodialysis unit is shown in Fig. 2.4.4.[2]

In electrodialysis, separation of an aqueous stream is achieved through the use of synthetic membranes and an electrical field. The membrane allows only one type of ion to pass through and may be chosen to remove either anions or cations. The electric field causes the positive and negative ions to move in opposite directions and hence produces one stream rich in a particular ion and one depleted of that ion. Both streams may subsequently be recycled or disposed. It is important that pH and concentration be controlled for efficient operation.

Capital cost for an electrodialysis process is ~20–25% of the total cost of a waste treatment facility. Capital and operating costs vary depending upon the initial ion concentration and water volume.

Evaporation

Evaporation is a common process used in both water desalination and the treatment of hazardous wastes. It may be used to treat a variety of waste streams: liquids, slurries, sludges, organic and inorganic streams, streams con-

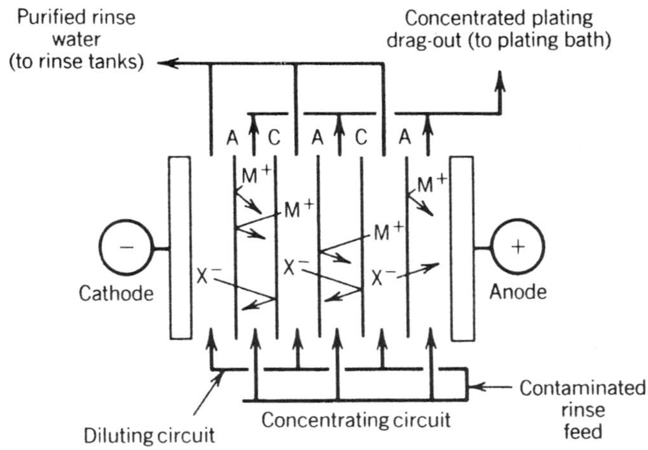

Figure 2.4.4 Schematic of an electrodialysis unit. C—cation selective membrane; A—anion selective membrane; M^+—cations; X^-—anions.

taining suspended or dissolved solids, and streams containing nonvolatile dissolved liquids. In waste treatment facilities, it has been used in processing radioactive wastes, wastes from paper mills and molasses distilleries, and wastes from trinitrotoluene (TNT) manufacturing.

The process and equipment used in evaporation are similar to that of distillation, except the vapor is not collected and condensed unless organic components are present. Usually the waste stream flows through metal pipes that are heated by low-pressure steam outside the pipe walls. Other modes of operation that have been used are the solar evaporation from ponds or the heating of open vessels. The process concentrates the hazardous wastes of the original feed stream and reduces its volume. Evaporation is an energy-intensive process and utility and equipment costs may be high.

Liquid–Liquid Extraction

Liquid–liquid extraction is used for the removal and recovery of organic solutes from aqueous and nonaqueous waste streams. Concentrations of solute in these streams range from a few hundred parts per million to a few percent. Most types of organic solutes may be removed by this process; though the technology is well developed, however, it is not commonly used in treating hazardous wastes. As environmental regulations on industrial wastewaters are made more stringent, its usage is likely to increase. Extraction has been used in removing and recovering phenols, oils, and acetic acid from aqueous streams and in removing and recovering freons and chlorinated hydrocarbons from organic solvent streams.

Liquid–liquid extraction may be performed in a single-stage or a multistage unit. In this process, the waste stream comes into contact with another liquid stream in which the waste stream solvent is immiscible. The organic solutes are soluble in the extracting solvent and are distributed between the two streams. Increased contact of the waste and solvent streams increases removal of the organic solutes. Recovery of the solute and solvent from the product stream is carried out by stripping or distillation. The recovered solute may be either treated and disposed, reused, or resold. Capital investment in this type of process depends on the particular waste stream to be processed.

Filtration

Filtration is a popular liquid–solid separation process commonly used in treating wastewater and sludges. In wastewater treatment, it is used to purify the liquid by removing suspended solids. This is usually followed by flocculation or sedimentation for further solids removal. In sludge treatment, it is used to remove the liquid (*sludge dewatering*) and concentrate the solid waste, thereby lowering the sludge volume. This method is highly competitive with other sludge dewatering processes. Filtration may also be used in treating nonaqueous liquid waste streams. It is likely to come into greater use as industrial wastewater regulations become more stringent.

In the filtration process, a liquid containing suspended solids is passed through a porous medium. The solids are trapped against the medium and the separation of solids from the liquid results. For large solid particles a thick barrier such as sand may be used. This is known as *granular media* filtration; Fig. 2.4.5 illustrates this process.[2] For smaller particles, a fine filter such as a filter cloth may be used. Fluid passage may be induced by gravity, positive pressure, or a vacuum. The filter is cleaned and the solids collected by passing a stream of water in the opposite direction of the waste stream flow; this is known as *backwashing*.

Flocculation, Precipitation, and Sedimentation

Flocculation, precipitation, and *sedimentation* are three processes used to separate hazardous waste streams that contain both a liquid and a solid phase. All three are well-developed, highly competitive processes, which are often used in the complete treatment of hazardous waste streams. They may also be used instead of, or in addition to, filtration. Some applications include the removal of suspended solid particles and soluble heavy metals from aqueous streams. Many industries use all three processes in the removal of pollutants from their wastewaters.

These processes work best when the waste stream contains a low concentration of the contaminating solids. Although they are applicable to a wide variety of aqueous waste streams, these processes are not generally used to treat nonaqueous waste streams or semisolid waste streams such as sludges and slurries.

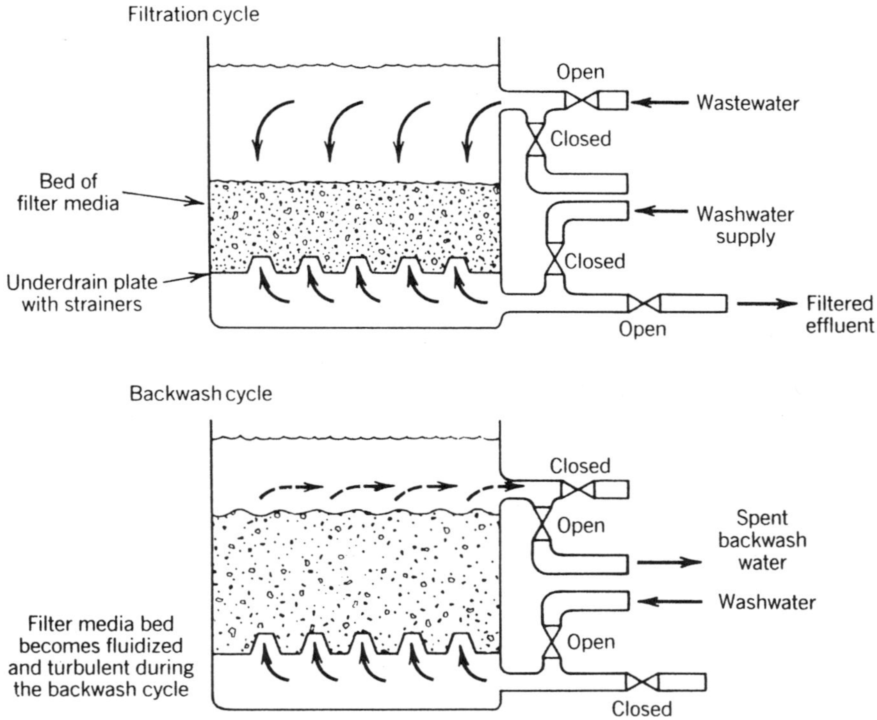

Figure 2.4.5 Granular media filtration.

Flocculation and precipitation are physicochemical processes. In *flocculation*, fine suspended particles, which are difficult to settle out of the liquid, are brought together by flocculating chemicals to form larger, more easily collected particles. In *precipitation*, metals soluble in the liquid are made insoluble by precipitating chemicals. *Sedimentation* is a physical process in which suspended solids are settled out through the use of gravitational and inertial forces acting on the solids. The solids removed by these processes form a sludge that may be disposed of or treated to recover any valuable material.

The operating costs for a hazardous waste treatment system using all three processes are dependent on the amount and type of flocculating and precipitating chemicals used. Systems using only sedimentation have much lower operating costs.

Flotation

Flotation, or dissolved air flotation, is basically a physical process used to remove organic or inorganic solids suspended in waste streams, such as wastewaters or slurries. The process is used to concentrate waste streams by forming a sludge to remove valuable or toxic solids. Since chemical reagents are sometimes used to remove specific inorganic solids, the process can be chemical as well as physical.

As a physical process, flotation has been used in removing oils and greases from waste streams. Flotation with chemical addition may be a future application for the removal of heavy-metal ions, cyanides, fluorides, and carbonyls from hazardous wastes. Flotation used to remove specific inorganics operates best when the desired product makes up <10% of the total solids in the stream.

The flotation process allows suspended solids to be removed from a hazardous waste stream through the use of agitation and, in some cases, reagents. The waste stream is agitated by air bubbles that reduce the density of the solid and carry the particles to the top of the liquid or slurry. A froth forms at the surface of the liquid or slurry, which is then removed by skimmers or scrapers. Individual materials or combinations of similar materials may be removed in this way if the proper reagents are used. Air emissions by the air bubbles cause no environmental problems. The sludge removed from the surface may be recovered or disposed. A typical air flotation system is shown in Fig. 2.4.6. Operating costs for flotation are dependent on the amount of waste treated, whether or not a reagent is used, and whether the solids are recovered and sold or disposed.

Freeze Crystallization

Freeze crystallization is a phase separation process not yet fully developed for commercial use in treating hazardous waste. Most of the development has been for use in water desalination. On a laboratory scale, this process has been used in the treatment of wastewater containing ammonium nitrate, paper mill bleach solutions, plating liquors, and arsenal redwater. The freeze crystallization process also has potential use in the removal of 1–10% total dissolved solids (TDS) in aqueous waste streams.

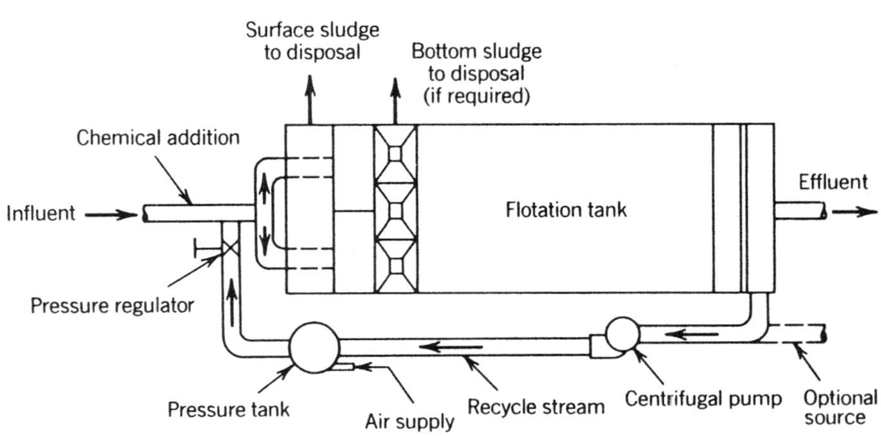

Figure 2.4.6 Dissolved air flotation system.

The contaminated wastewater is subcooled during freeze crystallization in order to form purified ice crystals. The remaining liquid is more concentrated in soluble organic or inorganic materials. The crystals are subsequently removed, washed, and melted to recover the water. The contaminated liquid could be freeze crystallized again to further concentrate the wastes and make waste disposal and treatment much easier.

High-Gradient Magnetic Separation

High-gradient magnetic separation (HGMS) is a phase and component separation process still in the development stage. It could be used to remove magnetic or nonmagnetic materials from a variety of hazardous waste streams: aqueous and nonaqueous liquids, slurries, sludges, and solids. Its potential use lies in the removal of ferromagnetic and paramagnetic particulates from liquids and slurries. It is being used on a small scale in wastewater treatment and coal desulfurization. The process works well when magnetic components make up only a small percentage of the total concentration, and solids make up 10–15% of the total waste stream.

The HGMS process uses a magnetized ferromagnetic filter to separate magnetic or paramagnetic particles from nonmagnetic particles. The filter captures the magnetic material, which is recovered when the filter is cleaned. By treating the waste stream with a magnetic seed, the filter can also be used to remove nonmagnetic material.

Capital investment may be low when only ferromagnetic material is removed but higher for the removal of nonmagnetic material. High-volume applications make the process more economically attractive. Full commercial development is not expected for at least 5 years.[1]

Ion Exchange

Ion exchange is generally used for the removal of dilute concentrations of heavy metals and anions from aqueous hazardous waste streams. Though the process is fully developed, it is not commonly used in industry. It has been used in recovering effluents from fertilizer manufacturing, the deionization of water, treating electroplating wastewaters, and looks promising for the removal of cyanides and selected heavy metals from waste streams.

Ion exchange is a two-step process. First, a solid material, the ion exchanger, collects specific ions after coming into contact with the aqueous waste stream. The exchanger is then exposed to another aqueous solution of a different composition that picks up the ions originally removed by the exchanger. The process is usually accomplished by sending the two aqueous streams through one or more fixed beds of exchangers. An illustration of this process is shown in Fig. 2.4.7. The ion-rich product stream may be recovered or disposed and the ion-poor stream is usually dilute enough for discharge to sewers.

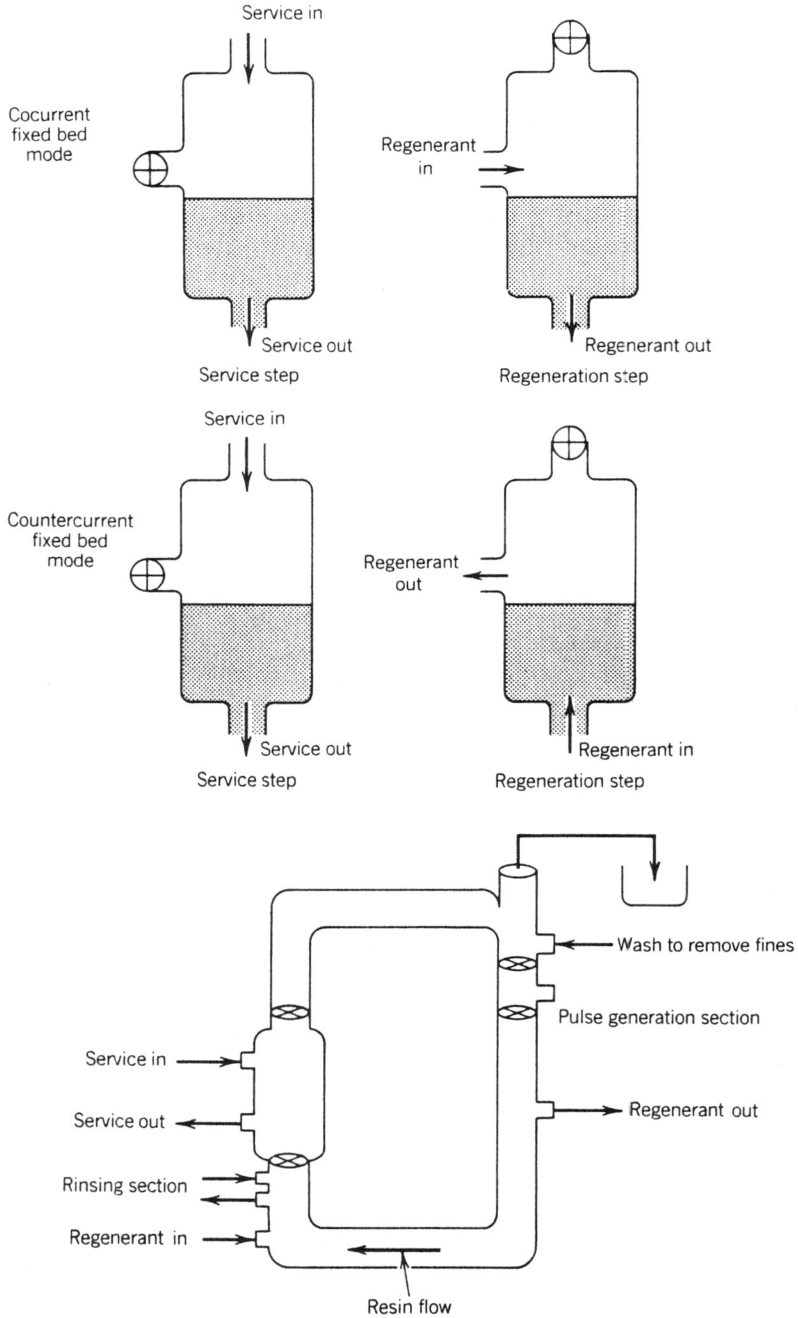

Countercurrent continuous mode (Higgins downflow type)

Figure 2.4.7 Operational modes for ion exchange.

Liquid Ion Exchange

Liquid ion exchange (LIE) serves the same purpose as ion exchange, the removal of heavy metals and anions from aqueous hazardous waste streams. However, LIE may be used for treating a higher concentration of contaminants. This process is also well-developed though not widely used in hazardous waste treatment. It has promising uses in the removal of cyanide from wastewaters and in treating hydroxide slimes produced in electroplating.

Liquid ion exchange uses an organic stream to carry out the transfer of ions from the aqueous waste stream to a second aqueous stream of different composition. The organic stream is immiscible in both aqueous streams. It removes the ionic or inorganic material from the waste feed stream and then passes it on to the second aqueous stream. The processing equipment is similar to that of liquid–liquid extraction. The ion-rich stream is either treated further for recovery or disposal. The purified aqueous stream usually contains a dilute concentration of organic solvent and may therefore require further treatment.

Capital investment is dependent upon the type of waste stream being treated. High-volume applications improve the economic aspects of this process.

Reverse Osmosis

Reverse osmosis is used in removing dissolved organic and inorganic materials from aqueous waste streams. Concentrations of these dissolved components are usually about 34,000 ppm or less. This process is more frequently used in water desalination but is becoming a more popular form of treating hazardous wastes. It is a good method for removing and concentrating waste streams, recovering valuable organic and inorganic material, and reducing wastewater volume.

Reverse osmosis employs a semipermeable membrane that allows passage of the solvent molecules, but not those of the dissolved organic and inorganic material. A pressure gradient is applied to cause separation of the solvent and solute. Any components that may damage or restrict the function of the membrane must be removed before the process is performed. Capital investment and operating costs depend on the waste stream composition.

Solidification

Solidification is a waste disposal process that transforms a hazardous waste into a nonhazardous solid product by fixation or encapsulation. In *fixation* a chemical or physical process and a solidifying agent are used to solidify the waste. *Encapsulation* is a process in which the waste is surrounded by a binder after it has been solidified by a chemical agent. Both produce a durable, impermeable, and supposedly environmentally safe product.

Solidification has only recently become a popular form of waste disposal. Previously, it had been used solely for treating radioactive wastes because of its high cost compared to other disposal means. Stricter regulations have made solidification a more popular choice of waste disposal. Solidification processes may be separated into five general categories depending on the solidifying agent used; these are silicate- and cement-based, lime-based, thermoplastic-based, organic polymer-based, and encapsulation techniques. All five of these processes are used to treat hazardous inorganic wastes. The cement- and lime-based methods are also used for sludges from stack-gas scrubbers after the sludges are chemically treated to precipitate out the metals. Solidification is unsuitable for treating organics, oxides, and toxic anions because these materials are difficult to solidify. Solidification has been used in treating wastes produced from steel mills, plating and lead-smelting plants, food production sludges, and sulfur residues.

The *silicate-* and *cement-based* methods may be used as batch or continuous processes. Portland cement and other additives are added to either a wet or sludgelike waste stream forming an impermeable, rocklike solid. The solidified product may then be used for land reclamation. The degree of solidification and strength of the product depends on the concentration of metals and organics present in the waste stream. This particular solidification process is the one most used in the United States. Some of its advantages are that the additives are not expensive, the process is well developed, equipment is available, and end product strength and permeability are easily controlled by the amount of cement used. The disadvantages are that waste weight and bulk are increased by the cement and pretreatment or special additives may be needed to ensure proper solidification.

The *lime-based* process is both chemical and physical. Lime, water, and siliceous material are added to the waste. A product known as *pozzolanic cement* is formed. Additives such as fly ash are also added to increase the strength of the end product. The degree of solidification depends on the reaction of the lime with the other components forming the cement. Once the product is formed, it is often used for landfill, mine reclamation fill, or capping material. The advantages and disadvantages of this process are similar to those of the cement-based process.

Thermoplastic solidification is a two-step process. Initially, the waste, which is first dried, is combined with paraffins, bitumen, and polyethylene at a temperature above 212°F; upon cooling, the mixture begins to solidify. In the second step, the solid waste is thermoplastically coated and then disposed. The thermoplasticized product is fairly permeable to most aqueous solutions. In addition to toxic inorganics, the process is also used in treating nuclear wastes. The equipment used in thermoplastic solidification is more specialized and therefore more costly than the equipment used in the cement-based or lime-based processes. Besides the end product's resistance to aqueous solutions, contaminant migrations are lower for this process than for the other four solidification methods. On the negative side, thermoplastics are flammable, and the waste stream must be dried before processing takes place.

In the *organic polymer* process, a monomer and a catalyst are combined with the waste stream, and the polymer is allowed to form. The product is then containerized and disposed. One popular technique is the urea-formaldehyde process that is used in treating nuclear wastes. Two advantages of the organic polymer process are that it may be used on wet or dry sludges and that the end product weighs less than those produced from the cement-based, lime-based, or thermoplastic-based techniques. The disadvantages are that some organic polymers are biodegradable, the end product must be containerized, and the end product may not be as durable as the end products produced from the other methods.

The *encapsulation* process is somewhat similar to the thermoplastic process. The dried waste is first chemically treated and then coated with a binder, usually polyethylene. The advantages of this process are that the encapsulated waste is very durable and resistant to water and deterioration, and the final product does not need to be stored in containers. The disadvantages are that the process is expensive, only small volumes of waste can be treated, and the sludge must be dried before processing.

The various solidification treatments may be performed on-site, which requires a large investment for the waste generator, or at regional facilities, which requires transportation costs. As with any other option, the choice and final cost of the solidification process will depend upon the type of waste involved, the desired end product, and the investment capabilities of the waste generator.[3]

Air Stripping

Air stripping is generally used in removing ammonia from biologically treated wastewater in order to reduce nitrogen concentrations. It may also be used to remove volatile organics and other gases. It has been used extensively at Superfund sites for soil remediation.

The packed tower (see Chapter 10) is the most compact and efficient device for air stripping. The wastewater in a lime slurry (for phosphate removal and pH control) is sent first to a mixing tank and then to a settling tank to settle out calcium phosphate and calcium carbonate. The clarified wastewater is introduced near the tops of two packed towers, while air is sent countercurrently through the towers to remove the ammonia. The concentration of ammonia in the air product should be about 3.7×10^{-7} lb/ft^3, which is below emission standards. The wastewater product is then sent to a recarbonation basin so that calcium carbonate may be precipitated, removed, and reused.

Steam Stripping

Steam stripping is used to remove dilute concentrations of ammonia, hydrogen sulfide, and other volatile components from aqueous waste streams. It has been used to recover sulfur from refinery waste and organics, such as phenol, from industrial wastes. Work is currently being done on its possible use in removing chlorinated light hydrocarbons from industrial wastewater.[1]

The steam stripping process is carried out in a distillation column, which may be either a packed or tray tower (see Chapter 10). Steam is sent up from the bottom of the column and is used to remove volatile components from the waste stream, which is flowing downward (see Fig. 2.4.8). The product stream, rich in volatile components, may be further treated to recover these components. If the volatiles contain sulfur, the stream may need treatment for sulfur dioxide emissions. The purified aqueous stream may be discharged if a sufficient amount of the volatiles has been removed.

Ultrafiltration

Ultrafiltration is a process used on aqueous wastewaters to remove both dissolved and suspended high-molecular-weight solids or colloids that are difficult to remove by other methods. This treatment can remove particles as small as 10^{-3}–10^{-2} microns (μm) and may be used to purify wastewaters by concentrating wastes or to fractionate waste streams. It is a fully developed process quite competitive with adsorption and evaporation in treating single-phase waste streams. It is currently being used in treating wastewaters produced by pharmaceutical, chemical, and electronics manufacturing companies. Its future uses may be in treating dye, pulpmill, and industrial laundry wastes.

Ultrafiltration uses a selective membrane to remove solutes or colloids from a pressurized waste stream passed through it (see Fig. 2.4.9).[2] The larger particles are retained by the membrane and the smaller particles pass through. The retained solids may be recovered or undergo further treatment before being

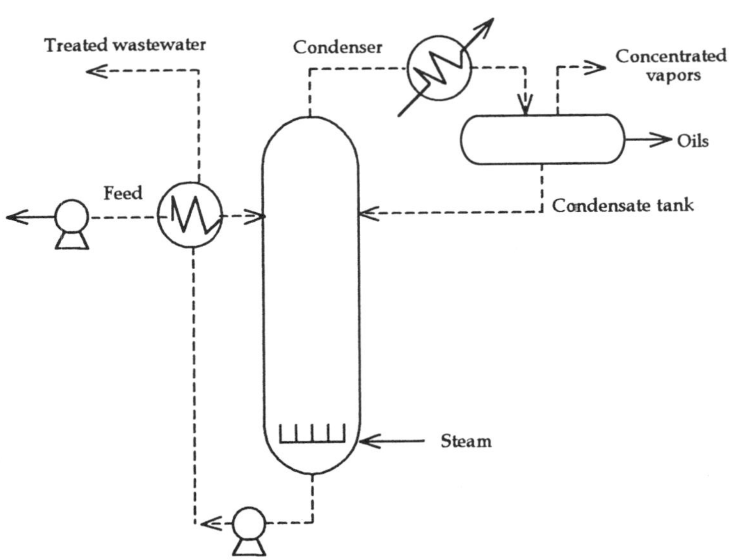

Figure 2.4.8 Steam stripping process.

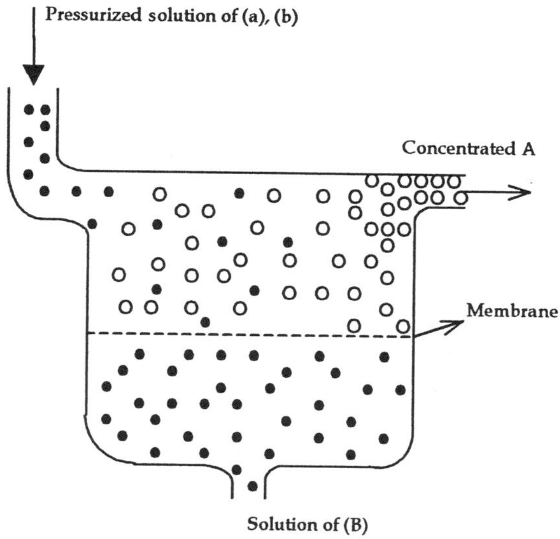

Figure 2.4.9 Membrane ultrafiltration.

disposed. Capital and operating costs depend on the capacity of the system, the type of wastewater being treated, and the type of membrane used.

2.5 ULTIMATE DISPOSAL

Ultimate disposal is described by many as the final process in the treatment and management of hazardous wastes.* Before disposal most wastes undergo the various treatments previously described (biological, chemical, and/or physical) in order to concentrate, detoxify, and reduce the volume of the wastes. This has not always been the case. Before the early 1970s, most wastes were haphazardly landfilled or ocean dumped with little concern to the environmental effects of these practices. The passage of certain environmental protection laws over the last two decades (see Chapter 3) has helped to eliminate the irresponsible handling of hazardous wastes by requiring that proper measures be taken before they are disposed.

Landfarming, deep-well injection, landfilling, and ocean dumping are four methods used in the so-called ultimate disposal of hazardous wastes. (Theodore[4,5] has recently added atmospheric dispersion to this list of four. However,

*The term *ultimate disposal methods* was coined by the U.S. Environmental Protection Agency (USEPA) and originally assigned to the four processes discussed in this section. The authors disagree with this nomenclature, preferring that *ultimate disposal* be reserved for a process in which the wastes are chemically or biologically rendered innocuous or a process that has been modified so that hazardous wastes are no longer generated.

this "solution to pollution by dilution" option does not realistically apply to hazardous wastes.) All four methods are discussed in this section. The latter three may be used on a variety of wastes; landfarming is used only on organic wastes. Landfarming is also a biological treatment method of treating wastes and could have been listed in Section 2.4. In this technique, nutrients in the soil convert the hazardous waste into nonhazardous materials that may enrich the soil. Wastes that are ocean dumped or landfilled also undergo some degree of biological conversion and degradation but not to the same extent as landfarmed wastes. The main purpose of ocean dumping, landfilling, and deep-well injection is the containment of the hazardous wastes.

Landfarming

Landfarming is one of several terms used to describe the process of disposing hazardous and nonhazardous wastes in the upper layer of the soil. The process is not a new one; it has been used for almost 30 years in the disposal of oily petroleum wastes. This application has increased in usage over the years and is now employed to treat up to 50% of these petroleum wastes as well as a number of other biodegradable wastes.

Landfarming has a number of advantages: It is an effective and low-cost disposal method; it is an environmentally safe and simple process not dependent upon processing equipment; and it is a natural form of waste disposal that can, in some cases, improve the fertility and nature of the soil.[6] There are also some limitations to landfarming. The bulk of the disposed waste must contain organic components. It is not recommended for use on inorganics, particularly when the pH of the waste is below 7. Wastes containing materials that could pollute the air, groundwater, or the soil itself are not candidates for this method. Some wastes may need pretreatment in order to make them suitable for landfarming; this additional treatment adds to the total cost of the waste disposal.

The mechanism behind landfarming is the breakdown of organic and inorganic material present in the waste by microorganisms residing in the soil. Once biodegraded, the material may then be used by other life forms. The upper layer is used in landfarming because it is more densely populated by microorganisms, resulting in a higher rate of biodegradation.

After its application to the soil, the waste may undergo decomposition, leaching of water-soluble components, volatilization, and finally, incorporation into the soil matrix.[6] Microorganisms, such as bacteria and yeasts, decompose the waste. Some of the carbon in the waste is released as carbon dioxide, and the remainder is converted to natural organic matter that benefits the soil. In order that maximum conversion take place, it is important that enough nitrogen and/or phosphates be present. It is also important that the degree of volatilization and leaching of the waste is low so that air and water are not polluted.

There are a number of factors that affect the degree of biodegradation of the wastes. These include: waste composition, microorganism and waste contact within the soil, temperature, pH, oxygen, inorganic nutrients (specifically nitrogen and phosphates), and moisture content.

- *Waste Composition* Waste composition is the most important factor affecting biodegradation. Earlier in this section it was mentioned that wastes consisting mostly of organic material (such as petroleum wastes) are the best candidates for landfarming and therefore biodegradation. Certain components are more readily biodegraded than others and their rate of biodegradation is therefore faster. By removing most nonorganic components before landfarming, the biodegradation rate may be increased.
- *Contact between Microorganism and Waste* The second factor that increases the degree of biodegradation is a larger contact area between the soil microorganisms and the waste. A mechanical device may be needed to achieve this in the case of sludges and semisolid wastes.
- *Soil Temperature* Soil temperature is one factor that is not easily controllable. Since biodegradation increases with increasing temperature, this process tends to work better in warmer climates and during the spring and summer months.
- *Soil pH* It was mentioned earlier that if the soil is too acidic (pH <7) excessive leaching of metals can occur, resulting in possible contamination of the groundwater. A pH between 7 and 9 is optimum for this process, and to achieve this, lime is often added to the soil.
- *Oxygen* Because the rate of aerobic biodegradation is greater than that of anaerobic biodegradation, an aerobic environment is preferred in the soil. If the soil is saturated with water and overloaded with biodegradable organic waste, it can become anaerobic. Control of the amount of waste and proper drainage are therefore required for this process to be successful.
- *Inorganic Nutrients* Sufficient amounts of nitrogen and phosphates are needed for a high degree of biodegradability. If not enough is present in the soil, these chemicals should be added in the form of fertilizer. Too much fertilizer, however, can cause groundwater contamination by the nitrates.
- *Moisture Content* The soil's moisture content is the final factor affecting the rate of biodegradation. A moisture content between 6 and 22 wt % is enough to produce good biodegradation.

The major steps involved in the landfarming process are as follows: site selection, site preparation, waste analysis, waste application, soil–waste blending, and post-waste-addition care. A landfarming site should be flat enough to prevent erosion problems but also slightly sloped to avoid flooding problems; the soil on the site should be uniform for better waste application; the waste should be well tested before application and pretreated if necessary; the waste should be applied to and mixed into the soil uniformly; and, after applying the waste, the land should be inspected periodically to make sure the previous factors are complied with.

Deep-Well Injection

Deep-well injection is an ultimate disposal method that transfers liquid wastes far underground and away from freshwater sources. Like landfarming, this disposal process has been used for many years by the petroleum industry and is used to dispose of salt waters in oil fields. When the method first came into use, very often the injected brine would eventually contaminate groundwater and freshwater sands because the site was poorly chosen. The process has been improved and laws such as the *Safe Drinking Water Act of 1974* ensure that sites for potential wells are better surveyed. Today, injection wells are placed as far away as possible from drinking water sources, usually at least $\frac{1}{4}$ mile.

Many factors go into the selection of a deep-well injection site. For example, the rock formation surrounding the disposal zone must be strong but permeable enough to absorb the liquid wastes, and the site must be far enough from drinking water sources to prevent contamination. Once a site is selected, it must be tested by drilling a pilot well. The performance data from the pilot well, besides testing permeability and water quality, also aid in the design of the final well and in determining the proper injection rate.[7]

The type of waste injected into the well plays a role in how deep the injection will be made. The more toxic the waste the farther down the disposal zone must be. Disposal zones have been classified into five different types:

1. *Zone of Rapid Circulation* This zone describes the area that runs from the soil surface down to only a few hundred feet. Injection is not done into this zone.
2. *Zone of Delayed Circulation* This zone contains circulating fresh water and may be used for certain wastewaters if properly monitored. The water circulation is slow enough so that residence times of a few decades to a few centuries can be achieved for the waste.
3. *Subzone of Lethargic Flow* The liquid flowing in this zone is very slow moving and saline. More concentrated hazardous wastes are injected here.
4. *Stagnant Subzones* The liquid contained in this region is hydrodynamically trapped, and the zone is generally several thousand feet below the soil surface. Highly toxic wastes may be injected here if the zone can properly accept and keep the waste.
5. *Dry Subzones* Salt beds fall under this classification. This zone does not contain water and is nearly impermeable. Because there is a possibility that liquid movement could occur through hydrofractures, this zone must be monitored. Otherwise, any waste injected here would be isolated from any water sources.

Salt beds have also been considered for storing nuclear wastes. One possibility is to inject water to remove some of the salt and thereby form an underground cavern. The cavern could then be filled with the waste, plus possibly a solidifying agent, and then cement sealed.

In order for wastewaters to be deep-well injected, the following criteria must apply: The wastewaters must have a low volume and high concentration of waste, must be difficult to treat by other methods, cannot cause an unfavorable reaction with material in the disposal zone, must be biologically inactive, and must be noncorrosive. Only wastes untreatable by other means are considered for deep-well injection because of the limited amount of available space for injection and its possible pollution of drinking waters.

Figure 2.5.1 shows an illustration of a typical injection well. The surface casing runs from the surface to about 200 ft below the freshwater table and is well sealed to prevent the degradation of freshwater sands. The protection cas-

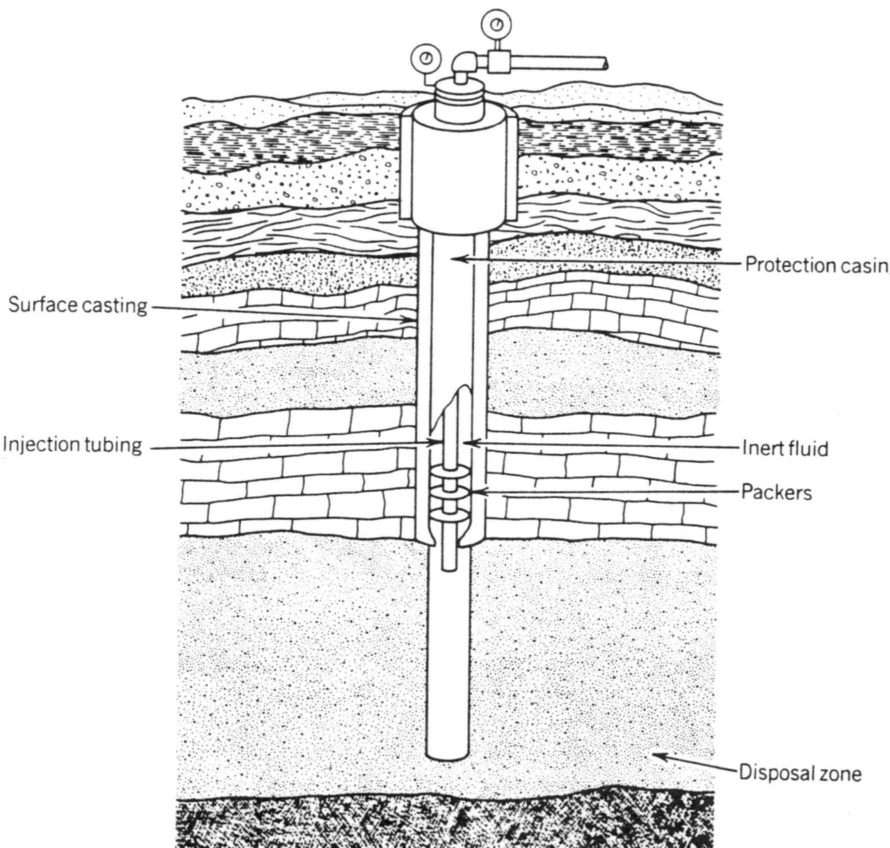

Figure 2.5.1 Injection disposal well.

ing is inside the surface casing and extends from ground level to the disposal zone; it is also sealed. The injection tubing is used to pass the waste stream through the well. It is sealed at the well head and just above the disposal zone. The annular space between the injection tubing and protection casing is filled with a noncorrosive fluid and kept at a pressure 0.5 psi (pounds per square inch) above the injection pressure.[7]

Landfilling

Landfilling is the third ultimate disposal method and is generally used on wastes in the form of sludges. There are two types of landfilling, area fill and trenching. *Area* fill is essentially accomplished above ground while *trenching* involves burying the waste. Trenching is the more established and popular form of the two. Since trenching requires excavation, however, area fill has the advantage that it requires less manpower and machinery. Area fill is also less likely to contaminate groundwater since the filling is above ground. Trenching, however, may be used for both stabilized and unstabilized sludges and makes more efficient use of the land. Both techniques require the use of lime and other chemicals to control odors and with both methods, cold and wet weather can cause problems. Both methods also produce gas, which can cause explosions or harm vegetation, and leachate, which can contaminate ground and surface water.

Most hazardous wastes must be subjected to one or more pretreatments such as solidification, degradation, volume reduction, and detoxification before being landfilled. This practice stabilizes the waste and helps decrease the amount of gas and leachate produced from the landfill. Landfilling is similar to landfarming in that both ultimate disposal methods combine wastes and soil. Landfarming, however, involves the biochemical reaction between soil nutrients and wastes to degrade and stabilize the waste; as a result, only specific types of wastes can be landfarmed. A larger variety of wastes may be handled by landfilling.

Area fill may be done in one of three ways: by mixing the waste with soil and forming a mound with the mixture and covering with soil; by spreading alternate layers of soil and soil–waste mixture over the area; or by filling a containment area surrounded by dikes with the waste and then covering with a soil layer. These forms of area fill are known as area fill mound, area fill layer, and dike containment, respectively. In *area fill mounds*, the sludge–soil mixture may be stacked in mounds as high as 6 ft. A soil covering, usually 3–5 ft, is placed on top of the mound. The amount of covering and soil-to-sludge ratio are dependent on the concentration of solids in the waste and on the degree of mound stability required. An earthen containment may be constructured for better mound stability. In an *area fill layer*, the sludge–soil mixture is spread out over an area in consecutive layers each having a thickness from 0.5 to 3 ft. Layers of soil 0.5–1 ft thick are often placed between the mixture layers. Final heights for these landfills will vary depending upon the amount of waste being disposed and the size of the site. Sites for these landfills are usually flat or slightly sloped. *Dike containment* landfills employ dikes to surround the four sides of the containment area. The site is often on a hill, which can serve as one of the containment sides. In this method, access roads are needed in order to bring the waste to the top of the landfill. Layers of soil, 1–3 ft thick, are usually placed at certain levels in the landfill, as well as a final layer on top, 3–5 ft thick. The final height of a containment area can range from 10 to 30 ft with widths of 50–100 ft and lengths of 100–200 ft.

Trenching involves placing the waste in an evacuated trench and covering it with one or more layers of soil. There are two types of trenching: *narrow* trenching operations, with trench widths of 10 ft or less, and *wide* trenching operations with trench widths >10 ft. The type of operation used depends upon the solids content of the waste. In either case, it is necessary that the width between trenches is large enough to achieve sidewall stability and space for soil stockpiles and equipment. There must be a soil thickness of 2–5 ft between the trench bottom and nearest groundwater level in order to provide adequate leachate control.

Wastes may be placed in *narrow* trenches in one application, with one soil layer placed on top of the waste. The soil removed from the trench may be used for the covering. Narrow trenches of 2–3 ft are best for sludges with low solids content (15–20%) because the support from the side walls prevents the soil layer from sinking to the bottom. Sludges with higher solids content should be placed in wider trenches.

In *wide trench* operations, excavating and filling equipment are placed within and at the top of the trench. The excavated soil is used to cover the sludge and to provide an intermediate soil layer in order to support the equipment. The sludge must have a fairly high solids content in order to support the equipment and soil covering at the top of the trench. If the trench is very wide, dikes may be used to confine the sludge to certain areas.

When a landfill site is to be selected, a number of factors must be evaluated; these factors fall into three categories: technical, economic, and public acceptance considerations. The following factors are technical considerations: haul distance, site size and life, topography, surface water, soils and geology, groundwater, soil quantity and suitability, vegetation, environmentally sensitive areas, archaeological or historical significance, site access, and land use. Economic considerations involve site capital cost, site operating cost, and hauling cost. Public acceptance considerations include local laws and public opinion; a public hearing is usually held in order to obtain local government and public input.

In comparing the costs of the different types of landfills, wide trench landfilling has the lowest capital and operating costs; area fill layer has the highest capital cost; and area fill mound has the highest operating costs.[8] The other types of landfills fall somewhere in between these extremes. Capital cost in this case involves land, site preparation, equipment purchase, and engineering. Overall costs for landfills are dependent upon the efficiency of the operation with respect to land use, equipment, and manpower. Hauling costs are independent of the landfill method and hence do not play a role in an economic comparison of the different techniques.

Once a landfill has been completed and closed, it must still be maintained in order to prevent soil erosion and other damaging effects of weather. It must also be monitored for gas and leachate as well as for any other potential environmental hazards.

Ocean Dumping

Ocean dumping is the last "ultimate" disposal method to be discussed. Although it is probably the simplest of the four techniques, its aftereffects and long-term consequences are more complex and less understood than those of other ultimate disposal methods. Regulations passed over the last 30 years have limited the types and amounts of wastes that may be dumped. Therefore, in some cases, ocean dumping may be no more attractive than any other ultimate disposal method, despite its simplicity.

Currently, only certain types of wastes may legally be ocean dumped. In some cases, such as industrial wastes, the disposal is regulated and may even require some treatment before disposal. *Uncontrolled* ocean dumping of untreated wastes occurs in the form of oil spills, rain carrying air pollutants, and runoff carrying land pollutants.

Until the early 1970s, ocean dumping in the United States was not strictly controlled; between 1945 and 1970, ocean dumping of wastes was used in ever-increasing amounts. This was done in order to avoid contaminating inland bodies of water. As a result of the passage of such regulations as the *Ocean Dumping Act* and the *Clean Water Act*, the practice of ocean dumping was severely restricted. The Ocean Dumping Act prohibited the dumping of any radioactive wastes and any biological or chemical warfare agents. It also gave the EPA and the Army Corps of Engineers the power to regulate ocean dumping of any wastes that were not dredge spoils.

Ocean dumping actually describes two forms of waste disposal: one into shallow offshore waters and the other into deep ocean waters. Each of these methods has advantages and disadvantages. The advantages of dumping offshore are more information and experience are available, transportation costs to the site are lower, and any resulting pollution is localized. The disadvantage is that dumping close to land could lead to problems with fishing industries and ruin offshore mineral deposits.[9]

Ocean dumping into deep waters offers the advantage of spreading the waste over large areas; the resulting lower waste concentration should have a less harmful effect on the marine environment. The disadvantage is that there is less information available and greater difficulty in monitoring the effects.

Ocean dumping is used not only as a final means of containment for hazardous wastes but also as a means of converting the hazardous waste to nonhazardous material. This is accomplished by dilution and biological conversion of the waste in the ocean. The mechanism for the biological degradation is not as well defined as that for landfarming, which is limited to specific types of hazardous wastes. Because such a variety of wastes are still ocean dumped and the long-term effects to the marine environment have not been well monitored, there is the possibility of very damaging, long-term consequences from this practice. The severity of the effects of ocean dumping are dependent on the type of waste, the input rate, the concentration of the waste in the ocean, the rate of dispersion, and the rate at which it undergoes biological conversion.

In recent years, criteria and monitoring procedures have been developed by the EPA that are used to measure the effects of ocean dumping, particularly for the offshore method. While this resulted in a better understanding of ocean dumping, more research is still needed to access the impact of present and possible future wastes (incineration residue, radioactive wastes) on the ocean environment.[9] There is still much to be learned about the various physical, chemical, and environmental processes that take place in the marine environment in order to understand and utilize ocean dumping fully.

2.6 PROCESS MODIFICATION/POLLUTION PREVENTION

As an alternative to treatments that change or dispose of a hazardous waste, the process that generates the waste can often be modified to reduce or eliminate the waste effluent. Although this method is being presented after the waste treatment and disposal methods (Sections 2.2–2.5), in practice it is usually the first line of action. The types of process modifications discussed here are pollution prevention options that include waste minimization/abatement, waste recycle/reuse, and waste exchange.

The term *pollution prevention*, in this text, is defined as that process or operation that attempts to reduce or eliminate the generation of wastes and/or pollutants that are emitted into the environment. Unfortunately, *pollution prevention* has come to mean different things to different people. Other terms—including waste minimization (a term that the EPA has discouraged because it focuses on hazardous wastes only), waste reduction, source reduction, and pollution minimization (a common term in industry)—have come, in a very general sense, to be used interchangeably with pollution prevention.

Applying pollution prevention strategies will not eliminate all wastes from all production processes. Rather, it offers a cost-effective means of minimizing the generation of waste. Source reduction is the first step in a hierarchy of options for reducing the pollution risks to human health and the environment. The next step in such a hierarchy is the responsible recycling of any wastes that cannot be reduced at the source. When recycling is conducted in an environmentally sound manner, it shares many of the same advantages as source reduction, such as conserving energy and other resources, reducing reliance on raw materials, and the need for end-of-pipe treatment or containment of wastes. Wastes that cannot be "feasibly" recycled should be treated in accordance with environmental standards that are designed to reduce both the hazard and volume of waste streams. Finally, any residues remaining from the treatment of waste should be disposed of safely in order to minimize their potential for release into the environment. This hierarchy establishes a set of presumptions rather than an ironclad rule since certain practices, such as treatment and proper disposal, can also be protective of the environment. Industries can be expected to balance costs and benefits when evaluating pollution prevention opportunities, considering such factors as savings in raw material and operating expenditures, avoided pollution control costs, reduced liability, and improved relations with local communities.[13] For more extensive detail on this topic the reader is directed to the literature.[7,10–14]

Waste Minimization–Abatement

In general, engineers are quite knowledgeable about the processes that generate wastes; few engineers are well informed on the systems and processes for the disposal of wastes. An option when considering alternatives for hazardous waste treatment is the redesign of the process that generates the waste. This can be accomplished either by substituting a new process for an old one or by revamping an old process. Some examples of such techniques are changing the solvent used from a hazardous to a nonhazardous solvent; changing the catalyst used from a hazardous to a nonhazardous catalyst; controlling the temperature and pressure of the reaction to inhibit hazardous side product formation; using higher quality raw materials that do not contain hazardous impurities; plus many others. Some specific examples where such methods have been used are:

In the manufacture of nitric acid, the air supply and temperature are carefully controlled to minimize the amount of NO_x emissions. (NO_x represents various oxides of nitrogen.)

In the pulp and paper industry, direct-contact evaporators are used in preference to the noncontact type to reduce particulate emissions.

In the purification of explosives where there are numerous washing steps, the effluent washes from one stage are used in later stages of the process.

The major drawback to this option is that most generators cannot afford to redesign the process; however, as disposal of hazardous waste becomes more difficult and more expensive, this option may gain in popularity.

Waste Exchange

The waste from one industry is often an acceptable raw material for another. Waste exchanges were first implemented and are now fairly common in Europe; there are few in the United States. A major stumbling block to widespread adoption of the method in this country has been the lack of an intermediary or an information bank to help match supply and demand. Even if a suitable industry (i.e., one that can use the waste) is found, it must be located fairly close to the waste generator. Transportation of the waste generally requires permitting and special handling considerations; this increases the cost.

Waste Recycle

In order to eliminate or significantly reduce waste discharge, recycle or reuse of waste streams can also be employed. In this method, however, treatment of the waste stream is usually required to regain the appropriate quality before the stream is used again. Recycle of waste streams is fairly common in many industries, for example: iron and steel manufacturing, explosive manufacturing, nonferrous metal manufacturing, steam and electric power plants, coal mining, organic acid and inorganic chemical manufacturing, aluminum forming, pharmaceutical manufacturing, battery manufacture, textiles, and paint-ink formulations. Usually, waste recycle is only used by large generators because of the high cost.

Methods of recycle are related to the particular reuse of the waste stream. The waste stream normally undergoes some processing to recover the liquid or solvent and to remove metal ions for disposal. Some examples of common recovery techniques are evaporation, reverse osmosis, acid absorption, and activated carbon adsorption. Whenever a recycle stream is used, the quality of the product produced should be carefully monitored.

Waste Minimization and Recycling Cannot Eliminate the Need for Incineration

Industrial wastes are generated from manufacturing processes that produce industrial and consumer products. Waste minimization (reducing the amount of waste produced) and recycling should be our first choices in managing wastes. Incorporating these concepts of *pollution prevention*[15,16] (preventing pollution at the source) are primary goals that should be rigorously pursued.

Advances in the technology of waste minimization and recycling have already resulted in a significant reduction in the amount of waste produced, but they will take time and resources to fully develop and implement. Even an aggressive waste reduction effort will leave substantial amounts of waste in need of final, safe disposal. The current USEPA definition of hazardous waste will be expanded in the future and more industrial wastes will be classified as hazardous. Existing contaminated sites and sediments in the beds of waterways will also require cleanup and waste treatment. Many of these remaining wastes can best be treated by incineration. The availability of a range of waste treatment options is critical to the long-term viability of many manufacturing items.[17]

2.7 THE SELECTION OF HAZARDOUS WASTE TREATMENT PROCESSES

If the hazardous waste cannot be eliminated through process modification, a viable hazardous waste management method must be chosen. The procedure used in this selection can be a long and complicated one. Before a specific treatment or means of disposal is chosen, the waste generator must determine how toxic and/or hazardous the waste is and if the treatment or disposal is to be done on- or off-site. Once these questions are answered, an appropriate method for treating the waste must be chosen; this choice is based not only on technical considerations but on economic and regulatory factors as well.

As described in Chapter 1, wastes may be classified in one of three ways: *highly toxic* or *highly hazardous* waste, which requires extra care in managing; *hazardous* waste, which may be managed by standard methods; and *nonhazardous* waste, which does not present much of a health or environmental threat. To determine the classification for a particular waste, federal (EPA) and state criteria should be used.

The next decision a waste generator must make is whether to manage the waste directly or have it done by a commercial firm. Companies that dispose of their wastes on-site do so because they have the technical and physical resources, prefer complete control over the waste management and its costs, and wish to avoid the specter of future liability associated with off-site disposal. According to the EPA, over 80% of all waste generators dispose of their wastes on-site for these reasons, or simply because there are no commercial firms within their area. If a waste generator decides to manage the waste directly, a method for treating and/or disposing of the waste must be chosen with regard to technical, economic, regulatory, and public opinion considerations.

Technical Considerations

Technical considerations include waste characteristics, technical suitability of the available treatments, and treatment objectives. The waste must be analyzed so that the proper form of treatment may be selected. Should the waste contain more than one type (e.g., a mixture of heavy metals and organics), more than one method of treatment may be needed. Waste treatment may be divided into five categories: physical, chemical, biological, thermal (incineration), and solidification or encapsulation (a subcategory of physical treatment). *Physical treatments* (Section 2.4) may be used alone or with the other types of treatment; their function is to concentrate wastes, reduce waste volume, and separate different waste components for continued treatment or disposal. *Chemical treatments* (Section 2.2) are used to convert hazardous wastes into other less hazardous forms; they are generally used on one substance or similar substances because the reactions involved are specific.[8] *Biological treatments* (Section 2.3) are more specific than the other types and are used mainly on organic wastes. *Solidification* and *encapsulation* (Section 2.4) are also treatments for specific types of waste; these methods work well on inorganic wastes but physical and/or chemical pretreatment is usually necessary.

Incineration is one of the most popular forms of waste treatment, especially when organic wastes are involved. It destroys most organic components in the waste, and the resulting sludge and ash are almost always less hazardous and easier to dispose of than the original waste. It is difficult to obtain a permit for incineration, however, and many stringent regulations govern its use. Table 2.7.1 shows the applicability of various treatment processes, excluding incineration, to a wide variety of different waste types.[1]

The *objective* of the treatment is another important factor that influences treatment selection. It may be desired, for example, to destroy the hazardous components in the waste, reduce the hazardous components and then isolate them, or merely to separate the hazardous components. The end product of the waste treatment must be compatible with the type of ultimate disposal to be used for the waste: sewage discharge, landfilling, deep-well injection, and so on. For example, if the waste is to be disposed of by landfarming, the end product of the waste treatment should consist of biodegradable organic components. Table 2.7.2 shows the output stream characteristics for many treatment processes, excluding incineration, as well as possible follow-up measures.[1] In Table 2.7.3, processes to perform various waste treatment tasks are compared. These tasks are (a) separation of heavy metals from liquid waste streams, (b) destruction of organics, (c) separation of toxic anions from liquid waste streams, (d) processes that can handle slurries or sludges, and (e) processes that can handle tars or solids. Not all of these processes are currently feasible for hazardous wastes; the second column describes the present state of the art for each process.

Economic Considerations

Economic feasibility is an obviously important factor to be considered in treatment selection. Cost is a strong function of the amount of waste to be treated and the simplicity and degree of commercial usage of the treatment process. The various costs include capital investment, operating costs, utility costs, and expense of the final disposal of the waste product.

Regulation Considerations

Environmental factors, in the form of state and local environmental regulations, also influence waste treatment selection. The EPA Resource Conservation and Recovery Act (RCRA) regulations especially control the type of waste treatment that may be used and the characteristics of the waste that is to finally be disposed. These regulations are the topic of Chapter 3.

Table 2.7.1 Applicability of treatment processes to various types of wastes[a]

Processes	Solid	Single Liquid Phase			Two Phases	
		Inorganic	Organic	Mixed	Slurry[b]	Sludge
Phase Separation						
Filtration	n	n	n	n	y	n
Sedimentation	n	n	n	n	y	n
Flocculation	n	n	n	n	n	n
Centrifugation	n	n	n	n	y	n
Distillation	n	n	n	n	y	y
Evaporation	n	n	n	n	y	y
Flotation	n	n	n	n	y	n
Ultrafiltration	n	n	n	n	n	n
HGMS	p	n	n	n	n	n
Precipitation	n	n	n	n	n	n
Component Separation						
Ion exchange	n	y	y	y	n	n
Liquid ion exchange	y	y	y	n	y	y
Freeze crystallization	n	y	y	n	y	y
Reverse osmosis	n	y	n	y	n	n
Carbon adsorption	n	y	n	y	n	n
Resin adsorption	n	y	n	y	n	n
Electrodialysis	n	y	n	y	n	n
Air stripping	n	n	y	y	y	n
Steam stripping	y	n	y	y	y	n
Ammonia Stripping	y	y	n	n	n	n
Ultrafiltration	n	y	y	y	n	n
Solvent Extraction	p	y	y	y	n	n
Distillation	n	n	y	y	y	y
Evaporation	n	y	y	y	y	y
Chemical Transformation						
Neutralization	n	y	y	y	y	y
Precipitation	n	y	y	y	n	n
Hydrolysis	p	n	y	y	y	y
Oxidation	n	y	y	y	n	n
Reduction	n	y	y	y	n	n
Ozonolysis	n	y	y	y	n	n
Calcination	y	y	y	y	y	y
Electrolysis	n	y	n	y	n	n
Microwave	n	n	y	y	n	n
Biological	n	n	y	y	y	y
Catalysis	n	y	y	y	n	n
Photolysis	n	n	y	y	n	n

[a]y = yes, workable; n = no; p = possible.
[b]Slurry is defined here as a pumpable mixture of solids and liquids.

Table 2.7.2 Characteristics of the end products of various waste treatments

Treatment		Output Streams			
Type	Process	Code Number	Form	Characteristics	Possible Follow-on Steps
Phase separation	Filtration	F1	Sludge	15–20% solids	Landfill, calcination
		F2	Liquid	500–5000 ppm total dissolved solids (TDS)	Component separation
	Sedimentation	S1	Sludge	2–15% solids	Decantation
	Centrifugation	S2	Liquid	10–200 ppm suspended solids (SS)	
	Flotation	FI1	Stabilized	Particle-bearing froth	Skimming
		FI2	Liquid	Solution	Component separation
	HGMS	H1	Slurry	Magnetic and paramagnetic particulates	Recovery
		H2	Liquid	Solution	Component separation
	Flocculation	Fc	Sludge or slurry	Flocculated particulates	Sedimentation, filtration, centrifugation
	Distillation	Dis1	Sludge	Still bottoms	Calcination
		Dis2	Liquid	Pure solvent	Sale
	Evaporation	E1	Solid		Resource recovery
			Liquid	Condensate	Recovery or disposal
Component separation					
Inorganics	Ion exchange	IE1	Liquid	Concentrated solution of hazardous components	Precipitation, recycle, electrolysis
		IE2	Liquid	Purified water with hazardous components at ppm levels	Discharge
	Liquid ion exchange	Similar to ion exchange			
	Carbon adsorption	CA1	Solid	Adsorbate on carbon	Chemical regeneration
		CA2	Liquid	Purified water	Discharge

(*Continued*)

Table 2.7.2 (*Continued*)

Treatment		Output Streams			
Type	Process	Code Number	Form	Characteristics	Possible Follow-on Steps
	Reverse osmosis	RO1	Liquid	Concentrated solutions of hazardous components	Precipitation, electrolysis, recycle
		RO2	Liquid	Purified solution, TDS > 5 ppm	To water treatment
	Electrodialysis	ED1	Liquid	Concentrated stream, 100–500 ppm salts	Precipitation, metal recovery
		ED2	Liquid	Dilute stream	To water treatment
	Freeze crystallization	FC1	Sludge	Concentrated brine	Recovery
		FC2	Liquid	Purified stream, ~100 TDS	To water treatment
Organics	Carbon adsorption	CA1	Solid	Adsorbate on carbon	Thermal or chemical regeneration
		CA2	Liquid	Purified water	Discharge
	Resin adsorption	RA1	Solid	Adsorbate on resin	Solvent regeneration
			Liquid	Purified water, <10 ppm organics	To water treatment
	Steam stripping	SS1	Liquid	Aqueous steam concentrated in volatile organics	Recovery; incineration
			Liquid	Dilute aqueous stream with 50–100 ppm organics	To water treatment
	Solvent extraction	SE1	Liquid	Concentrated solution of hazardous components in extraction solvent	Recovery of extraction solvent
		SE2	Liquid	Purified liquid; hazardous component concentration <10 ppm	Recycle or discharge
	Distillation	Dis1	Sludge	Still bottoms	Incineration
		Dis2	Liquid	Pure liquid	Sale
Chemical transformation					
	Neutralization	N	Unchanged	Stream of altered pH	Component separation
	Precipitation	Ppt1	Sludge		Landfill, calcination

	Precipitation	Ppt2	Liquid	Supernatant with concentrations governed by solubility of precipitate	Depends on product stream composition
	Oxidation	O	Liquid	CO_2, H_2O and other oxidation products	Depends on product stream
	Ozonation				composition
	Calcination	Cal1	Solid	Oxide and/or other residue	Landfill or recovery
		Cal2	Gas	Volatiles (CO_2, NO_x, SO_x, hydrocarbons, fine particles)	Wet scrubbing
	Hydrolysis	Hy	Liquid, slurry, or sludge	Mixture of products that may or may not be toxic	Resource recovery
	Electrolysis	El1		Cathode reaction products	Often a recovered metal
		El2		Anode reaction products	Depends on nature of products
	Photolysis	P	Liquid	Solution of photodecomposition products	Depends on nature of products
	Microwave discharge	M	Gas	Similar to incinerator emissions	Wet scrubbing
	Catalysis	Depends on reaction catalyzed			
Biological treatment					
	Activated sludge	AC1	Liquid	Clean water	Discharge
		AC2	Sludge	Heavy metals and refractory organics	Calcination
	Aerated lagoon				
	Trickling filter				
	Waste stabilization pond	W	Liquid	Clear water	Discharge
	Anaerobic digestion	An1	Sludge		Incineration
		An2	Gas	CO_2, methane	Fuel recovery
	Composting	Com1	Sludge	Concentrated in metals	Calcination or recovery
		Com2	Liquid	Leachate solution of partially degraded organics	Component separation

Table 2.7.3 Comparison of waste treatment processes

Process	Required Feed Stream Properties[a]	Characteristics of Output Stream(s)
(a) Processes That Separate Heavy Metals from Liquid Waste Streams		
Physical removal		
Ion exchange	Concentration <4000 ppm; aqueous solutions, low SS	One concentrated in heavy metals; one purified
Reverse osmosis	Concentration >400 ppm; aqueous solution; controlled pH; low SS; no strong oxidants	One concentrated in heavy metals; one with heavy metal concentrations >5 ppm
Electrodialysis	Aqueous solutions; neutral or slightly acidic; Fe and Mn <0.3 ppm; Cu <400 ppm	One with 1000–5000 ppm heavy metals; one with 100–500 ppm heavy metals
Liquid ion exchange	Aqueous solutions; no concentration limits; no surfactants; SS <0.1%	Extraction solvent concentrated in heavy metals; purified water or slurry
Freeze crystallization	Aqueous solutions; TDS ~10%	Concentrated brine or sludge; purified water, TDS ~100 ppm
Chemical removal		
Precipitation	Aqueous or low viscosity nonaqueous solutions; no concentration limits	Precipitated heavy-metal sulfides, hydroxides, oxides, etc.; solvent with TDS governed by solubility products of precipitates
Reduction	Aqueous solutions; concentrations of heavy metals <1%; controlled pH	Acidic solutions with reagent (oxidized $NaBH_4$ or Zn); metallic precipitates
Electrolysis	Aqueous solutions; heavy metal concentrations <10%	Recovered metals; solution with 2–10 ppm heavy metals
(b) Processes That Destroy Organics		
Biodegradation	Dilute aqueous streams with soluble organics	Pure water
Oxidation	Dilute aqueous solutions of phenols, organic sulfur compounds, chlorinated hydrocarbons, etc.	Oxidation products in aqueous solutions
Ozonation	Aqueous solutions; concentrations <1%	Oxidation products in aqueous solutions

Table 2.7.3 (*Continued*)

Process	Required Feed Stream Properties[a]	Characteristics of Output Stream(s)
Calcination	Organics and inorganics that decompose thermally	Solid oxides; volatile emissions
Hydrolysis	Aqueous or nonaqueous streams; no concentration limits	Hydrolysis products
Photolysis	Aqueous streams; transparent to light; components that absorb radiation	Photolysis products
Chlorinolysis	Chlorinated hydrocarbon waste streams; low sulfur; low oxygen; can contain benzene and other aromatics; no solids; no tars	CCl_4; HCl, and phosgene
Microwave discharge	Organic liquids or vapors	Discharge products; not accurately predictable
(c) Processes That Separate Toxic Anions from Liquid Waste Streams		
Physical removal		
Ion exchange	Inorganic or organic anions in aqueous solution	Concentrated aqueous solutions
Liquid ion exchange	Inorganic or organic anions in aqueous solution	Concentrated solutions in extraction solvent
Electrodialysis	Aqueous stream with 1000–5000 ppm inorganic salts; and pH; Fe and Mn <0.3 ppm	Concentrated aqueous stream (10,000 ppm salts); dilute stream (100–500 ppm salts)
Reverse osmosis	Aqueous solutions with up to 34,000 ppm total dissolved solids	Dilute solution (~5 ppm TDS); concentrated solution of hazardous components
Freeze crystallization	Aqueous salt solutions	Purified water; concentrated brine
Chemical removal		
Oxidation	Aqueous solutions of cyanides, sulfides, sulfites etc.; concentrations <1%	Oxidation products
Ozonation	Aqueous solutions of cyanides; concentrations <1%	Cyanate solutions

Table 2.7.3 (*Continued*)

Process	Required Feed Stream Properties[a]	Characteristics of Output Stream(s)
Electrolysis	Alkaline aqueous solutions of cyanides or concentrated HCl solutions (>20%)	Cyanides to ammonium and carbonate salt solutions; HCl to Cl_2 gas
(d) Processes That Can Accept Slurries or Sludges		
Calcination	Waste stream with components that decompose by volatilization (hydroxides, carbonates, nitrates, sulfites, sulfates)	Solid greatly reduced in volume; volatiles
Freeze crystallization	Low-viscosity aqueous slurry or sludge	Brine sludge; purified water
HGMS	Magnetic or paramagnetic particles in slurry	Particles absorbed on magnetic filter
Liquid ion exchange	Solvent-extractable inorganic component	Solution in extraction solvent
Flotation	Flotable particles in slurry	Froth
Hydrolysis	Hydrolyzable component	Hydrolysis products
Anaerobic digestion	Aqueous slurry; <7% solids; no oils or greases; no aromatics or long-chain hydrocarbons	Sludges; methane and CO_2
Composting	Aqueous sludge; <50% solids	Sludge; leachate
Steam distillation	Sludge or slurry with volatile organics	Volatile, solid residue
Solvent extraction	Solvent extractable organic	Solution of extracted components; residual sludge
(e) Processes That Can Accept Tars or Solids		
Calcination	Tars or solids that can be volatilized	Volatiles; char and/or metal oxides
Hydrolysis	Tars or solid powders	Hydrolysis products
Steam distillation	Solids contaminated with volatile organics	Purified solids; condensed organics
Dissolution	Tars, solids, or solid powders that will dissolve in some reagent	Liquid solution for further treatment; solid residue
Crushing and grinding	Bulk solid	Powdered solid
Cryogenics	Tars, bulk solid	Reduced particle size

[a]SS = suspended solids; TDS = total dissolved solids.

Public Opinion

The attitude of local residents and politicians also has an influence on hazardous waste management. As described earlier, the NIMBY (not in my back yard) syndrome often arises when a hazardous waste facility is about to be constructed, in spite of adherence to EPA regulations that have been established to protect the community. If the waste generator works with the community early in the planning stages for a new treatment facility, there is likely to be less public opposition later on. A review of these factors is available in the literature.[18]

REFERENCES

1. National Technical Information Service, U.S. Department of Commerce, *Physical, Chemical, and Biological Treatment Techniques for Industrial Wastes*, Vol. 1, PB-275-054, 1977.
2. Office of Research and Development, U.S. Environmental Protection Agency, *Treatability Manual: Volume 1/1. Technologies for Control/Removal of Pollutants*, EPA-600/8-80-042c, July 1980.
3. R. B. Pojasek, "Solid-Waste Disposal: Solidification," *Chem. Eng.*, 140–145, August 13 (1979).
4. R. Dupont, K. Ganesan, and L. Theodore, *Pollution Prevention*, 2nd ed., CRC Press, Boca Raton, FL, 2000.
5. R. Theodore: personal notes.
6. R. L. Huddleston, "Solid-Waste Disposal: Landfarming," *Chem. Eng.*, 119–124, February 26 (1979).
7. G. W. Dawson and B. W. Mercer, *Hazardous Waste Management*, Wiley, New York, 1986.
8. M. Sittig, *Landfill Disposal of Hazardous Wastes and Sludges*, Noyes Data Corporation, Park Ridge, NJ, 1979.
9. B. H. Ketchum, D. R. Kester, and P. K. Park, *Ocean Dumping of Industrial Wastes*, Plenum, New York, 1981.
10. J. A. Mackie and K. Niesen, "Hazardous-Waste Management: The Alternatives," *Chem. Eng.*, 50–64, August 6 (1984).
11. R. E. Kenson, "Recovery and Reuse of Solvents for VOC Air Emissions," *Environmental Progress*, **4**(3), 161–164, August (1985).
12. H. Bhatt, R. Sykes, and T. Sweeney, *Management of Toxic and Hazardous Wastes*, Lewis Chelsea, MI, 1985.
13. K. Noll, C. Haas, C. Schmidt, and P. Kodukula, *Recovery. Recycle, and Reuse of Industrial Wastes,* Lewis Chelsea, MI, 1985.
14. L. L. Tavlarides, *Process Modifications for Industrial Source Reduction*, Lewis, Chelsea, MI, 1985.
15. R. Dupont, L. Theodore, and K. Ganesan, "Pollution Prevention," Lewis Publishers, CRC Press, Boca Raton, FL, 1999.

16. L. Theodore and R. Allen, "Pollution Prevention," *A Theodore Tutorial*, Air and Waste Management Association Bookstore, Pittsburgh, PA, 1996.
17. "Hazardous Waste Incineration: What Engineering Experts Say," A.S.M.E.-CRTD-Vol. 32, Pg. 2–3, 1994.
18. J. J. Santoleri, L. Theodore, and J. Lauber, "Facts or Myths: The Burning Issue of Incineration"—AWMA-86th Annual Meetings, Denver, CO, June 13–18, 1993.

3

STANDARDS AND REGULATIONS

Contributing author: S. Heather McHale;
Partial contribution: Michael L. Budin

3.1 INTRODUCTION

The emergence of hazardous waste as a major environmental issue has forced society to reexamine the ultimate ends of the whole pollution control enterprise. The hazardous waste problem is a by-product of this nation's rapid economic growth over the last 40 or 50 years. As industry has expanded, we have developed more and more reliance on a broader range of chemicals and chemical by-products in plastics, automobiles, and other major manufacturing sectors. When chemicals are used to develop new products, however, wastes are generated, and unfortunately the phenomenal growth in the production of waste was not mirrored by growth in the field of waste management. Much of the waste produced made its way into the environment where it poses a serious threat to ecological systems and public health. It was not until the mid-1970s that Congress enacted the *Resource Conservation and Recovery Act* (RCRA), the first major hazardous waste regulatory document to address waste generation and disposal effectively.

This chapter is organized in historical fashion. In Section 3.2, hazardous waste legislation prior to the RCRA of 1976—what little there was—is discussed. Section 3.3 examines the 1976 RCRA with particular emphasis on Subtitle C of the act, which lays the groundwork for the current standards and regulations pertaining to the management of hazardous waste. In Section 3.4, the 1984 *Hazardous and Solid Waste Amendments* (HSWA) to the RCRA, also known as the *New RCRA*, are discussed. Along with key elements of the amendments, a discussion of Subtitle 1, which expands the regulation domain of Subtitle C, is included. Section 3.5 covers the *permitting process* for incinerators, treatment/storage tanks, and land treatment, with special emphasis on incinerators. Section 3.6 covers recent activities by the U.S. Environmental Protection Agency (USEPA) in assisting and expediting the permit processes for incinerators and other hazardous waste combustors including concerns regarding emissions of toxic metals, hydrogen chloride and chlorine and their control, control of products of incomplete combustion (PICs), and measurement techniques for all emissions of concern. Section 3.7 covers the release and implementation of the boiler and industrial furnace regulations. Section 3.8 covers the USEPA hazardous waste reduction and combustion strategy in the 1993–1994 era under the Clinton/Gore/Browner administration in Washington. Included in this section are the activities initiated by the USEPA beginning on May 18, 1993, with the moratorium on new incinerator construction for an 18-month period followed by public meetings in February and November, 1993. These hearings openly discussed waste minimization and eventual issuance of the Maximum Achievable Control Technology (MACT) rules for combustion of hazardous wastes in incinerators, cement kilns, and lightweight aggregate kilns. Section 3.9 discusses the Combustion Emissions Technical Resources Document (CETRED), issued in May, 1994. Section 3.10 covers the MACT rules initially introduced in April, 1996, following many reviews by industry, industrial associations, peer review groups, and public groups. It received final approval by the Office of Management and Budget (OMB) and was released in final form on September 30, 1999. Section 3.11 discusses the future of incineration in light of economic factors, pollution prevention, waste minimization, recycling and reuse, and restrictions on land disposal. In addition, the future of incineration is discussed and compared with other available alternatives.

3.2 EARLY LEGISLATION

Prior to the passage of RCRA, little solid waste legislation existed. Practices that permitted technologies to remove pollutants efficiently from air and water streams, only to dispose of them indiscriminantly on land, were commonplace. Legislative authority to control solid waste disposal dates as far back as the 1899 *River and Harbor Act*, the main concern of which was keeping the waterways *navigable*, rather than keeping them *clean.* The following is a quotation from the *Refuse Act*, which comprises Section 13 of the River and Harbor Act[1]:

> . . . it shall not be lawful to throw, discharge or deposit, or cause, suffer or procure to be thrown, discharged or deposited either from or out of any ship, barge, or other floating craft of any kind, or from the shore, wharf, manufacturing establishment, or mill of any kind, any refuse matter of any kind or description whatever other than that flowing from streets and sewers and passing therefrom in a

> liquid state, into any tributary of any navigable water from which the same shall float or be washed into such navigable water; and it shall not be lawful to deposit, or cause, suffer, or procure to be deposited material of any kind in any place on the bank of any navigable water, or on the bank of any tributary of any navigable water, where the same shall be liable to be washed into such navigable water; either by ordinary or high tides, or by storms or floods, or otherwise, whereby navigation shall or may be impeded or obstructed: provided that nothing herein contained shall extend to, apply to, or prohibit the operations in connection with the improvement of navigable waters or construction of public works, considered necessary and proper by the United States officers supervising such improvement or public work: and provided further, that the Secretary of War, whenever in the judgment of the Chief of Engineers, anchorage and navigation shall not be injured thereby, may permit the deposit of any material above mentioned in navigable waters within limits to be defined and under conditions to be prescribed by him, provided application is made to him prior to depositing such material; and whenever any permit is so granted the conditions thereof shall be strictly complied with, and any violation thereof shall be unlawful.

The first specific solid waste legislation aimed at protecting the environment did not pass through Congress until the mid-1960s. The *Solid Waste Disposal Act* of 1965 fostered a research program concerned with utility-generated solid waste and the recovery of valuable fractions of municipal solid waste. This research and development (R&D) program concentrated on the improvement of collection, transport, recycling, reuse, processing, and disposal methods of solid waste. The program was jointly administered by the Department of Interior (Bureau of Mines) and the Department of Health, Education, and Welfare (Bureau of Solid Waste Management).

The *Resource Recovery Act* of 1970 provided a new directive to the 1965 Solid Waste Disposal Act. While the 1965 act was limited mainly to R&D, the 1970 act expanded the federal role to the planning of solid waste management programs and training; and while the 1965 act made no mention of *hazardous wastes*, the 1970 act registered concern for the adverse health effects of such wastes and directed the Department of Health, Education, and Welfare to study the problem of storage and disposal of hazardous wastes.

The 1970 act initiated a change in the nation's attitude toward waste disposal. For one thing, the emphasis shifted from simply waste disposal to include waste management and resource recovery; for another, it became evident that the existing regulations were inadequate to assure the proper management of hazardous wastes. The result of this latter concern was the passage by Congress in 1976 of the RCRA, an amendment to the Resource Recovery Act of 1970. See Section 3.3 below for additional details.

In 1976, the Toxic Substances and Control Act prohibited the further manufacture of polychlorinated biphenyls (PCBs) after July 2, 1979, established limits on PCB use in commerce, and established regulations for proper disposal of PCBs. In 1980, the Comprehensive Environmental Response, Compensation and Liability Act (CERCLA) established a National Fund (Superfund) to assist in remedial actions of cleanup of the uncontrolled waste sites created by the poor disposal practices of the past. In 1984, the Hazardous and Solid Waste Act (HSWA) amended the RCRA and established a strict timeline for restricting untreated hazardous waste from land disposal. This is covered in more detail in Section 3.4. The Superfund Amendments and Reauthorization Act (SARA) of 1986 not only reauthorized the Superfund program, but greatly expanded the provision and funding of the initial act. In 1988, the Medical Waste Tracking Act established a 2-year demonstration program to track regulated medical waste from the point of generation to the point of disposal in five states. The Clean Air Act Amendments (CAAA) of 1990, as well as numerous state and county regulations, regulated emissions from various incineration sources.

3.3 RESOURCE CONSERVATION AND RECOVERY ACT OF 1976

The RCRA of 1976 completely replaced the previous language of the Solid Waste Disposal Act of 1965 to address the enormous growth in the production of waste. It provided "cradle to grave" provisions for controlling the storage, transport, treatment, and disposal of hazardous waste. The objectives of this act were to promote the protection of health and the environment and to conserve valuable materials and energy resources by[1,2]:

- Providing technical and financial assistance to state and local governments and interstate agencies for the development of solid waste management plans (including resource recovery and resource conservation systems) that promote improved solid waste management techniques (including more effective organizational arrangements), new and improved methods of collection, separation and recovery of solid waste, and the environmentally safe disposal of nonrecoverable residues.
- Providing training grants in occupations involving the design, operation, and maintenance of solid waste disposal systems.
- Prohibiting future open dumping on the land and requiring the conversion of existing open dumps to facilities that do not pose danger to the environment or to health.
- Regulating the treatment, storage, transportation, and disposal of hazardous wastes that have adverse effects on health and environment.
- Providing for the promulgation of guidelines for solid waste collection, transport, separation, recovery, and disposal practices and systems.
- A national research and development program for improved solid waste management and resource conservation techniques, more effective organizational arrangements, new and improved methods of collection, separation, recovery, and recycling of solid wastes and environmentally safe disposal of nonrecoverable residues.

- Promoting the demonstration, construction, and application of solid waste management, resource recovery, and resource conservation systems that preserve and enhance the quality of air, water and land resources.
- Establishing a cooperative effort among federal, state, and local governments and private enterprises in order to recover valuable materials and energy from solid waste.

Structurewise, the RCRA is divided into eight subtitles. These subtitles are (A) General Provisions; (B) Office of Solid Waste; Authorities of the Administrator; (C) Hazardous Waste Management; (D) State or Regional Solid Waste Plans; (E) Duties of the Secretary of Commerce in Resource and Recovery; (F) Federal Responsibilities; (G) Miscellaneous Provisions; and (H) Research, Development, Demonstration, and Information. Portions of Subtitles A and C are discussed here; in Subtitle A, key words and terms used in the act are defined; Subtitle C lays the framework for the RCRA hazardous waste management program.

Subtitle A: General Provisions

In Section 1004, legal definitions for many key terms used in waste management are presented. Some of the more pertinent of these are[3]:

> *Hazardous waste:* A solid waste or combination of solid wastes, which because of its quantity, concentration, or physical, chemical or infectious characteristics may (a) cause, or significantly contribute to an increase in mortality or an increase in serious irreversible, or incapacitated reversible illness, or (b) pose a substantial present or potential hazard to human health or the environment when improperly treated, stored, transported or disposed of, or otherwise managed.
>
> *Hazardous waste generation:* The act or process of producing hazardous waste.
>
> *Hazardous waste management:* The systematic control of collection, source separation, storage, transportation, processing, treatment and recovery and disposal of hazardous waste.
>
> *Solid waste:* Any garbage, refuse, sludge from a waste treatment plant, water supply treatment plant, or air pollution control facility and other discarded material, including solid, liquid, semisolid or contained gaseous material resulting from industrial, commercial, mining and agricultural operations and from community activities, but does not include dissolved material in domestic sewage or solid or dissolved material in irrigation return flows or industrial discharges which are point sources subject to permits under the federal *Water Pollution Control Act* or source, special nuclear, or byproduct material as defined by the *Atomic Energy Act.*
>
> *Storage*: When used in connection with hazardous waste, means the containment of hazardous waste, either on a temporary basis, or for a period of years, in such a manner as not to constitute disposal of such hazardous waste.

Disposal: The storage, deposit, injection, dumping, spilling, leaking or placing of any solid waste or hazardous waste into or in any land or water so that such solid waste or hazardous waste or any constituent thereof may not enter the environment or be emitted into the air or discharged into any waters including ground waters.

Treatment: When used in connection with hazardous waste, means any method, technique or process, including neutralization designed to change the physical, chemical or biological character or composition of any hazardous waste so as to neutralize such waste or as to render such waste non-hazardous, safer for transport, amenable for recovery, or reduced in volume. Such terms include any activity of processing designed to change the physical form or chemical composition of hazardous waste so as to render it non-hazardous.

Subtitle C: Hazardous Waste Management

Subtitles C and D generate the framework for regulatory control programs for the management of hazardous and solid nonhazardous wastes, respectively. The hazardous waste program outlined under Subtitle C is the one most people associate with the term *RCRA*. It is divided into 11 subsections, each of which is summarized here, using much of the language contained in the original document.[2,4]

1. *Section 3001. Identification and Listing of Hazardous Waste* Section 3001 required EPA to promulgate the criteria to be used in determining how hazardous a particular waste is. Factors such as flammability (ignitability), corrosivity, toxicity, persistence, and degradability in the environment as well as potential for accumulation in tissue are to be considered in defining the criteria. In addition, other hazardous characteristics may be added to the list in the future if the need arises.
2. *Section 3002. Standards Applicable to Generators of Hazardous Waste* The administrator (EPA) is required to promulgate regulations, after consultation with appropriate federal and state agencies, establishing standards for generators of hazardous waste. These standards are to establish record-keeping practices that accurately identify the waste stream's constitutents and quantities; the disposition of the waste; labeling the waste for storage, transport, or disposal identification; use of appropriate containers; furnishing information to treatment facilities, storage, or disposal facilities on the chemical composition of the waste; use of a manifest system to reconcile the disposition of the waste and to assure waste only proceeds to a permitted facility; submitting reports to EPA regarding the quantities of waste generated during this reporting period; and the disposition of all hazardous wastes during that same period.
3. *Section 3003. Standards Applicable to Transporters of Hazardous Waste* Under Section 3003, standards for the transporters of hazardous wastes are established. The standards involve record keeping of the waste trans-

ported, including source and delivery points; proper labeling of the wastes; compliance with the prescribed manifest system; and limiting transportation of all such hazardous waste only to permitted facilities designated on the manifest form. This section also stipulates that EPA must coordinate its efforts with the Department of Transportation in those cases where the hazardous wastes fall under the Hazardous Materials Transportation Act.

4. *Section 3004. Standards Applicable to Owners and Operators of Hazardous Waste Treatment, Storage and Disposal Facilities* This section provides for the establishment of standards and regulations for facilities that treat, store, and dispose of hazardous wastes. These standards and regulations pertain to maintaining records of all hazardous waste handled with respect to treatment, storage, and disposal; reporting, monitoring, and inspection of manifests for wastes received to assure compliance; maintaining and operating facilities in accordance with procedures satisfactory to EPA; designating the location, design, and construction of the facility; maintaining contingency plans to minimize unanticipated damage; maintaining the operation of facilities to provide training, continuity of ownership, and to provide standards for financial responsibility.
5. *Section 3005. Permits for Treatment, Storage, and Disposal of Hazardous Waste* This section covers permits for the treatment, storage, or disposal of hazardous waste. A permit application must include estimates of the quantities, compositions, and concentrations of the waste anticipated, as well as the time, frequency, or rate of receiving and handling the waste, and description of the site. The permit process is discussed in detail in Section 3.5.
6. *Section 3006. Authorized State Hazardous Waste Program* State hazardous waste programs under Section 3006 will receive approval from EPA unless the state program (a) is not equivalent to the federal programs, (b) is not consistent with federal or state programs of other states, or (c) does not provide adequate enforcement of compliance. Any action taken under a hazardous waste program authorized under this section has the same force and effect as actions taken by EPA under this subtitle.
7. *Section 3007. Inspections* This section provides one of the means of monitoring compliance. Through it, EPA is given the right to gain entry to facilities and records where hazardous wastes are generated, treated, stored, or disposed of for the purpose of inspection and sampling. Records, reports, and information resulting from these inspections are available to the public, unless the operator can demonstrate that release of these data would divulge information privileged to protection.
8. *Section 3008. Federal Enforcement* When EPA determines that a facility is in violation of the regulations, a failure-to-comply notification is issued. If the violation extends beyond 30 days after notification, EPA

may (a) issue a second notification requiring compliance within a specified time limit, (b) assess civil penalties, or (c) revoke the RCRA permit. Criminal penalties may be assessed to transporters delivering to nonpermitted facilities, persons disposing of waste without a permit, or persons found to have falsified information on permit applications.

9. *Section 3009. Retention of State Authority* In order to make the federal program the minimum standard for hazardous waste regulation, states are prohibited through Section 3009 from enacting regulations less stringent than those promulgated under RCRA.
10. *Section 3010. Effective Date* Section 3010 requires that, 90 days after the promulgation of Section 3001, all generators, transporters, and disposers must have completed the hazardous waste notification procedure. This section also sets a period of 6 months between effective date and promulgation date of Subtitle C regulations.
11. *Section 3011. Authorization of Assistance to States* This section authorizes funding to assist states in the development and implementation of hazardous waste programs at the state level.

The RCRA regulations are continuously being developed and published according to the following procedure. When a regulation is first proposed, it is published in the *Federal Register* as a *proposed* regulation, along with a discussion of EPAs rationale for the regulation. For a period of time, normally 60 days, public reaction is invited. After this period, EPA revises the proposed regulation and finalizes it by again publishing it in the *Federal Register.* Annually, all such regulations are compiled and placed in the *Codes of Federal Regulations* (CFR); this process is called *codification.* Most of RCRA has been codified in this manner and can be found in Volume 40 of the CFR.[4]

3.4 HAZARDOUS AND SOLID WASTE AMENDMENTS OF 1984

Since 1976 the RCRA has been twice amended by Congress, once in 1980 and again in 1984. Thus it has, since its inception, undergone a process of continuous evaluation in order to keep pace with society's changing needs. The 1984 amendments, known by the title *Hazardous and Solid Waste Amendments* (HSWA or the New RCRA) significantly expanded the scope of RCRA and has resulted in numerous changes in EPA's policy and practice. These amendments charted a new course for hazardous waste management—away from historic land disposal practices and toward much greater use of treatment technologies prior to disposal. HSWA also further articulated the national policy to emphasize pollution prevention as the first and primary goal for the waste management program.

As a result of the above change, combustion of hazardous waste in incinerators, boilers and industrial furnaces (BIFs) began to increase substantially.

Concurrent with increased use of combustion as a form of waste management, public concerns began to be voiced about the safety and reliability of combustion facilities. In addition, citizens began to ask whether an overabundance of combustion capacity serves to undercut reduction of waste generated at industrial facilities. Section 3.8 covers the activities, initiated by the EPA in the late 1980s and early 1990s, to take a closer look at the safety and reliability of hazardous waste combustion. Key elements of these HSWA amendments are given here[4]:

1. *Landfilling of Liquids* The placement of bulk or noncontainerized liquid hazardous waste or hazardous waste containing free liquids in any permitted or interim status landfills is banned after May 8, 1985. As of November 8, 1985, the placement of nonhazardous liquids in permitted or interim status landfills is also banned; however, an owner or operator may obtain an exemption from the ban on nonhazardous liquids by demonstrating to EPA's satisfaction that the only reasonably available alternative for these nonhazardous liquids is a landfill or outlined surface impoundment that already contains hazardous waste, and that the disposal of the nonhazardous liquids in the owner's or operator's landfill will not present a risk of contamination to any underground source of drinking water.
2. *Minimum Technological Requirements* This provision establishes minimum standards for hazardous waste management facilities such as double liners or their performance equivalent for new land disposal facilities or interim status facilities.
3. *Corrective Action* Any permit issued after November 8, 1984, must require corrective action for all releases of hazardous waste or constituents from any solid waste management unit regardless of when the waste was placed in such a unit; financial assurance for the completion of such corrective action is required as well. This provision applies to any solid waste management unit, including inactive units, at any treatment, storage, or disposal facility seeking a permit under RCRA.
4. *Cleanup Beyond a Facility's Property Boundary* The owner or operator of a permitted facility must institute corrective action beyond the facility boundary whenever necessary to protect human health and the environment, unless the owner or operator shows that permission to undertake such action on adjacent property cannot be obtained.
5. *Groundwater Monitoring Variances* The new regulations allow surface impoundments, waste piles, and landfills to waiver groundwater monitoring requirements if such units are fitted with a double liner, leak detection system, leachate collection system (for landfills and waste piles), and are located entirely above the seasonal high water table. This section also exempts any lined waste pile from groundwater monitoring requirements even if the pile does not have leak detection and leachate

collection systems as long as the pile is located entirely above the seasonal high water table and the waste is removed periodically so that the liner can be examined.

6. *Groundwater Monitoring Variance for Certain Engineered Structures* This variance is applicable on a case-by-case basis, only to an engineered structure that (a) does not receive or contain liquid waste (or waste containing free liquid), (b) is designed and operated to exclude liquid from precipitation, or other runoff, (c) has multiple leak detection systems within the outer layer of containment that are operated and maintained throughout the life of the unit, including the closure and postclosure care periods, and (d) prevents the migration of hazardous constitutents beyond the outer layer of containment prior to the end of the postclosure period.
7. *Salt Dome Formations, Salt Bed Formations, Underground Mines, and Caves* For all noncontainerized (or bulk) liquid hazardous waste, the placement of waste in these four settings is prohibited until (a) EPA has determined and promulgated performance and permitting standards and (b) a permit has been issued for the facility. For containerized liquid and all other nonliquid hazardous waste, the placement of such waste in the four settings is prohibited until a permit has been issued to the facility.
8. *Dust Suppression* This provision prohibits the use of waste oil, used oil, or other material that is contaminated with dioxin or any other hazardous waste (other than a waste that is identified as hazardous solely on the basis of ignitability) for dust suppression or road treatment.
9. *Underground Injection* The injection of hazardous waste into or above any underground formation that contains an underground source of drinking water within $\frac{1}{4}$ mile of the injection well is prohibited. This statutory ban on so-called Class IV wells is effective, automatically, 6 months from the date of enactment of the amendments, in any state that does not have identical or more stringent prohibitions in effect under an applicable underground injection control (UIC) program. The ban does not apply to the injection of contaminated groundwater into the aquifer from which it was withdrawn if (a) the injection is part of a federally supervised cleanup action under RCRA, (b) contaminated groundwater is treated to substantially reduce hazardous constituents prior to injection, and (c) such cleanup, when completed, will be sufficient to protect human health and the environment.
10. *Small-Quantity Generators* Small-quantity generators (between 100 and 1000 kg/month) of hazardous waste must: (a) treat, store, or dispose of the waste at a permitted facility, (b) dispose of the waste at a facility authorized by a state to manage municipal or industrial solid waste, (c) store no more than 1000 kg on-site, or (d) recycle the waste. These

requirements remain in effect until EPA promulgates new standards applicable to small-quantity generators.

11. *Preconstruction Ban/TSCA Exception* Construction of a new hazardous waste management facility without an RCRA permit is prohibited. However, this statute makes exception for any facility with EPA approval under the *Toxic Substances Control Act* (TSCA) for the incineration of PCBs. Under TSCA regulation, any PCB article or PCB container may be stored for disposal for up to 1 year. The facility need not obtain a permit for construction of the storage unit but must ensure that the unit meets certain criteria specified in the regulation.
12. *Permit Life* Any permit for a treatment, storage, or disposal facility is for a fixed term, not to exceed 10 years. Each permit for a land disposal facility is to be reviewed 5 years after the date of issuance or reissuance and modified as necessary to assure that the facility complies with currently applicable requirements necessary to protect human health and the environment.
13. *Authority to Add Conditions* The intent of this amendment is to authorize the agency (EPA) to impose permit conditions beyond those mandated by the regulations, where necessary to protect human health and the environment.
14. *Expansion of Interim Status for Newly Regulated Units* (The term *interim* is defined in detail in Section 3.5. A facility that has been granted such status is temporarily regarded by EPA as being permitted and allowed to operate even though Part B of the two-part permit application has not yet received EPA approval.) A facility may qualify for interim status if it is in existence on the effective date of a statutory or regulatory change under RCRA that requires the facility to have a permit (e.g., a facility that treats, stores, or disposes of a newly listed hazardous waste).
15. *Loss of Interim Status for Failure to Submit Part B* Interim status for owners and operators of land disposal facilities terminates within a 12-month period unless the owner or operator submits Part B (see previous paragraph) prior to that date and demonstrates compliance with the applicable groundwater monitoring and financial responsibility requirements. Interim status for owners and operators of incinerators terminates by November 8, 1989, unless Part B of the application was submitted by November 8, 1986. For all other facilities, interim status terminates by November 8, 1992, unless Part B is submitted by November 8, 1988.
16. *Ban on Hazardous Waste in Certain Cement Kilns* Cement kilns located in incorporated cities with populations >500,000 are prohibited from burning hazardous waste, or any fuel containing a hazardous waste, unless these kilns comply with the regulations applicable to hazardous waste incinerators. This prohibition remained in effect until EPA developed standards for cement kilns burning hazardous waste in 1991 under the BIF rules.

17. *Labeling of Hazardous Waste Fuels* Any fuel containing a hazardous waste, or a hazardous waste burned directly as a fuel, must include a warning label in the invoice or bill of sale for the fuel. The warning label must state that the fuel contains hazardous wastes and must list the hazardous wastes contained therein. The statute contains three exceptions to the labeling requirements; these are for hazardous-waste-derived petroleum coke and for two types of hazardous-waste-derived fuels from petroleum refining operations.
18. *Household Wastes* A resource recovery facility generating energy by burning municipal waste is not considered to be managing hazardous waste, as long as the facility receives and burns only household waste or solid waste from other sources that contains no hazardous waste. Such a facility may not accept hazardous wastes from any nonhousehold sources.
19. *Minimum Technological Requirements for Incinerators* Incinerators that receive permits after enactment of the amendments must meet a minimum destruction and removal requirement; namely, that the principal organic hazardous constitutents (POHCs) of the waste feed be incinerated with a minimum destruction and removal efficiency (DRE) of 99.99%.
20. *Exposure Information and Health Assessments* Part B permit applications for landfills and surface impoundments (described in detail in Section 3.5) must be accompanied by information on the potential for exposure of the public to hazardous wastes through releases related to the unit.
21. *Additional Criteria and Delisting Procedures* In evaluating a delisting petition, EPA must consider factors other than those for which the waste was listed, if there is reason to believe that such additional factors could cause the waste to be a hazardous waste.
22. *Research, Development, and Demonstration Permits* The HSWA provide EPA with authority to issue permits for research, development, and demonstration treatment activities for which permit standards have not been established. As authorized, EPA reserves the right to waive or modify procedures to expedite permitting as long as human health and the environment are protected. The permit is to be issued for a maximum of 360 days of operation and may be renewed up to three times.
23. *Hazardous Waste Exports* In general, the export of hazardous waste out of the country is prohibited unless: (a) the person exporting such waste provides notification to EPA, (b) the government of the receiving country has consented to accept the waste, (c) a copy of the receiving country's written consent is attached to the manifest that accompanies the shipment, and (d) the shipment conforms to the terms of the consent. In lieu of meeting these requirements, a person may export hazardous waste if the United States and the government of the receiving country

have entered into an agreement and the shipment conforms to the terms of the agreement.

24. *Waste Minimization* Except for small-quantity generators, waste generators must certify that waste volume and/or quantity and toxicity have been minimized and that the generator's proposed treatment, storage, or disposal method minimizes the threat to human health and the environment.

The HSWA amendments differ from previous hazardous waste management legislation in that Congress not only placed *explicit* requirements in the statute, but also instructed EPA in general language to develop *specific* regulations. Another factor that distinguishes the HSWA is the ambitious timetable they established; many of the HSWA provisions were already in place in 1984; others went into effect shortly thereafter. The HSWA also established what are called *hammer* provisions—statutory requirements ultimately banning the landfilling of hazardous wastes that go into effect automatically if EPA fails to issue regulations by certain dates.[5]

3.5 PERMIT PROCESS

The permitting of any hazardous or toxic waste management site or process is a highly complex, lengthy, and costly procedure. The RCRA imposes legally enforceable requirements on owners or operators of *TSD* facilities (TSD is an acronym for the treatment, storage, or disposal of solid wastes considered hazardous under RCRA). A permit is granted only after compliance with federal standards has been demonstrated. Under this permit, the facility has legal authorization to treat hazardous waste within the limits specified in the permit.

An RCRA permit application has been composed of two parts: A and B. To satisfy application requirements for Part A, Consolidated Permit Application Forms 1 (EPA Form 3510-1) and 3 (EPA Form 3510-3) must be completed and submitted to the appropriate federal or state agency. To satisfy application requirements for Part B, a wide variety of procedures to safeguard human health and the environment, for example, groundwater monitoring, contingency plans, emergency response plans, and risk assessment must be described in detail. The EPA has provided listings and general descriptions of what must be included in this part. However, because there is no specific form to complete, the success of the Part B application depends on how well the applicant can follow and interpret EPA directives to the satisfaction of the agency (federal, state, and/or local) that has RCRA permitting authority over the facility.[6]

Definitions of Terms

Definitions of selected terms associated with the RCRA permit application process are given here.[7] Again, most of the language of the original RCRA document is used.

- *Closure* The process in which the owner/operator of a hazardous waste management facility discontinues active operation by treating, removing from the site, or disposing of on-site, all hazardous wastes in accordance with an approved closure plan. *Closure* entails specific financial guarantees and technical tasks that are included in a permittee's closure plan and that the permittee must implement as part of his/her permit obligations. Under RCRA, the facility must be closed in such a way that (1) minimizes the need for further facility maintenance and (2) controls, minimizes, or eliminates postclosure escapes of hazardous waste, its constitutents, or byproducts to ground or surface waters or to the atmosphere. Generally, closure must occur within 90 days after the facility receives its final shipment of hazardous wastes.
- *Closure Plan* A plan, subject to approval by authorized regulatory agencies, which must be submitted by the owner/operator of a hazardous waste management facility as part of the RCRA permit application. The approved plan becomes part of the permit conditions subsequently imposed on the applicant. The plan identifies steps required to (1) *completely* or *partially close* the facility at any time during its intended operating life and (2) *completely close* the facility at the end of its intended operating life.
- *Acutely Hazardous Wastes* Commercial chemical products and manufacturing intermediates that are listed (under their generic names) by RCRA as *acutely hazardous substances.* Also included are off-specification commercial chemical products and manufacturing chemical intermediates that, if they met specifications, would have the generic names listed; and any residue or contaminated soil, water, or other debris resulting from the cleanup of a spill of any of these substances.
- *Postclosure Care Period* The period following closure of a hazardous waste management facility during which the owner/operator must continue to secure the facility sufficiently to protect human health and the environment. The postclosure care period entails specified and guaranteed financial obligations plus the performance of certain technical tasks such as groundwater monitoring. Typically, this period continues for 30 years after the date of completing closure of a TSD facility for which postclosure care is required.
- *Postclosure Plan* Similar to the closure plan, except that it identifies activities (monitoring, maintenance, etc.) to be performed after closure of a TSD facility.
- *TSD Facility* Any facility that treats, stores, and/or disposes of hazardous wastes.
- *Delisting* The process of exclusion (or petitioning for exclusion) of a solid waste from the definition of *hazardous waste* even though that waste is listed as *hazardous* under RCRA. To have a waste delisted, the owner/operator of a facility must petition for a regulatory amendment, the grant-

ing of which is contingent on demonstrating to EPA and/or state authorities that the waste does not meet any of the criteria under which the waste was originally listed as hazardous. The amendment, if granted, excludes only the waste generated at the specific facility and for which the applicant has provided sufficient demonstration by way of sampling, testing, and other procedures required by RCRA.

- *Existing TSD Facility* A treatment, storage, or disposal facility that began operation or construction on or before November 19, 1980; such a facility may qualify for *interim status.* (See definition of *interim status* below.)
- *New TSD Facility* One that began operation or construction after November 19, 1980; such a facility cannot qualify for interim status.
- *Final Authorization* A final permission granted by federal RCRA authorities for a state program to supersede the federal RCRA program. In such a case, all hazardous waste activities in the state are subject only to the state and not to the federal program.
- *Notification or Notification Form* A form that notifies the regulatory agencies of any hazardous waste management activities. Owners/operators of existing TSD facilities must complete and submit this form to the authorized federal or state RCRA regulatory authorities. New facilities are not required to file notification, but both Parts A and B of the RCRA permit application must be filed before these facilities can legally begin operation. Existing facilities that handle hazardous wastes identified in RCRA regulations must file notification no later than 90 days after the waste is listed in the *Federal Register.*
- *Interim Status* The period during which the owner/operator of an existing TSD facility is treated as having been issued an RCRA permit even though Part B of the permit application has not yet been submitted. Owners/operators of a new facility cannot, by definition, qualify for interim status. An existing facility may automatically qualify for interim status if the owner/operator files the notification form with Part A. Interim status continues until Part B is requested by regulation officials, submitted by the applicant, and processed. The applicant has at least 6 months to submit Part B after it is requested by the authorities. Interim status should not be confused with *interim authorization* (see below), a term that relates specifically to state programs and has nothing to do with permit applications.
- *Interim Authorization* Temporary authorization granted by federal RCRA authorities to a state that allows some components of the state requirements to apply in place of federal RCRA requirements. However, federal requirements still apply for those components not covered by the interim authorization.
- *Finally Effective RCRA Permit* A set of legally enforceable requirements imposed on the owner/operator of a TSD facility whose completed RCRA permit application (Parts A and B) has been approved and duly processed by authorized regulatory agencies. In return for complying with permit

conditions, the permittee receives legal authorization, within the limits specified in the permit, to handle hazardous waste. The *finally effective permit* is generally issued for a fixed period not to exceed 10 years, after which the permittee must reapply for a new one at least 180 days before the current permit expires. A finally effective permit may also be subject to updating (via an application to be filed by the permittee) if changes are made at the permitted facility.

Application for a Permit

The specific steps that must be taken to secure an RCRA permit depends on (a) whether the facility is existing or new, (b) its location, and (c) its operational characteristics. The facility is considered by RCRA to be an *existing facility* if it was operating or under construction on or before November 19, 1980. Submission of both notification and Part A of the application automatically qualifies such a facility for interim status; this facility is treated by EPA as if it were permitted, until a final determination is made on the permit application. If the facility is *new,* that is, one that began operation or construction after November 19, 1980, both Parts A and B of the RCRA application must be filed with the appropriate agencies at least 180 days before construction of the facility may begin. To operate the new facility, both parts of the application must first be accepted and a finally effective permit issued.

Since the completion of Part A is normally simple and routine, it will not be discussed further. Part B is much more complicated and warrants more complete discussion.

Part B

Two types of information are required for Part B of the application.[6] The first type is *general information* that is required on all RCRA permit applications; the second type covers *facility-specific information* on the particular type of treatment, storage, and disposal operations employed at the facility. These may include either one or a combination of the following methods: landfill, surface impoundment, tanks, containers, incinerators, or land treatment. The effort involved in completing the facility-specific section of Part B can be relatively easy or difficult depending on the specific methods of disposal employed. For example, if only temporary storage facilities such as tanks or containers are involved, the task is fairly simple; if treatment such as incineration or land disposal is involved, the task is quite complex.

General Information The general information requirements for Part B are fairly extensive and involve a variety of tasks and documentation such as engineering studies, chemical, hydrogeological, and geological analysis, and planning and financial analysis. A public advisement program through the mechanism of public hearings is also necessary. Other pertinent general requirements imposed on virtually all RCRA permit applicants are general facility description, chemical and physical waste characterization, waste analysis plan, security procedures, general inspection schedule, preparedness and prevention requirements, contingency plan, traffic patterns and traffic control procedures, facility location information, outline of personnel training programs designed to meet RCRA requirements, closure plans (and where applicable, postclosure plans), closure cost estimate and related financial assurance mechanism (and where applicable, postclosure cost estimate plus related financial assurance mechanism), proof of coverage by a state financial mechanism (where appropriate), a topographic map for the facility including the land area defined by a 1000-ft margin perimeter of the groundwater monitoring and ground-waste installations (where applicable).

Facility-Specific Information Besides the general information section, which is required on virtually all RCRA permit applications, a detailed section for *facility-specific information* must also be submitted for Part B. The type of information requested varies from facility to facility and depends on whether or not the facility (1) uses tanks to store or treat hazardous waste, (2) stores containers of hazardous waste, (3) incinerates hazardous waste, (4) uses land treatment to dispose of hazardous waste, (5) disposes of hazardous waste in landfills, (6) stores, treats, or disposes of hazardous waste in surface impoundments, or (7) treats or stores hazardous waste piles.

If it can be demonstrated that the requested information cannot be provided, at least to the extent required, federal EPA or authorized state agencies may modify Part B application requirements. For example, in a case where seasonal treatment monitoring information is required, the allotted 6-month period may not be sufficient. In such a case, agencies may grant a permit that includes special provisions for continued monitoring, analysis, and review of the treatment technology.

Permitting of Hazardous Waste Incinerators

The permitting of hazardous waste incinerators is a complex multifaceted program conducted simultaneously on the federal, state, and local levels. Because of the variety of state and local regulations for the handling, transportation, treatment, and disposal of hazardous wastes, as well as those concerning the operation of incinerators, each start-up has a unique set of permit requirements.

Generally speaking, hazardous waste incinerators require the following permits: federal RCRA, state RCRA, TSCA (for PCBs), federal air quality (Title V), state air quality, state wastewater discharge, and federal wastewater discharge permits. A variety of local permits may also be necessary. Each of these require data substantiating an incinerator's operation at or above performance levels determined by environmental legislation. Each, as well, requires a public hearing and discussion of environmental impacts.

Trial Burn

As part of the RCRA Part B permitting process, an incineration facility is required to perform a *trial burn*. A trial burn is performed to demonstrate conformity with the regulations. In the case of a hazardous waste incinerator, the specific regulations are as follows:

- 40 CFR § 264.343(a)(1), *Standards to Control Organic Emissions*, require a destruction and removal efficiency (DRE) of selected principal organic hazardous constituents (POHCs) of 99.99 percent (four nines DRE). The DRE for an incinerator/air pollution control system is defined by the following formula:

$$\text{DRE}(\%) = (W_{\text{in}} - W_{\text{out}})/W_{\text{in}} \times 100$$

where DRE = destruction and removal efficiency, percent
W_{in} = mass feed rate of POHC fed to the incinerator (lb/h)
W_{out} = mass emission rate of POHC to the atmosphere (lb/h)

- 40 CFR § 264.343(b), *Standards to Control Hydrogen Chloride*, requires the control of hydrogen chloride emissions such that the rate of emission is no greater than the larger of either 1.8 kg/h (4 lb/h) or 1% of the hydrogen chloride in the stack gas prior to entering any pollution control equipment.
- 40 CFR § 264.343(c), *Standards to Control Particulate Matter*, mandates that the incinerator may not emit particulate matter in excess of 180 mg per dry standard cubic meter (mg/dscm) or 0.08 grains per dry standard cubic foot (gr/dscf) after correction to 7% O_2 dry volume in the stack gas. The correction factor (CF) is defined as follows:

$$\text{Correction factor (CF)} = (21 - 7)/(21 - Y) = 14/(21 - Y)$$

where Y = measured O_2 concentration in the stack gas on a dry basis (expressed as percentage).

When systems operate with excess air that results in stack O_2 above 7% d.v. (dry volume), the CF is less than one. If the system operates with stack O_2 less than 7% d.v., the CF is greater than one. This impacts systems that dilute the combustion gases with excess air. This CF is used when air (21% O_2 by volume) is used as the combustion air. If enriched air (>21% O_2 up to 100% O_2) is used as combustion air, this CF does not apply.

A *trial burn plan* must be included as part of the engineering description section of the RCRA Part B permit application. The trial burn plan describes the goals and procedures of the trial burn. The plan includes discussions on the waste characteristics and the system equipment, including controls and monitoring devices, descriptions of the conditions that will be demonstrated during the test, and descriptions of the sampling and analysis methods that will be used to collect the data.

To verify compliance with the DRE performance standard during the trial burn, it is not required that the incinerator DRE for every POHC identified in the waste be measured. Typically, two or three compounds are chosen as representative POHCs for the waste stream. Interestingly, the designated trial burn POHCs do not have to be actually present in the normal waste; they do, however, have to be considered more difficult to incinerate than any POHCs in the normal waste. The substitute POHCs in this strategy are referred to as *surrogate* POHCs. The surrogate POHCs would be added to the waste stream during the trial burn testing.

Two primary ranking hierarchies have been used as criteria in the selection of POHCs to ensure that the POHCs chosen represent the widest range of compounds expected to be burned. The first ranking is the USEPA's incinerability list, included in USEPA's *Guidance Manual for Hazardous Waste Incinerator Permits*, July 1983 (EPA/SW-966), which ranks compounds by heat of combustion. This approach to POHC selection is based on thermodynamic and kinetic equilibrium theories that claim that the primary concern in evaluating the difficulty of destruction for a compound is the amount of energy necessary to complete the combustion process. The second POHC ranking is the thermal stability index, which is included in the USEPA's handbook *Guidance on Setting Permit Conditions and Reporting Trial Burn Results*, January 1989 (EPA/625/6-89/019). This ranking of 40 CFR Part 60, Appendix VIII, compounds is based on their thermal stability, with the most stable being considered the most difficult to burn. These two ranking systems are used with other factors, such as waste characteristics and physical and chemical characteristics of compounds, to choose an appropriate POHC for the trial burn demonstration. The EPA permit review personnel work with the owners/operators of the incinerator facility in determining which POHCs in a given waste should be designated for sampling and analysis during the trial burn.

The trial burn plan also describes the operating condition(s) to be demonstrated during the testing. The test conditions should be developed to demonstrate the worst-case operating conditions of the system. Factors that should be considered include the following:

- Maximum capacity (heat release) of the system
- Design capacity for each waste stream
- Worst-case composition of each waste stream, for example, maximum chlorine and ash concentrations
- Minimum combustion temperature and minimum combustion gas oxygen concentration to be demonstrated

The complexity of a trial burn will depend on the number of operating limits that must be demonstrated during the test. More than one test condition may be required to demonstrate all limits.

The trial burn plan is submitted with the Part B permit application or, occasionally, will be submitted prior to the application. This is done to give the agency sufficient time to review the plan and provide comments to the facility. After the agency has approved the trial burn plan, the facility must prepare for and execute the trial burn. In some cases, it is advisable to plan and conduct a pretrial burn to ensure owner/operator of proper performance before final trial burn.

In conducting the trial burn, only approved EPA sampling and analysis methods or their equivalents may be employed. The current EPA sampling methods, Volatile Organic Sampling Train (VOST) (SW-846 Method 0030) and Modified Method 5 (MM-5) (SW-846 Method 0010) are commonly used for sampling of POHCs. USEPA Method 5 sampling protocol is used to sample for particulate, and SW-846 Method 0050 is used to sample for hydrogen chloride. Analytical methods for hazardous organic compounds include *gas chromatography* (GC), *flame ionization detection* (FID), *photoionization detection* (PID), *Hall electrolytic conductivity detection* (HECD), *electron capture detection* (ECD), *gas chromatography/mass spectrometry* (GC/MS), *high-performance liquid chromatography* (HPLC), and *atmospheric pressure chemical ionization mass spectrometry* (APCI-MS).

The POHC compounds normally have to be concentrated from large amounts of the flue gas sample on special polymer resins; a 3-h timeframe is typical. The extraction and concentration techniques required to prepare the sample for analysis are complicated and time consuming. As a result, trial burns are costly, and it usually requires 30–60 days before the final laboratory analysis results are known.

Because of the complexity of the sampling and analysis, trial burns can be very costly. The costs associated with the trial burn normally include expenditures for surveying, agency meetings, equipment, sampling, analysis, cleanup, and report preparation. These details follow.

- *Site Survey* These costs include professional services and travel for inspection of the facility to be tested and discussion of plans for trial burns. Specific characteristics of wastes to be incinerated are assumed known or provided by the waste generator. Information on the compatibility of these wastes with the specific facility characteristics is assumed available.
- *Meetings* Meetings with owner/operator of the facility, state or federal agency personnel, trial burn consultants, sampling and analysis contractors to discuss trial burn plan, wastes, and schedule. This provides an opportunity for all parties to resolve any questions or concerns before the final trial burn commences. If a pretrial burn is planned, this is also discussed at this meeting.
- *Waste Feed Preparation* Waste feeds for all injection points whether in solid, slurry, sludge, or liquid form must be prepared and stored prior to the final trial burn. Quantities of each, including amounts of POHCs suf-

ficient for pretrial burn and final trial burn, must be made available. This information should be included in trial burn plans. The plan should also include process operations that provide the number of hours needed for stabilization of temperature in the combustion zones. For solids feeds, amount needed to completely fill the primary combustion chamber must be calculated prior to start of sampling.[8]

- *Quality Assurance/Quality Control (QA/QC)* Because every test represents a unique situation, appropriate QA/QC procedures must be employed throughout both the sampling and analytical phases of any test. For example, all measuring devices should be properly calibrated, all solvents and resins should be checked for purity before the sampling program is begun, sampling trains should be all glass and thoroughly cleaned to remove all sealant greases and any other residues, and ground glass plugs and aluminum foil should be used to cap off the adsorbent trap and other sample-exposed portions of the trains to minimize the potential for contamination. Costs associated with QA/QC include professional services and sampling and testing equipment.
- *Equipment Preparation* Sampling equipment may have to be leased by the facility or provided by a consulting firm. Certain costs are incurred in calibrating, loading of instruments, and transport to facility site.
- *Equipment Setup and Takedown* Installation of equipment includes the erection of scaffolding, securing of ports, and placement of sampling instruments at facility stacks for the proper monitoring and collecting of trial burn samples.
- *Stack Sampling* A minimum of three tests involving 3–6 h per test is normal. Costs for testing, instrumentation, and adjustments associated with a permit application are considered separate costs not attributable to the trial burn. Development of sampling procedures and verification of test methods are presumed available and accomplished prior to the trial burns.
- *Sample Analysis* Laboratory analysis costs include the preservation and transporting of samples to an off-site laboratory. Compounds to be analyzed generally include any potential air pollutants. Gas chromatography (GC) and mass spectrometry (MS) testing is conducted for the chlorinated hydrocarbons.
- *Equipment Cleanup* These costs are the routine expenditures incurred for cleaning and storing various sampling and analysis equipment.
- *Report Preparation* The written report displays and interprets the trial burn results. Costs for writing, typing, artwork, reproduction, and so on are included here.

Phases of Permitting

A new facility is required to submit a complete Part B permit application prior to construction of an incinerator. After submittal of the application, the agency reviews the documents for completeness. If the documentation is incomplete, the agency will issue a *notice of deficiency* requesting further information. When the agency determines that the permit application is complete, a public hearing is held and a temporary or draft permit is issued, allowing the owner or operator to build the incinerator and initiate the trial burn procedure.

The temporary permit covers four phases of operation. During the first phase, immediately following construction, the unit is operated for *shakedown* purposes to identify possible mechanical deficiencies and to assure its readiness for the trial burn procedures. There is no limit on the time for this shakedown as long as hazardous waste is not injected into the incineration system. The trial burn phase of the permit is limited to 720 h of operation using hazardous waste feed and is conducted during the second phase. This is the most critical component of the permitting process since it demonstrates the incinerator's ability to meet the performance standards. As discussed above, a pretrial burn may also be conducted during this phase to ensure proper performance before the final trial burn. In addition, performance data collected during the trial burn phase are reviewed by the permitting official and become the basis for setting the conditions of the facility permit. These conditions include: (1) allowable waste analysis procedures, (2) allowable waste feed composition (including acceptable variations in the physical or chemical properties of the waste feed), (3) acceptable operating limits for carbon monoxide in the stack, (4) waste feed rate, (5) combustion chamber temperatures, (6) combustion gas flow rate, and (7) allowable variations in incinerator design and operating procedures (including a requirement for automatic waste feed cutoff during startup, shutdown, and at any time when conditions of the permit are violated).

The third phase consists of completing and submitting the trial burn results. This phase can last several weeks to several months, during which the incinerator is allowed to operate under specified conditions. The data to be reported to regulatory agencies after the burn include: (1) a quantitative analysis of the POHCs in the waste feed, (2) a determination of the concentration of particulates, POHCs, oxygen, carbon monoxide (CO), and HCl in the exhaust gas, (3) a quantitative analysis of any scrubber water, ash residues, and other residues to determine the fate of the POHCs, (4) a computation of the DRE of the POHCs, (5) a computation of the HCl emission rate and removal efficiency, (6) a computation of particulate emissions, (7) the identification of sources of fugitive emissions and their means of control, (8) a measurement of average, maximum, and minimum temperatures and combustion gas velocities (gas flows), (9) a continuous measurement of CO in the exhaust gas, and (10) any other information the agency may require to determine compliance.

Provided that performance standards are met in the trial burn, the facility can begin its fourth and final phase, which continues through the duration of the permit. In the event that the trial burn results do not demonstrate compliance with standards, the temporary permit must be modified to allow for a second trial burn.[7]

An existing facility will follow most of the same procedures as a new facility when renewing the permit. The facility is required to submit a complete application for a renewal permit at least 180 days prior to the expiration of the existing permit. This renewal application follows the same agency review process. After approval of the facility's trial burn plan, the facility can begin with the trial burn phase of the process. The existing facility is allowed the 720 h of operation for the trial burn using hazardous waste feed. During this time, the facility may exceed its existing permit limits in order to test the incinerator to establish new operating limits, providing it requested this increase from the regulatory agency.

From start to finish, the permitting of a hazardous waste incinerator usually takes from 18 to 36 months. This process can be considerably longer if such obstacles as *notice of deficiencies*, inconclusive trial burn results, or intense public opposition occur.

3.6 EPA GUIDANCE MANUALS

In order to expedite the permitting process covered in Section 3.5, the EPA developed a series of guidance manuals for EPA and state permit writers and industry's permit applicants to assist in executing all necessary procedures leading to a final operating permit for a hazardous waste incinerator.[9]

A guidance manual is a supplemental document to EPA regulations. It provides a detailed procedure for EPA permit writers on how to implement regulations and for industry on how to comply with regulations.[10] These guidance manuals are listed below:

1. Volume 1—*Guidance Manual for Hazardous Waste Incinerator Permits*, SW-966, NTIS PB84-100577, July 1983. This describes the overall incinerator permitting process, highlights the specific guidance provided by other manuals, and addresses permitting issues not covered in the other manuals such as treatment of data in lieu of a trial burn. It can be viewed as a road map and a good summary of all permitting issues.
2. Volume 2—*Guidance on Setting Permit Conditions and Reporting Trial Burn Results*, EPA/625/6-89/019, January, 1989. This provides guidance on selecting key operating parameters, translating trial burn data into permit operating conditions, and reporting trial burn results. It also discusses planning a trial burn to achieve workable permit limits.
3. Volume 3—*Hazardous Waste Incineration Measurement Guidance Manual*, EPA/1989. This provides general guidance to permit writers in reviewing the measurement aspects of incineration permit applications and trial burn plans. It is oriented to show how measurements are made, not what measurements to make. The guidance deals with commonly required measurement parameters and measurement methods for process monitor-

ing, sampling and analysis aspects of trial burns and subsequent operation of the incinerator, and QA/QC associated with these activities.

4. Volume 4—*Guidance on Metals and Hydrogen Chloride Controls for Hazardous Waste Incinerators*, EPA/530-SW-90-007, August, 1989. This provides specific guidance on limiting metals emissions from incinerators.
5. Volume 5—*Guidance on PIC Controls for Hazardous Waste Incinerators*, EPA—1989. This provides detailed permit requirements for CO and total hydrocarbon (THC) emissions from hazardous waste incinerators in the RCRA system. Emission limits for CO and THC and the rationale for their selection are discussed.
6. Volume 6—*Methods for Measurements of CO, O_2, THC, HCl, and Metals at Hazardous Waste Incinerators*, EPA—1989. This provides measurement methods for the above parameters including performance specifications for continuous CO monitors.
7. Volume 7—*Quality Assurance/Quality Control Procedures for Hazardous Waste Incineration.* (This provides procedures that ensure measurements and sampling follow guidance and methods outlined in Volumes 1–6.)

3.7 BOILER AND INDUSTRIAL FURNACE (BIF) REGULATIONS[11]

Boilers and industrial furnaces burning hazardous waste were initially exempt from the RCRA incinerator requirements. Under RCRA, these units are considered to be "resource recovery" units, which do not fall under the definition of incinerators. As burning hazardous waste in boilers and industrial furnaces became common practice, the USEPA studied these units and determined that there was a need for regulations. These units were burning the same wastes as permitted hazardous waste incineration facilities but were not required to meet any of the same performance standards.

In 1991, the final rules governing the operation of boilers and industrial furnaces (BIF) burning hazardous waste were promulgated. These rules took effect on August 21, 1991. The BIF regulations are found in the Code of Federal Regulations, Chapter 40, Part 266. These rules cover the burning of hazardous wastes in boilers, industrial furnaces, and kilns for cement, lime, and aggregate. The rule requires that these facilities obtain RCRA Part B permits to burn hazardous waste in resource recovery units. Details on these rules are provided below.

Performance Standards

BIFs are required to meet performance standards similar to those of RCRA incinerators. In addition, the BIF regulations set limits on other pollutants, such as heavy metals and chlorine, which are not regulated by RCRA.

Destruction and Removal Efficiency A boiler or industrial furnace burning hazardous waste must achieve DRE of 99.99% for all organic hazardous constituents in the waste feed. To demonstrate conforming with this requirement, 99.99% DRE must be demonstrated during a trial burn for each designated waste feed stream.

BIF units burning hazardous waste containing (or derived from) EPA Hazardous Wastes Nos. F020, F021, F022, F023, F026, or F027 must achieve a DRE of 99.9999% for each designated POHC. This performance must be demonstrated on POHCs that are more difficult to burn than tetra-, penta-, and hexachlorodibenzo-*p*-dioxins and dibenzofurans. DRE is determined using the same calculation method as that for incinerators.

Carbon Monoxide/Hydrocarbons The stack gas concentration of carbon monoxide (CO) from a BIF unit must not exceed 100 parts per million by volume (ppm_v) on an hourly rolling average basis (i.e., over any 60-min period), continuously corrected to 7% O_2, on a dry gas basis. The CO and O_2 must be continuously monitored.

As an alternate to the CO standard, the CO concentration from a BIF unit burning hazardous waste may exceed the 100 ppm_v limit provided that stack gas concentrations of hydrocarbons (HC) do not exceed 20 ppm_v. The HC limits must be on an hourly rolling average basis (i.e., over any 60-min period), reported as propane and continuously corrected to 7% O_2, dry gas basis.

If the alternative standard is used, HC emissions must be continuously monitored. CO and O_2 must also be continuously monitored. The alternative CO standard is established based on CO data during the trial burn or the compliance test. The alternative CO standard is the average over all valid runs of the highest hourly average CO level for each run. The CO limit is implemented on an hourly rolling average basis and continuously corrected to 7% O_2, dry gas basis.

Dioxins and Furans BIF units that are equipped with a dry particulate matter control device that operates within the temperature range of 450–750°F, and industrial furnaces operating under an alternative hydrocarbon limit must conduct a site-specific risk assessment to demonstrate that emissions of polychlorinated dibenzo-*p*-dioxins (PCDD) and polychlorinated dibenzofurans (PCDF) do not result in an increased lifetime cancer risk to the hypothetical maximum exposed individual (MEI) exceeding 1 in 100,000.

Particulate Matter A BIF unit may not emit particulate matter in excess of 180 mg/dscm (0.08 gr/dscf) after correction to a stack gas concentration of 7% O_2. The method of calculating the corrected concentration is identical to that for incinerators.

Hydrogen Chloride and Chlorine Hydrogen chloride (HCl) and chlorine gas (Cl_2) emissions are regulated using a tiered set of standards. The standards have been developed to ensure that emissions of HCl and Cl_2 do not exceed ambient health-based levels. The standards involve the use of dispersion modeling or screening limits based on worst-case facilities. Typically, a facility has the option to choose any of the four tiers to demonstrate compliance. For some facilities, Tier III or Adjusted Tier I must be used because of site-specific conditions such as neighboring building heights, ground level rise within 5 km of stack, bodies of water, and so forth.

TIER I Tier I is the simplest approach to determine HCl and Cl_2 limits. This approach also produces the most conservative feed rate limits. Feed rate limits are determined using a set of tables developed by the USEPA. The feed rate limits are a function of the stack height, surrounding land use, and the flue gas flow rate. By limiting the HCl and Cl_2 feed rates, the facility can eliminate the need for stack testing. However, with this approach, no credit can be taken for any HCl or Cl_2 removed from the gas stream by downstream air pollution control equipment.

TIER II Tier II also involves the use of USEPA-developed tables. The Tier II tables set limits on stack emission rates that are also functions of stack height, surrounding land use, and the flue gas flowrate. In order to demonstrate compliance with Tier II, a facility is required to perform a stack test. By measuring the HCl and Cl_2 at the stack, the facility can take credit for any removal efficiency of the air pollution control system. For this reason, Tier II can be less restrictive than Tier I if a facility utilizes a HCl/Cl_2 control device.

TIER III Tier III involves the use of site-specific risk assessment. With this tier, a facility can choose to perform their own dispersion modeling to set emission limits. In a Tier III risk assessment, the calculated ambient concentration of HCl and Cl_2 is compared to the allowable reference air concentration (RAC). The RACs are concentration levels that are considered to pose an acceptable level of risk to the surrounding population. Typically, a facility will use a Tier III analysis to back-calculate from the RAC to determine the maximum allowable emission rate for the unit. This tier is usually the least restrictive for a facility. Stack sampling is required to demonstrate compliance.

ADJUSTED TIER I Adjusted Tier I is a derivative of the Tier III procedure. This involves all the same procedures as the Tier III risk assessment and includes a final step to back-calculate from stack emission rate limits to feed rate limits. With this tier, no stack testing would be required to demonstrate compliance. However, as with Tier I, there is no allowance in this tier for any removal efficiency associated with a control device.

Metals Emissions Standards Metals emissions for BIF units are limited using the same tiered approach as is used for HCl and Cl_2. There are 12 metals regulated under the BIF regulations. Four are considered to be carcinogens: arsenic, beryllium, cadmium, and chromium. The other 8 are noncarcinogenic: antimony, barium, lead, mercury, nickel, selenium, silver, and thallium.

Tables have been developed for Tiers I and II, just as for HCl and Cl_2. Tier III and Adjusted Tier I involve site specific dispersion modeling. For the non-carcinogenic metals, a facility is required to limit the emissions (or feed rate) below the values in the tables for Tiers I and II. For Tier III and Adjusted Tier I, the facility is required to maintain the emission rates of the metals at a level that would produce ambient concentrations below the individual RACs of the metals.

For the carcinogenic metals, the process is slightly different. For Tiers I and II, the facility is required to maintain the emission rates (or feed rates) at some fraction of the values given in the tables. The sum of the individual fractions must be less than or equal to one. For example, the limit for each of the four metals could be set to 25% of the values given in the table. In Tier III, each carcinogen has been assigned a unit risk factor (URF). The BIF regulations state that the aggregate risk from emissions of carcinogenic metals must be less than 1 in 100,000. To determine risk, the ambient concentration is multiplied by the URF. The four individual risks are then summed to determine the total aggregate risk. This value must be less than 1 in 100,000.

Permitting Requirements

Existing facilities subject to the new BIF standards operated under interim status after the BIF promulgation in 1991. Existing facilities were defined as those that were burning hazardous waste on or before August 21, 1991, or commenced construction on or before August 21, 1991. To attain interim status, facilities were required to submit a *certification of precompliance* that included information to demonstrate that the unit was capable of operating within the allowable limits of the standard. Facilities then had one year to conduct an initial compliance test, analyze the test results, and submit a *certification of compliance* to the agency. Facilities were then required to conduct additional compliance tests and submit a *recertification of compliance* to the agency every 3 years.

In the periodic compliance tests, facilities are required to demonstrate compliance with the particulate, chlorine, and metals standards of the BIF regulations. These tests are similar to RCRA trial burns, but typically less complicated.

In recent years, agencies have begun calling BIF facilities to begin the process to obtain their Part B permits. These facilities have been required to submit Part B permit applications, complete with trial burn plans. Many facilities have performed trial burns, although, to date, there have been very few Part B permits issued.

Effects of BIF Regulations on Incinerators

The promulgation of the BIF regulations also had an effect on the permitting of hazardous waste incinerators under RCRA. After 1991, RCRA incinerators were required to comply with BIF emission standards for metals and HCl/Cl_2 upon renewal of the permit. Although there had been no regulations promulgated that required incinerators to meet the limits set in the BIF regulations, the EPA or state regulatory agency could make such restrictions on a facility by enforcing what is known as *omnibus authority*. In the Code of Federal Regulations, Chapter 40, Part 270.32, federal and state agencies are given the right to include any terms and conditions in a facility's permit that they determine to be necessary to protect human health and the environment. Because of this authority, agencies are able to set permit limits, such as metals emissions limitations, which are more restrictive than current regulations.

Waste Fuels: A Source of Energy

The major operating cost in the production of cement, lime, and aggregate is energy. Owners and operators of these facilities have continuously searched for means of improving the energy efficiency of these systems. In the late 1970s, many of these facilities determined that waste solvents normally burned in on-site incinerators at chemical, agricultural, and pharmaceutical processing plants and off-site commercial disposal incinerators could replace their purchased fossil fuels (coal, oil, and gas) with a fuel that could be burned at no cost to the facility. In fact, some of these plants were paid to dispose of these waste solvents. Many industrial plant operations were also using these waste fuels to supplement base fuels fired in boilers and process heaters. In these applications, there was a concern regarding the corrosive effect of combustion products containing halogens, sulfur, and phosphorous to the heat transfer surface of the boilers and process furnaces. In the case of the production kilns, the raw materials processed would, in most cases, aid in neutralizing acid gases that are formed in the combustion process.

Since these facilities utilized a hazardous waste fuel to generate energy, thereby replacing fossil fuels, they were not regulated by the existing RCRA regulations. RCRA established the standards for the treatment, storage, and disposal (TSD) of hazardous wastes. Hazardous materials (wastes) that were recycled or used to provide energy to a boiler, process furnace, and/or cement kiln were not covered by these regulations. Only the storage and transportation of these waste materials were covered.

Why the BIF Regulations?

As a result of emission test data obtained from boilers and kilns burning hazardous wastes, the USEPA and many environmentally concerned groups began to have second thoughts about exempt burning of hazardous waste. Collection and resale of ignitable wastes was becoming a thriving business. In 1984 the Hazardous and Solid Waste Amendments (HSWA) mandated regulation of waste fuels by November 1986. Rules for storage and transportation of these wastes were issued in May 1986 and emission standards proposed in May 1987.

Subsequent notices were drafted in the *Federal Register* in October 1989 and again in April 1990. These regulations were designed to control emission of metals using both risk assessments and emission limits. Up until this point, the only emissions controls by RCRA were those for particulates (0.08 g/dscf) and hydrochloric acid (99% removal or 4 lb/hr). The USEPA Science Advisory Board (SAB) "Report on the Incineration of Liquid Hazardous Wastes by the Environmental Effects, Transport and Fate Committee" (issued in April, 1985) stated concerns over the risks to the public created by emissions from hazardous waste combustion devices.[12] In "Report of the Products of Incomplete Combustion" the SAB, in October 1989, restated these concerns.[13] The final rule was promulgated December 31, 1990, and published on February 21, 1991, with an effective compliance date 6 months later, August 21, 1991.[14]

With the advent of the BIF regulations, the burning of all hazardous waste had effectively been put on an equal basis. The final rule, as published in 40 CFR 261, 265, and 266 on February 21, 1991, with technical amendments as late as August 16, 1991, was issued on August 21, 1991.

Initial Effect of BIF Regulations

As of December 1990, approximately 148 hazardous waste incinerators were regulated under the RCRA by the USEPA and states with authorization. At the same time, approximately 1300 installations were burning hazardous wastes as fuels in boilers, industrial furnaces, acid furnaces, and the like. The kiln operations were burning up to 1 million tons annually, while another 125 large boilers were consuming an additional 1 million tons per year.

The USEPA estimated that some 200 industrial facilities would stop burning hazardous wastes as fuel because of the new rule. Nearly 600 small boilers burning less than one percent of hazardous waste as fuel would continue operating under an exemption in the new rule. Smelters reclaiming metals from hazardous wastes and coke ovens receiving certain soot-making by-products also could continue operating under exemptions from the final rule.

Under the BIF regulations, existing facilities were those that were burning hazardous waste on or before August 21, 1991, or commenced construction on or before August 21, 1991. The facility attained interim status (IS) by submitting the precompliance certification prior to the August 21 date. Even if a facility owner was not sure whether or not it would continue burning hazardous wastes in BIFs after August 21, a precompliance certificate was required.

Certification of Compliance

The BIF regulations required that a compliance test be conducted, the results reported and analyzed, and a certificate of compliance sent to the regional administrator prior to August 21, 1992.

Results of BIF Regulations

The above discussion of BIF regulations has pointed out the implications of the regulations to owner/operators of these facilities. Of prime importance was the need to determine the cost–benefit of continued waste burning. Proper staffing of plant management and operations was necessary. A corporate environmental department with necessary legal and technical assistance was essential. Environmental managers were needed at each facility to ensure daily compliance and also to maintain communications with state and federal regulators. This resulted in continuing and increasing personnel costs to the operation. Major expenditures included laboratory facilities, air pollution control (APC) system additions, induced-draft (ID) fan modifications with possible horsepower (HP) increase, stack modifications, process and flue gas monitors, and instrumentation. In addition, other costs included operator training programs, increased number of maintenance and instrumentation personnel, and staffing of the laboratory with qualified and trained personnel.

Companies had to deal with the new laws and maintain contact with regulatory agencies. A complete understanding of the many types of BIFs that existed resulted in an intensive training program for the agency personnel similar to that conducted on incinerators in the mid-1980s.

3.8 EPA HAZARDOUS WASTE REDUCTION AND COMBUSTION STRATEGY

Incinerator Moratorium

On May 18th, 1993, the USEPA administrator, Carol Browner, announced a new hazardous waste reduction and combustion strategy. The agency took a series of immediate actions related to hazardous waste incinerators and industrial furnaces including:

- Freezing new permits for incinerators
- Requiring health risk assessments, including those for indirect exposure, in all new permits to ensure that they are based on the best scientific evidence
- Requiring new permits to include dioxin emission standards
- Requiring new permits to include more stringent control for metals
- Calling for hazardous waste producers to commit to waste reduction programs
- Calling for a national review of the relative roles of waste combustion and waste reduction in hazardous waste management
- Calling for greater public involvement opportunities
- Requiring permits for those within interim status, including 171 industrial furnaces and 30 cement kilns

In addition, the EPA announced an overhaul of the federal rules governing waste combustion. These actions were justified "to protect human health and the environment by reducing the amount of hazardous waste produced in this country and strengthening federal controls governing hazardous waste incinerators and furnaces."[15]

EPA Roundtable of BIF Regulations

The above action followed the issuance of the BIF compliance certificates to many interim status operations in August 1992. A national roundtable was conducted by EPA in Crystal City, Virginia, on February 24–25, 1993. Representatives from industry, industry trade associations, environmental and local citizen groups, states, and EPA regional offices met to discuss issues related to regulations of boilers and industrial furnaces that burn hazardous waste.

The discussions centered on these topics:

- Hazardous waste burning vs. waste minimization
- Siting requirements
- Public participation
- Technology vs. risk-based emission standards
- Review exclusion of residues
- Organic emissions from cement kilns
- Continuous emission monitoring
- Consistency of EPA regional office interpretations/actions
- Trial burns
- Record keeping
- Regulatory exemptions
- Applicability of HCl and Cl_2 emission limits
- Limit on excess oxygen
- Combustion chamber temperature
- Feed-stream sampling frequency
- Opacity monitoring

The outcome of this meeting in February was the May 18, 1993, announcement by Carol Browner described above.

3.9 COMBUSTION EMISSIONS TECHNICAL RESOURCE DOCUMENT[16]

The Combustion Emissions Technical Resource Document (CETRED) contained the initial technical analysis by the USEPA concerning emissions of dioxins/furans and particulate matter from certain types of devices that burn hazardous waste: cement kilns, lightweight aggregate kilns, incinerators, and industrial boilers. CETRED represented the first preliminary step in the development of regulations under the RCRA and the Clean Air Act (CAA) to impose upgraded standards on hazardous waste combustors (HWCs). CETRED also represented a major effort toward implementing the commitment made by EPA Administrator Carol M. Browner in the Draft Hazardous Waste Minimization and Combustion Strategy, released on May 18, 1993, to upgrade the technical standards governing emissions from HWCs.

EPA's intention in releasing CETRED was to give the regulated community and other interested individuals the earliest possible opportunity to understand the direction that EPA was pursuing. CETRED may be regarded as a preliminary technical analysis of certain HWCs and their emissions of particulate matter (PM) and dioxins/furans. CETRED represented the current state of analysis of EPA's technical staff in the Office of Solid Waste regarding the emission levels of PM and dioxins/furans achievable by the best controlled sources.

CETRED did not contain a characterization of emissions for toxic metals and other hazardous air pollutants from the HWCs studied. EPA initiated a technical analysis to characterize these emissions in the 1994–1995 time period. EPA made the results of that analysis available to the public for review prior to the time that any regulatory proposal would be developed.

Hazardous Waste Combustor Systems

The HWCs affected by CETRED are both the hazardous waste incinerators (HWIs) and the BIFs. Included in the HWIs are rotary kiln, liquid injection, fluid bed, and fixed hearth units. The BIFs include cement kilns, lightweight aggregate kilns, and boilers.

Air Pollution Control Systems

Incinerators use many types of air pollution control devices to control emissions of particulate matter, metals, and acid gases. In general, the control systems can be grouped into the following three categories: (1) wet systems, where a wet scrubber is used for both particulate and acid gas control; (2) dry systems, where a fabric filter (FF) or electrostatic precipitator (ESP) is used for particulate matter (PM) control, sometimes in combination with dry scrubbing for acid gas control; and (3) hybrid wet/dry systems where a dry technique (ESP or FF) is used for PM (and possibly acid gas control with use of dry scrubbing) followed by a wet technique (venturi or packed-bed scrubber) for acid gas control.

Cement kilns are normally equipped with either an ESP or FF to control emissions of PM. Most light-weight aggregate kilns burning hazardous waste are equipped with FFs to control PM emissions.

PM Emission Levels

EPA evaluated PM emission levels because controlling PM will control emissions of most toxic metals and toxic organic compounds absorbed in the PM. PM levels are evaluated for several different source categories (or subsets thereof), as well as various degrees of aggregation of these source categories. For example, commercial incinerators are evaluated separately from on-site incinerators as a crude attempt to determine if the best controlled of the generally smaller on-site incinerators are achieving PM emission levels substantially different from the best controlled commercial incinerators. Furthermore, PM emissions from the best controlled incinerators, boilers, and industrial furnaces are evaluated as an aggregated group representing the hazardous waste combustor source category. While the agency was interested in examining the appropriateness and implementability of establishing standards that apply across all types of hazardous waste combustors, EPA recognized that in doing so there may be technical and policy determinations that had not yet been fully illuminated or explored. The agency emphasized that it had not decided if and which source categories would be grouped in determining PM emission limits.

For the study, EPA evaluated PM emission data that facilities had submitted to the EPA regional offices and the states as part of certifications of compliance under the agency's boiler and industrial furnace (BIF) rule, and from incinerator trial burns used to consider issuance of operating permits. To identify the best controlled sources, the sources are ranked in order of ascending PM emission levels, considering the average emission level for the source and the variability of emissions data. The best 12% of the sources (or best five sources, whichever is greater) would be selected as (potentially) the best controlled sources. Each of the sources in the selected pool is evaluated to determine if it achieves low PM emissions because it uses advanced control techniques or because it simply burns wastes or other materials with low levels of ash. Sources with low PM levels because of low ash feed rates are screened out of the pool as being unrepresentative of best controlled sources, and the next best controlled sources are brought into the pool as replacements.

Once the pool of the best controlled sources was established, their emissions data were analyzed statistically to predict a PM emission level that could be achieved under two alternative approaches. Under option 1, a PM emission level is identified so that all of the sources in the best controlled source pool could be expected to comply with 99% of the time, with 95% confidence. Under option 2, a PM emission level is identified so that a source with the average emission level of the best controlled source pool could be expected to meet 99% of the time, with 95% confidence, and assuming that the variability of the source's emissions data is similar to the variability of the emissions data from the sources in the pool. Under option 1, all of the sources in the best controlled source pool should be able to achieve the PM level without modifications, while under option 2, some of the sources in the pool may not be able to achieve the PM level without modifications to the facility design or operation.

Table 3.9.1 PM emission levels achievable by the best controlled sources (gr/dscf @ 7% O_2)[16]

Source category	Option 1[a]	Option 2[b]
Cement kilns	0.033	0.010
Lightweight aggregate kilns	0.022	0.0077
Commercial incinerators	0.010	0.0049
On-site incinerators	0.015	0.0075
All HW incinerators	0.0057	0.004
Boilers	0.021	0.011
All hazardous waste combustors	0.0086	0.0052

[a]Under option 1, a PM emission level is identified that all of the sources in the best controlled source pool could be expected to meet 99% of the time, with 95% confidence.
[b]Under option 2, a PM emission level is identified that a source with emissions equivalent to the average for the pool of the best controlled sources could be expected to meet 99% of the time, with 95% confidence, and assuming that the variability of the source's emissions is similar to the variability of emissions for the sources in the pool.

Table 3.9.1 presents the PM emissions levels achievable by the best controlled sources for the source categories and groups of source categories, and for the alternative analytical approaches.

Dioxin/Furan Emission Levels

Dioxin is a term used to describe a large group of chemical compounds having a similar basic structure. The most common subgroup of dioxin compounds (of which there are 75 different varieties) comprises those that include chlorine atoms. A few are toxic; the most toxic is 2,3,7,8-tetrachlorodibenzo(para) dioxin (2,3,7,8-TCDD). There are also 135 polychlorinated-dibenzofuran compounds. These are known as "furans."

These compounds are chlorinated tricyclic aromatic compounds. Each of these compounds has a triple-ring structure consisting of two benzene rings interconnected to each other by, respectively, two or one oxygen atoms. The number of chlorine atoms can vary between 1 and 8. The PCDD and PCDF compounds are referred to as "congeners"—a specific member of a group of structurally related compounds. The chemical 2,3,7,8-TCDD is a colorless crystalline solid, is slightly soluble in water, and binds strongly to solids and particulate matter.[17] The known natural sources of PCDDs and PCDFs are related to fires and combustion processes. Thus, these toxic compounds may be produced from forest fires, by lightning or volcanic action. The formation of PCDD/PCDFs is very dependent on the presence of oxygen, carbon, chlorine, and heat. Since fire produces PCDD/PCDFs, combustion devices may have emissions of these compounds. These include municipal waste incinerators, industrial waste incinerators, medical waste combustors, sewage sludge incineration, wood and coal combustion, oil combustion including motor vehicle engines and accidental fires. Recent testing has shown the high-efficiency burners produce minimal to zero PCDD/PCDFs in the burning zones, but formation has been found due to low-temperature (250–350°C) reformation in the waste heat boilers and air pollution control systems used with the incineration systems. These lower temperature zones cause the formation of PCDD/PCDFs from the combination of chlorine, ash particles, and hydrocarbons at these critical temperatures when exposed for specific time cycles. The chlorine results from the incomplete combustion of the chlorinated hydrocarbons in the wastes or fuels used in the incinerator. Wastes that contain salts (e.g., chlorides) will not produce the free chlorine found to be the source of PCDD/PCDFs when combined with ash and/or unburned hydrocarbons in the lower temperature zones beyond the combustor.

Dioxin/furan emission levels are evaluated for hazardous waste burning incinerators, cement kilns, lightweight aggregate kilns, and boilers in the aggregate [i.e., as a hazardous waste combustor (HWC) source category]. EPA was not aware of any overriding technical reason why all HWCs should not be able to meet the emission level achievable by the best controlled sources.

The dioxin/furan emissions data used for the analysis were obtained from BIF certifications of compliance and incinerator trial burn results. The agency used the same methodology for PM to identify the best controlled sources, except that the sources in the (potentially) best controlled source pool were screened for three criteria. First, the pool sources were tested to determine if they have low dioxin/furan emissions because their feed streams contain insignificant levels of chlorine. Then, two additional screening criteria were used because of concern about the potential conflict between minimizing dioxin/furan and PM emissions at the same time. Given that dioxins/furans may be formed in HWCs operating under good combustion practices by surface-catalyzed reactions, dioxin/furan emissions can be affected by PM emissions. Thus, these screening criteria ensure that the dioxin/furan emissions from the pool sources were representative of emissions from sources that use best operating practices to control both PM and dioxin/furans. The first of these criteria screened out of the pool sources that had low PM emissions because they are feeding materials with low ash content. These sources may have low dioxin emissions simply because there was little PM to promote surface-catalyzed formation. The second criterion screened out sources that have high PM emissions. These sources may have low dioxin emissions simply because PM are not removed and thereby providing an attenuated residence time (e.g., in an ESP or FF) for surface-catalyzed reactions to take place.

After the pool of best controlled and representative sources was identified, the same statistical approaches used to analyze PM emissions were used to analyze dioxin/furan emissions. Table 3.9.2 presents the dioxin/furan emission levels achievable by the best controlled HWC sources under the alternative analytical approaches.

Note that EPA believes that it is appropriate to control dioxin/furan emissions for HWCs based on toxicity equivalents (TEQs). The first congener iden-

Table 3.9.2 Dioxin/furan emission levels achievable by the best controlled sources (ng/dscm @ 7% O_2)[16]

Basis	Option 1[a]	Option 2[b]
TEQ, ng/dscm	0.17	0.12
Total tetra through octa congeners	9.4	5.4

[a]Under option 1, an emission level is identified that all of the sources in the best controlled source pool could be expected to meet 99% of the time, with 95% confidence.
[b]Under option 2, an emission level is identified that a source with emissions equivalent to the average for the pool of the best controlled sources could be expected to meet 99% of the time, with 95% confidence, and assuming that the variability of the sources emissions is similar to the variability of emissions for the sources in the pool.

tified as specifically responsible for a dermal disorder called "chloracne" among production workers in trichlorophenol and trichlorophenoxy herbicide plants was 2,3,7,8-TCDD. Other congeners were noted to have quantitatively different acute toxic effects. In the early 1980s, there was a mandatory need to quantify the toxicological potential of mixtures of PCDDs and PCDFs. Scientists developed procedures for quantitative evaluation based on available toxicological data. This formed the basis of expressing the toxic potential of each congener by a factor ("toxicity equivalency factor" or TEF). The total TEQ of a given environmental sample is calculated by multiplying the amount of each congener present in the sample by its specific TEF and adding up the total. Under this approach, weighting functions know as toxicity equivalence factors are assigned to the various dioxin and furan congeners to account for their toxicity (TEQ) relative of 2,3,7,8-TCDD. [16]

European Emission Regulations

On March 23, 1992, the European Economic Community (EEC) issued a proposal for a Council Directive on the Incineration of Hazardous Waste. The directive required all its member states to establish laws, regulations, and administrative procedures to comply with the directive by June 30, 1994. The directive stipulated that any new incinerator must comply immediately and existing facilities by June 30, 1997.

The regulatory approach adopted in the 1992 EEC directive established a wide array of continuous emission monitoring requirements, including continuous monitors for carbon monoxide and total dust emission levels, and monthly measurements for metals, dioxins, and furans. A summary of European guidelines and limits for PM and dioxins/furans is presented in Table 3.9.3.

Waste incineration has been in use in Europe longer than in North America. The air pollution control device (APCD) systems are similar. However, because the majority of the European facilities have undergone retrofits and have faced more stringent emission standards, design differences exist. Incinerators in Eu-

Table 3.9.3 Emission guidelines/limits for waste incineration in Europe[16]

Pollutant (daily average)	EEC guideline	Netherlands limit	Germany limit
Total dust (mg/m^3)	5	3	5
Dioxin/furan ($ngTEQ/m^3$)[a]	0.1	0.1	0.1

[a]It is important to note that the European guideline or limit of 0.1 TEQ is corrected to 11% oxygen, and compliance is based on daily averaging. EPA requires that dioxin/furan emissions be corrected to a stack gas oxygen level of 7%. A 0.1 limit at a 11% oxygen correction factor is equivalent to a 0.14 limit at a 7% correction factor. Further, EPA requires hazardous waste burning devices operating under RCRA regulations to comply with emissions standards generally on a hourly rolling averaging period. The European guidelines/limits are based on daily averaging, a less stringent approach in terms of operation variability. Finally, RCRA regulations require a facility to comply with the emissions standard for each of three triplicate runs during a Trial Burn or Compliance Test. Compliance with the European guideline/limits is based on the average of test runs.

rope currently incorporate some sort of dust control device, such as wet and dry electrostatic precipitators (ESP) or fabric filters (FF). Most facilities have added multistage wet and dry scrubbers or spray drying and dry absorption processes for controlling acid gas and heavy-metal emissions. The future trend is expected to be toward wet scrubbers, even though all APCD systems must be zero liquid discharge systems. Some new technologies that are emerging include adding selective catalytic reduction $DeNO_x$ reactors, activated carbon filters, and gas suspension absorbers. As new options arise, it appears that the general practice in Europe is to continue to retrofit facilities with new APCDs in series with existing equipment.[16]

3.10 HAZARDOUS WASTE COMBUSTOR MAXIMUM ACHIEVABLE CONTROL TECHNOLOGY (MACT) STANDARDS FOR HAZARDOUS WASTE INCINERATORS, CEMENT KILNS, AND LIGHTWEIGHT AGGREGATE KILNS

On April 19, 1996, the EPA proposed hazardous air pollutant (HAP) emissions standards for hazardous waste incinerators (HWIs), cement kilns (CKs), and lightweight aggregate kilns (LWAKs) under joint authority of the Clean Air Act Amendments of 1990 (CAAA) and the Resource Conservation and Recovery Act (RCRA) (61 FR 17358). The proposed MACT standards limit emissions from both new and existing facilities in each equipment category. Pollutants regulated under the rule would be dioxins and furans (D/F), mercury (Hg), total chlorine (HCl/Cl_2), semivolatile metals (SVM) including lead and cadmium, low-volatility metals (LVM) including antimony, arsenic, beryllium, and chromium, particulate matter (PM), carbon monoxide (CO), and hydrocarbons (HC). In May 1997, revised proposed emission standards were presented in a Notice of Data Availability (62 FR 24212).

Fast Track Portion of the Rule

On June 19, 1998, USEPA promulgated a portion of the proposed MACT standards for HWCs (63 FR 33820). This portion of the rule, known as the *fast track* portion, contains provisions for:

- Exclusions from RCRA requirements for hazardous waste derived fuels that are comparable to fossil fuels
- Expedited RCRA permit modification process for facilities making modifications to meet the MACT standards
- Allowances for extension of compliance dates to promote the use of pollution prevention technologies
- Notification requirements for facilities subject to the final HWC MACT standards

The flow diagram in Fig. 3.10.1 depicts the timeline to comply with the HWC MACT standards. Within one year of the effective date of the emission standards of the MACT rule, facilities will be required to submit a final *Notification of Intent to Comply* (NIC). The NIC must include information on the facility location, waste minimization technique(s), emission control technique(s), and emission monitoring technique(s) being considered, and a statement of the facility's intention to comply.

Within 10 months of the effective date of the emission standards of the MACT rule, facilities will be required to hold at least one informal meeting with the public to discuss anticipated activities required to achieve compliance with the MACT standards.

Within 2 years of the effective date of the emission standards, facilities must submit a progress report. The report must include information on the engineering design for any modifications needed to achieve compliance, the construction applications to be submitted to the applicable regulatory authority, proof of a binding contractual commitment to purchase, fabricate, and install any equipment, devices, and ancillary structures needed to achieve compliance. The progress report must also include a schedule for the system modification and the projected date by which compliance will be achieved.

The rule also allows facilities to request a 1-year extension to install pollution prevention or waste minimization controls. The request must be received in writing not later than 12 months before the facility's nonextended compliance date.

The proposed standards were submitted to the Office of Management and Budget (OMB) on Monday, March 22, 1999. OMB had 90 days to review the regulation. The rule was signed on July 30, 1999, and appeared in the *Federal Register* on September 30, 1999.

Site-Specific Risk Assessments

In the April 19, 1996, preamble to the proposed MACT standards for hazardous waste combustors (61 FR 17358), the USEPA recommended that site-specific risk assessments be conducted as part of the RCRA permitting process for hazardous waste combustors. Specifically, these site-specific risk assessments are intended to address potential concerns about hazardous air pollutants, including dioxins, furans, metals, and nondioxin products of incomplete combustion (PICs). These risk assessments consider the direct pathway (inhalation), as well as indirect pathways such as ingestion of water, soil, vegetation, and animal tissues. Site-specific risk assessments have been performed for hazardous waste combustors since mid-1993 as a result of EPA's *hazardous waste minimization and combustion strategy.*

Most hazardous waste combustors permitted in the mid-1990s had to incorporate a site-specific risk assessment into the permitting processes. For hazardous waste combustors that have not performed a site-specific risk assessment, the EPA recommends that the agency require a risk assessment as part of the MACT performance testing, if appropriate. When a site-specific risk assessment is determined appropriate, the final permit would take into account both the MACT standards and any relevant site-specific considerations resulting from the risk assessment.

The EPA is incorporating risk assessments in the permitting process to ensure that emissions from RCRA combustors are safe. This does not mean that every combustor must undergo a risk assessment. There are factors that the EPA recommends agencies use to determine if a risk assessment is warranted for a certain facility. Some factors include:

- Site-specific considerations such as proximity to receptors, unique dispersion patterns, etc.
- PICs most likely to be found and to pose significant risk
- Off-site sources of pollutants that could influence risk
- Significant ecological considerations
- Volume and types of wastes being burned
- Potentially sensitive receptors such as schools, hospitals, nursing homes, etc.
- Other on-site sources of hazardous air pollutants that would impact risk
- Public concerns

The regulating agency has the final word on whether or not to require a site-specific risk assessment. Most hazardous waste combustors currently in the process of obtaining initial or renewal permits have been required to perform some risk assessment as part of the permitting process.

Numerous guidance documents detailing the requirements of a site-specific risk assessment are available from state to state. The EPA has published two

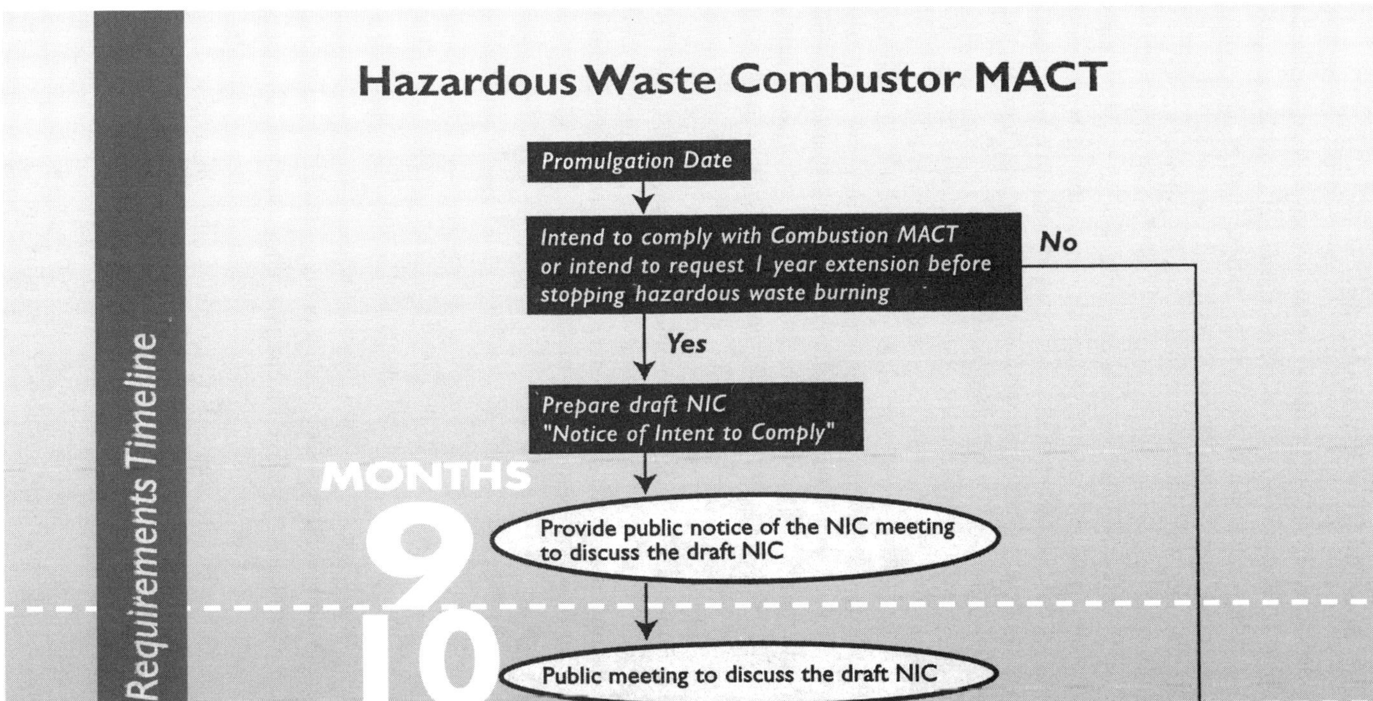
Requirements Timeline
Hazardous Waste Combustor MACT
Promulgation Date
Intend to comply with Combustion MACT or intend to request 1 year extension before stopping hazardous waste burning
No
Yes
Prepare draft NIC "Notice of Intent to Comply"
MONTHS
9
Provide public notice of the NIC meeting to discuss the draft NIC
10
Public meeting to discuss the draft NIC

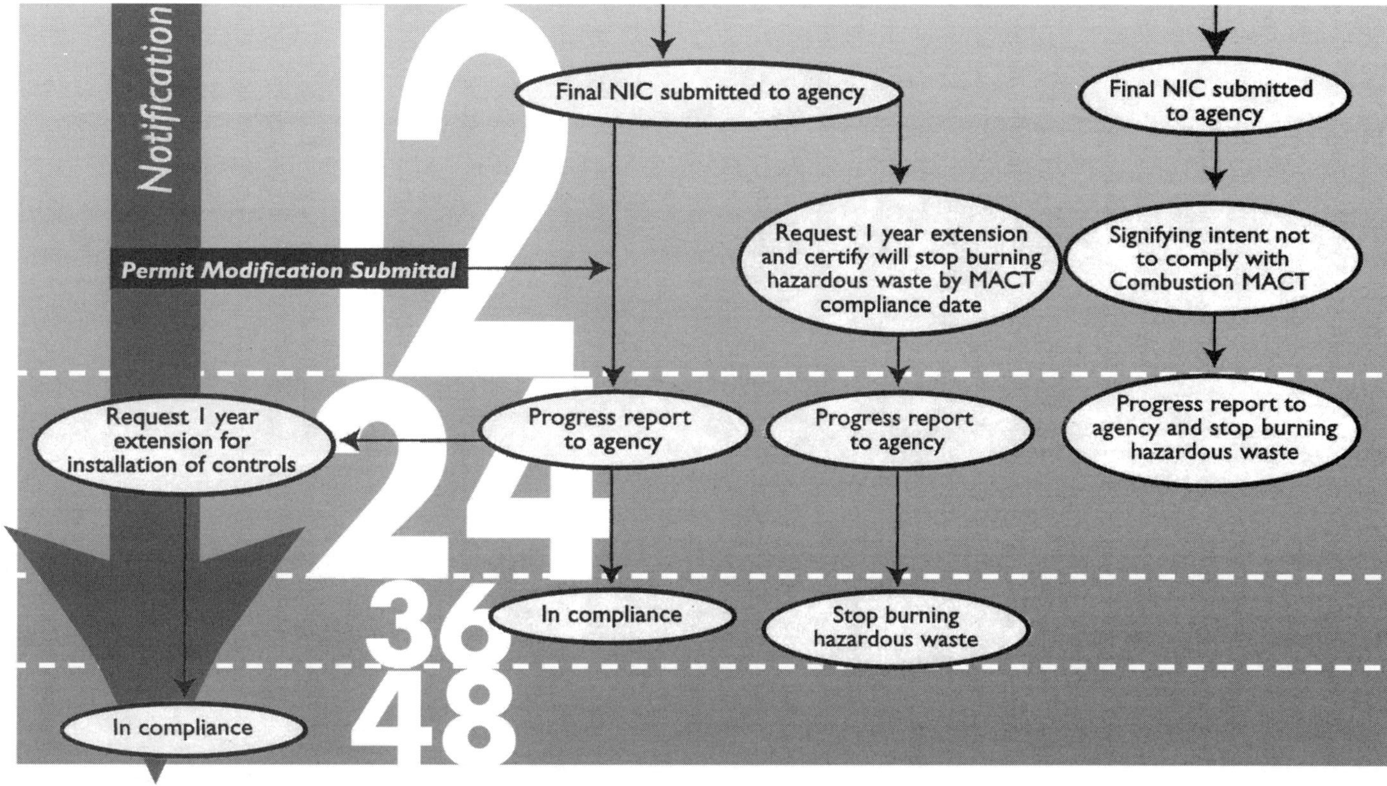

Figure 3.10.1 MACT flow diagram (provided by RMT/Four Nines of RMT, Inc.).

guidance documents in draft form. *Human Health Risk Assessment Protocol for Hazardous Waste Combustion Facilities* (July 1998, EPA530-D-98-001A) recommends a set of procedures for performing risk assessments. Many states are using this draft document as the official guidance for site-specific risk assessments. The EPA has also recently released *Guidance on Collection of Emissions Data to Support Site-Specific Risk Assessments at Hazardous Waste Combustion Facilities* (August 1998, EPA530-D-98-002). This draft document provides guidance on collecting emissions data in trial burns or risk burns. Both documents are available on the EPA web site (www.epa.gov/epaoswer/hazwaste/combust.htm).

EPA is expected to release a guidance document on ecological risk assessments for hazardous waste combustors. To date, most site-specific risk assessments have not addressed ecological risk. With this new guidance, facilities will most likely be required to include ecological risk in future site-specific risk assessments.

Most facilities must submit a Risk Assessment Protocol prior to performing a site-specific risk assessment. The Risk Assessment Protocol must be reviewed and approved by the agency prior to developing the actual risk assessment. The protocol document details the methodologies to be used in the risk assessment. It should include items such as details on pollutants of concern, the site-specific pathways and input parameters to be used, and the dispersion modeling to be performed for the risk assessment.

The identification of compounds of potential concern (COPCs) is an important facet of protocol development. These are the compounds analyzed in the risk burn testing and ultimately evaluated in the risk assessment. The facility's objective is to provide the agency with a list of compounds that is defensible to the public. Public opinion and safety are usually key factors in the RCRA permitting process. The EPA has provided guidance on selecting COPCs for risk assessments in both guidance documents noted above. The compounds included in the EPA suggested list are typically considered to contribute to risk. These compounds should be considered when developing the COPC list for the facility. However, it is also very important to review the chemistry of the facility's waste streams. There is no one definitive COPC list because a compound that is a concern at one facility may not be a concern at another. In addition, the COPCs from a combustion unit will vary depending on the type of combustion unit, the type of waste being burned, and the type of air pollution control equipment used. The COPC list should be site specific, taking into consideration the operating conditions of the hazardous waste combustor.

After agency approval of the Risk Assessment Protocol, the site-specific risk assessment is performed using data collected during a risk burn. A risk burn can be a separate test condition or combined with another test, such as the MACT performance test.

After completing the risk burn, the collected data are used to perform the risk assessment. Risk assessments are not typically pass/fail situations but are rather developed in numerous stages to obtain acceptable risk values. A risk assessment would begin with very conservative assumptions to obtain an acceptable result with minimal effort. If the conservative assumptions result in unacceptable risk levels, the process is refined by providing more site-specific information. Real data, such as farm locations, population density, and water body locations, would replace conservative assumptions to provide a more realistic risk estimate. This staged process provides a facility with a good chance of obtaining an acceptable risk level from the risk assessment.

Incorporating site-specific, risk-based permit conditions into a permit is not anticipated to be necessary for the vast majority of hazardous waste combustors. Rather, such conditions would be necessary only if compliance with the MACT requirements is insufficient to protect human health and the environment and if the resulting risk-based conditions are more stringent than those required under the MACT standards.

The results of recent risk assessments provide insight to the process. The main drivers appear to be emissions of dioxins/furans, polycyclic aromatic hydrocarbons (PAHs), and heavy metals such as chromium, mercury, beryllium, and cadmium. This implies that facilities burning materials with high chlorine content or metals content will be most affected by risk assessment results. Based on these results, the agency may choose to limit facility emissions to rates lower than those required by regulations.

MACT Standards Promulgated on September 30, 1999

The final Maximum Achievable Control Technology (MACT) standards for hazardous waste combustors were promulgated on Thursday September 30, 1999. The complete rule can be found in the *Federal Register*, Volume 64, No. 189/Thursday, September 30, 1999.

The promulgated MACT standards limit emissions from both new and existing facilities in each equipment category. Pollutants regulated under the rule include dioxins and furans (D/F), mercury (Hg), total chlorine (HCl/Cl_2), semivolatile metals (SVM) including lead and cadmium, low-volatility metals (LVM) including arsenic, beryllium, and chromium, particulate matter (PM), carbon monoxide (CO), and hydrocarbons (HC). The standards also require continuous emissions monitoring systems (CEMS) and performance testing.

MACT Standards for Hazardous Waste Incinerators

The MACT standards include emission limits for both new and existing hazardous waste incinerators (Table 3.10.1). The standards require a minimum *destruction and removal efficiency* (DRE) of 99.99% for each POHC and a minimum DRE of 99.9999% for dioxin-listed wastes. These requirements are identical to the existing hazardous waste incinerator performance standards found in 40 CFR 264 Subpart O, except that the POHCs are to be chosen from the hazardous air pollutants established by the CAAA instead of the hazardous constituents provided in 40 CFR 261 Appendix VIII.

Table 3.10.1 Emissions limits from incinerators[a]

Pollutant	Units	Existing	New
DF	ng TEQ/dscm	0.2 or 0.4[b]	0.2
Mercury	μg/dscm	130	45
SVM	μg/dscm	240	24
LVM	μg/dscm	97	97
HCl/Cl_2	ppm_v (dry)	77	21
PM	mg/dscm (gr/dscf)	34 (0.015)	34 (0.015)
CO	ppm_v (dry)	100	100
HC	ppm_v (dry)	10	10

[a]All emission rates corrected to 7% oxygen dry volume.
[b]If the gas temperature at the inlet of the initial particulate control device is maintained below 400°F.

Cement Kiln Standards

Cement kilns have the same DRE requirements as incinerators under the MACT standards (Table 3.10.2). Other emission standards are similar to those for incinerators with some exceptions due to differences in equipment and operations.

In addition to the requirements above, new facilities, which began kiln construction after April 19, 1996, at a plant site where a cement kiln had not previously existed, must continuously monitor HC and meet an HC limit of 50 ppm by volume, over a 30-day block average, dry basis, corrected to 7% oxygen.

Lightweight Aggregate Kiln Standards

LWAKs have the same DRE requirements as incinerators under the MACT standards. Other emission standards are similar to those for incinerators with some exceptions due to differences in equipment and operations (see Table 3.10.3).

Continuous Emission Monitoring

Continuous emission monitoring systems (CEMS) are required to demonstrate and monitor compliance with the CO and HC standards. Facilities have the option of using CEMS for either CO or HC. If CO is continuously monitored, compliance with the HC limit needs to be demonstrated only during DRE testing. If HC is continuously monitored, compliance with the CO limit need not be demonstrated. An oxygen CEMS is also required to continuously correct the CO and HC levels to 7% oxygen. Facilities will also be required to install particulate matter CEMS to demonstrate compliance with the PM standards. However, all requirements relating to the PM CEMS are deferred pending further rulemaking.

Table 3.10.2 Emissions limits from cement kilns[a]

Pollutant	Units	Existing	New
D/F	ng TEQ/dscm	0.2 or 0.4[b]	0.2 or 0.4[b]
Mercury	μg/dscm	120	56
SVM	μg/dscm	240	180
LVM	μg/dscm	56	54
HCl/Cl_2	ppm_v (dry)	130	86
PM	kg/Mg	0.15	0.15
Opacity	%	20	20
CO	ppm_v (dry)	100[c]	100[c]
HC[a]	ppm_v (dry)	10[d] or 20[e]	10[d] or 20[e]

[a]All emission rates corrected to 7% oxygen dry volume.
[b]If the gas temperature at the inlet of the dry particulate control device is maintained below 400°F.
[c]In the by-pass duct or main stack.
[d]In the by-pass duct.
[e]In the main stack.

Performance Testing

A facility must conduct comprehensive performance tests to demonstrate compliance with the emission standards, establish limits for the operating parameters, and demonstrate compliance with the performance specifications for continuous monitoring systems. Comprehensive performance tests are to be conducted every 5 years. The initial comprehensive performance test must be conducted no later than 6 months after the compliance date.

A facility must also conduct confirmatory performance tests to demonstrate compliance with the D/F emission standard when the source operates under normal operating conditions. A performance evaluation of the continuous monitoring systems required for compliance assurance with the D/F emission stan-

Table 3.10.3 Emission limits from LWAKs[a]

Pollutant	Units	Existing	New
D/F	ng TEQ/dscm	0.2 or 0.4[b]	0.2 or 0.4[b]
Mercury	μg/dscm	47	33
SVM	μg/dscm	250	43
LVM	μg/dscm	110	110
HCl/Cl_2	ppm_v (dry)	150	41
PM	mg/dscm	57	57
CO	ppm_v (dry)	100	100
HC	ppm_v (dry)	20	20

[a]All emission rates corrected to 7% oxygen dry volume.
[b]If the combustion gas is rapidly quenched to 400°F or lower.

dard must also be conducted. Confirmatory performance tests shall be conducted approximately midway between comprehensive performance tests.

3.11 FUTURE OF INCINERATION

The choice of a hazardous waste disposal method is limited to basically three alternatives: (1) waste reduction through process modification (elimination, substitution, recycle/reuse, or minimization), (2) conversion of the hazardous waste to a less hazardous waste, or (3) placement of the waste or residuals in the environment. These three are not necessarily exclusive of each other and, more often than not, more than one are used. The first of these, waste reduction, is usually preferable in terms of economic return as well as short- and long-term protection of the environment and public health. The third approach, the placement of waste and waste residuals in the environment, includes secure landfills, above-ground landfills, deepwell injection, ocean disposal, geological isolation, and seabed implacement. The 1984 amendments (HSWA), coupled with expected future prohibition and financial liability, place severe limitations on these options.

The second approach includes all forms of waste treatment. In the 1976 RCRA, treatment is described as any method, technique, or process, including neutralization, designed to change the physical, chemical, or biological character or composition of any hazardous waste so as to neutralize such waste or as to render such waste less hazardous, safer for transport, amenable for recovery, amenable for storage, or reduced in volume. With the exception of incineration, these physical, chemical, and biological methods were discussed in some detail in Chapter 2. Of all of these options, incineration is becoming a more and more attractive alternative. This method has the capability of safely converting large volumes of primarily organic hazardous waste to much smaller volumes of ash and other residues plus nontoxic gaseous emissions. Moreover, process incineration can often provide an optimum permanent solution to management of hazardous waste with minimal long-term ecological burden. Based on the 1984 RCRA amendments and the more recent MACT rules, waste combustion in HWCs still appears (from a regulatory point of view) to be the most viable means of waste treatment for organics. The demand for incineration is expected to be maintained at the capacity of the mid-1990s due, in large extent, to restrictions established in the 1984 RCRA amendments regarding land disposal options.

3.12 ILLUSTRATIVE EXAMPLES

The illustrative examples provided here pertain to standards and regulations. Those readers not familiar with technical calculations should bypass these examples and the problems at the conclusion of this chapter until Part II has been reviewed. The reader may then choose to return to this problem set.

Example 3.12.1 A packed column is operating at an efficiency of 85.5% for the removal of HCl from a waste flue gas. Since the recent RCRA regulations require that no more than 4.0 lb/h be emitted into the atmosphere, what is the maximum inlet HCl rate in the flue gas to operate within the regulations? In order to meet the latest MACT regulations of 21 ppm$_v$(dry), determine compliance if the stack volume is 8000 dscfm and if the HCl rate is 20 lb/h.

SOLUTION By definition,

$$E = \frac{(w_{\text{in}} - w_{\text{out}})}{w_{\text{in}}}$$

where E = efficiency (expressed as a fraction)
w_{in} = inlet mass flow rate
w_{out} = outlet mass flow rate

This equation may be rearranged and solved for w_{in}:

$$w_{\text{in}} = \frac{w_{\text{out}}}{1 - E}$$

For this problem, w_{out} is the maximum allowable emission rate (4 lb/h) and the fractional efficiency (E) is 0.855:

$$w_{\text{in}} = \frac{4.0}{1.0 - 0.855} = 27.58 \text{ lb/h}$$

This represents the maximum allowable mass flow rate of HCl in the inlet flue gas.

$$\text{Volume of HCl} = 4 \text{ lb/h} \times \frac{1 \text{ hr}}{60 \text{ min}} \times \frac{1}{36.5 \text{ lb/mol}} \times 379 \text{ ft}^3 \text{ lb mol} = 0.6922 \text{ cfm}$$

If the stack volume is 8000 dscfm, then

$$\text{Concentration} = \frac{0.6922}{8000} = 86.53 \text{ ppm}$$

This indicates noncompliance for the existing incinerators. In order to attain compliance as a new incinerator (21 ppm_v), the emission level must be reduced to

$$w_{out} = \frac{21}{86.5} \times \left(4\ \frac{lb}{h}\right)$$

$$= 0.97 \text{ lb/h}$$

The fractional efficiency, E, must then be increased to

$$E = \frac{w_{in} - w_{out}}{w_{in}} = \frac{20 - .97}{20} = 95.15\%$$

Example 3.12.2 A quench tower operates at a HCl removal efficiency of 65%. This is then followed by a packed tower absorber. What is the minimum collection efficiency of the absorber if an overall HCl removal efficiency of 99.0% is required?

SOLUTION Select as a basis, 100 lb/h of HCl entering the unit. The mass of HCl leaving spray tower is calculated from the following equation:

$$E = \frac{w_{in} - w_{out}}{w_{in}}$$

$$w_{out} = w_{in}(1 - E) = 100(1.0 - 0.65)$$

$$= 35 \text{ lb/h HCl leaving the spray tower}$$

Use the overall efficiency to calculate the mass flow rate of HCl leaving the packed tower absorber:

$$w_{out} = w_{in}(1 - E) = 100(1.0 - 0.990)$$

$$= 1.0 \text{ lb/h leaving the packed tower}$$

The efficiency of the packed tower can now be calculated:

$$E = \frac{w_{in} - w_{out}}{w_{in}} = \frac{35.0 - 1.0}{35.0}$$

$$= 0.971 = 97.1\%$$

Example 3.12.3 A small quantity of a waste mixture is incinerated at 2000°F using 5% ash coal to provide energy necessary for combustion. If 300 ft^3 of the flue gas (measured at 2000°F) is produced for every pound of coal burned at an O_2 concentration of 10% dry volume basis, what is the maximum effluent particulate loading in gr/ft^3 (at 2000°F) corrected to 7% O_2 dry volume. Assume no contribution to the particulates from the waste.

SOLUTION Based on the problem statement, assume the volume of gas produced on incinerating the waste is negligible. Select as a basis 1.0 lb of coal burned.

$$\text{Mass of particulates} = (1.0\text{ lb})(0.05) = 0.05\text{ lb}$$

The maximum particulate loading is given by:

$$W = \frac{0.05}{300} = 1.667 \times 10^{-4}\text{ lb/ft}^3$$

where W = particulate loading (mass of particulates/total volume of flue gas).

To convert to gr/ft^3, note that there are 7000 gr/lb. Therefore,

$$W = (1.667 \times 10^{-4})(7000) = 1.167\text{ gr/ft}^3$$

Thus the correction factor is

$$\text{Correction factor} = \frac{14}{21 - 10} = 1.272$$

Corrected value is $1.167 \times 1.272 = 1.4899$ gr/ft^3 at 7% dry volume.

Example 3.12.4 With reference to Example 3.12.3, what dilution factor or particulate collection efficiency is required to achieve the ambient air quality standard of 75 $\mu g/m^3$ if stack O_2 is 10% dry volume? What efficiency is required to achieve an outlet loading of 0.015 gr/ft^3 at 7% O_2 dry volume?

SOLUTION First convert W_{in} to micrograms per cubic meter ($\mu g/m^3$).

$$W_{in} = 1.667 \times 10^{-4}\text{ lb/ft}^3 = 2.6727 \times 10^6\ \mu\text{g/m}^3$$

where W_{in} = inlet loading (mass particulate/volume gas).

The dilution factor (DF) may now be calculated:

$$\mathrm{DF} = \frac{w_{\text{in}}}{w_{\text{std}}} = \frac{2.6727 \times 10^6}{75} = 3.564 \times 10^4$$

where W_{std} = ambient air quality standard.

The particulate collection efficiency (E) is given by:

$$E = \frac{w_{\text{in}} - w_{\text{std}}}{w_{\text{in}}} = \frac{2.6727 \times 10^6 - 75}{2.6727 \times 10^6}$$

$$= 0.99997 = 99.997\%$$

Note: This definition of E is equivalent to that given in Example 3.12.1 in terms of mass flow rates (w).

The required efficiency to achieve an outlet loading of .015 gr/ft^3 at 7% O_2 dry volume is

$$E = \frac{1.484 - 0.015}{1.484}$$

$$= 0.9878 = 98.78\%$$

Example 3.12.5 The waste flow rate into an incinerator is 100 lb/h. Calculate the waste flow rate leaving the unit to achieve a destruction efficiency of **(1)** 95%, **(2)** 99%, **(3)** 99.9%, **(4)** 99.99%, **(5)** 99.9999%.

SOLUTION Since one may rearrange this equation and solve for w_{out},

$$w_{\text{out}} = w_{\text{in}} \left[1 - \left(\frac{\text{DRE}}{100}\right)\right]$$

1. For a DRE of 95%:

$$w_{\text{out}} = 100 \left[1 - \left(\frac{95}{100}\right)\right]$$

$$= 5 \text{ lb/h}$$

2. For a DRE of 99%:

$$w_{\text{out}} = 100 \left[1 - \left(\frac{99}{100}\right)\right]$$

$$= 1 \text{ lb/h}$$

3. For a DRE of 99.9%:

$$w_{out} = 0.1 \text{ lb/h}$$

4. For a DRE of 99.99%:

$$w_{out} = 0.01 \text{ lb/h}$$

5. For a DRE of 99.9999%:

$$w_{out} = 0.0001 \text{ lb/h}$$

Example 3.12.6 A waste mixture consisting of toluene and chlorobenzene is presently being incinerated at 2050°F with a 2.2-s residence time. The componential waste inlet feed rate and the stack discharge rate (including particulate catch) are given. The stack gas flow rate has been determined to be 13,250 dscfm (dry standard cubic feet per minute) with an O_2 concentration of 4% O_2 dry volume. Is the unit in compliance? Components (1) and (2) are the designated POHCs.

	Compound	Inlet (lb/h)	Outlet (lb/h)
1	C_7H_8 (toluene)	953	0.082
2	C_6H_5Cl (chlorobenzene)	337.5	0.022
	HCl	—	2.2
	Particulates	—	4.0

SOLUTION Since

$$\text{DRE}_i = \frac{(w_{in})_i - (w_{out})_i}{(w_{in})_i} 100$$

$$\text{DRE}_{toluene} = \frac{953 - 0.082}{953} 100 = 0.99.991\%$$

$$\text{DRE}_i = \frac{337.5 - 0.022}{337.5} 100 = 99.993\%$$

Regulations require that the DRE for each designated POHC be at least 99.99%. In this case, the designated POHCs are toluene and chlorobenzene. Both POHCs exhibit a DRE of >99.99%. Therefore, the unit is in compliance with current regulations.

The reaction of chlorobenzene in the incinerator may be shown as the following equation

$$C_6H_5Cl + \text{Air} \rightarrow 6CO_2 + HCl + 2H_2O$$

Assume all the chlorine in the feed is converted to HCl. The molar feed rate of chlorobenzene $\dot{n}_3$ is

$$\dot{n}_2 = \frac{w_2}{(\text{MW})_2} = \frac{337.5}{112.5}$$

$$= 3.0 \text{ lbmol/h}$$

where n_i = molar flow rate of component i
$(\text{MW})_i$ = molecular weight of component i

Note: Molecular weights, moles, and the units of lbmol and gmol are discussed in Chapter 4.

Each molecule of chlorobenzene contains one atom of chlorine. Therefore,

$$\dot{n}_{\text{HCl}} = \dot{n}_2$$

$$= 3.0 \text{ lbmol/h}$$

$$w_{\text{HCl}} = 109.5 \text{ lb/h}$$

The removal efficiency (RE) for HCl is

$$\text{RE} = \frac{w_{\text{in}} - w_{\text{out}}}{w_{\text{in}}} 100 = \frac{109.5 - 2.2}{109.5} 100 = 97.99\%$$

Previous RCRA regulations require that HCl emissions must either achieve a 99% HCl scrubbing efficiency or emit <4 lb/h of hydrogen chloride. In this case, emissions are <4 lb/h and therefore in compliance with past regulations. To determine if the emissions meet the new MACT rules of September 30, 1999, the outlet emission must not exceed 21 ppm corrected to 7% O_2 dry volume.

The stack flow is given at 13,250 dscfm. The HCl emissions are 2.2 lb/h. The HCl volume is calculated as follows:

$$\text{Volume, cfm} = w_3 \times \frac{1}{365} \times \frac{1 \text{ h}}{60 \text{ min}} \times 379 \frac{\text{cu ft}}{\text{mol}}$$

$$= .3807 \text{ cfm}$$

$$\text{Concentration} = \frac{\text{vol. HCl}}{\text{vol. flue gases}}$$

$$= \frac{.3807}{13{,}250}$$

$$= 28 \text{ ppm at } 4\% \text{ O}_2 \text{ dry volume}$$

$$\text{Corrected concentration} = 0.8235 \times 28$$

$$= 23.06 \text{ ppm}$$

Removal efficiencies must increase to

$$\frac{21}{23.06}(1 - .9799) = 98.2\%$$

The outlet loading W_{out} of the particulates is

$$\text{PM} = \frac{\text{PM}_{out}(7000)}{\text{dscfm} \times 60} = \frac{(40)(7000)}{(13{,}250)(60)}$$

$$= 0.0352 \text{ gr/dscf}$$

$$\text{Correction factor} = \frac{14}{21 - 4} = 0.8235$$

$$\text{Corrected PM}_{out} = \text{PM} \times \text{CF}$$

$$= 0.0352 \times 0.8235$$

$$= 0.029 \text{ gr/dscf}$$

New MACT regulations require that particulate emissions must be <0.015 gr/dscf corrected to 7% O_2 dry volume. It is therefore in compliance with past regulations at <0.08 gr/dscf but will not meet the new MACT standards for particulate emissions.

The unit is therefore in compliance with the new MACT regulations for all but PM_{out}.

Example 3.12.7 During a trial burn, an incinerator was operated at a waste feed rate of 5000 lb/h and 7% O_2 dry volume in the stack gases. The gas flow rate measured in the stack was 19,200 dscfm. Under these conditions, the measured concentrations of the principal organic hazardous components were:

Trichloroethylene (1) = 4.9 μg/dscf
1,1,1-Trichloroethane (2) = 1.0 μg/dscf
Methylene chloride (3) = 49 μg/dscf
Perchloroethylene (4) = 490 μg/dscf

Each hazardous component listed here constitutes approximately 5% of the total waste feed rate. Calculate the destruction and removal efficiency of the POHCs. Is the unit in compliance with present regulations?

SOLUTION Each hazardous component constitutes 5% of the total waste feed. Therefore,

$$(w_{in})_I = (0.05)(\text{total feed rate}) = (0.05)(5000 \text{ lb/h})$$

$$= 250 \text{ lb/h}$$

The mass flow rate of each hazardous component in the stack is given by:

$$(w_{\text{out}})_i = \frac{Q_s C_i}{7.57 \times 10^6}$$

where Q_s = volumetric gas flow rate (dscfm)
C_i = concentration of component i (μg/dscf)
7.57×10^6 = conversion factor that converts lb/h to g/min

Therefore,

$$(w_{\text{out}})_1 = \frac{(19{,}200)(4.9)}{7.57 \times 10^6} = 0.01243 \text{ lb/h} \qquad (w_{\text{out}})_2 = \frac{(19{,}200)(1.0)}{7.57 \times 10^6} = 0.00254 \text{ lb/h}$$

$$(w_{\text{out}})_3 = \frac{(19{,}200)(49)}{7.57 \times 10^6} = 0.1243 \text{ lb/h} \qquad (w_{\text{out}})_4 = \frac{(19{,}200)(490)}{7.57 \times 10^6} = 1.243 \text{ lb/h}$$

The destruction and removal efficiency for component i, DRE_i, is given by:

$$\text{DRE}_i \left(\frac{(w_{\text{in}})_i - (w_{\text{out}})_i}{w_{\text{in}}} \right) 100$$

$$\text{DRE}_1 = \left(\frac{250 - 0.01243}{250} \right) 100 = 99.995\% \qquad \text{DRE}_2 = \left(\frac{250 - 0.00254}{250} \right) 100 = 99.999\%$$

$$\text{DRE}_3 = \left(\frac{250 - 0.1243}{250} \right) 100 = 99.95\% \qquad \text{DRE}_4 = \left(\frac{250 - 1.243}{250} \right) 100 = 99.50\%$$

Present regulation dictates that 99.99% destruction and removal efficiency be achieved. In this case, only two of the principal organic hazardous components (1 and 2) achieved this status. Assuming that either component 3 or 4 (or both) was a designated POHC, the unit would be out of compliance.

PROBLEMS

1. A proposed incineration facility design requires that a spray tower and a packed column are to be used in series for the removal of HCl from the flue gas. The spray tower is operating at an efficiency of 65% and the packed column at an efficiency of 98%. Calculate the mass flow rate of HCl leaving the spray tower, the mass flow rate of HCl entering the packed tower, and the overall efficiency of the removal system if 76.0 lb of HCl enters the system every hour.

2. Two packed columns operating in series are used for removing HCl from a flue gas. If the first column is operating at 65% efficiency and the second column is operating at 99.5% efficiency, calculate the overall efficiency and mass flow rate of HCl leaving the system.

3. Two spray columns operating in series are used for removing HCl from a flue gas coming from a hazardous waste incinerator. An overall efficiency of 99.0% is required. What is the required minimum collection efficiency of the second column if the first column is operating at an efficiency of 70%?

4. A hazardous waste incinerator is burning an ash coal to provide the energy necessary for combustion. Calculate the maximum ash content of the coal. A collection efficiency of 98.60% can be achieved for a maximum required outlet load of 0.001 lb/lb coal. Also calculate the maximum particulate loading in gr/ft^3 if 325 ft^3 of total flue gas is produced for every pound of coal burned.

5. A hazardous waste incinerator is burning anthracite coal containing 7.1% ash to provide the necessary energy for combustion. If 300 ft^3 of total flue gas are produced for every pound of coal burned, what is the maximum effluent particulate loading in gr/ft^3? Assume no contribution to the particulates from the waste. The secondary ambient air quality standard for particulates is 75 μg/m^3. What dilution factor or particulate collection efficiency is required to achieve this standard?

6. A proposed incinerator is designed to destroy a hazardous waste at 2100°F and 1 atm. Current regulations dictate that a minimum destruction efficiency of 99.99% must be achieved. The waste flow rate into the unit is 960 lb/h while that flowing out of the unit is measured as 0.08 lb/h. Is the unit in compliance?

7. A hazardous waste incinerator is burning a waste mixture consisting of toluene (1), chlorobenzene (2), and dichlorobenzene (3) with 50% excess air at 2100°F with a 2.5-s residence time. Analyses of the feed rate composition and stack discharge rate are given. The stack gas flow rate has been determined to be 14,280 dscfm. Is the unit in compliance with present regulations? Compounds 2 and 3 are the designated POHCs.

	Compound	Inlet (lb/h)	Outlet (lb/h)
1	C_7H_8 (toluene)	867	0.112
2	C_6H_5Cl (chlorobenzene)	442	0.015
3	$C_6H_4Cl_2$ (dichlorobenzene)	297	0.018
	HCl	—	4.1
	Particulates	—	9.65

Note: In most trial burn plans, chlorobenzene would be selected as the only POHC since dichlorobenzene can generate chlorobenzene as a PIC.

8. Analysis of a trial burn performance established emission concentrations of the principal organic components as follows:

1,1,1-Trichloroethane (1) = 2.1) μg/dscf
Perchloroethylene (2) = 10 μg/dscf
Methylene chloride (3) = 6.0 μg/dscf

The incinerator was operated at a waste feed rate of 6200 lb/h and 50% excess air. The gas flow rate measured in the stack was 21,100 dscfm. Each hazardous component constituted approximately 6% of the total waste feed rate. Calculate the destruction and removal efficiency of the POHCs. Is the unit in compliance with present regulations?

REFERENCES

1. P. N. Cheremisinoff and F. Ellerbusch, *Solid Waste Legislation, Resource Conservation & Recovery Act, A Special Report*, Washington, DC, 80th ed., CRC Press, 1999.
2. "Resource Conservation and Recovery Act of 1976," *International Environmental Reporter*, Bureau of National Affairs, Washington, DC, Oct. 21, 1976.
3. "Resource Conservation and Recovery Act, 1976," *Congressional Record*, U.S. Govt., Washington, DC, Oct. 3 and 11, 1984.
4. "Hazardous Waste Management System; Final Codification Rule," *Federal Register*, 40 CFR, Washington, DC, Monday, July 15 (1985).
5. T. Shen, U. Choi, and L. Theodore, *Hazardous Waste Incineration Manual*, U.S. EPA Air Pollution Training Institute, Research Triangle Park, NC, 1985.
6. J. Casler, *On Permitting the Resource Conservation and Recovery Act*, Environmental Research and Technology, Inc., Concord, MA, 1984.
7. E. D. Cooper, "Incineration Permitting," *Hazardous Materials and Waste Management,* **4**, 10, Jan./Feb. (1986).
8. J. J. Santoleri "RCRA Trial Burn Planning—Importance of Mini-Burns," AWMA, Atlantic City, NJ, October 5, 1989.
9. C. C. Lee, *Environmental Engineering Dictionary*, Govt. Institute, Inc., Rockville, MD, 1999.
10. C. C. Lee and George L. Huffman, U.S. EPA, "Minimization of Combustion By-Products: Regulatory Framework," *Hazardous Waste & Hazardous Materials*, **8**(4), (1991).
11. J. J. Santoleri and R. A. Carnes, "New RCRA Regulations for Industrial Boilers, Furnaces and Incinerators," AWMA, Orlando, FL, Mar. 16–19, 1992.
12. U.S. EPA Science Advisory Board, "Report on Incineration of Liquid Wastes by the Environmental Effects, Transport and Fate Committee," U.S. EPA, Washington, DC, April (1985).
13. U.S. EPA Science Advisory Board, "Report of the Products of Incomplete Combustion Subcommittee of the Science Advisory Board," U.S. EPA SAB-EC-90-004, October (1989).

14. "Burning of Hazardous Wastes in Boilers and Industrial Furnaces," 40 CFR 261, 265, 166, Washington, DC, August 21, 1991.
15. U.S. EPA, "Draft Hazardous Waste Minimization and Combustion Strategy," Washington, DC, May 18, 1993.
16. U.S. EPA Solid Waste and Emergency Response, *Combustion Emissions Technical Resource Document (CETRED)*, EPA530-R-94-014, May (1994).
17. European Centre for Ecotoxicology (ECOTEC), "Exposure of Man to Dioxins: A Perspective on Industrial Waste Incineration," Technical Report No. 49, Brussels, Belgium, December, 1992.

II

INCINERATION PRINCIPLES

From the authors' point of view, Chapters 4–6 are the "heart" of this text. While Part III presents much valuable information on the description, design, and operation of equipment used in the incineration of hazardous wastes and Part IV investigates overall incineration plant design (including economic considerations), Part II contains the core calculations on which incineration design is based.

The purpose of Chapter 4 is to review certain basic topics with which the reader may or may not already be familiar. Among these are definitions and explanations of concepts and properties that pertain to incineration: thermal conductivity, heat capacity, heating value, viscosity, density, Reynolds number, ideal gas law, psychrometric chart, steam tables, and so on. Chapter 5 discusses a number of scientific principles, primarily from the fields of chemistry, physics, and chemical engineering: topics such as stoichiometry, thermochemistry, chemical equilibrium, and kinetics. Although these topics are pertinent to hazardous waste incineration, they are treated in a general fashion without any attempt to apply them to specific problems. This latter is the subject of Chapter 6, where the same topics studied in Chapter 5 are applied directly to the incineration process. Important calculational procedures are presented in this chapter; the more important among these are methods for predicting (a) incinerator operating temperature based on the properties of the waste or waste–fuel mixture and the amount of excess air employed during combustion, and (b) flue products produced when burning a given waste–fuel mixture in stoichiometric or excess air.

4

BASIC CONCEPTS

4.1 INTRODUCTION

This chapter is a review of some basic concepts from physics, chemistry, and engineering in preparation for material that is covered in later chapters. These basic concepts include: units and dimensions, some physical and chemical properties of substances, the ideal gas law, the Reynolds number, and phase equilibria. All of these topics are vital to hazardous waste incineration applications. Because many of these topics are unrelated to each other, this chapter admittedly lacks the cohesiveness that chapters covering a single topic might have. This is usually the case when basic material from such widely differing areas of knowledge as physics, chemistry, and engineering is surveyed. Though these topics are widely divergent and covered with varying degrees of thoroughness, all of them will find later use in this text. If additional information on these review topics is needed, the reader is directed to the literature in the reference section of this chapter.

4.2 FUNDAMENTALS OF MEASUREMENT

Units and Dimensions

The units used in this text are consistent with those adopted by the engineering profession in the United States. For engineering work, SI (*Système International*) and *English* units are most often employed; in the United States, the *English engineering* units are generally used, although efforts are still underway to obtain universal adoption of SI units for all engineering and science applications. The SI units have the advantage of being based on the decimal system, which allows for more convenient conversion of units within the system. There are other systems of units; some of the more common of these are shown in Table 4.2.1. The English engineering units will primarily be used here. Tables 4.2.2 and 4.2.3 present units for the English and SI systems, respectively. Some of the more common prefixes for SI units are given in Table 4.2.4. Conversion factors between SI and English units and additional details on the SI system are provided in the appendix.

Conversion of Units

Converting a measurement from one unit to another can conveniently be accomplished by using *unit conversion factors*; these factors are obtained from the simple equation that relates the two units numerically. For example, from

$$1 \text{ foot (ft)} = 12 \text{ inches (in.)} \tag{4.2.1}$$

the following conversion factor can be obtained:

$$12 \text{ in.}/1 \text{ ft} = 1 \tag{4.2.2}$$

Since this factor is equal to unity, multiplying some quantity (e.g., 18 ft) by the factor cannot alter its value. Hence

$$18 \text{ ft } (12 \text{ in.}/1 \text{ ft}) = 216 \text{ in.} \tag{4.2.3}$$

Note that in Eq. (4.2.3), the old units of *feet* on the left-hand side cancel out leaving only the desired units of *inches.*

Physical equations must be dimensionally consistent. For the equality to hold, each term in the equation must have the same dimensions. This condition can be and should be checked when solving engineering problems. Throughout the text, and in particular in Part II, great care is exercised in maintaining the dimensional formulas of all terms and the dimensional homogeneity of each equation. Equations will generally be developed in terms of specific units rather than general dimensions (e.g., feet, rather than length). This approach should help the reader to more easily attach physical significance to the equations presented in these chapters.

Significant Figures and Scientific Notation

Significant figures provide an indication of the precision with which a quantity is measured or known. The last digit represents, in a qualitative sense, some degree of doubt. For example, a measurement of 8.32 in. implies that the actual quantity is somewhere between 8.315 and 8.325 in. This applies to calculated and measured quantities; quantities that are known exactly (e.g., pure integers) have an infinite number of significant figures.

Table 4.2.1 Common systems of units

System	Length	Time	Mass	Force	Energy	Temperature
SI	meter	second	kilogram	newton	joule	kelvin, degree Celsius
egs	centimeter	second	gram	dyne	erg, joule, or calorie	kelvin, degree Celsius
fps	foot	second	pound	poundal	foot poundal	degree Rankine, degree Fahrenheit
American engineering	foot	second	pound	pound (force)	British thermal unit, horsepower hour	degree Rankine, degree Fahrenheit
British engineering	foot	second	slug	pound (force)	British thermal unit, foot pound (force)	degree Rankine, degree Fahrenheit

Table 4.2.2 English engineering units

Physical Quantity	Name of Unit	Symbol for Unit
Length	foot	ft
Time	second	s
Mass	pound (mass)	lb
Temperature	degree Rankine	°R
Temperature (alternate)	degree Fahrenheit	°F
Moles	pound-mole	lbmol
Energy	British thermal unit	Btu
Energy (alternate)	horsepower-hour	hp-h
Force	pound (force)	lb_f
Acceleration	foot per second square	ft/s^2
Velocity	foot per second	ft/s
Volume	cubic foot	ft^3
Area	square foot	ft^2
Frequency	cycles per second, hertz	cycles/s, Hz
Power	horse power, Btu per second	hp, Btu/s
Specific heat capacity	British thermal unit per (pound mass-degree Rankine)	Btu/lb-°R
Density	pound (mass) per cubic foot	lb/ft^3
Pressure	pound (force) per square inch	psi
	pound (force) per square foot	psf
	atmospheres	atm
	bar	bar

The significant digits of a number are the digits from the first nonzero digit on the left to either (a) the last digit (whether it is nonzero or zero) on the right if there is a decimal point, or (b) the last nonzero digit of the number if there is no decimal point. For example:

370	Has 2 significant figures
370.	Has 3 significant figures
370.0	Has 4 significant figures
28,070	Has 4 significant figures
0.037	Has 2 significant figures
0.0370	Has 3 significant figures
0.02807	Has 4 significant figures

Whenever quantities are combined by multiplication and/or division, the number of significant figures in the result should equal the lowest number of significant figures of any of the quantities. In long calculations, the final result should be rounded off to the correct number of significant figures. When quantities are combined by addition and/or subtraction, the final result cannot be more precise than any of the quantities added or subtracted. Therefore, the

Table 4.2.3 SI units

Physical Unit	Name of Unit	Symbol for Unit
Length	meter	m
Mass	kilogram, gram	kg, g
Time	second	s
Temperature	kelvin	K
Temperature (alternate)	degree Celsius	°C
Moles	gram-mole	gmol
Energy	joule	J, kg-m^2/s^2
Force	newton	N, kg-m/s^2, J/m
Acceleration	meters per square second	m/s^2
Pressure	pascal, newton per square meter	Pa, N/m^2
Pressure (alternate)	bar	bar
Velocity	meters per second	m/s
Volume	cubic meter, liters	m^3, L
Area	square meter	m^2
Frequency	hertz	Hz, cycles/s
Power	watt	W, kg-m^2-s^3, J/s
Specific heat capacity	joule per kilogram-kelvin	J/kg-K
Density	kilogram per cubic meter	kg/m^3
Angular velocity	radians per second	rad/s

Table 4.2.4 Prefixes for SI units

Multiplication Factors	Prefix	Symbol
1,000,000,000,000,000,000 = 10^{18}	exa	E
1,000,000,000,000,000 = 10^{15}	peta	P
1,000,000,000,000 = 10^{12}	tera	T
1,000,000,000 = 10^{9}	giga	G
1,000,000 = 10^{6}	mega	M
1,000 = 10^{3}	kilo	k
100 = 10^{2}	hecto	h
10 = 10^{1}	deka	da
0.1 = 10^{-1}	deci	d
0.01 = 10^{-2}	centi	c
0.001 = 10^{-3}	milli	m
0.000 001 = 10^{-6}	micro	μ
0.000 000 001 = 10^{-9}	nano	n
0.000 000 000 001 = 10^{-12}	pico	p
0.000 000 000 000 001 = 10^{-15}	femto	f
0.000 000 000 000 000 001 = 10^{-18}	atto	a

position (relative to the decimal point) of the last significant digit in the number that has the lowest degree of precision is the position of the last permissible significant digit in the result. For example, the sum of 3702., 370, 0.037, 4, and 37. should be reported as 4110 (without a decimal). The least precise of the five numbers is 370, which has its last significant digit in the *tens* position. The answer should also have its last significant digit in the *tens* position.

Unfortunately, engineers and scientists rarely concern themselves with significant figures in their calculations. However, it is recommended that—at least for this chapter—the reader attempt to follow the calculational procedure set forth in this subsection.

In the process of performing engineering calculations, very large and very small numbers are often encountered. A convenient way to represent these numbers is to use *scientific notation.* Generally, a number represented in scientific notation is the product of a number (<10 but > or = 1) and 10 raised to an integer power. For example,

$$28{,}070{,}000{,}000 = 2.807 \times 10^{10}$$

$$0.000\ 002\ 807 = 2.807 \times 10^{-6}$$

A nice feature of using scientific notation is that only the significant figures need appear in the number.

4.3 CHEMICAL AND PHYSICAL PROPERTIES

Temperature

Whether in the gaseous, liquid, or solid state, all molecules possess some degree of kinetic energy; that is, they are in constant motion—vibrating, rotating, or translating. The kinetic energies of individual molecules cannot be measured, but the combined effect of these energies in a very large number of molecules can. This measurable quantity is known as *temperature*; it is a macroscopic concept only and as such does not exist on the molecular level.

Temperature can be measured in many ways; the most common method makes use of the expansion of mercury (usually encased inside a glass capillary tube) with increasing temperature. (On incineration systems, however, thermocouples or thermistors are more commonly employed.) The two most commonly used temperature scales are the Celsius (or Centigrade) and Fahrenheit scales. The Celsius is based on the boiling and freezing points of water at 1-atm pressure; to the former, a value of 100°C is assigned, and to the latter, a value of 0°C. On the older Fahrenheit scale, these temperatures correspond to 212 and 32°F, respectively. Equations (4.3.1) and (4.3.2) show the conversion from one scale to the other.

$$°F = 1.8(°C) + 32 \tag{4.3.1}$$

$$°C = (°F - 32)/1.8 \tag{4.3.2}$$

where °F = a temperature on the Fahrenheit scale
°C = a temperature on the Celsius scale

Experiments with gases at low-to-moderate pressures (up to a few atmospheres) have shown that, if the pressure is kept constant, the volume of a gas and its temperature are linearly related (Charles' law) and that a decrease of 0.3663% or (1/273) of the initial volume is experienced for every temperature drop of 1°C. These experiments were not extended to very low temperatures, but if the linear relationship were extrapolated, the volume of the gas would *theoretically* be zero at a temperature of approximately −273°C or −460°F. This temperature has become known as *absolute zero* and is the basis for the definition of two *absolute* temperature scales. (An *absolute* scale is one which does not allow negative quantities.) These absolute temperature scales are the kelvin (K) and Rankine (°R) scales; the former is defined by shifting the Celsius scale by 273°C so that 0 K is equal to −273°C; Eq. (4.3.3) shows this relationship:

$$K = °C + 273 \tag{4.3.3}$$

The Rankine scale is defined by shifting the Fahrenheit scale 460°, so that

$$°R = °F + 460 \tag{4.3.4}$$

The relationships among the various temperature scales are shown in Fig. 4.3.1.

Pressure

In the gaseous state, the molecules possess a high degree of translational kinetic energy, which means they are able to move quite freely throughout the body of the gas. If the gas is in a container of some type, the molecules are constantly bombarding the walls of the container. The macroscopic effect of this bombardment by a tremendous number of molecules—enough to make the effect measurable—is called *pressure.* The natural units of pressure are force per unit area. In the example of the gas in a container, the *unit area* is a portion of the inside solid surface of the container wall and the *force,* measured perpendicularly to the unit area, is the result of the molecules hitting the unit area and giving up momentum during the sudden change of direction.

There are a number of different methods used to express a pressure measurement. Some of them are natural units, that is, based on a force per unit area, for example, pound (force) per square inch (abbreviated $lb_f/in.^2$ or psi) or dyne per square centimeter (dyn/cm^2). Others are based on a fluid height, such

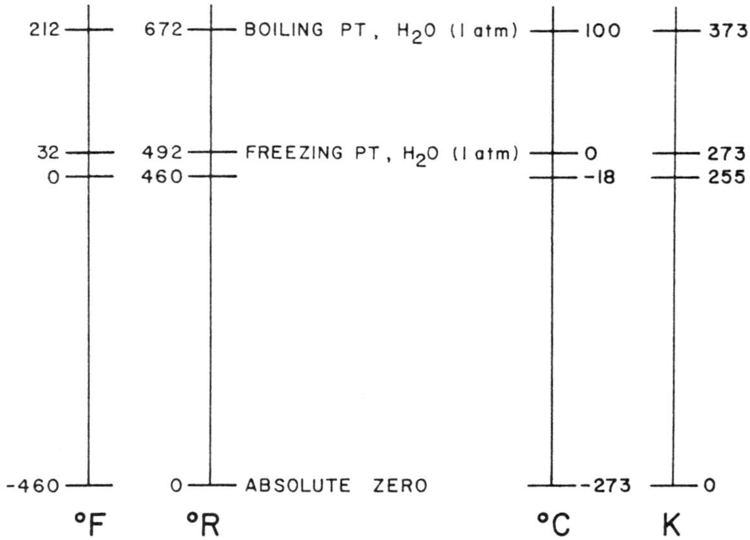

Figure 4.3.1 Temperature scales.

as inches of water (in. H_2O) or millimeters of mercury (mm Hg); units such as these are convenient when the pressure is indicated by a difference between two levels of a liquid as in a *manometer* or *barometer*. *Barometric pressure* and *atmospheric pressure* are synonymous and measure the ambient air pressure. *Standard barometric pressure* is the average atmospheric pressure at sea level, 45° north latitude at 32°F. It is used to define another unit of pressure called the atmosphere (atm). Standard barometric pressure is 1 atm and is equivalent to 14.696 psi and 29.921 in. Hg. As one might expect, barometric pressure varies with weather and altitude.

Measurements of pressure by most gauges indicate the difference in pressure either above or below that of the atmosphere surrounding the gauge. *Gauge pressure* is the pressure indicated by such a device. If the pressure in the system measured by the gauge is greater than the pressure prevailing in the atmosphere, the gauge pressure is expressed positively; if lower than atmospheric pressure, the gauge pressure is a negative quantity; the term *vacuum* designates a negative gauge pressure. Gauge pressures are often identified by the letter *g* after the pressure unit; for example, psig (pounds per square inch gauge) is a gauge pressure in psi units.

Since gauge pressure is the pressure relative to the prevailing atmospheric pressure, the sum of the two gives the *absolute pressure*, indicated by the letter *a* after the unit [e.g., psia (pounds per square inch absolute)]:

$$P = P_a + P_g \tag{4.3.5}$$

where P = absolute pressure (psia)
P_a = atmospheric pressure (psia)
P_g = gauge pressure (psig)

The absolute pressure scale is absolute in the same sense that the absolute temperature scale is absolute; that is, a pressure of zero psia is the lowest possible pressure theoretically achievable—a perfect vacuum.

Moles and Molecular Weights

An atom consists of protons and neutrons in a nucleus surrounded by electrons. An electron has such a small mass relative to that of the proton and neutron that the weight of the atom (called the *atomic weight*) is approximately equal to the sum of the weights of the particles in its nucleus. Atomic weight may be expressed in *atomic mass units (amu) per atom* or *in grams per gram-atom*. One gram-atom contains 6.02×10^{23} atoms (Avogadro's number). The atomic weights of the elements are listed in Table 4.3.1.

The *molecular weight* (MW) of a compound is the sum of the atomic weights of the atoms that make up the molecule. Units of atomic mass units per molecule (amu/molecule) or grams per gram-mole (g/gmol) are used for molecular weight. One gram-mole (gmol) contains an Avogadro number of molecules. For the English system, a pound-mole (lbmol) contains $454 \times 6.023 \times 10^{23}$ molecules.

Molal units are used extensively in pollutant control calculations as they greatly simplify material balances where chemical (including combustion) reactions are occurring. For mixtures of substances (gases, liquids, or solids), it is also convenient to express compositions in mole fractions or mole percentages instead of mass fractions. The mole fraction is the ratio of the number of moles of one component to the total number of moles in the mixture.

Mass and Volume

The *density* (ρ) of a substance is the ratio of its mass to its volume and may be expressed in units of pounds per cubic foot (lb/ft^3), kilograms per cubic meter (kg/m^3), and so on. For solids, density can be easily determined by placing a known mass of the substance in a liquid and determining the displaced volume. The density of a liquid can be measured by weighing a known volume of the liquid in a volumetric flask. For gases, the ideal gas law, discussed in Section 4.4, can be used to calculate the density from the pressure, temperature, and molecular weight of the gas.

Densities of pure solids and liquids are relatively independent of temperature and pressure and can be found in standard reference books.[1] The *specific volume* (v) of a substance is its volume per unit mass (ft^3/lb, m^3/kg, etc.) and is, therefore, the inverse of its density.

Table 4.3.1 Atomic weights of the elements[a,b]

Element	Symbol	Atomic Weight
Actinium	Ac	227.0278
Aluminum	Al	26.9815
Americium	Am	(243)
Antimony	Sb	121.75
Argon	Ar	39.948
Arsenic	As	74.9216
Astatine	At	(210)
Barium	Ba	137.34
Berkelium	Bk	(247)
Beryllium	Be	9.0122
Bismuth	Bi	208.980
Boron	B	10.811
Bromine	Br	79.904
Cadmium	Cd	112.40
Calcium	Ca	40.08
Californium	Cf	(251)
Carbon	C	12.01115
Cerium	Ce	140.12
Cesium	Cs	132.905
Chlorine	Cl	35.453
Chromium	Cr	51.996
Cobalt	Co	58.9332
Copper	Cu	63.546
Curium	Cm	(247)
Dysprosium	Dy	162.50
Einsteinium	Es	(252)
Erbium	Er	167.26
Europium	Eu	151.96
Fermium	Fm	(257)
Fluorine	F	18.9984
Francium	Fr	(223)
Gadolinium	Gd	157.25
Gallium	Ga	69.72
Germanium	Ge	72.59
Gold	Au	196.967
Hafnium	Hf	178.49
Helium	He	4.0026
Holmium	Ho	164.930
Hydrogen	H	1.00797
Indium	In	114.82
Iodine	I	126.9044
Iridium	Ir	192.2
Iron	Fe	55.847
Krypton	Kr	83.80
Lanthanum	La	138.91
Lawrencium	Lr	(260)

Table 4.3.1 (*Continued*)

Element	Symbol	Atomic Weight
Lead	Pb	207.19
Lithium	Li	6.939
Lutetium	Lu	174.97
Magnesium	Mg	24.312
Manganese	Mn	54.9380
Mendelevium	Md	(258)
Mercury	Hg	200.59
Molybdenum	Mo	95.94
Neodymium	Nd	144.24
Neon	Nc	20.183
Neptunium	Np	237.0482
Nickel	Ni	58.71
Niobium	Nb	92.906
Nitrogen	N	14.0067
Nobelium	No	(259)
Osmium	Os	190.2
Oxygen	O	15.9994
Pallasium	Pd	106.4
Phosphorus	P	30.9738
Platinum	Pt	195.09
Plutonium	Pu	(244)
Polonium	Po	(209)
Potassium	K	39.102
Praseodymium	Pr	140.907
Promethium	Pm	(145)
Protactinium	Pa	231.0359
Radium	Ra	226.0254[c]
Radon	Rn	(222)
Rhenium	Re	186.2
Rhodium	Rh	102.905
Rubidium	Rb	84.57
Rutherium	Ru	101.07
Samarium	Sm	150.35
Scandium	Sc	44.956
Selenium	Se	78.96
Silicon	Si	28.086
Silver	Ag	107.868
Sodium	Na	22.9898
Strontium	Sr	87.62
Sulfur	S	32.064
Tantalum	Ta	180.948
Technetium	Tc	(98)
Tellurium	Te	127.60
Terbium	Tb	158.924
Thallium	Tl	204.37
Thorium	Th	232.038

(*continued*)

Table 4.3.1 ***(Continued)***

Element	Symbol	Atomic Weight
Thulium	Tm	168.934
Tin	Sn	118.69
Titanium	Ti	47.90
Tungsten	W	183.85
Uranium	U	238.03
Vanadium	V	50.942
Xenon	Xe	131.30
Ytterbium	Yb	173.04
Yttrium	Y	88.905
Zinc	Zn	65.37
Zirconium	Zr	91.22

[a]Atomic weights apply to naturally occurring isotopic compositions and are based on an atomic mass of ^{12}C = 12.
[b]For radioactive elements a value given in parenthesis is the atomic mass number of the isotope of longest known half-life.
[c]Geologically exceptional samples are known in which the element has an isotopic composition outside the limits for normal material.

The *specific gravity* (SG) is the ratio of the density of a substance to the density of a reference substance at a specific condition:

$$\text{SG} = \rho/\rho_{\text{ref}} \tag{4.3.6}$$

The reference most commonly used for *solids* and *liquids* is water at its maximum density, which occurs at 4°C; this reference density is 1.000 g/cm^3, 1000 kg/m^3, or 62.43 lb/ft^3. Note that, since the specific gravity is a ratio of two densities, it is dimensionless. Therefore, any set of units may be employed for the two densities as long as they are consistent. The specific gravity of *gases* is used only rarely; when it is, air at the same conditions of temperature and pressure as the gas is usually employed as the reference substance.

Another dimensionless quantity related to density is the API (American Petroleum Institute) gravity, which is often used to indicate densities of fuel oils and some liquid hazardous wastes. The relationship between the API scale and the specific gravity is

$$\text{degrees API} = \frac{141.5}{\text{SG}(60/60°\text{F})} - 131.5 \tag{4.3.7}$$

where SG(60/60°F) = specific gravity of the liquid at 60°F using water at 60°F as the reference.

Viscosity

Viscosity is a property associated with a fluid's resistance to flow; more precisely, this property accounts for energy losses, which result from shear stresses that occur between different portions of the fluid, which are moving at different velocities. The *absolute viscosity* (μ) has units of mass per length-time; the fundamental unit is the *poise*, which is defined as 1 g/cm-s. This unit is inconveniently large for many practical purposes and viscosities are frequently given in *centipoises* (0.01 poise), which is abbreviated cP. The viscosity of pure water at 68.6°F is 1.00 cP. In English units, absolute viscosity is expressed either as pounds (mass) per foot-second (lb/ft-s) or pounds per foot-hour (lb/ft-h). The absolute viscosity depends primarily on temperature and to a lesser degree on pressure. The *kinematic viscosity* (ν) is the absolute viscosity divided by the density of the fluid and is useful in certain fluid flow problems; the units for this quantity are length squared per time, for example, square foot per second (ft^2/s) or square meters per hour (m^2/h). A kinematic viscosity of 1 cm^2/s is called a *stoke*. For pure water at 70°F, $\nu = 0.983$ cS (centistokes). Because fluid viscosity changes rapidly with temperature, a numerical value of viscosity has no significance unless the temperature is specified.

Liquid viscosity is usually measured by the amount of time it takes for a given volume of liquid to flow through an orifice. The *Saybolt universal viscometer* is the most widely used device in the United States for the determination of the viscosity of fuel oils and liquid wastes. It should be stressed that Saybolt viscosities, which are expressed in *Saybolt seconds (SSU)*, are not even approximately proportional to absolute viscosities except in the range above 200 SSU; hence, converting units from Saybolt seconds to other units requires the use of special conversion tables. As the time of flow decreases, the deviation becomes more marked. In any event, viscosity is an important property because of potential flow problems with liquid wastes and/or fuel oil, particularly with liquid injection and rotary kiln incinerators.

The viscosities of air at atmospheric pressure and water are presented in Tables 4.3.2 and 4.3.3, respectively, as functions of temperature. Viscosities of other substances are available in the literature.[1]

Heat Capacity

The *heat capacity* of a substance is defined as the quantity of heat required to raise the temperature of that substance by 1°; the *specific heat capacity* is the heat capacity on a unit mass basis. The term *specific heat* is frequently used in place of *specific heat capacity*. This is not strictly correct, because traditionally specific heat has been defined as the ratio of the heat capacity of a substance to the heat capacity of water. However, since the specific heat of water is approximately 1 cal/g-°C or 1 Btu/lb-°F, the term *specific heat* has come to imply heat capacity per unit mass. For gases, the addition of heat to cause the 1° temperature rise may be accomplished either at constant pressure

Table 4.3.2 Viscosity of air at 1 atmosphere[a]

T (°C)	Viscosity, Micropoise (μP)
0	170.8
18	182.7
40	190.4
54	195.8
74	210.2
229	263.8

[a] 1 P = 100 cP = 10^6 μP; 1 cp = 6.72×10^{-4} lb/ft-s.

or at constant volume. Since the amounts of heat necessary are different for the two cases, subscripts are used to identify which heat capacity is being used —c_P for constant pressure and c_V for constant volume. For liquids and solids, this distinction does not have to be made since there is little difference between the two. Values of specific heats are available in the literature.[1]

Heat capacities are often used on a *molar* basis instead of a *mass* basis, in which case the units become cal/gmol-°C or Btu/lbmol-°F. To distinguish between the two bases, uppercase letters (C_P, C_V) will be used in this text to represent the molar-based heat capacities, and lowercase letters (c_P, c_V) will be used for the mass-based heat capacities or specific heats.

Table 4.3.3 Viscosity of water

T (°C)	Viscosity, Centipoise (cP)
0	1.792
5	1.519
10	1.308
15	1.140
20	1.000
25	0.894
30	0.801
35	0.723
40	0.656
50	0.594
60	0.469
70	0.406
80	0.357
90	0.317
100	0.284

Heat capacities are functions of both the temperature and pressure, although the effect of pressure is generally small and is neglected in almost all engineering calculations. The effect of temperature on C_P can be described by

$$C_P = \alpha + \beta T + \gamma T^2 \tag{4.3.8}$$

or

$$C_P = a + bT = cT^{-2} \tag{4.3.9}$$

Average or *mean* heat capacity data over specific temperature ranges are also available. Values for α, β, γ, and and a, b, c, as well as average heat capacity information are provided in tabular form in Chapter 5. These will find extensive application in Chapters 5 and 6, as well as some of the later chapters. Properties such as enthalpy and heat of vaporization are also discussed in Chapter 5.

Thermal Conductivity

Experience has shown that when a temperature difference exists across a solid body, energy in the form of heat will transfer from the high-temperature region to the low-temperature region until thermal equilibrium (same temperature) is reached. This mode of heat transfer in which vibrating molecules pass along kinetic energy through the solid is called *conduction.* Liquids and gases may also transport heat in this fashion. (For a more comprehensive discussion of heat transfer mechanisms, the reader is directed to Chapter 9.) The property of *thermal conductivity* provides a measure of how fast (or how easily) heat flows through a substance. It is defined as the amount of heat that flows in unit time through a unit surface area of unit thickness as a result of a unit difference in temperature. Typical units for conductivity are Btu-ft/h-ft^2-°F or Btu/h-ft-°F. Thermal conductivities for some liquids and gases are given in Table 4.3.4, and for some insulating solids in Table 8.2.1.

With regard to hazardous waste incineration systems, this particular property finds application in designing heat exchangers and waste heat boilers (see Chapter 8), as well as in selecting insulating material for the devices within the system, particularly the incinerator.

Reynolds Number

The Reynolds number, Re, is a dimensionless number that indicates whether a fluid flowing is in the laminar or turbulent mode. *Laminar* flow is characteristic of fluids flowing slowly enough so that there are no eddies (whirlpools) or macroscopic mixing of different portions of the fluid. *(Note:*. In any fluid, there is always *molecular* mixing due to the thermal activity of the molecules; this is distinct from *macroscopic* mixing due to the swirling motion of different portions of the fluid.) In laminar flow, a fluid can be imagined to flow like a

Table 4.3.4 Thermal conductivities of gases and liquids

Gases and Vapors (1 atm)	Thermal Conductivity k^a	
	32°F	212°F
Acetone	0.0057	0.0099
Acetylene	0.0108	0.0172
Air	0.0140	0.0184
Ammonia	0.0126	0.0192
Benzene	0.0052	0.0103
Carbon dioxide	0.0084	0.0128
Carbon monoxide	0.0134	0.0176
Carbon tetrachloride		0.0052
Chlorine	0.0043	
Ethane	0.0106	0.0175
Ethyl alcohol		0.0124
Ethyl ether	0.0077	0.0131
Ethylene	0.0101	0.0161
Helium	0.0818	0.0988
Hydrogen	0.0966	0.1240
Methane	0.0176	0.0255
Methyl alcohol	0.0083	0.0128
Nitrogen	0.0139	0.0181
Nitrous oxide	0.0088	0.0138
Oxygen	0.0142	0.0188
Propane	0.0087	0.0151
Sulfur dioxide	0.0050	0.0069
Water vapor		0.0136
Liquids	**Temperature (°F)**	**k^a**
Acetic acid	68	0.909
Acetone	86	0.102
Ammonia (anhydrous)	5–86	0.29
Aniline	32–68	0.100
Benzene	86	0.092
n-Butyl alcohol	86	0.097
Carbon bisulfide	86	0.093
Carbon tetrachloride	32	0.107
Chlorobenzene	50	0.083
Ethyl acetate	68	0.101
Ethyl alcohol (absolute)	68	0.105
Ethyl ether	86	0.080
Ethylene glycol	32	0.153
Gasoline	86	0.078
Glycerine	68	0.164
n-Heptane	86	0.081
Kerosene	68	0.086
Methyl alcohol	68	0.124
Nitrobenzene	86	0.095
n-Octane	86	0.083
Sulfur dioxide	5	0.128
Sulfuric acid (90%)	86	0.21
Toluene	86	0.086
Trichloroethylene	122	0.080
o-Xylene	68	0.090

[a]Units of k: Btu/h-ft-°F.

deck of cards, with adjacent layers sliding past one another. *Turbulent* flow is characterized by eddies and macroscopic currents. In practice, moving gases are generally in the turbulent region. For flow in a pipe, a Reynolds number above 2100 is an indication of turbulent flow.

The Reynolds number is dependent on the fluid velocity, density, viscosity, and some *length* characteristic of the system or conduit; for pipes, this characteristic length is the inside diameter:

$$\mathrm{Re} = vD\rho/\mu = vD/\nu \tag{4.3.10}$$

where Re = Reynolds number
D = inside diameter of the pipe (ft)
v = fluid velocity (ft/s)
ρ = fluid density (lb/ft^3)
μ = fluid viscosity (lb/ft-s)
ν = fluid kinematic viscosity (ft^2/s)

Any consistent set of units may be used with Eq. (4.3.10).

pH

An important chemical property of an aqueous solution is its pH. The pH measures the acidity or basicity of the solution. In a neutral solution, such as pure water, the hydrogen (H^+) and hydroxyl (OH^-) ion concentrations are equal. At ordinary temperatures, this concentration is

$$C_{\mathrm{H}^+} = C_{\mathrm{OH}^-} = 10^{-7} \text{ g-ion/L} \tag{4.3.11}$$

where C_{H^+} = hydrogen ion concentration
C_{OH^-} = hydroxyl ion concentration

The unit g-ion stands for gram-ion, which represents an Avogadro number of ions. In all aqueous solutions, whether neutral, basic, or acidic, a chemical equilibrium or balance is established between these two concentrations, so that

$$K_{\mathrm{eq}} = C_{\mathrm{H}^+}C_{\mathrm{OH}^-} = 10^{-14} \tag{4.3.12}$$

where K_{eq} = equilibrium constant.

The numerical value for K_{eq} given in Eq. (4.3.12) holds for room temperature and only when the concentrations are expressed in gram-ion per liter (g-ion/L). In acid solutions, C_{H^+} is > C_{OH^-}; in basic solutions, C_{OH^-} predominates. The pH is a direct measure of the hydrogen ion concentration and is defined by

$$\mathrm{pH} = -\log C_{\mathrm{H}^+} \tag{4.3.13}$$

Thus, an acidic solution is characterized by a pH below 7 (the lower the pH, the higher the acidity); a basic solution, by a pH above 7; and a neutral solution by a pH of 7. It should be pointed out that Eq. (4.3.13) is not the exact definition of pH but is a close approximation to it. Strictly speaking, the *activity* of the hydrogen ion, a_{H^+}, and not the ion concentration belongs in Eq. (4.3.13). For a discussion of chemical activities, the reader is directed to the literature.[2]

Vapor Pressure

This section concludes with a discussion of *vapor pressure,* which is an important property of liquids, and, to a much lesser extent, of solids. If a liquid is allowed to evaporate in a confined space, the pressure in the vapor space increases as the amount of vapor increases. If there is sufficient liquid present, a point is eventually reached at which the pressure in the vapor space is exactly equal to the pressure exerted by the liquid at its own surface. At this point, a dynamic equilibrium exists in which vaporization and condensation take place at equal rates and the pressure in the vapor space remains constant. The pressure exerted at equilibrium is called the vapor pressure of the liquid. The magnitude of this pressure for a given liquid depends on the temperature, but not on the amount of liquid present. Solids, like liquids, also exert a vapor pressure. Evaporation of solids (called *sublimation*) is noticeable only for those with appreciable vapor pressures.

4.4 IDEAL GAS LAW

Observations based on physical experimentation often can be synthesized into simple mathematical equations called *laws.* These laws are never perfect and hence are only an approximate representation of reality. The *ideal gas law* (IGL) was derived from experiments in which the effects of pressure and temperature on gaseous volumes were measured over moderate temperature and pressure ranges. This law works well in the pressure and temperature ranges that were used in taking the data; extrapolations outside of the ranges have been found to work well in some cases and poorly in others. As a general rule, this law works best when the molecules of the gas are far apart, that is, when the pressure is low and the temperature is high. Under these conditions, the gas is said to behave *ideally*, that is, its behavior is a close approximation to the so-called *perfect* or *ideal gas*, a hypothetical entity that obeys the ideal gas law perfectly. For engineering calculations, and specifically for hazardous waste incinerator problems, the ideal gas law is almost always assumed to be valid, since it generally works well (usually within a few percent of the correct result) up to the highest pressures and down to the lowest temperatures used in hazardous waste incinerator applications.

The two precursors of the ideal gas law were *Boyle's* and *Charles'* laws. Boyle found that the volume of a given mass of gas is inversely proportional to the absolute pressure if the temperature is kept constant:

$$P_1V_1 = P_2V_2 \tag{4.4.1}$$

where V_1 = volume of gas at absolute pressure P_1 and temperature T
V_2 = volume of gas at absolute pressure P_2 and temperature T

Charles found that the volume of a given mass of gas varies directly with the absolute temperature at constant pressure:

$$\frac{V_1}{T_1} = \frac{V_2}{T_2} \tag{4.4.2}$$

where V_1 = volume of gas at pressure P and absolute temperature T_1
V_2 = volume of gas at pressure P and absolute temperature T_2

Boyle's and Charles' laws may be combined into a single equation in which neither temperature nor pressure need be held constant:

$$\frac{P_1V_1}{T_1} = \frac{P_2V_2}{T_2} \tag{4.4.3}$$

For Eq. (4.4.3) to hold, the mass of gas must be constant as the conditions change from (P_1, T_1) to (P_2, T_2). This equation indicates that for a given mass of a specific gas, PV/T has a constant value. Since, at the same temperature and pressure, volume and mass must be directly proportional, this statement may be extended to

$$\frac{PV}{mT} = C \tag{4.4.4}$$

where m = mass of a specific gas
C = constant that depends on the gas

Moreover, experiments with different gases showed that Eq. (4.4.4) could be expressed in a far more generalized form. If the number of moles (n) is used in place of the mass (m), the constant is the same for all gases:

$$\frac{PV}{nT} = R \tag{4.4.5}$$

where R = universal gas constant.

Equation (4.4.5) is called the ideal gas law. Numerically, the value of R depends on the units used for P, V, T, and n (see Table 4.4.1). In this text, all gases are assumed to approximate ideal gas behavior. As is generally the case in engineering practice, the ideal gas law is assumed to be valid for all illus-

Table 4.4.1 Values of *R* in various units

R	Temperature Scale	Units of *V*	Units of *n*	Units of *P*	Unit of *PV* (energy)
10.73	°R	ft^3	lbmol	psia	—
0.7302	°R	ft^3	lbmol	atm	—
21.85	°R	ft^3	lbmol	in. Hg	—
555.0	°R	ft^3	lbmol	mm Hg	—
297.0	°R	ft^3	lbmol	in. H_2O	—
0.7398	°R	ft^3	lbmol	bar	—
1545.0	°R	ft^3	lbmol	psfa	—
24.75	°R	ft^3	lbmol	ft H_2O	—
1.9872	°R	—	lbmol	—	Btu
0.0007805	°R	—	lbmol	—	hp-h
0.0005819	°R	—	lbmol	—	kW-h
500.7	°R	—	lbmol	—	cal
1.314	K	ft^3	lbmol	atm	—
998.9	K	ft^3	lbmol	mm Hg	—
19.32	K	ft^3	lbmol	psia	—
62.361	K	L	gmol	mm Hg	—
0.08205	K	L	gmol	atm	—
0.08314	K	L	gmol	bar	—
8314	K	L	gmol	Pa	—
8.314	K	m^3	gmol	Pa	—
82.057	K	cm^3	gmol	atm	—
1.9872	K	—	gmol	—	cal
8.3144	K	—	gmol	—	J

trative and assigned problems. If a case is encountered in practice where the gas behaves in very nonideal fashion, for example, a high-molecular-weight gas (such as a chlorinated organic) under high pressures, one of the many *real gas* correlations found in the literature[3] should be used.

Other useful forms of the ideal gas law are shown in Eqs. (4.4.6) and (4.4.7). Equation (4.4.6) applies to gas flow rather than to gas confined in a container:[4]

$$PQ = \dot{n}RT \tag{4.4.6}$$

where Q = gas volumetric flow rate (ft^3/h)
P = absolute pressure (psia)
$\dot{n}$ = molar flow rate (lbmol/h)
T = absolute temperature (°R)
R = 10.73 psia-ft^3/lbmol°R

Equation (4.4.7) combines n and V from Eq. (4.4.5) to express the law in terms of density:

$$PM = \rho RT \tag{4.4.7}$$

where M = molecular weight of gas (lb/lbmol)
ρ = density of gas (lb/ft^3)

Volumetric flow rates are often given not at the actual conditions of pressure and temperature but at arbitrarily chosen standard conditions (*STP, standard temperature and pressure*). To distinguish between flow rates based on the two conditions, the letters *a* and *s* are often used as part of the unit. The units acfm and scfm stand for actual cubic feet per minute and standard cubic feet per minute, respectively. The ideal gas law can be used to convert from *standard* to *actual* conditions, but, since there are many standard conditions in use, the STP being used must be known. Standard conditions most often used are shown in Table 4.4.2. The reader is cautioned on the incorrect use of acfm and/or scfm. The use of standard conditions is a convenience; when predicting the performance of or designing equipment, the actual conditions must be employed. Designs based on standard conditions can lead to disastrous results, with the unit usually underdesigned. For example, for a flue gas stream at 2140°F, the ratio of acfm to scfm (standard temperature = 60°F) for an incinerator application is *5.0*. Equation (4.4.8), which is a form of Charles' law, can be used to correct flow rates from standard to actual conditions:

$$Q_a = Q_s(T_a/T_s) \tag{4.4.8}$$

where Q_a = volumetric flow rate at actual conditions (ft^3/h)
Q_s = volumetric flow rate at standard conditions (ft^3/h)
T_a = actual absolute temperature (°R)
T_s = standard absolute temperature (°R)

The reader is again reminded that absolute temperatures and pressures must be employed in all ideal gas law calculations.

Table 4.4.2 Common standard conditions

System	Temperature	Pressure	Molar Volume
SI	273 K	101.3 kPa	22.4 m^3/kmol
Universal scientific	0°C	760 mm Hg	22.4 L/gmol
Natural gas industry	60°F	14.7 psia	379 ft^3/lbmol
American engineering	32°F	1 atm	359 ft^3/lbmol
Hazardous waste	60°F	1 atm	379 ft^3/lbmol
incinerator industry	70°F	1 atm	387 ft^3/lbmol

Partial Pressure and Partial Volume

In engineering practice, mixtures of gases are more often encountered than single or pure gases. The ideal gas law is based on the *number* of molecules present in the gas volume; the *kind* of molecules is not a significant factor, only the number. This law applies equally well to mixtures and pure gases alike. Dalton and Amagat both applied the ideal gas law to mixtures of gases. Since pressure is caused by gas molecules colliding with the walls of the container, it seems reasonable that the total pressure of a gas mixture is made up of pressure contributions due to each of the component gases. These pressure contributions are called *partial pressures.* Dalton defined the partial pressure of a component as the pressure that would be exerted if the same mass of the component gas occupied the same total volume alone at the same temperature as the mixture. The sum of these partial pressures equals the total pressure:

$$P = P_a + P_b = P_c + \cdots + P_n = \sum_{i=1}^{n} P_i \tag{4.4.9}$$

where P = total pressure
n = number of components
P_i = partial pressure of component i

Equation (4.4.9) is known as *Dalton's law*. Applying the ideal gas law to one component (a) only,

$$P_aV = n_aRT \tag{4.4.10}$$

where n_a = number of moles of component a.

Eliminating R, T, and V between Eq. (4.4.5) and (4.4.10) yields

$$\frac{P_a}{P} = \frac{n_a}{n} = y_a$$

or

$$P_a = y_aP \tag{4.4.11}$$

where y_a = mole fraction of component a.

Amagat's law is similar to Dalton's. Instead of considering the total pressure to be made up of partial pressures where each component occupies the total container volume, Amagat considered total volume to be made up of partial volumes in which each component is exerting the total pressure. The definition of the *partial volume* is therefore the volume occupied by a component gas alone at the same temperature and pressure as the mixture:

$$V = V_a + V_b + V_c + \cdots + V_n + \sum_{i=1}^{n} V_i \tag{4.4.12}$$

Applying Eq. (4.4.5), as before,

$$\frac{V_a}{V} = \frac{n_a}{n} = y_a \tag{4.4.13}$$

where V_a = partial volume of component *a*.

It is common in the hazardous waste incineration industry to describe low concentrations of pollutants in gaseous mixtures in parts per million (ppm) by volume. Since partial volumes are proportional to mole fractions, it is only necessary to multiply the mole fraction of the pollutant by 1 million to obtain the concentration in parts per million. [For liquids and solids, parts per million (ppm) is also used to express concentration, although it is usually on a *mass* basis rather than a *volume* basis. The terms ppmv and ppmw are sometimes used to distinguish between the volume and mass bases.]

4.5 PHASE EQUILIBRIUM

The term *phase*, for a pure substance, indicates a *state of matter*, that is, solid, liquid, or gas. For mixtures, however, a more stringent connotation must be used, since a totally liquid or solid system may contain more than one phase (a mixture of oil and water, e.g.). A phase is characterized by uniformity or *homogeneity;* the same composition and properties must exist throughout the phase region. At most temperatures and pressures, a pure substance normally exists as a single phase. At certain temperatures and pressures, two or perhaps even three phases can coexist in equilibrium. This is shown on the phase diagram for water (see Fig. 4.5.1). Regarding the interpretation of this diagram, the following points should be noted:

- The line between the gas and liquid phase regions is the *boiling point* line and represents equilibrium between the gas and liquid.
- The boiling point of a liquid is the temperature at which its vapor pressure is equal to the *external* pressure. The temperature at which the vapor pressure is equal to 1 atm is the *normal* boiling point.
- The line between the solid and gas phase regions is the *sublimation point* line and represents equilibrium between the solid and gas.
- The line between the solid and liquid phase regions is the *melting point* or *freezing point* line and represents equilibrium between the liquid and solid.

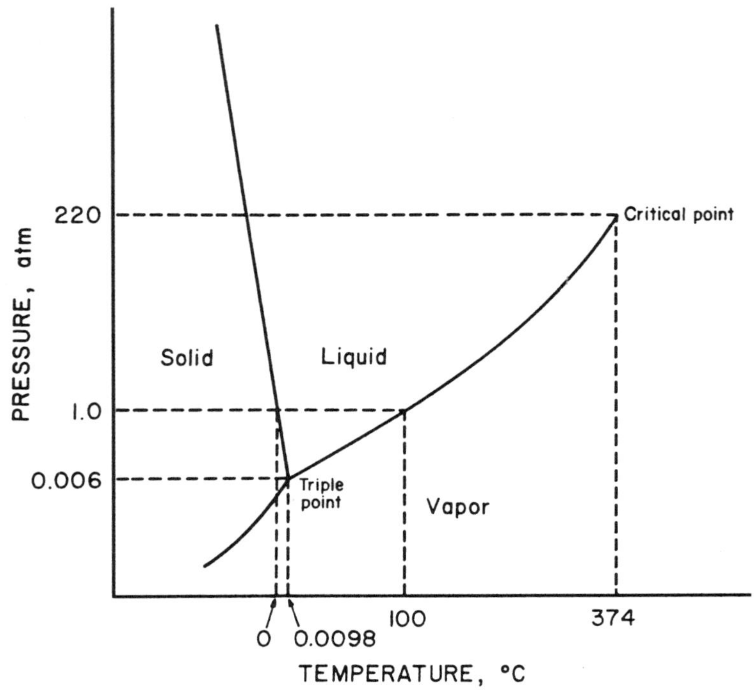

Figure 4.5.1 Phase diagram for water.

- The point at which all three equilibrium lines meet (i.e., the one pressure and temperature where solid, liquid, and gas phases can all coexist) is the *triple point.*
- The liquid–gas equilibrium line is bounded on one end by the triple point and the other end by the *critical* point. The *critical temperature* (the temperature coordinate of the critical point) is defined as the temperature above which a gas or vapor cannot be liquefied by the application of pressure alone.

The term *vapor*, strictly speaking, is used only for a *condensable* gas, that is, a gas below its *critical* temperature, and should not be applied to a noncondensable gas. It should also be pointed out that the phase diagram for water (Fig. 4.5.1) differs from that of other substances in one respect—the freezing point line is *negatively* sloped; for other materials, the slope of this line is *positive.* This is a consequence of the fact that liquid water is denser than ice, and isothermal (constant temperature) compression of the liquid can result in a transformation to the solid.

Incineration calculations rarely involve single (pure) components, however. Phase equilibria for multicomponent systems are considerably more complex, mainly because of the addition of *composition* variables; for example, in a *ternary* or three-component system, the mole fractions of two of the components are pertinent variables along with temperature and pressure. In a single-component system, dynamic equilibrium between two phases is achieved when the rate of molecular transfer from one phase to the second equals that in the opposite direction. In multicomponent systems, the equilibrium requirement is more stringent—the rate of transfer of *each component* must be the same in both directions.

In hazardous waste incineration applications, the most important equilibrium phase relationship is that between liquid and vapor. Raoult's and Henry's laws theoretically describe liquid–vapor behavior and under certain conditions are applicable in practice. *Raoult's law* is sometimes useful for mixtures of components of similar structure and states that the partial pressure of any component in the vapor is equal to the product of the vapor pressure of the *pure* component and the mole fraction of that component in the liquid, that is,

$$P_i = P_i' x_i \tag{4.5.1}$$

where P_i = partial pressure of component i in the vapor
P_i' = vapor pressure of pure i at the same temperature
x_i = mole fraction of component i in the liquid

This expression may be applied to all components. If the gas phase is ideal, Eq. (4.5.1) becomes

$$y_i = (P_i'/P)x_i \tag{4.5.2}$$

where y_i = mole fraction component i in the vapor
P = total system pressure

Thus, the mole fraction of water vapor in a gas that is saturated, that is, in equilibrium contact with pure water ($x = 1.0$), is given simply by the ratio of the vapor pressure of water at the system temperature divided by the system pressure. This finds application in many quench and absorber calculations (see Part III).

Unfortunately, relatively few mixtures follow Raoult's law. *Henry's law* is a more empirical relationship used for representing data on many systems:

$$P_i = H_i x_i \tag{4.5.3}$$

where H_i = Henry's law constant for component i (in units of pressure).

If the gas behaves ideally, Eq. (4.5.3) may be written as

$$y_i = m_i x_i \tag{4.5.4}$$

where m_i = constant (dimensionless).

This is a more convenient form of Henry's law to work with. The constant m_i (or H_i) has been determined experimentally for a large number of compounds and is usually valid at low concentrations.

In other applications, mixtures of condensable vapors and noncondensable gases must be handled. A common example is water vapor and air; a mixture of organic vapors and air is another such example that often appears in air pollution problems. Condensers can be used to control organic emissions to the atmosphere by lowering the temperature of the gaseous stream, although an increase in pressure will produce the same result. The calculation for this is accomplished using the phase equilibrium constant (K_i). This constant has been referred to in industry as a *componential split factor* since it provides the ratio of the mole fractions of a component in two equilibrium phases. The defining equation is

$$K_i = y_i/x_i \tag{4.5.5}$$

where K_i = phase equilibrium constant for component i (dimensionless).

This equilibrium constant is a function of the temperature, pressure, the other components in the system, and the mole fractions of these components. However, as a first approximation, K_i is generally treated as a function only of the temperature and pressure. For this condition, K_i may be approximated by

$$K_i = P_i/P \tag{4.5.6}$$

Another phase equilibrium application involves the psychrometric or humidity chart (see Fig. 4.5.2). A humidity chart is used to determine the properties of moist air and to calculate moisture content in air. The ordinate of the chart is the absolute humidity $\mathcal{H}$, which is defined as the mass of water vapor per mass of bone-dry air. (Some charts base the ordinate on moles instead of mass.)

Based on this definition, Eq. (4.5.7) gives $\mathcal{H}$ in terms of moles and also in terms of partial pressure:

$$\mathcal{H} = \frac{18 n_{H_2O}}{29(n_t - n_{H_2O})} = \frac{18 P_{H_2O}}{29(P - P_{H_2O})} \tag{4.5.7}$$

where n_{H_2O} = number of moles of water vapor
n_T = total number of moles in gas
P_{H_2O} = partial pressure of water vapor
P = total system pressure

Curves showing the *relative humidity* (ratio of the mass of the water vapor in the air to the maximum mass of water vapor the air could hold at that temperature, that is, if the air were saturated) of humid air also appear on the charts. The curve for 100% relative humidity is also referred to as the *saturation curve*. The abscissa of the humidity chart is air temperature, also known

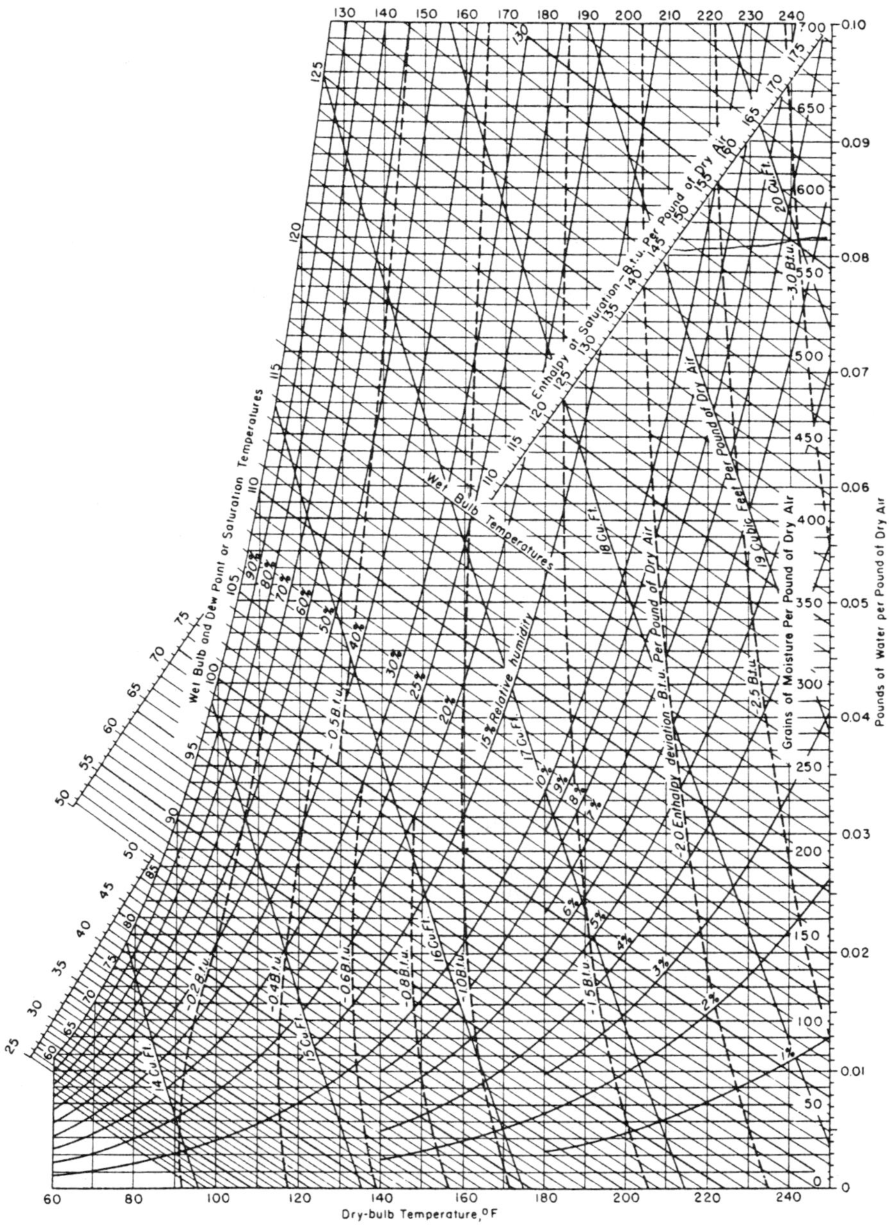

Figure 4.5.2 Psychrometric chart. (From R. Perry and D. Green, *Perry's Chemical Engineering Handbook*, 7th ed., © 1997, McGraw-Hill, New York; reproduced with permission of McGraw-Hill.)

as the *dry-bulb* temperature (T_{DB}). The *wet-bulb* temperature (T_{WB}) is another measure of humidity; it is the temperature at which a thermometer with a wet wick wrapped around the bulb stabilizes. As water evaporates from the wick to the ambient air, the bulb is cooled; the rate of cooling depends on how humid the air is. No evaporation occurs if the air is saturated with water, hence T_{WB} and T_{DB} are the same. The lower the humidity, the greater the difference between these two temperatures. On the psychrometric chart, constant wet-bulb temperature lines are straight with negative slopes. The value of T_{WB} corresponds to the value of the abscissa at the point of intersection of this line with the saturation curve.

The *humid volume* is the volume of wet air per mass of dry air and is linearly related to the humidity. (In Fig. 4.5.2, this quantity is used as an alternate ordinate. Note the straight parallel lines labeled with units of cubic feet.) The product of the humid volume and the absolute humidity gives the volume of the moist air. The *humid enthalpy* (also called *humid heat*) is the enthalpy of the moist air on a bone-dry air basis. The term *enthalpy* is a measure of the energy content of the mixture and is defined in Chapter 5. The enthalpy for *saturated* air can be read from the chart by extending the approximate wet-bulb temperature line upwards to the diagonal scale labeled *enthalpy of saturation.*

The following are some helpful points on the use of psychrometric charts:

- Heating or cooling at temperatures above the *dew point* (temperature at which the vapor begins to condense) corresponds to a horizontal movement on the chart. As long as no condensation occurs, the absolute humidity stays constant.
- If the air is cooled, the system follows the appropriate horizontal line to the left until it reaches the saturation curve and follows it thereafter.
- In problems involving use of the humidity chart, it is convenient to choose the *mass of dry air* as a basis, since the chart uses this basis.

Steam tables, charts, and diagrams are often employed in hazardous waste incineration calculations. A somewhat abbreviated steam table is given in Table 4.5.1; more comprehensive versions are available in the literature.

4.6 ILLUSTRATIVE EXAMPLES

Example 4.6.1 Convert units of acceleration in cm/s^2 to miles/yr^2.

SOLUTION

$$\frac{1\ \text{cm}}{\text{s}^2} \times \frac{3600^2\text{s}^2}{1\ \text{h}^2} \times \frac{24^2\ \text{h}^2\ 365^2\ \text{day}^2}{1\ \text{day}^2\ \ 1\ \text{yr}^2} \times \frac{1\ \text{in.}}{2.54\ \text{cm}} \times \frac{1\ \text{ft}}{12\ \text{in.}} \times \frac{1\ \text{mile}}{5280\ \text{ft}}$$

$$= 6.18 \times 10^9\ \text{miles/yr}^2$$

Thus, 1.0 cm/s^2 is equal to 6.18×10^9 miles/yr^2.

Table 4.5.1 Properties of saturated steam and water[a]

Temp. T, °F	Vapor press. P_A $lb_f/in.^2$	Specific vol., ft^3/lb Liquid v_x	Sat. vapor v_y	Enthalpy, Btu/lb Liquid h_x	Vaporization λ	Sat. vapor h_y
32	0.08854	0.01602	3,306	0.00	1075.8	1075.8
35	0.09995	0.01602	2,947	3.02	1074.1	1077.1
40	0.12170	0.01602	2,444	8.05	1071.3	1079.3
45	0.14752	0.01602	2,036.4	13.06	1068.4	1081.5
50	0.17811	0.01603	1,703.2	18.07	1065.6	1083.7
55	0.2141	0.01603	1,430.7	23.07	1062.7	1085.8
60	0.2563	0.01604	1,206.7	28.06	1059.9	1088.0
65	0.3056	0.01605	1,021.4	33.05	1057.1	1090.2
70	0.3631	0.01606	867.9	38.04	1054.3	1092.3
75	0.4298	0.01607	740.0	43.03	1051.5	1094.5
80	0.5069	0.01608	633.1	48.02	1048.6	1096.6
85	0.5959	0.01609	543.5	53.00	1045.8	1098.8
90	0.6982	0.01610	468.0	57.99	1042.9	1100.9
95	0.8153	0.01612	404.3	62.98	1040.1	1103.1
100	0.9492	0.01613	350.4	67.97	1037.2	1105.2
110	1.2748	0.01617	265.4	77.94	1031.6	1109.5
120	1.6924	0.01620	203.27	87.92	1025.8	1113.7
130	2.2225	0.01625	157.34	97.90	1020.0	1117.9
140	2.8886	0.01629	123.01	107.89	1014.1	1122.0
150	3.718	0.01634	97.07	117.89	1008.2	1126.1
160	4.741	0.01639	77.29	127.89	1002.3	1130.2
170	5.992	0.01645	62.06	137.90	996.3	1134.2
180	7.510	0.01651	50.23	147.92	990.2	1138.1
190	9.339	0.01657	40.96	157.95	984.1	1142.0
200	11.526	0.01663	33.64	167.99	977.9	1145.9
210	14.123	0.01670	27.82	178.05	971.6	1149.7
212	14.696	0.01672	26.80	180.07	970.3	1150.4
220	17.186	0.01677	23.15	188.13	965.2	1153.4
230	20.780	0.01684	19.382	198.23	958.8	1157.0
240	24.969	0.01692	16.323	208.34	952.2	1160.5
250	29.825	0.01700	13.821	218.48	945.5	1164.0
260	35.429	0.01709	11.763	228.64	938.7	1167.3
270	41.858	0.01717	10.061	238.84	931.8	1170.6
280	49.203	0.01726	8.645	249.06	924.7	1173.8
290	57.556	0.01735	7.461	259.31	917.5	1176.8
300	67.013	0.01745	6.466	269.59	910.1	1179.7
310	77.68	0.01755	5.626	279.92	902.6	1182.5
320	89.66	0.01765	4.914	290.28	894.9	1185.2
330	103.06	0.01776	4.307	300.68	887.0	1187.7
340	118.01	0.01787	3.788	311.13	879.0	1190.1
350	134.63	0.01799	3.342	321.63	870.7	1192.3
360	153.04	0.01811	2.957	332.18	862.2	1194.4
370	173.37	0.01823	2.625	342.79	853.5	1196.3
380	195.77	0.01836	2.335	353.45	844.6	1198.1
390	220.37	0.01850	2.0836	364.17	835.4	1199.6
400	247.31	0.01864	1.8633	374.97	826.0	1201.0
410	276.75	0.01878	1.6700	385.83	816.3	1202.1
420	308.83	0.01894	1.5000	396.77	806.3	1203.1
430	343.72	0.01910	1.3499	407.79	796.0	1203.8
440	381.59	0.01926	1.2171	418.90	785.4	1204.3
450	422.6	0.0194	1.0993	431.1	774.5	1204.6

[a]From J. H. Keenan and F. G. Keyes, *Thermodynamic Properties of Steam*, John Wiley & Sons, New York, 1936.

Example 4.6.2 If a 55-gal tank contains 20.0 lb of water, (1) how many pound-moles of water does it contain; (2) how many gram-moles does it contain; and (3) how many molecules does it contain?

SOLUTION The molecular weight of the water is

$$\text{MW} = (2)(1.008) = (15.999) = 18.015 \text{ g/gmol}$$
$$= 18.015 \text{ lb/lbmol}$$

1. $20.0 \text{ lb} \times \dfrac{\text{lbmol}}{18.015 \text{ lb}} = 1.11 \text{ lbmol water}$
2. $20.0 \text{ lb} \times \dfrac{453.593 \text{ g}}{1 \text{ lb}} \times \dfrac{\text{gmol}}{18.015 \text{ g}} = 503.6 \text{ gmol water}$
3. $503.6 \text{ gmol} \times \dfrac{6.023 \times 10^{23} \text{ molecules}}{1 \text{ gmol}} = 3.033 \times 10^{26} \text{ molecules}$

Example 4.6.3 Perform the following temperature conversions:

1. Convert 55°F to (a) Rankine, (b) Celsius, and (c) Kelvin
2. Convert 55°C to (a) Fahrenheit, (b) Rankine and (c) Kelvin

SOLUTION

1. (a) $°\text{R} = °\text{F} + 460 = 55 + 460 = 515$
   (b) $°\text{C} = \frac{5}{9}(°\text{F} - 32) = \frac{5}{9}(55 - 32) = 12.8$
   (c) $\text{K} = \frac{5}{9}(°\text{F} + 460) = \frac{5}{9}(55 + 460) = 286$
2. (a) $°\text{F} = 1.8(°\text{C}) + 32 = 1.8(55) + 32 = 131$
   (b) $°\text{R} = 1.8(°\text{C}) + 492 = 1.8(55) + 492 = 591$
   (c) $\text{K} = °\text{C} + 273 = 55 + 273 = 328$

Example 4.6.4 Consider the following pressure calculations.

1. A liquid weighing 100 lb held in a cylindrical column with a base area of 3 in.^{2} exerts how much pressure at the base in lb_f/ft^2?
2. If the pressure is 35 psig (pounds per square inch gauge), what is the absolute pressure?

SOLUTION

1. $F = mg/g_c = 100 \text{ lb}(1 \text{ lb}_f/\text{lb})$
   $= 100 \text{ lb}_f$

   Note: g_c is a conversion factor equal to 32.16 $\text{lb-ft/lb}_f\text{-s}^2$; g is the gravitational acceleration, which is equal, or close to, 32.16 ft/s^2 on Earth's surface:

$$P = F/\text{area} = 100\ \text{lb}_f/3\ \text{in.}^2$$

$$= 33.33\ \text{lb}_f/\text{in.}^2$$

$$= 4800\ \text{lb}_f/\text{ft}^2$$

2. P absolute = P gauge + P atmosphere = 35 + 14.7
   = 49.7 psia

Example 4.6.5

1. What is the density of air at 75°F and 14.7 psia? The molecular weight of air is 29.
2. Calculate the volume (in ft^3) of 1.0 lbmol of any ideal gas at 60°F and 14.7 psia.
3. Calculate the density of a gas (MW = 29) in g/cm^3 at 20°C and 1.2 atm using the ideal gas law.
4. Data from an incinerator indicate a volumetric flow rate of 30,000 acfm (60°F, 1 atm). If the operating temperature and pressure of the unit are 1100°F and 1 atm, respectively, calculate the flow rate in standard cubic feet per minute.
5. The exhaust gas flow rate from a facility is 1000 scfm. All of the gas is vented through a small stack that has an inlet area of 1.0 ft^2. The exhaust gas temperature is 300°F. What is the velocity of the gas through the stack inlet in feet per second? Assume standard conditions to be 70°F and 1.0 atm. Neglect the pressure drop across the stack.

SOLUTION This example is solved using the ideal gas law:

$$PV = nRT = \frac{m}{M}RT$$

1. $$\rho = \frac{PM}{RT} = \frac{(14.7\ \text{psia})(29\ \text{lb/lbmol})}{(10.73\ \text{ft}^3 - \text{psi/lbmol} - {}^\circ\text{R}(75 + 460)}$$

   $$= 0.0743\ \text{lb/ft}^3$$

2. Solve the ideal gas law for V and calculate the volume:

   $$V = \frac{nRT}{P} = \frac{(1)(10.73)(60 + 460)}{14.7}$$

   $$= 379\ \text{ft}^3$$

   This result is an important number to remember in combustion calculations—1 lbmol of any (ideal) gas at 60°F and 1 atm occupies 379 ft^3.

3. Calculate the density of the gas again using the ideal gas law:

$$PV = nRT = \frac{m}{M}RT$$

$$\frac{m}{V} = \rho = \frac{PM}{RT} = \frac{(1.2)(29)}{(82.06)(20 + 273)}$$

$$= 0.00145 \text{ g/cm}^3$$

The effects of pressure, temperature, and molecular weight on density can be obtained directly from the ideal gas law equation. Increasing the pressure and molecular weight increases the density; increasing the temperature decreases the density.

4. Since the pressure remains constant, calculate the standard cubic feet per minute using Charles' law:

$$Q_s = Q_a\left(\frac{T_a}{T_s}\right) = 30{,}000\left(\frac{60 + 460}{1100 + 460}\right)$$

$$= 10000 \text{ scfm}$$

The reader is again cautioned on the use of acfm and/or scfm. Predicting the performance of and designing equipment should always be based on *actual* conditions. Designs based on standard conditions can lead to disastrous results, with the unit underdesigned. The reader is also reminded that absolute temperatures and pressures must be employed in all ideal gas law calculations.

5. Calculate the actual flow rate, in acfm, using Charles' law:

$$Q_a = Q_s\left(\frac{T_a}{T_s}\right) = 1000\left(\frac{460 + 300}{460 + 70}\right)$$

$$= 1434 \text{ acfm}$$

Note that since the gas in vented through the stack to the atmosphere, the pressure is 1.0 atm. Calculate the velocity of the gas:

$$v = \frac{Q_a}{A} = \frac{1434}{1.0}$$

$$= 1434 \text{ ft/min}$$

Example 4.6.6 The exhaust to the atmosphere from an incinerator has a CO concentration of 0.15 mm Hg partial pressure. Calculate the parts per million of CO in the exhaust.

SOLUTION First calculate the mole fraction (y). By definition,

$$y = P_{\text{CO}}/P$$

Since the exhaust is discharged to the atmosphere, the atmospheric pressure (760 mm Hg) is the total pressure (P):

$$y = (0.15)/(760) = 1.97 \times 10^{-4}$$

$$\begin{aligned}\text{ppm} &= (y)(10^6) = (1.97 \times 10^{-4})(10^6)\\ &= 197 \text{ ppm}\end{aligned}$$

Note: Since the concentration of *gases* is involved, it is understood that ppm is on a volume basis.

Example 4.6.7 What is the kinematic viscosity of a gas, the specific gravity and absolute viscosity of which are 0.8 and 0.02 cP, respectively?

SOLUTION

$$\frac{0.02 \text{ cP}}{1} \times \frac{6.720 \times 10^{-4} \text{ lb/ft-s}}{1 \text{ cP}} = 1.344 \times 10^{-5} \text{ lb/ft-s}$$

$$\begin{aligned}\rho &= (\text{SG})(\rho_{\text{ref}}) = (0.08)(62.43 \text{ lb/ft}^3) = 49.94 \text{ lb/ft}^3\\ \nu &= \mu/\rho = (1.344 \times 10^{-5} \text{ lb/ft-s})/(49.94 \text{ lb/ft}^3)\\ &= 2.691 \times 10^{-7} \text{ ft}^2\text{/s}\end{aligned}$$

Example 4.6.8 If the heat capacity of liquid water is 1.0 Btu/lb-°F at 17°C, what is the change in temperature if 50 Btu of heat is added to 3 lb of water at constant pressure, assuming no loss of heat?

SOLUTION From the definition of heat capacity (see Chapter 5 for additional details),

$$Q = mc_P\Delta T$$

where Q = amount of heat (Btu)
m = mass (lb)
$\Delta T = Q/mc_P$ = (50 Btu)/(3 lb)(1 Btu/lb-°F)
= 16.67°F

Example 4.6.9 What is the heat flux (Btu/h-ft^2) through a copper plate 0.3 in. thick if one side is at 100°C and the other side is 400°C? The thermal conductivity of copper is 214 Btu/h-ft-°F.

SOLUTION From the definition of thermal conductivity (see Chapter 8 for additional details),

$$\frac{q}{A} = \frac{k\Delta T}{x} = (214 \text{ Btu/h-ft-°F}[(400 - 100)(1.8)])\text{°F}/(0.3/12)\text{ft}$$

$$= 4.62 \times 10^6 \text{ Btu/h-ft}^2$$

where q/A = heat flux
x = thickness

Note: The direction of heat transfer is from the high-temperature region to the low-temperature region.

Example 4.6.10 Calculate the Reynolds number for a liquid waste flowing through a 5-in.-diameter pipe at 10 fps (feet per second) with a density of 50 lb/ft^3 and a viscosity of 0.65 cP? Is the flow turbulent or laminar?

SOLUTION By definition

$$\text{Re} = \rho v D/\mu$$

Substitution yields

$$\text{Re} = \frac{50 \text{ lb}}{\text{ft}^3} \times \frac{10 \text{ ft}}{\text{s}} \times \frac{5 \text{ in.}}{1} \times \frac{1 \text{ ft}}{12 \text{ in.}} \times \frac{1}{0.65 \text{ cP}} \times \frac{1 \text{ cP}}{6.720 \times 10^{-4} \text{ lb/ft-s}}$$

$$= (50 \text{ lb/ft}^3)(10 \text{ ft/s})(5/12 \text{ ft})/(0.65 \times 6.72 \times 10^{-4} \text{ lb/ft-s})$$

$$= 477{,}000$$

The Reynolds number is >2100; therefore, this flow is turbulent.

Example 4.6.11 Perform the following ideal gas law calculations:

1. What is the final volumetric flow rate of a gas that is heated from 100 to 300°F if its initial flow is 3500 scfm (60°F, 1 atm)?
2. What is the volumetric flow rate of the gas (100°F, 1 atm) if it is compressed isothermally (constant temperature) to 3 atm?
3. What is the final volumetric flow rate if the gas is compressed to 3 atm and heated from 100 to 300°F with an initial volumetric flow rate of 3500 scfm (60°F, 1 atm)?

SOLUTION

1. Converting the initial standard conditions to actual conditions using Charles' law,

$$Q_{\mathrm{a}} = Q_{\mathrm{s}}\left(\frac{T_{\mathrm{a}}}{T_{\mathrm{s}}}\right) = \frac{(3500 \text{ scfm}(100 + 460)}{60 + 460}$$
$$= 3769 \text{ acfm}$$

Using Charles' law again to obtain the final volumetric flow rate,

$$Q_{\mathrm{a2}} = Q_{\mathrm{a}}\left(\frac{T_2}{T_1}\right) = \frac{(3769)(760)}{560}$$
$$= 5115 \text{ acfm}$$

2. Using Boyle's law,

$$Q_{\mathrm{a2}} = Q_{\mathrm{a1}}\left(\frac{P_1}{P_2}\right) = \frac{(1)(3769)}{3}$$
$$= 1256 \text{ acfm}$$

3. Using the combined gas law,

$$Q_{\mathrm{a2}} = Q_{\mathrm{a1}}\left(\frac{P_1}{P_2}\right)\left(\frac{T_2}{T_1}\right) = \frac{(1)(3769)(760)}{(560)(3)}$$
$$= 1705 \text{ acfm}$$

Example 4.6.12 Refer to the psychometric chart in Fig. 4.5.2 to answer the following:

1. List key properties for humid air at a dry-bulb temperature of 160°F and a wet-bulb temperature of 100°F.
2. A stream of moist air is cooled and humidified adiabatically from T_{DB} of 100°F and T_{WB} of 70°F to a T_{DB} of 80°F. How much moisture is added per pound of dry air?

SOLUTION

1. If the air were to be cooled until the moisture just begins to condense, the dew point would be reached. This is represented by a horizontal line at constant humidity intersecting the saturation curve at a dew point of 87.5°F.

   The relative humidity is approximately 14% (interpolating between the 15 and 10% relative humidity lines). The absolute humidity is the horizontal line extended to the right intersecting the ordinate at a humidity of 0.0285 lb H_2O/lb dry air.

The humid volume is approximately 16.3 ft^3 moist air/lb dry air (interpolating between 16 and 17 ft^3 moist air volume.)

The enthalpy for saturated air at a T_{WB} of 100°F is 71.8 Btu/lb dry air. For the unsaturated air, the enthalpy deviation is −1.0 Btu/lb dry air; therefore, the actual enthalpy for the moist air at a T_{WB} of 100°F and a T_{DB} of 160°F is 70.8 Btu/lb dry air.

2. Adiabatic cooling follows the wet-bulb temperature line upwards (toward the saturation curve). The difference in the final and initial humidities is the added moisture.

	$\mathcal{H}$
Initial	0.0090
Final	0.0133

$$\Delta\mathcal{H} = 0.0133 - 0.0090 = 4.3 \times 10^{-3} \text{ lb H}_2\text{O/lb dry air}$$

Example 4.6.13 A flue gas is discharged at 120°F from an HCl absorber in a hazardous waste incinerator (HWI) facility in which carbon tetrachloride (CCl_4) is being incinerated. If 9000 lb/h (MW = 30) of gas enters the absorber essentially dry (negligible water) at 560°F, calculate the moisture content, the mass flow rate, and the volumetric flow rate of the discharge gas. The discharge gas from the absorber may safely be assumed to be saturated with water vapor.

SOLUTION From Fig. 4.5.2, the discharge humidity of the flue gas is approximately

$$\mathcal{H}_{out} = 0.0814 \text{ lb H}_2\text{O/lb bone dry air}$$

This represents the moisture content of the gas at outlet conditions in lb H_2O/lb dry air. If the gas is assumed to have the properties of air, the discharge water vapor rate is

$$W_{H_2O} = (0.0814)(9000)$$
$$= 733 \text{ lb/h}$$

The total flow rate leaving the absorber is

$$W_{total} = 733 + 9000$$
$$= 9733 \text{ lb/h}$$

The volumetric (or molar) flow rate can only be calculated if the molecular weight of the gas is known. The average molecular weight of the discharge flue gas must first be calculated from the mole fraction of the flue gas (fg) and water vapor (wv):

$$y_{fg} = \frac{9000/30}{(9000/30) + (733/18)} = 0.88$$

$$y_{wv} = \frac{733/18}{(733/18) + (9000/30)} = 0.12$$

$$\overline{MW} = (0.88)(30) + (0.12)(18) = 28.6 \text{ lb}$$

The ideal gas law is employed to calculate the volumetric flow rate, Q_a:

$$PQ_a = \frac{w}{\overline{MW}} RT$$

$$Q_a = \frac{9733}{28.6} \frac{(0.73)(460 + 140)}{1.0}$$

$$= 1.49 \times 10^5 \text{ ft}^3\text{/h}$$

PROBLEMS

Use the correct number of significant figures and scientific notation whenever possible.

1. Convert **(a)** 8.03 yr to s, **(b)** 150 miles/h to yd/h, **(c)** 100.0 m/s^2 to ft/min^2, and **(d)** 0.03 g/cm^3 to lb/ft^3.

2. **(a)** What is the molecular weight of nitrobenzene ($C_6H_5O_2N$)?
   (b) How many moles are there in 50.0 g of nitrobenzene?
   (c) If the specific gravity is 1.203, what is the density in g/cm^3?
   (d) What is the volume occupied by 50.0 g of nitrobenzene in cm^3, in ft^3, and in in.^{3}?
   (e) If the nitrobenzene is held in a cylindrical container with a base of 1 in. in diameter, what is the pressure at the base? What is it in gauge pressure?
   (f) How many molecules are contained in 50.0 g of nitrobenzene?

3. Convert the following temperature: **(a)** 20°C to °F, K, °R and **(b)** 20°F to °C, K, °R.

4. **(a)** What is the density of air at 115°F and 2 atm?
   (b) What volume does 100 g of air occupy at the condition of (a)?
   (c) What is the new volume if the temperature is changed to 200°F and the pressure is 1 atm?

5. A liquid with a viscosity of 0.78 cP and a density of 1.50 g/cm^3 flows through a 1-in.-diameter pipe at 20 cm/s. Calculate the Reynolds number. What region (laminar or turbulent) is it in?

6. A heat pump takes in 3500 gpm of water at a temperature of 38°F and discharges back to the lake at 36.2°F. How many Btus are removed from the water per day? (C_p for H_2O = 75.4 J/gmol-°C, ρ = 62.4 lb/ft^3)

7. One wall of an oven has a 3-in. insulation cover. The temperature on the inside of the wall is at 400°F; the temperature on the outside is at 25°C. What is the heat flux (heat flow rate per unit area) across the wall if the insulation is made of glass wool (k = 0.022 Btu/h-ft-°F)?

8. A container holds a gaseous mixture comprised of 30% CO_2, 5% CO, 5% H_2O, 50% N_2, and 10% O_2, by volume. What is the partial pressure of each component if the total pressure is 2 atm? What are their partial volumes if the total volume is 10 ft^3? What are their concentrations in ppm (parts per million)?

9. The exhaust to the atmosphere from an incinerator shows a CO concentration with a partial pressure of 0.19 mm Hg. What is this concentration in ppm?

10. The inlet gas to a spray tower is at 1600°F. It is piped through a 3.0-ft inside diameter duct at 25 ft/s to the spray tower. The scrubber cools the gas to 500°F. In order to maintain the velocity of 25 ft/s, what size duct would be required at the outlet of the unit? Neglect the pressure across the spray tower and any moisture considerations.

11. **(a)** What is the density of air at 60°F and 14.7 psia?
(b) Calculate the volume (in ft^3) of 1.0 lbmol of any ideal gas at 77°F and 14.7 psia.
(c) Calculate the density of a gas (MW = 32) in g/cm^3 at 20°C and 1.0 atm using the ideal gas law.
(d) Data from an incinerator indicate a volumetric flow rate of 10,000 scfm (60°F, 1 atm). If the operating temperature and pressure of the unit are 1950°F and 1 atm, respectively, calculate the actual flow rate in cubic feet per minute.

REFERENCES

1. R. C. Weast, ed., *CRC Handbook of Chemistry and Physics*, 80th ed., CRC Press, Boca Raton, FL, 1999.
2. S. Maron and C. Prutton, *Principles of Physical Chemistry*, 4th ed., MacMillan, New York, 1970.
3. J. M. Smith and H. C. Van Ness, *Introduction to Chemical Engineering Thermodynamics*, 4th ed., McGraw-Hill, New York, 1992.
4. R. Perry and D. Green, *Perry's Chemical Engineering Handbook*, 7th ed., McGraw-Hill, New York, 1997.

5

STOICHIOMETRIC AND THERMODYNAMIC CONSIDERATIONS

5.1 INTRODUCTION

In order to understand completely the design as well as the operation and performance of waste combustors, it is necessary first to understand the chemical theory underlying this technology. How can one predict what products will be emitted from effluent streams? At what temperature must the combustor be operated to ensure compliance with the *four nines* (99.99% destruction and/or removal efficiency)? How much energy in the form of heat is given off during combustion? Is it economically feasible to recover this heat? Is the waste feed high enough in heating value, or must additional fuel be added to assist in the combustion process? If so, how much fuel must be added? The answers to these questions are rooted in the various theories of thermodynamics, thermochemistry, and chemical reaction equilibrium. Answers will be provided in the following sections of this chapter and Chapter 6.

5.2 CONSERVATION LAWS

Conservation of Mass

The *conservation law* for mass can be applied to any process or system. The general form of this law is given by Eq. (5.2.1):

$$\text{Mass in} - \text{mass out} + \text{mass generated} = \text{mass accumulated}$$

or on a time rate basis by

$$\begin{aligned}&\text{rate of mass in} - \text{rate of mass out} + \text{rate of mass generated}\\&\quad = \text{rate of mass accumulated}\end{aligned} \tag{5.2.1}$$

This equation may be applied either to the total mass involved or to a particular species, on either a mole or mass basis. In combustion processes, it is often necessary to obtain quantitative relationships by writing mass balances on the various elements in the system. This law can be applied to steady-state or unsteady-state (transient) processes and to batch or continuous systems. In order to isolate a system for study, it is separated from the surroundings by a boundary or envelope. This boundary may be real (e.g., the walls of a combustor) or imaginary. Mass crossing the boundary and entering the system is part of the *mass in* term in Eq. (5.2.1), while that crossing the boundary and leaving the system is part of the *mass out* term. Equation (5.2.1) may be written for any compound whose quantity is not changed by chemical reaction and for any chemical element whether or not it has participated in a chemical reaction. It may be written for one piece of equipment, around several pieces of equipment, or around an entire process. It may be used to calculate an unknown quantity directly, to check the validity of experimental data, or to express one or more of the independent relationships among the unknown quantities in a particular problem situation.

A *steady-state* process is one in which there is no change in conditions (pressure, temperature, composition, etc.) or rates of flow with time at any given point in the system. The accumulation term in Eq. (5.2.1) is then zero. (If there is no chemical or nuclear reaction, the generation term is also zero.) All other processes are *unsteady state.* In a *batch* process, a given quantity of reactants is placed in a container, and by chemical and/or physical means, a change is made to occur. At the end of the process, the container (or adjustment containers to which material may have been transferred) holds the product or products. In a *continuous* process, reactants are continuously fed to a piece of equipment or to several pieces in series, and products are continuously removed from one or more points. A continuous process may or may not be steady state. A coal-fired power plant, for example, operates continuously. However, because of the wide variation in power demand between peak and slack periods, there is an equally wide variation in the rate at which the coal is fired. For this reason, power plant problems may require the use of average data over long periods of time. Most waste combustor operations are assumed to be steady state and continuous.

As indicated previously, Eq. (5.2.1) may be applied to the total mass of each stream (referred to as an *overall* or *total material balance*) or to the individual component(s) of the streams (referred to as a *componential* or *component material balance*). Often the primary task in preparing a material balance in waste combustion calculations is to develop the quantitative relationships among the streams. The primary factors, therefore, are those that *tie* the streams together. An element, compound, or unreactive mass (ash, e.g.) that enters or leaves in a single stream or passes through a process unchanged is so convenient for this purpose that it may be considered a *key* to the calculations. If sufficient data are given about this component, it can be used in a component balance to determine the total masses of the entering and exiting streams. Such a component is sometimes referred to as a *key component.* Since a key component does not react in a process, it must retain its identity as it passes through the process. Obviously, except for nuclear reactions, elements may always be used as key components because they do not change identity even though they may undergo a chemical reaction. Thus CO (carbon monoxide) may be used as a key component only when it does not react, but C (carbon) may always be used as a key component. A component that enters the system in only one stream and leaves in only one stream is usually the most convenient choice for a key component.

Four important processing concepts are bypass, recycle, purge, and makeup. With *bypass,* part of the inlet stream is diverted around the equipment to rejoin the (main) stream after the unit (see Fig. 5.2.1). This stream effectively moves in parallel with the stream passing through the equipment. In *recycle*, part of the product stream is sent back to mix with the feed. If a small quantity of nonreactive material is present in the feed to a process that includes recycle, it may be necessary to remove the nonreactive material in a *purge* stream to prevent its building up above a maximum tolerable value. This can also occur in a process without recycle; if a nonreactive material is added in the feed and not totally removed in the products, it will accumulate until purged. The purging process is sometimes referred to as *blowdown. Makeup*, as its name implies, involves adding or making up part of a stream that has been removed from a process. Makeup may be thought of, in a final sense, as the opposite of purge and/or blowdown.

Conservation of Energy

A presentation of the conservation law for energy would be incomplete without a brief review of some introductory thermodynamic principles. *Thermodynamics* is defined as that science that deals with the relationships among the various forms of energy. A system may possess energy due to its:

1. Temperature
2. Velocity
3. Position

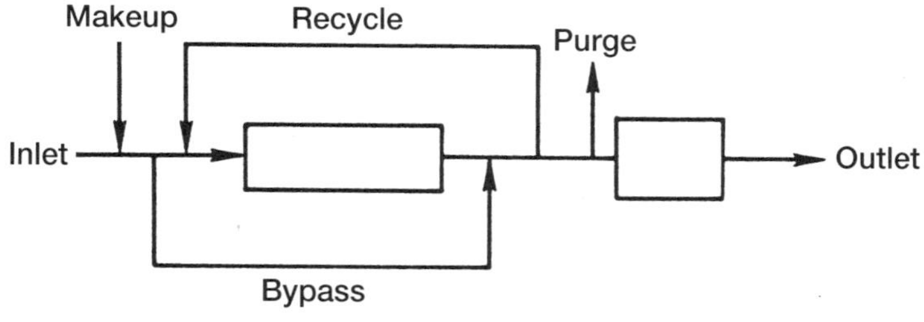

Figure 5.2.1 Recycle, bypass, and purge.

4. Molecular structure
5. Surface, and so on

The energies corresponding to these states are, for example:

1. Internal
2. Kinetic
3. Potential
4. Chemical
5. Surface

Engineering thermodynamics is founded on three basic laws. Energy, like mass and momentum, is conserved. Application of the conservation law for energy gives rise to the first law of thermodynamics. This law, in steady-state equation form for batch and flow processes, is presented here.

For *batch* processes:

$$\Delta E = Q - W \tag{5.2.2}$$

For flow processes:

$$\Delta H = Q - W_s \tag{5.2.3}$$

where potential, kinetic, and other energy effects have been neglected and

Q = energy in the form of heat transferred across the boundaries of the system
W = energy in the form of work transferred across the boundaries of the system
W_s = energy in the form of mechanical work transferred across the boundaries of the system
E = internal energy of the system
H = enthalpy of the system
ΔE, ΔH = changes in the internal energy and enthalpy, respectively, during the process

The internal energy and enthalpy in Eqs. (5.2.2) and (5.2.3), as well as the other equations in this section, may be on a *mass* basis (i.e., for 1 gal or 1 lb of material), on a *mole* basis (i.e., for 1 gmol or 1 lbmol of material), or represent the total internal energy and enthalpy of the entire system. As long as these equations are dimensionally consistent, it makes no difference. For the sake of clarity, the same convention that was used for the heat capacities will be employed throughout this text—uppercase letters (e.g., H, E, C_P) represent properties on a mole basis, while lowercase letters (e.g., h, e, c_P) represent properties on a mass basis. Properties for the entire system will rarely be used and therefore require no special symbols.

Perhaps the most important thermodynamic function the engineer works with is the *enthalpy*. The enthalpy is defined by:

$$H = E + PV$$

where P = pressure of the system
V = volume of the system

The terms E and H (and free energy G, to be discussed later) are *state* or *point* functions. By fixing a certain number of variables on which the function depends, this automatically fixes the numerical value of the function; that is, it is single valued (see Section 4.5). For example, fixing the temperature and pressure of a one-component single-phase system immediately specifies the enthalpy and internal energy.

The change in enthalpy as it undergoes a change in state from (T_1, P_1) to (T_2, P_2) is given by

$$\Delta H = H_2 - H_1 \tag{5.2.4}$$

Note that H and ΔH are independent of the path. This is a characteristic of all state or point functions, that is, the state of the system is independent of the path by which the state is reached. The terms Q, W, and W_s in Eqs. (5.2.2) and (5.2.3) are *path* functions; their values depend on the path used between the two states; unless a process or change of state is occurring, path functions have no value.

For a mathematical representation of this thermodynamic point function, the following can be written:

$$H = H(T, P)$$

By the rules of partial differentiation, a differential change in H is given by

$$dH = \left(\frac{\partial H}{\partial T}\right)_P dT + \left(\frac{\partial H}{\partial P}\right)_T dP \tag{5.2.5}$$

The term $(\partial H/\partial P)_T$ is assumed to be negligible in most engineering applications. It is exactly zero for an ideal gas and is small for solids and liquids and gases near ambient conditions. The term $(\partial H/\partial P)_T$ is defined as the heat capacity at constant pressure (see Chapter 4 for additional details):

$$C_P = \left(\frac{\partial H}{\partial P}\right)_P \tag{5.2.6}$$

Equation (5.2.5) may now be written as

$$dH = C_p \, dT \tag{5.2.7}$$

If average molar heat capacity data are available, this equation may be integrated to yield

$$\Delta H = \bar{C}_p \, \Delta T \tag{5.2.8}$$

where $\bar{C}_P$ = average value of C_p in the temperature range ΔT.

Average molar heat capacity data are provided in Table 5.2.1. Thus, the calculation of enthalpy change(s) associated with a temperature change may be

Table 5.2.1 Mean molar heat capacities of gases for the temperature range 0°F to T[a]

					$\overline{C_P}$ = Btu/lbmol-°F						
T (°F)	N_2	O_2	H_2O	CO_2	H_2	CO	CH_4	SO_2	NH_3	HCl	NO
0	6.94	6.92	7.93	8.50	6.86	6.92	8.25	9.9	8.80	6.92	7.1
200	6.96	7.03	8.04	9.00	6.89	6.96	8.42	10.0	8.85	6.96	7.2
400	6.98	7.14	8.13	9.52	6.93	7.00	9.33	10.3	9.05	7.01	7.2
600	7.02	7.26	8.25	9.97	6.95	7.05	10.00	10.6	9.40	7.05	7.3
800	7.08	7.39	8.39	10.37	6.97	7.13	10.72	10.9	9.75	7.10	7.3
1000	7.15	7.51	8.54	10.72	6.98	7.21	11.45	11.2	10.06	7.15	7.4
1200	7.23	7.62	8.69	11.02	7.01	7.30	12.13	11.4	10.43	7.19	7.5
1400	7.31	7.71	8.85	11.29	7.03	7.38	12.78	11.7	10.77	7.24	7.6
1600	7.39	7.80	9.01	11.53	7.07	7.47	13.38	11.8		7.29	7.7
1800	7.46	7.88	9.17	11.75	7.10	7.55		12.0		7.33	7.7
2000	7.53	7.96	9.33	11.94	7.15	7.62		12.1		7.38	7.8
2200	7.60	8.02	9.48	12.12	7.20	7.68		12.2		7.43	7.8
2400	7.66	8.08	9.64	12.28	7.24	7.75		12.3		7.47	7.9
2600	7.72	8.14	9.79	12.42	7.28	7.80		12.4		7.52	8.0
2800	7.78	8.19	9.93	12.55	7.33	7.86		12.5		7.57	8.0
3000	7.83	8.24	10.07	12.67	7.38	7.91		12.5		7.61	8.1
3200	7.87	8.29	10.20	12.79	7.43	7.95					
3400	7.92	8.34	10.32	12.89	7.48	8.00					
3600	7.96	8.38	10.44	12.98	7.53	8.04					
3800	8.00	8.42	10.56	13.08	7.57	8.08					
4000	8.04	8.46	10.67	13.16	7.62	8.11					
4200	8.07	8.50	10.78	13.23	7.66	8.14					
4400	8.10	8.54	10.88	13.31	7.70	8.18					
4600	8.13	8.58	10.97	13.38	7.75	8.20					
4800	8.16	8.62	11.08	13.44	7.79	8.23					

[a]From E. T. Williams and R. C. Johnson, *Stoichiometry for Chemical Engineers*, McGraw-Hill, New York, 1958.

accomplished through the application of either Eq. (5.2.4) or (5.2.8). Actual enthalpy data as provided in Table 5.2.2 must be available for use of Eq. (5.2.4). If the heat capacity is a function of temperature, for example,

$$C_P = \alpha + \beta T + \gamma T^2 \tag{5.2.9}$$

Equation (5.2.7) may be integrated directly between some reference or standard temperature (T_0) and the final temperature (T_1):

$$\begin{aligned}\Delta H &= H_1 - H_0 \\ &= \alpha(T_1 - T_0) + \frac{\beta}{2}(T_1^2 - T_0^2) + \frac{\gamma}{3}(T_1^3 - T_0^3)\end{aligned} \tag{5.2.10}$$

Equation (5.2.7) may also be integrated if the heat capacity as a function of temperature is of the form

Table 5.2.2 Molar enthalpies of combustion gases[a] in Btu/lbmol[b]

T (°F)	N_2	Air (MW 28.97)	CO_2	H_2O
32	0	0	0	0
60	194.9	194.6	243.1	224.2
77	312.2	312.7	392.2	360.5
100	473.3	472.7	597.9	545.3
200	1,170	1,170	1,527	1,353
300	1,868	1,870	2,509	2,171
400	2,570	2,576	3,537	3,001
500	3,277	3,289	4,607	3,842
600	3,991	4,010	5,714	4,700
700	4,713	4,740	6,855	5,572
800	5,443	5,479	8,026	6,460
900	6,182	6,227	9,224	7,364
1,000	6,929	6,984	10,447	8,284
1,200	8,452	8,524	12,960	10,176
1,500	10,799	10,895	16,860	13,140
2,000	14,840	14,970	23,630	18,380
2,500	19,020	19,170	30,620	23,950
3,000	23,280	23,460	37,750	29,780

[a] K. A. Kobe and E. G. Long, "Thermochemistry for the Petroleum Industry," *Pet. Refiner,* **28**(11), 127–132, (1949).
[b] To convert values to enthalpy in Btu/lb, divide by the molecular weight.

$$C_P = a + bT + cT^{-2} \tag{5.2.11}$$

The enthalpy change is then given by

$$\Delta H = a(T_1 - T_0) + \frac{b}{2}(T_1^2 - T_0^2) - c\left(\frac{1}{T_1} - \frac{1}{T_0}\right) \tag{5.2.12}$$

Tabulated values of α, b, γ and a, b, c for a host of compounds (including some chlorinated organics) are given in Tables 5.2.3 and 5.2.4, respectively. The reader is cautioned that the use of Eqs. (5.2.9)–(5.2.12) required the use of absolute temperature in kelvins.

Most waste combustion facilities operate in a steady-state flow mode with no significant mechanical or shaft work added (or withdrawn) from the system. For this condition, Eq. (5.2.3) reduces to

$$Q = \Delta H \tag{5.2.13}$$

This equation is routinely used in many waste combustion calculations. If a unit or system is operated adiabatically, $Q = 0$ and Eq. (5.2.13) becomes

$$\Delta H = 0 \tag{5.2.14}$$

Although the topics of material and energy balances have been covered separately in this section, it should be emphasized that this segregation does not exist in reality. Waste combustion is invariably accompanied by heat effects, and one must work with both energy and material balances simultaneously.

5.3 STOICHIOMETRY

The term *stoichiometry* has come to mean different things to different people. In a loose sense, stoichiometry involves the balancing of an equation for a chemical reaction that provides a quantitative relationship among the reactants and products. In the simplest stoichiometric situation, exact quantities of pure reactants are available, and these quantities react completely to give the desired product(s). In an industrial process, the reactants usually are not pure, one reactant is usually in excess of what is needed for the reaction, and the desired reaction may not go to completion because of a host of other considerations. For example, complete combustion of pure hydrocarbons yields carbon dioxide and water as the reaction products. Consider the combustion of methane in oxygen:

$$CH_4 + O_2 \rightarrow CO_2 + H_2O \tag{5.3.1}$$

In order to balance this reaction, two molecules of oxygen are needed. This

Table 5.2.3 Molar heat capacities[a,b]

Compound	Formula	α	$\beta \times 10^3$	$\gamma \times 10^6$
	Normal Paraffins (gases)			
Methane	CH_4	3.381	18.044	−4.300
Ethane	C_2H_6	2.247	38.201	−11.049
Propane	C_3H_8	2.410	57.195	−17.533
n-Butane	C_4H_{10}	3.844	73.350	−22.655
n-Pentane	C_5H_{12}	4.895	90.113	−28.039
n-Hexane	C_6H_{14}	6.011	106.746	−33.363
n-Heptane	C_7H_{16}	7.094	123.447	−38.719
n-Octane	C_8H_{18}	8.163	140.217	−44.127
Increment per C atom above C_8	—	1.097	16.667	−5.338
	Normal Monoolefins (gases) (1-alkenes)			
Ethylene	C_2H_4	2.830	28.601	−8.726
Propylene	C_3H_6	3.253	45.116	−13.740
1-Butene	C_4H_8	3.909	62.848	−19.617
1-Pentene	C_5H_{10}	5.347	79.990	−24.733
1-Hexene	C_6H_{12}	6.399	95.752	−30.116
1-Heptene	C_7H_{14}	7.488	112.440	−35.462
1-Octene	C_8H_{16}	8.592	129.076	−40.775
Increment per C atom above C_8	—	1.097	16.667	−5.338
	Miscellaneous Gases			
Acetaldehyde[c]	C_2H_4O	3.364	35.722	−12.236
Acetylene	C_2H_2	7.331	12.622	−3.889
Ammonia	NH_3	6.086	8.812	−1.506
Benzene	C_6H_6	−0.409	77.621	−26.429
1,3-Butadiene	C_4H_6	5.432	53.224	−17.649
Carbon dioxide	CO_2	6.214	10.396	−3.545
Carbon monoxide	CO	6.420	1.665	−0.196
Chlorine	Cl_2	7.576	2.424	−0.965
Cyclohexane	C_6H_{12}	−7.701	125.675	−41.584
Ethyl alcohol	C_3H_6O	6.990	39.741	−11.926
Hydrogen	H_2	6.947	−0.200	0.481
Hydrogen chloride	HCl	6.732	0.433	0.370
Hydrogen sulfide	H_2S	6.662	5.134	−0.854
Methyl alcohol	CH_4O	4.394	24.274	−6.855
Nitric oxide	NO	7.020	−0.370	2.546
Nitrogen	N_2	6.524	1.250	−0.001
Oxygen	O_2	6.148	3.102	−0.923
Sulfur dioxide	SO_2	7.116	9.512	3.511
Sulfur trioxide	SO_3	6.077	23.537	−0.687
Toluene	C_7H_8	0.576	93.493	−31.227
Water	H_2O	7.256	2.298	0.283

(*continued*)

Table 5.2.3 (*Continued*)

Compound	Formula	α	$\beta \times 10^3$	$\gamma \times 10^6$
	Chlorinated Organic Liquids			
Methyl chloride	CH_3Cl	8.994	10.280	−21.218
Dichloromethane	CH_2Cl_2	13.672	7.616	−33.272
Chloroform	$CHCl_3$	18.453	4.932	−37.923
Carbon tetrachloride	CCl_4	23.777	2.224	−38.558
Ethyl chloride	C_2H_5Cl	15.267	17.060	−48.798
1,1-Dichloroethane	$C_2H_4Cl_2$	20.014	14.315	−54.917
1,1,2,2-Tetrachloroethane	$C_2H_2Cl_4$	28.397	10.338	−66.739
n-Propyl chloride	C_3H_7Cl	20.703	24.070	−69.351
1,3-Dichloropropane	$C_3H_6Cl_2$	24.840	21.810	−68.214
n-Butyl chloride	C_4H_9Cl	26.258	31.208	−89.482
1-Chloropentane	$C_5H_{11}Cl$	31.812	38.345	−109.612
1-Chloroethylene	C_2H_3Cl	14.058	10.939	−40.586
trans-1,2-Dichloroethylene	$C_2H_6Cl_2$	18.050	8.845	−42.981
Trichloroethylene	C_2HCl_3	22.410	6.394	−46.257
Tetrachloroethylene	C_2Cl_4	26.595	4.106	−46.208
3-Chloro-1-propene	C_3H_5Cl	18.728	18.514	−56.495
Chlorobenzene	C_6H_5Cl	28.558	25.786	−116.150
p-Dichlorobenzene	$C_6H_5Cl_2$	33.265	23.025	−116.848
Hexachlorobenzene	C_6Cl_6	50.759	13.286	−115.620

[a]Selected mainly from values given by H. M. Spencer, J. Justice, and G. Flanagan, *J. Am. Chem. Soc.*, **56**, 2311 (1934); **64**, 2511 (1942); **67**, 1859 (1945); *Ind. Eng. Chem.* **40**, 2152 (1948); also from personal notes: L. Theodore and J. Reynolds, 1986.
[b]Constants for the equation $C_P = \alpha + \beta T + \gamma T^2$, where T is in K and C_P is in Btu/lbmol-°F or cal/gmol-°C.
[c]Applicable range 298–1000 K.

requires that there be four oxygen atoms on the right side of the reaction. This is satisfied by introducing two molecules of water as product. The final balanced reaction becomes

$$CH_4 + 2O_2 \rightarrow CO_2 + 2H_2O \tag{5.3.2}$$

Thus, two molecules (or moles) of oxygen are required to completely combust one molecule (or mole) of methane to yield one molecule (or mole) of carbon dioxide and two molecules (or moles) of water. Note that the numbers of carbon, oxygen, and hydrogen atoms on the right-hand side of this reaction are equal to those on the left-hand side. The reader should verify that the total mass (obtained by multiplying the number of each molecule by its molecular weight and summing) on each side of the reaction is the same. [The term *moles* here may refer to either gram-moles (gmol) or pound-moles (lbmol); it makes no difference.]

Table 5.2.4 Molar heat capacities[a,b]

Compound	Formula	Temperature Range (K)	a	$b \times 10^3$
	Inorganic Gases			
Ammonia	NH_3	298–1800	7.11	6.00
Bromine	Br_2	298–3000	8.92	0.12
Carbon monoxide	CO	298–2500	6.79	0.98
Carbon dioxide	CO_2	298–2500	10.57	2.10
Carbon disulfide	CS_2	298–1800	12.45	1.60
Chlorine	Cl_2	298–3000	8.85	0.16
Hydrogen	H_2	298–3000	6.52	0.78
Hydrogen sulfide	H_2S	298–2300	7.81	2.96
Hydrogen chloride	HCl	298–2000	6.27	1.24
Hydrogen cyanide	HCN	298–2500	9.41	2.70
Nitrogen	N_2	298–3000	6.83	0.90
Nitrous oxide	N_2O	298–2000	10.92	2.06
Nitric oxide	NO	298–2500	7.03	0.92
Nitrogen dioxide	NO_2	298–2000	10.07	2.28
Nitrogen tetroxide	N_2O_4	298–1000	20.05	9.50
Oxygen	O_2	298–3000	7.16	1.00
Sulfur dioxide	SO_2	298–2000	11.04	1.88
Sulfur trioxide	SO_3	298–1500	13.90	6.10
Water	H_2O	298–2750	7.30	2.46
	Chlorinated Organic Liquids			
Methyl chloride	Ch_3Cl	298–1800	3.44	23.52
Dichloromethane	CH_2Cl_2	298–1800	5.34	27.00
Chloroform	$CHCl_3$	298–1800	8.91	27.18
Carbon tetrachloride	CCl_4	298–1800	1.39	24.97
Ethyl chloride	C_2H_5Cl	298–1800	2.69	46.81
1,1-Dichloroethane	$C_2H_4Cl_2$	298–1800	5.99	47.31
1,1,2,2-Tetrachloroethane	$C_2H_2Cl_4$	298–1800	1.15	49.83
n-Propyl chloride	C_3H_7Cl	298–1800	2.88	66.18
1,3-Dichloropropane	$C_3H_6Cl_2$	298–1800	7.35	63.07
n-Butyl chloride	C_4H_9Cl	298–1800	3.22	85.65
1-Chloropentane	$C_5H_{11}Cl$	298–1800	3.57	10.51
1-Chloroethylene	C_2H_3Cl	298–1800	3.70	35.30
trans-1,2-Dichloroethylene	$C_2H_6Cl_2$	298–1800	7.13	34.46
Trichloroethylene	C_2HCl_3	298–1800	10.76	33.59
Tetrachloroethylene	C_2Cl_4	298–1800	14.99	31.13
3-Chloro-1-propene	C_3H_5Cl	298–1800	4.26	52.61
Chlorobenzene	C_6H_5Cl	298–1800	−1.12	95.63
p-Dichlorobenzene	$C_6H_5Cl_2$	298–1800	3.49	92.98
Hexachlorobenzene	C_6Cl_6	298–1800	21.53	81.64

[a]Selected from K. Kelley, *U.S. Bur. Mines Bull.* **584** (1960); also from personal Theodore and J. Reynolds, 1986.

[b]Constants for the equation $C_P = a\ bT + cT^{-2}$, where T is in K and C_P is in Btu/lb-°F or cal/gmol-°C.

The terms used to describe a reaction that does not involve stoichiometric ratios of reactants must be carefully defined in order to avoid confusion. If the reactants are not present in formula or stoichiometric ratio, one reactant is said to be *limiting*; the others are said to be in *excess*. Consider the following reaction:

$$CO + \tfrac{1}{2}O_2 \rightarrow CO_2 \tag{5.3.3}$$

If the starting amounts are 1 mol of CO and 3 mol of oxygen, CO is the limiting reactant, with O_2 present in excess. There are 2.5 mol of excess O_2, because only 0.5 mol is required to combine with the CO. Thus there is 500% excess oxygen present. The percentage of excess must be defined in relation to the amount of the reactant necessary to react completely with the limiting reactant. Thus, if for some reason only part of the CO actually reacts, this does not alter the fact that the oxygen is in excess by 500%. However, there are often several possible products. For instance, the reactions

$$C + O_2 \rightarrow CO_2 \quad \text{and} \quad C + \tfrac{1}{2}O_2 \rightarrow CO \tag{5.3.4}$$

can occur simultaneously. In this case, if there are 3 mol of oxygen present per mole of carbon, the oxygen is in excess. The extent of this excess, however, cannot be definitely fixed. It is customary to choose one product (e.g., the desired one) and specify the excess reactant in terms of this product. For this case, there is 200% excess oxygen for the reaction going to CO_2, and there is 500% excess oxygen for the reaction going to CO. The discussion on excess oxygen can be extended to excess air using the same approach. Stoichiometric or theoretical oxygen (or air) is defined as 0% excess oxygen (or air). This is an important concept since waste combustors operate with excess air. For example, approximately 25–50% excess air is employed with liquid injection incinerators and waste-fired boilers and process furnaces burning liquid wastes. Minimum excess air requirements for rotary kilns are approximately 50–80% if only bulk solids are burned and 80–120% if containerized wastes are incinerated.

Consider the combustion of 1 mol of ethane. The reaction for complete combustion may be written

$$C_2H_6 + 3.5O_2 \rightarrow 2CO_2 + 3H_2O \tag{5.3.5}$$

Thus, 2 mol of CO_2 and 3 mol of H_2O will be formed from the complete combustion of 1 mol of C_2H_6. The oxygen required is 3.5 mol. If 60% excess is used, this means an additional 2.1 mol, or a total of 5.6 mol of oxygen is required. Equation (5.3.5) describes a gas-phase reaction. In accordance with Charles' law, one may also interpret this equation as follows: When 1 ft^3 of C_2H_6 reacts with 3.5 ft^3 of O_2, 2 ft^3 of CO_2 and 3 ft^3 of H_2O will form. Similar balanced stoichiometric reactions may be written for other hydrocarbons and organics to determine oxygen (or air) requirements and the products of combustion. However, this somewhat tedious calculation may be bypassed by use of Table 5.3.1. The reader should carefully review this table before proceeding to the illustrative examples at the end of this chapter.

This discussion may now be extended to reaction systems that involve the combustion of carbon and/or carbonaceous hazardous wastes and fuels. To simplify matters, it will be assumed that both the air and waste–fuel mixture are dry. Throughout this text, air is assumed to contain 21% O_2 by volume. This is perhaps a bit high by a few hundredths of a percent. The remaining 79% is assumed to be inert and to consist of nitrogen (and a trace of the noble gases). The usual combustion products of hydrocarbons are CO_2 and H_2O. The combustion products of other organics may contain additional compounds; for example, an organic chloride will also produce chlorine and/or hydrochloric acid. Combustion products from certain fuels will yield sulfur dioxide and nitrogen. So-called complete combustion of an organic or a fuel involves conversion of all the elemental carbon to carbon dioxide, hydrogen to water, sulfur to sulfur dioxide, and nitrogen to its elemental form of N_2. Thus, the *theoretical* or *stoichiometric* oxygen described for a combustion reaction is the amount of oxygen required to burn all the carbon to carbon dioxide, all the hydrogen to water, and so on; excess oxygen (or excess air) is the oxygen furnished in excess of the theoretical oxygen required for combustion. Since combustion calculations assume dry air to contain 21% oxygen and 79% nitrogen on a mole or volume basis, 4.76 mol of air consists of 1.0 mol of oxygen and 3.76 mol of nitrogen. One mole of O_2 equals 32 lb; 3.76 mol of N_2 equals 105.28 lb. Therefore, on a weight basis, dry air contains 23.3% O_2 by weight (32/137.28) and 77% N_2 by weight (105.28/137.28). The temperature of the products that results from complete combustion under adiabatic conditions is defined as the *adiabatic flame temperature;* this will be discussed in more detail in Section 5.4.

For the combustion of 1.0 lbmol of chlorobenzene (C_6H_5Cl) in stoichiometric oxygen, the balanced reaction is

$$C_6H_5Cl + 7O_2 \rightarrow 6CO_2 + 2H_2O + HCl \tag{5.3.6}$$

Since air, not oxygen, is employed in incineration processes, this reaction with air becomes

$$C_6H_5Cl + 7O_2 + 26.3N_2 \rightarrow 6CO_2 + 2H_2O + HCl + 26.3N_2 \tag{5.3.7}$$

where the nitrogen in the air has been retained on both sides of the equation since it does not participate in the combustion reaction. The moles and masses involved in this reaction, based on the stoichiometric combustion of 1.0 lbmol of C_6H_5Cl, are given here:

Table 5.3.1 Combustion constants[a]

Compound	Density	Specific volume	Heat of combustion (Btu/ft^3)^c		Heat of combustion (Btu/lb)		For 100% total air (mol/mol or ft^3/ft^3 of combustible): Required for combustion			For 100% total air (mol/mol or ft^3/ft^3 of combustible): Flue products			For 100% total air (lb/lb of combustible): Required for combustion			For 100% total air (lb/lb of combustible): Flue products			Flammability limit (% by volume)	
	lb/ft^{3c}	ft^3/lb^c	Gross (high)	Net (low)	Gross (high)	Net (low)	O_2	N_2	Air	CO_2	H_2O	N_2	O_2	N_2	Air	CO_2	H_2O	N_2	Lower	Upper
Carbon, C[b]	—	—	—	—	14,093	14,093	1.0	3.76	4.76	1.0	—	3.76	2.66	8.86	11.53	3.66	—	8.86	—	—
Hydrogen, H_2	0.0053	187.723	325	275	61,100	51,623	0.5	1.88	2.38	—	1.0	1.88	7.94	26.41	34.34	—	8.94	26.41	4.00	74.20
Oxygen, O_2	0.0846	11.819	—	—	—	—	—	—	—	—	—	—	—	—	—	—	—	—	—	—
Nitrogen (atm), N_2	0.0744	13.443	—	—	—	—	—	—	—	—	—	—	—	—	—	—	—	—	—	—
Carbon monoxide, CO	0.0740	13.506	322	322	4,347	4,347	0.5	1.88	2.38	1.0	—	1.88	0.57	1.90	2.47	1.57	—	1.90	12.50	74.20
Carbon dioxide, CO_2	0.1170	8.548	—	—	—	—	—	—	—	—	—	—	—	—	—	—	—	—	—	—
Paraffin Series																				
Methane, CH_4	0.0424	23.565	1013	913	23,879	21,520	2.0	7.53	9.53	1.0	2.0	7.53	3.99	13.28	17.27	2.74	2.25	13.28	5.00	15.00
Ethane, C_2H_6	0.0803	12.455	1792	1641	22,320	20,432	3.5	13.18	16.68	2.0	3.0	13.18	3.73	12.39	16.12	2.93	1.80	12.39	3.00	12.50
Propane, C_3H_8	0.1196	8.365	2590	2385	21,661	19,944	5.0	18.82	23.82	3.0	4.0	18.82	3.63	12.07	15.70	2.99	1.68	12.07	2.12	9.35
n-Butane, C_4H_{10}	0.1582	6.321	3370	3113	21,308	19,680	6.5	24.47	30.97	4.0	5.0	24.47	3.58	11.91	15.49	3.03	1.55	11.91	1.86	8.41
Isobutane, C_4H_{10}	0.1582	6.321	3363	3105	21,257	19,629	6.5	24.47	30.97	4.0	5.0	24.47	3.58	11.91	15.49	3.03	1.55	11.91	1.80	8.44
n-Pentane, C_5H_{12}	0.1904	5.252	4016	3709	21,091	19,517	8.0	30.11	38.11	5.0	6.0	30.11	3.55	11.81	15.35	3.05	1.50	11.81	—	—
Isopentane, C_5H_{12}	0.1904	5.252	4008	3716	21,052	19,478	8.0	30.11	38.11	5.0	6.0	30.11	3.55	11.81	15.35	3.05	1.50	11.81	—	—
Neopentane, C_5H_{12}	0.1904	5.252	3993	3693	20,970	19,396	8.0	30.11	38.11	5.0	6.0	30.11	3.55	11.81	15.35	3.05	1.50	11.81	—	
n-Hexane, C_6H_{14}	0.2274	4.398	4762	4412	20,940	19,403	9.5	35.76	45.26	6.0	7.0	35.76	3.53	11.74	15.27	3.06	1.46	11.74	1.18	7.40

Olefin Series																				
Ethylene, C_2H_4	0.0746	13.412	1614	1513	21,644	20,295	3.0	11.29	14.29	2.0	2.0	11.29	3.42	11.39	14.81	3.14	1.29	11.39	2.75	28.60
Propylene, C_3H_6	0.1110	9.007	2336	2186	21,041	19,691	4.5	16.94	21.44	3.0	3.0	16.94	3.42	11.39	14.81	3.14	1.29	11.39	2.00	11.10
1-Butene, C_4H_8	0.1480	6.756	3084	2885	20,840	19,496	6.0	22.59	28.59	4.0	4.0	22.59	3.42	11.39	14.81	3.14	1.29	11.39	1.75	9.70
Isobutene, C_4H_8	0.1480	6.756	3068	2869	20,730	19,382	6.0	22.59	28.59	4.0	4.0	22.59	3.42	11.39	14.81	3.14	1.29	11.39	—	—
1-Pentene, C_5H_{10}	0.1852	5.400	3836	3586	20,712	19,363	7.5	28.23	35.73	5.0	5.0	28.23	3.42	11.39	14.81	3.14	1.29	11.39	—	—
Aromatic Series																				
Benzene, C_6H_6	0.2060	4.852	3751	3601	18,210	17,480	7.5	28.23	35.73	6.0	3.0	28.23	3.07	10.22	13.50	3.38	0.69	10.22	1.40	7.10
Toluene, C_7H_8	0.2431	4.113	4484	4284	18,440	17,620	9.0	33.88	42.88	7.0	4.0	33.88	3.13	10.40	13.53	3.34	0.78	10.40	1.27	6.75
Xylene, C_8H_{10}	0.2803	3.567	5230	4980	18,650	17,760	10.5	39.52	50.02	8.0	5.0	39.52	3.17	10.53	13.70	3.32	0.85	10.53	1.00	6.00
Miscellaneous Gases																				
Acetylene, C_2H_2	0.0697	14.344	1499	1448	21,500	20,776	2.5	9.41	11.91	2.0	1.0	9.41	3.07	10.22	13.30	3.38	0.69	10.22	—	—
Napthalene, $C_{10}H_8$	0.3384	2.955	5854	5654	17,298	16,708	12.0	45.17	57.17	10.0	4.0	45.17	3.00	9.97	12.96	3.43	0.56	9.97	—	—
Methyl alcohol, CH_3OH	0.0846	11.820	868	768	10,259	9,078	1.5	5.65	7.15	1.0	2.0	5.65	1.50	4.98	6.48	1.37	1.13	4.98	6.72	36.50
Ethyl alcohol, C_2H_5OH	0.1216	8.221	1600	1451	13,161	11,929	3.0	11.29	14.29	2.0	3.0	11.29	2.08	6.93	9.02	1.92	1.17	6.93	3.28	18.95
Ammonia, NH_3	0.0456	21.914	441	365	9,668	8,001	0.75	2.82	3.57	—	1.5	3.32	1.41	4.69	6.10	—	1.59	5.51	15.50	27.00
Sulfur, S[b]	—	—	—	—	3,983	3,983	1.0	3.76	4.76	SO_2 1.0	—	3.76	1.00	3.29	4.29	SO_2 2.00	—	3.29	—	—
Hydrogen sulfide, H_2S	0.0911	10.979	647	596	7,100	6,545	1.5	5.65	7.15	1.0	1.0	5.65	1.41	4.69	6.10	1.88	0.53	4.69	4.30	45.50
Sulfur dioxide, SO_2	0.1733	5.770	—	—	—	—	—	—	—	—	—	—	—	—	—	—	—	—	—	—
Water vapor, H_2O	0.0476	21.017	—	—	—	—	—	—	—	—	—	—	—	—	—	—	—	—	—	—
Air	0.0766	13.063	—	—	—	—	—	—	—	—	—	—	—	—	—	—	—	—	—	—
Gasoline	—	—	—	—	—	—	—	—	—	—	—	—	—	—	—	—	—	—	1.40	7.60

[a]Adapted from "Fuel Flue Gases." *Combustion Flame and Explosions of Gases*, American Gas Association, New York, NY, 1951.

[b]Carbon and sulfur are considered as gases for molal calculations only.

[c]All gas volumes corrected to 60°F and 30 in. Hg dry.

	C_6H_5Cl +	$7O_2$ +	$26.3N_2$ →	$6CO_2$ +	$2H_2O$ +	HCl +	$26.3N_2$	
Moles	1	7	26.3	6	2	1	26.3	
MW	112.5	32	28	44	18	36.5	28	
Mass	112.5	224	736.4	264	36	36.5	736.4	(5.3.8)

Initial mass = 1072.9 initial number of moles = 34.3

Final mass = 1072.9 final number of moles = 35.3

Note that, in accordance with the conservation law for mass, the initial and final masses balance. The number of moles, as is typically the case in waste combustor calculations, do not balance. The concentrations of the various species may also be calculated. For example,

$$\%CO_2 \text{ by weight} = (264/1072.9)100\% = 24.61\%$$

$$\%CO_2 \text{ by mol (or volume)} = (6/35.3)100\% = 17.0\%$$

$$\%CO_2 \text{ by weight (dry basis)} = (264/1036.9)100\% = 25.46\%$$

$$\%CO_2 \text{ by mol (dry basis)} = (6/33.3)100\% = 18.0\%$$

The air requirement for this reaction is 33.3 lbmol. This is stoichiometric or 0% excess air (EA). For 100% EA (100% above stoichiometric), one would use (33.3) (2.0) or 66.6 lbmol of air. For 50% EA one would use (33.3)(1.5) or 50 lbmol of air; for this condition 16.7 lbmol excess or additional air is employed.

If 100% excess air is employed in the combustion of 1.0 lbmol of C_6H_5Cl, the combustion reaction would become

$$C_6H_5Cl + 14O_2 + 52.6N_2 \rightarrow 6CO_2 + 2H_2O + HCl + 7O_2 + 52.6N_2 \quad (5.3.9)$$

The reader is left with the exercise of verifying these results:

Total final number of moles = 68.6

$$\%O_2 \text{ by mol (or volume)} = (7/68.6)100\% = 10.2\%$$

$$\%HCl \text{ by mol} = (1/68.6)100\% = 1.46\%$$

$$\%H_2O \text{ by mol} = (2/68.6)100\% = 2.92\%$$

Since waste combustors are operated at essentially atmospheric pressure, then

$$\text{Partial pressure } O_2 = 0.102 \text{ atm}$$

$$\text{Partial pressure HCl} = 0.0146 \text{ atm}$$

$$\text{Partial pressure } H_2O = 0.0292 \text{ atm}$$

If the C_6H_5Cl waste contains 0.5% sulfur (S) by mass, then

$$\text{Weight of S} = (0.005)(112.5) = 0.5625 \text{ lb}$$

$$\text{Number of lbmol of S} = 0.5625/32 = 0.0176 \text{ lbmol}$$

$$\text{Number of lbmol of } SO_2 \text{ formed} = \text{number of lbmol of S} = 0.0176 \text{ lbmol}$$

For this condition (approximately),

$$\%SO_2 \text{ by mol} = (0.0176/68.6)100\% = 0.0257\%$$

$$\text{Partial pressure } SO_2 = 2.57 \times 10^{-4} \text{ atm}$$

5.4 THERMOCHEMISTRY

Consider now the energy effects associated with a chemical reaction. To introduce this subject, the reader is reminded that the engineer and applied scientist are rarely concerned with the *magnitude* or *amount* of the energy in a system; the primary concern is with *changes* in the amount of energy. In measuring energy changes for systems, the enthalpy (H) has been found to be the most convenient term to work with. There are many different types of enthalpy effects; these include:

Sensible (temperature)
Latent (phase)
Reaction (chemical)
Dilution (with water), for example, HCl with H_2O
Solution (nonaqueous), for example, HCl with a solvent other than H_2O

The *sensible* enthalpy change was reviewed in Section 5.2. The *latent* enthalpy change finds application in hazardous waste incineration calculations in determining the heat (enthalpy) of condensation or vaporization of water. Steam tables (or the equivalent) are usually employed for this determination. The *dilution* and *solution* enthalpy effects are often significant in some industrial absorber calculations but may safely be neglected in hazardous waste incineration calculations. The *heat of reaction* is defined as the enthalpy change of a system undergoing chemical reaction. If the reactants and products are at the same temperature and in their standard states, the heat of reaction is termed the *standard heat of reaction*. For engineering purposes, the *standard state* of a chemical may be taken as the *pure* chemical at 1 atm pressure. A superscript zero is often employed to identify a standard heat of reaction, for example, ΔH^0. A T subscript (ΔH_T^0) is sometimes used to indicate the temperature; standard heat of reaction data are meaningless unless the temperatures are specified. ΔH_{298}^0 data (i.e., for 298 K or 25°C) for many reactions are available in the literature.[1,2]

As described earlier, the first law of thermodynamics provides that, in a steady-flow process with no mechanical work,

$$Q = \Delta H \tag{5.4.1}$$

Since enthalpy is a point function, it is independent of the path for any process. If heat of reaction determinations are made in a flow reactor, the energy (in the form of heat) transferred across the reactor boundary or surface is exactly equal to the heat of reaction. This is not the case for batch or nonflow systems. For this reason, *heat of reaction* is a misleading term. More recently it has been referred to as the *enthalpy of reaction,* although a more accurate term would be *enthalpy change* of reaction.

The *heat of formation* (ΔH_f^0) is defined as the enthalpy change occurring during a chemical reaction where 1 mol of a product is formed from its elements. The *standard* heat of formation (ΔH_f^0) is applied to formation reactions that occur at constant temperature with each element and the product in its standard state.

Consider the formation reaction for CO_2 at standard conditions at 25°C:

$$\begin{array}{lccc} & C + O_2 & \xrightarrow{\Delta H^0_{f\,298}} & CO_2 \\ \text{Enthalpy:} & 0 \quad\; 0 & & H_{CO_2} \end{array} \tag{5.4.2}$$

Once again, this reaction reads: "1 mol of carbon and 1 mol of oxygen react to form 1 mol of carbon dioxide." The enthalpies of reactants and products are printed below the symbols in the reaction, and the enthalpy change for the reaction, ΔH_f^0, placed above the arrow. Note that the enthalpies of elements in their standard states (pure, 1 atm) at 25°C have arbitrarily been set to zero. The enthalpy change accompanying this reaction is the *standard heat of formation* and is given by

$$\Delta H^0_{f\,298} = H_{CO_2} - (\cancel{H_C^0} + \cancel{H_{O_2}^0}) = H_{CO_2} \tag{5.4.3}$$

The *standard heat of combustion* at temperature T is defined as the enthalpy change during a chemical reaction where 1 mol of material is burned in oxygen, where all reactants and products are in their standard states. This quantity finds extensive application in calculating enthalpy changes for waste combustion reactions and is often given in the literature for 60°F (16°C). Although much of the literature data on standard heats of reaction are given for 25°C (76°F), there is little sensible enthalpy change between these two temperatures, and the two sets of data may be considered compatible.

Note that this formation reaction for CO_2 is a chemical reaction; the heat of formation, in this case, is therefore equal to the heat of combustion. Since a combustion reaction is one type of chemical reaction, the development to follow will concentrate on chemical reactions in general.

Chemical (stoichiometric) equations may be combined by addition or subtraction. The standard heat of reaction (ΔH^0) associated with each equation may likewise be combined to give the standard heat of reaction associated with the resulting chemical equation. This is possible, once again, because enthalpy is a point function, and these changes are independent of path. In particular, formation equations and standard heats of formation may always be combined by addition and subtraction to produce any desired equation and its accompanying standard heat of reaction. This desired equation cannot itself be a formation equation. Thus, the enthalpy change for a chemical reaction is the same whether it takes place in one or several steps. This is referred to as the *law of constant enthalpy summation* and is a direct consequence of the first law of thermodynamics.

Consider the general reaction

$$aA + bB + \cdots \rightarrow cC + dD + \cdots \tag{5.4.4}$$

where A, B = formulas for the reactants (r)
C, D = formulas for the products (P)
a, b, c, d = stoichiometric coefficients of the balanced reaction

To simplify the presentation that follows, Eq. (5.4.4) is shortened to

$$aA + bB \rightarrow cC + dD \tag{5.4.5}$$

(Although this presentation plus those in later sections of this chapter and Chapter 6 will deal with the hypothetical species, A, B, C, etc., application to real systems can be found in the illustrative examples at the end of this and Chapter 6.) This reaction reads: "*a* moles of A react with *b* moles of B to form *c* moles of C and *d* moles of D." The heat of reaction for this chemical change is given by

$$\Delta H^0 = cH_C^0 + dH_D^0 - aH_A^0 - bH_B^0 \tag{5.4.6}$$

where H_C^0 = enthalpy of C in its standard state
C, D, A, B = subscripts indicating chemical species C, D, A, and B

If the temperature is 25°C, the enthalpies of the elements at standard state are, by convention, equal to zero. Therefore,

$$(\Delta H_f^0)_{i\,298} = H_{i\,298}^0 \tag{5.4.7}$$

Substituting Eq. (5.4.7) for each component in Eq. (5.4.6) yields

$$\Delta H_{298}^0 = c(\Delta H_f^0)_C + d(\Delta H_f^0)_D - a(\Delta H_f^0)_A - b(\Delta H_f^0)_B$$

or (5.4.8)

$$\Delta H_{298}^0 = \sum_p n_p(\Delta H_f^0)_p - \sum_p n_r(\Delta H_f^0)_r$$

where p = products
r = reactants
n_p, n_r = coefficients from the chemical equation

The standard heat of a reaction is obtained by taking the difference between the standard heat of formation of the products and that of the reactants. If the standard heat of reaction or formation is negative (*exothermic*), as is the case with most incineration reactions, then energy is liberated as a result of the reaction. Energy is absorbed if ΔH is positive (*endothermic*). Standard heat of formation and standard enthalpy of combustion data at 25°C are provided in Table 5.4.1. Both of these heat (or enthalpy) effects find extensive application in incinerator calculations.

The reader is cautioned on the use of the published heat of combustion data in the literature for chlorinated organics. It appears that the laboratory combustion tests used to generate these data are conducted in bomb calorimeters using excess oxygen. Burning carbon tetrachloride under these conditions yields carbon dioxide and chlorine oxides. Based on the published heat of combustion data, calculations suggest that chlorine monoxide or chlorine dioxide are the likely products, but there may be an equilibrium mixture of several species. The important lesson to be learned is that the published heat of combustion data for halogenated species may be useless since the products of reaction in an incinerator are acid gas (e.g., HCl, Cl_2) not halogen oxides. The use of heat of formation data is therefore recommended; these can be used to determine the heat of reaction for the burning of a particular waste that produces specific products. This avoids the pitfalls of using published data that may not apply to the reaction occurring in the waste combustion.[3]

Other tables of heat of formation, combustion, and reaction are available in the literature (particularly thermodynamics text/reference books) for a wide variety of compounds.[1,2] It is important to note that these are valueless unless the stoichiometric equation, temperature, and the state of the reactants and products are included. However, *heat of reaction* is a term rarely employed in air pollution and/or waste combustion calculations. The two terms most often used in this field are the gross (or higher) heating value and the net (or lower) heating value. The former is designated by HHV or HV_G and the latter by NHV or HV_N. The *gross heating value* represents the enthalpy change or heat re-

Table 5.4.1 Standard heats of formation and combustion at 25°C in calories per gram-mole[a,b]

Compound	Formula	State	ΔH^0_{f298}	$-\Delta H^0_{c298}$
	Normal Paraffins			
Methane	CH_4	g	−17,889	212,800
Ethane	C_2H_6	g	−20,236	372,820
Propane	C_3H_8	g	−24,820	530,600
n-Butane	C_4H_{10}	g	−30,150	687,640
n-Pentane	C_5H_{12}	g	−35,000	845,160
n-Hexane	C_6H_{14}	g	−39,960	1,002,570
Increment per C atom above C_6	—	g	−4,925	157,440
	Normal Monoolefins (1-alkenes)			
Ethylene	C_2H_4	g	12,496	337,230
Propylene	C_3H_6	g	4,879	491,990
1-Butene	C_4H_8	g	−30	649,450
1-Pentene	C_5H_{10}	g	−5,000	806,850
1-Hexene	C_6H_{12}	g	−9,960	964,260
Increment per C atom above C_6	—	g	−4,925	157,440
	Miscellaneous Organic Compounds			
Acetaldehyde	C_2H_4O	g	−39,760	
Acetic acid	$C_2H_4O_2$	l	−116,400	
Acetylene	C_2H_2	g	54,194	310,620
Benzene	C_6H_6	g	19,820	789,080
Benzene	C_6H_6	l	11,720	780,980
1,3-Butadiene	C_4H_6	g	26,330	607,490
Cyclohexane	C_6H_{12}	g	−29,430	944,790
Cyclohexane	C_6H_{12}	l	−37,340	936,880
Ethanol	C_2H_6O	g	−56,240	
Ethanol	C_2H_6O	l	−66,356	
Ethylbenzene	C_8H_{10}	g	7,120	1,101,120
Ethylene glycol	$C_2H_6O_2$	l	108,580	
Ethylene oxide	C_2H_4O	g	−12,190	
Methanol	CH_4O	g	−48,100	
Methanol	CH_4O	l	−57,036	
Methylcyclohexane	C_7H_{14}	g	−36,990	1,099,590
Methylcyclohexane	C_7H_{14}	l	−45,450	1,091,130
Styrene	C_8H_8	g	35,220	1,060,900
Toluene	C_7H_8	g	11,950	943,580
Toluene	C_7H_8	l	2,870	934,500
	Miscellaneous Inorganic Compounds			
Ammonia	NH_3	g	−11,040	
Calcium carbide	CaC_2	s	−15,000	

(*continued*)

Table 5.4.1 (*Continued*)

Compound	Formula	State	ΔH^0_{f298}	$-\Delta H^0_{c298}$
Calcium carbonate	$CaCO_3$	s	−288,450	
Calcium chloride	$CaCl_2$	s	−190,000	
Calcium chloride	$CaCl_2 \cdot 6H_2O$	s	−623,150	
Calcium hydroxide	$Ca(OH)_2$	s	−235,800	94,052
Calcium oxide	CaO	s	−151,900	
Carbon	C	Graphite	—	67,636
Carbon dioxide	CO_2	g	−94,052	
Carbon monoxide	CO	g	−26,416	68,317
Hydrochloric acid	HCl	g	−22,063	
Hydrogen	H_2	g	—	
Hydrogen sulfide	H_2S	g	−4,815	
Iron oxide	FeO	s	−64,300	
Iron oxide	Fe_3O_4	s	−267,000	
Iron oxide	Fe_2O_3	s	−196,500	
Iron sulfide	FeS_2	s	−42,520	
Lithium chloride	$LiCl$	s	−97,700	
Lithium chloride	$LiCl \cdot H_2O$	s	−170,310	
Lithium chloride	$LiCl \cdot 2H_2O$	s	−242,100	
Lithium chloride	$LiCl \cdot 3H_2O$	s	−313,500	
Nitric acid	$HN\dot{O}_3$	l	−41,404	
Nitrogen oxides	NO	g	21,600	
	NO_2	g	8,041	
	N_2O	g	19,490	
	N_2O_4	g	2,309	
Sodium carbonate	Na_2Co_3	s	−270,300	
Sodium carbonate	$Na_2CO_3 \cdot 10H_2O$	s	−975,600	
Sodium chloride	$NaCl$	s	−98,232	
Sodium hydroxide	$NaOH$	s	101,990	
Sulfur dioxide	SO_2	g	−70,960	
Sulfur trioxide	SO_3	g	−94,450	
Sulfur trioxide	SO_3	l	−104,800	
Sulfuric acid	H_2SO_4	l	−193,910	
Water	H_2O	g	−57,798	
Water	H_2O	l	−68,317	
	Chlorinated Organic Compounds			
Methyl chloride	CH_3Cl	l	−20,630	
Dichloromethane	CH_2Cl_2	l	−22,800	
Chloroform	$CHCl_3$	l	−24,200	36,900
Carbon tetrachloride	CCl_4	l	−24,000	
Ethyl chloride	C_2H_5Cl	l	−26,700	
1,1-Dichloroethane	$C_2H_4Cl_2$	l	−31,050	
1,1,2,2-Tetrachloroethane	$C_2H_2Cl_4$	l	−36,500	
n-Propyl chloride	C_3H_7Cl	l	−31,100	
1,3-Dichloropropane	$C_3H_6Cl_2$	l	−38,600	

Table 5.4.1 (*Continued*)

Compound	Formula	State	ΔH^0_{f298}	$-\Delta H^0_{c298}$
n-Butyl chloride	C_4H_9Cl	1	−35,200	
1-Chloropentane	$C_5H_{11}Cl$	1	−41,800	
1-Chloroethylene	C_2H_3Cl	1	−8,400	
trans-1,2-Dichloroethylene	$C_2H_6Cl_2$	1	−1,000	
Trichloroethylene	C_2HCl_3	1	−1,400	510,000
Tetrachloroethylene	C_2Cl_4	1	−3,400	782,000
3-Chloro-1-propene	C_3H_5Cl	1	−150	1,600,000
Chlorobenzene	C_6H_5Cl	1	12,390	
p-Dichlorobenzene	$C_6H_5Cl_2$	1	5,500	
Hexachlorobenzene	C_6Cl_6	1	−8,100	
Benzylchloride	C_7H_7Cl	1		
1,1,1-trichloro-2,2-bis (*p*-chlorophenyl)ethane (DDT)	$C_{14}H_9Cl_5$	1		

[a]Selected mainly from F. D. Rossini, ed., "Selected Values of Physical and Thermodynamic Properties of Hydrocarbons and Related Compounds," *American Petroleum Institute Research Project 44*, Carnegie Institute of Technology, Pittsburgh, PA, 1953; F. D. Rossini, D. D. Wagman, W. H. Evans, S. Levine, and I. Jaffe, "Selected Values of Chemical Thermodynamic Properties," *Natl. Bur. Stand. Circ. 500*, 1952; also from personal notes, L. Theodore and J. Reynolds, 1985.
[b]For combustion reactions the products are H_2O (l) and CO_2 (g).

leased when a compound is stoichiometrically reacted (combusted) at a reference temperature with the final (flue) products also at the reference temperature and any water present in the liquid state. Most of these data are available at a reference temperature of 60°F. The *net heating value* is similar to the gross heating value except the water is in the vapor state. The difference (if any) between the two values represents the energy necessary to vaporize any water present. Thus, the standard heat of reaction and the gross and/or net heating values employed in the waste combustion industry both represent the same phenomenon. Gross and net heating values for a number of hydrocarbons are presented in Table 5.3.1. In addition, the net heating value may be approximated by a form of Dulong's equation that includes the chlorine content:

$$\text{NHV} = 14{,}000m_C + 45{,}000(m_H - \tfrac{1}{8}m_O) - 760m_{Cl} + 4500m_S \qquad (5.4.9)$$

where NHV = net heating value of waste–fuel mixture (Btu/lb)
m_C = mass fraction of C (carbon) in the waste–fuel mixture
C, H, O, Cl, S = subscripts indicating carbon, hydrogen, oxygen, chlorine, and sulfur, respectively

Another common term employed in waste combustion calculations is the *available heat*, usually designated as HA_T. The available heat at any temperature T is the gross heating value minus the amount of heat ($\Sigma\Delta H$) required to take the product(s) of combustion (flue gas) from the reference temperature to that temperature T. Thus

$$\mathrm{HA}_T = \mathrm{HHV} - \Sigma\Delta H$$

If all the heat liberated by the reaction goes into heating up the products of combustion (the flue gas), the temperature achieved is defined as the *flame temperature*. If the combustion process is conducted adiabatically, with no heat transfer to the surroundings, the final temperature achieved by the flue gas is defined as the *adiabatic flame temperature*. If the combustion process is conducted with theoretical or stoichiometric air (0% excess), the resulting temperature is defined as the *theoretical adiabatic flame temperature*. It is rare that a waste combustor operates at the adiabatic flame temperature. This is never the case in a typical combustion process due to losses through the burner, refractories and insulation. The actual flame temperature is usually a few hundred degrees below the calculated or theoretical adiabatic furnace temperature. Again, most combustion processes whether batch or continuous are conducted with air in excess of that required for theoretical combustion. It is rare in actual applications that the "theoretical adiabatic flame temperature" is ever achieved.

Heat losses (conduction and convection) through the enclosure (furnace walls) due to construction materials and external shell conditions (air cooled, water cooled, or natural radiation and convection to surrounding atmosphere) as well as heat losses by radiation to downstream equipment (boiler tubes, spray quench chamber, etc.) will affect the final temperature conditions of the reaction (combustion) products.

To calculate fuel requirements, operating temperature, and (excess) air requirements for a waste combustor operation, one must apply the conservation laws for mass and energy in conjunction with thermochemical principles. This is an extremely involved, rigorous calculation. An enthalpy balance is applied around the waste combustor following comprehensive overall and componential material balances. The enthalpy balance must account for all temperature changes of both the feed and fuel (reactants) as well as the flue gas (products). Latent (phase) and combustion (reaction) enthalpy effects must also be included in the analysis. Fortunately, algorithms are available to perform these detailed thermodynamic calculations; these can be found in Chapter 6.

Effect of Temperature on Heats of Reaction

The heat of reaction is a function of temperature because the heat capacities of both the reactants and products vary with temperature. Smith and Van Ness[2] have described this effect mathematically in the following manner:

$$\Delta H_T^0 = \Delta H_{298}^0 + \int_{298}^{T} \Delta C_P \, dT \tag{5.4.10}$$

where T is the absolute temperature (K), and

$$\Delta C_P = \sum_{\text{products}} nC_P - \sum_{\text{reactants}} nC_P$$

If the heat capacity for each product and reactant is expressed by

$$C_P = \alpha + \beta T + \gamma T^2$$

then

$$\Delta C_P = \Delta\alpha + (\Delta\beta)T + (\Delta\gamma)T^2 \tag{5.4.11}$$

where

$$\Delta\alpha = \sum_{\text{products}} n\alpha - \sum_{\text{reactants}} n\alpha$$

with similar definitions for $\Delta\beta$ and $\Delta\gamma$. Combining Eqs. (5.4.10) and (5.4.11), the following expression for the standard heat of reaction at temperature T is obtained:

$$\Delta H_T^0 = \Delta H_{298}^0 + \int_{298}^{T} [\Delta\alpha + (\Delta\beta)T + (\Delta\gamma)T^2] \, dT$$

or

$$\Delta H_T^0 = \Delta H_{298}^0 + \Delta\alpha(T - 298) + \tfrac{1}{2}\Delta\beta(T^2 - 298^2) + \tfrac{1}{3}\Delta\gamma(T^3 - 298^3) \tag{5.4.12}$$

If all the constant terms in this equation are collected and lumped together into a constant designated ΔH_0, the result is

$$\Delta H_T^0 = \Delta H_0 + \Delta\alpha T + \tfrac{1}{2}(\Delta\beta)T^2 + \tfrac{1}{3}(\Delta\gamma)T^3 \tag{5.4.13}$$

where ΔH_T^0 = standard heat of reaction at temperature T
ΔH_0 = constant = $\Delta H_{298}^0 - 298\Delta\alpha - \tfrac{1}{2}(298)^2\,\Delta\beta - \tfrac{1}{3}(298)^3\,\Delta\gamma$

Note: the use of the equations in this subsection requires that the temperature T be expressed in kelvins.

If the standard heat of reaction is known at a single temperature, for example, 25°C, ΔH_T^0 can be calculated. The constant ΔH_0 may then be calculated directly from Eq. (5.4.13).

The reader should attempt to rederive the equivalent of Eqs. (5.4.12) and (5.4.13) if heat capacity variation with temperature is given by

$$C_P = a + bT + cT^{-2}$$

One application of this calculation is that associated with the adiabatic flame temperature (to be discussed in Chapter 6). It represents the maximum temperature the products of combustion (flue) can achieve if the reaction is conducted adiabatically. For this condition, all the energy liberated on combustion at or near standard conditions (ΔH_c^0) appears as sensible heat in heating up the flue products, ΔH_p, that is,

$$\Delta H_c^0 + \Delta H_p = 0 \tag{5.4.14}$$

where ΔH_c^0 = standard heat of combustion at 25°C
ΔH_p = enthalpy change of the products as the temperature increases from 25°C to the theoretical flame temperature

The right term of Eq. (5.4.10), in conjunction with Eq. (5.4.11), should be applied to the product flue gases in evaluating ΔH_p. Details of this calculation are presented in the illustrative examples provided in Section 5.7.

5.5 CHEMICAL REACTION EQUILIBRIUM

With regard to chemical reactions, there are two important questions that are of concern to the engineer: (1) how *far* will the reaction go and (2) how *fast* will the reaction go? Chemical thermodynamics provides the answer to the first question; however, it tells nothing about the second. Reactor rates fall within the domain of chemical kinetics and will be treated in Section 5.6. To illustrate the difference and importance of both questions on an engineering analysis of a chemical reaction, consider the following process. Substance A, which costs 1 cent/ton, can be converted to B, which costs \$1 million/lb, by the reaction A $\leftrightarrow$ B. Chemical thermodynamics will provide information on the maximum amount of B that can be formed. If 99.99% of A can be converted to B, the reaction would then appear to be economically feasible, from a *thermodynamic* point of view. However, a *kinetic* analysis might indicate that the reaction is so slow that, for all practical purposes, its rate is vanishingly small. For example, it might take 10^6 years to obtain a 10^{-6}% conversion of A. The reaction is then economically unfeasible. Thus, it can be seen that both equilibrium and kinetic effects must be considered in an overall engineering analysis of a chemical reaction. The same principle applies to control of gaseous pollutants by combustion and incineration of hazardous wastes.

A rigorous, detailed presentation of this topic is beyond the scope of this text. However, a superficial treatment is presented in the hope that it may provide at least a qualitative introduction to chemical reaction equilibrium. This topic does find application in waste combustor calculations involving SO_3 and Cl_2 discharges and will hopefully explain, in part, the role of reaction equilibria in these calculations.

Chemical reaction equilibrium calculations are structured around another thermodynamic term referred to as *free energy.* This so-called *free energy* (G) is a thermodynamic property that cannot be easily defined without some basic grounding in thermodynamics. No attempt will be made to define it here, and the interested reader is again directed to the literature for further development of this topic. In the opinion of the authors, the fact that standard free energy data are available in the literature and can be used to calculate chemical equilibrium constants warrants the inclusion of this property in the discussion to follow. Free energy has the same units as enthalpy and internal energy and may be used on a mole or total mass basis. The uppercase G signifies a mole basis.

Consider the equilibrium reaction

$$aA + bB = cC + dD \tag{5.5.1}$$

where A, B, C, D = chemical formulas of the reactant and product species
a, b, c, d = stoichiometric coefficients

and the equal sign is a reminder that the reacting system is at equilibrium. For this reaction (as with enthalpy),

$$\Delta G^0 = cG^0_C + dG^0_D - aG^0_A - bG^0_B \tag{5.5.2}$$

where ΔG^0 represents the free energy change for this reaction when reactants and products are in their standard states. Note that ΔG^0 may be calculated and obtained in a manner similar to ΔH^0. Note further that, at 25°C, the free energy of an element in its standard state is arbitrarily set at zero. Therefore,

$$(\Delta G^0_f)_{i\,298} = G^0_{i\,298}$$

so that

$$\Delta G^0_{298} = c(\Delta G^0_f)_C + d(\Delta G^0_f)_D - a(\Delta G^0_f)_A - b(\Delta G^0_f)_B \tag{5.5.3}$$

According to Eq. (5.5.3), at 25°C, the standard free energy of reaction ΔG^0 may be calculated from standard free energy of formation data. Some of this information is presented in Table 5.5.1. The following equation is used to calculate the chemical reaction equilibrium constant K at a temperature T:

$$(\Delta G_T^0) = -RT \ln K \tag{5.5.4}$$

The value of this equilibrium constant depends on the temperature at which the equilibrium is established. The effect of temperature on K must now be examined.

Effect of Temperature on the Equilibrium Constant

The dependence of ΔG^0 on temperature is given by[2]

$$\frac{d(\Delta G^0/RT)}{dT} = \frac{\Delta H^0}{RT^2} \tag{5.5.5}$$

According to Eq. (5.5.4)

$$\frac{\Delta G^0}{RT} = \ln K$$

Substitution yields

$$\frac{d \ln K}{dT} = \frac{\Delta H^0}{RT^2} \tag{5.5.6}$$

Equation (5.5.6) shows the effect of temperature on the equilibrium constant, and hence on the equilibrium yield. If the reaction is exothermic, ΔH^0 is negative and the equilibrium constant decreases with an increase in temperature; for an endothermic reaction, the equilibrium constant increases with an increase in temperature.

If the term ΔH^0, which is the standard enthalpy change (heat of reaction), is assumed to be constant with temperature, Eq. (5.5.6) can be integrated between the temperatures T and T_1 to give

$$\ln \frac{K}{K_1} = -\frac{\Delta H^0}{R}\left(\frac{1}{T} - \frac{1}{T_1}\right) \tag{5.5.7}$$

This approximate equation may be used to determine the equilibrium constant at a temperature T from the known value at some other temperature T_1 if the difference between the two temperatures is small. If the standard heat of reaction is known as a function of temperature, however, Eq. (5.5.6) can be integrated rigorously:

Table 5.5.1 Standard free energy of formation at 25°C in calories per gram-mole[a]

Compound	Formula	State	ΔG^0_{f298}
	Normal Paraffins		
Methane	CH_4	g	−12,140
Ethane	C_2H_6	g	−7,860
Propane	C_3H_8	g	−5,614
n-Butane	C_4H_{10}	g	−4,100
n-Pentane	C_5H_{12}	g	−2,000
n-Hexane	C_6H_{14}	g	−70
*n*Heptane	C_7H_{16}	g	1,920
n-Octane	C_8H_{18}	g	3,920
Increment per C atom above C_8		g	2,010
	Normal Monoolefins (1-alkenes)		
Ethylene	C_2H_4	g	16,282
Propylene	C_3H_6	g	14,990
1-Butene	C_4H_8	g	17,090
1-Pentene	C_5H_{10}	g	18,960
1-Hexene	C_6H_{12}	g	20,940
Increment per C atom above C_6		g	2,010
	Miscellaneous Organic Compounds		
Acetaldehyde	C_2H_4O	g	−31,960
Acetic acid	$C_2H_4O_2$	l	−93,800
Acetylene	C_2H_2	g	50,000
Benzene	C_6H_6	g	30,989
Benzene	C_6H_6	l	29,756
1,3-Butadiene	C_4H_6	g	36,010
Cyclohexane	C_6H_{12}	g	7,590
Cyclohexane	C_6H_{12}	l	6,370
Ethanol	C_2H_6O	g	−40,130
Ethanol	C_2H_6O	l	−41,650
Ethylbenzene	C_8H_{10}	g	31,208
Ethylene glycol	$C_2H_6O_2$	l	−77,120
Ethylene oxide	C_2H_4O	g	−2,790
Methanol	CH_4O	g	−38,810
Methanol	CH_4O	l	−39,850
Methylcyclohexane	C_6H_{14}	g	6,520
Methylcyclohexane	C_6H_{14}	l	4,860
Styrene	C_8H_8	g	51,100
Toluene	C_7H_8	g	29,228
Toluene	C_7H_8	l	27,282
	Miscellaneous Inorganic Compounds		
Ammonia	NH_3	g	−3,976
Ammonia	NH_3	aq	−6,370

(continued)

Table 5.5.1 (*Continued*)

Compound	Formula	State	ΔG^0_{f298}
Calcium carbide	CaC_2	s	−16,200
Calcium carbonate	$CaCO_3$	s	−269,780
Calcium chloride	$CaCl_2$	s	−179,300
Calcium chloride	$CaCl_2$	aq	−194,880
Calcium hydroxide	$Ca(OH)_2$	s	−214,330
Calcium hydroxide	$Ca(OH)_2$	aq	−207,370
Calcium oxide	CaO	s	−144,400
Carbon dioxide	CO_2	g	−94,258
Carbon monoxide	CO	g	−32,781
Hydrochloric acid	HCl	g	−22,778
Hydrogen sulfide	H_2S	g	−7,892
Iron oxide	Fe_3O_4	s	−242,400
Iron oxide	Fe_2O_3	s	−177,100
Iron sulfide	FeS_2	s	−39,840
Nitric acid	HNO_3	l	−19,100
Nitric acid	HNO_3	aq	−26,410
Nigrogen oxides	NO	g	20,690
	NO_2	g	12,265
	N_2O	g	24,933
	N_2O_4	g	23,395
Sodium carbonate	Na_2CO_3	s	−250,400
Sodium chloride	$NaCl$	s	−91,785
Sodium chloride	$NaCl$	aq	−93,939
Sodium hydroxide	$NaOH$	s	−90,600
Sodium hydroxide	$NaOH$	aq	−100,184
Sulfur dioxide	SO_2	g	−71,790
Sulfur trioxide	SO_3	g	−88,520
Sulfuric acid	H_2SO_4	aq	−177,340
Water	H_2O	g	−54,635
Water	H_2O	l	−56,690
	Chlorinated Organics		
Methyl chloride	CH_3Cl	l	−15,030
Dichloromethane	CH_2Cl_2	l	−16,460
Chloroform	$CHCl_3$	l	−16,380
Carbon tetrachloride	CCl_4	l	−13,920
Ethyl chloride	C_2H_5Cl	l	−14,340
1,1-Dichloroethane	$C_2H_4Cl_2$	l	−17,470
1,1,2,2-Tetrachloroethane	$C_2H_2Cl_4$	l	−20,480
n-Propyl chloride	C_3H_7Cl	l	−12,110
1,3-Dichloropropane	$C_3H_6Cl_2$	l	−19,740
n-Butyl chloride	C_4H_9Cl	l	−9,270
1-Chloropentane	$C_5H_{11}Cl$	l	−8,940
1-Chloroethylene	C_2H_3Cl	l	−12,310
trans-1,2-Dichloroethylene	$C_2H_6Cl_2$	l	−6,350

Table 5.5.1 ***(Continued)***

Compound	Formula	State	ΔG^0_{f298}
Trichloroethylene	C_2HCl_3	l	−4,750
Tetrachloroethylene	C_2Cl_4	l	−4,900
3-Chloro-1-propene	C_3H_5Cl	l	10,420
Chlorobenzene	C_6H_5Cl	l	23,700
p-Dichlorobenzene	$C_6H_5Cl_2$	l	18,440
Hexachlorobenzene	C_6Cl_6	l	10,560

[a]Selected mainly from F. D. Rossini, ed., "Selected Values of Physical and Thermodynamic Properties of Hydrocarbons and Related Compounds," *American Petroleum Institute Research Project 44*, Carnegie Institute of Technology, Pittsburgh, PA, 1953; F. D. Rossini, D. D. Wagman, W. H. Evans, S. Levine, and I. Jaffee, "Selected Values of Chemical Thermodynamic Properties," *Natl. Bur. Stand. Circ.* **500** (1952); and also from personal notes, L. Theodore and J. Reynolds, 1986.

$$\ln K = \frac{1}{R}\int \frac{\Delta H^0}{T^2}\, dT + I \tag{5.5.8}$$

where I is a constant of integration.

In the previous section, it was shown that, if the molar heat capacity for each chemical species taking part in the reaction is known and can be expressed as a power series in T (kelvin),

$$C_P = \alpha + \beta T + \gamma T^2$$

then ΔH^0 at a given temperature T becomes

$$\Delta H^0_T = \Delta H_0 + \Delta\alpha T + \frac{\Delta\beta T^2}{2} + \frac{\Delta\gamma T^3}{3} \tag{5.4.13}$$

In this equation, the constant ΔH_0 can be calculated from the standard heat of reaction at 25°C [see Eq. (5.4.13)]. With ΔH_0 determined ΔH^0_T can be substituted into Eq. (5.5.8) and integrated to obtain

$$\ln K = -\frac{\Delta H_0}{RT} + \frac{\Delta\alpha}{R}\ln T + \frac{\Delta\beta}{2R} + \frac{\Delta\gamma}{6R}T^2 + I \tag{5.5.9}$$

Here the constant I may be evaluated from a knowledge of the equilibrium constant at one temperature. This is usually obtained from standard free energy of formation data at 25°C. Equation (5.5.4) is employed to obtain K at this temperature. A similar equation for ΔG^0_T may be obtained by combining Eqs. (5.5.4) and (5.5.9):

$$\Delta G_T^0 = \Delta H_0 - (\Delta\alpha)T \ln T - \left(\frac{\Delta\beta}{2}\right) T^2 - \left(\frac{\Delta\gamma}{6}\right) T^3 - IRT \quad (5.5.10)$$

The reader is left the exercise of developing companion equations to Eqs. (5.5.9) and (5.5.10) if the heat capacity variation with temperature is given by:

$$C_P = a + bT + cT^{-2}$$

The problem that remains is to relate K to understandable physical quantities. For gas-phase reactions, as waste combustor operation, the term K for Eq. (5.5.4) may be approximately represented in terms of the partial pressures of the components involved. This functional relationship is given in Eqs. (5.5.11) and (5.5.12):

$$K = K_P \quad (5.5.11)$$

where K is an equilibrium constant based on partial pressures;

$$K_P = \frac{P_C^c P_D^d}{P_A^a P_B^b} \quad (5.5.12)$$

where P_A = partial pressure of component A, etc.

This definition of K_P obviously applies to the reaction of Eq. (5.5.1). Assuming a K value is available or calculable, this equation may be used to determine the partial pressures of the participating components at equilibrium. It is important to note that the component partial pressures (P_i) are *equilibrium* values. For product gases, P_i usually represents the *maximum* values that can ultimately be achieved; for reactant gases, P_i represents *minimum* values. For liquid-phase reactions, K is approximately given by

$$K = K_C \quad (5.5.13)$$

where $K_C = C_C^c C_D^d / C_A^a C_B^b$

C_C = concentration of component C (gmol/L)

C, D, A, B = subscripts indicating chemical species, C, D, A, and B

5.6 CHEMICAL KINETICS

Chemical kinetics involves the study of reaction rates and the variables that affect these rates. It is a topic that is critical for the analysis of incinerator/combustion systems. The objective of this section is to develop a working understanding of this subject in order to apply chemical kinetics principles to waste combustion applications. The topic is treated from an engineering point of view, that is, in terms of physically measurable quantities. The rate of a chemical reaction can be described in any of several different ways. The most commonly used definition involves the time rate of change in the amount of one of the components participating in the reaction; this rate is usually based on some arbitrary factor related to the reacting system size or geometry, such as volume, mass, and interfacial area. The definition shown in Eq. (5.6.1), which applies to homogeneous reactions (all reactants and products are in the same phase), is a convenient one from an engineering point of view and is adopted for use in this text.

$$R_A = \frac{1}{V}\left(\frac{dn_A}{d\theta}\right) \tag{5.6.1}$$

where R_A = reaction rate based on component A
V = volume of reacting system
n_A = number of moles of A at time θ
θ = time

If the volume term is constant, one may write Eq. (5.6.1) as

$$\begin{aligned} R_A &= \frac{d(n_A/V)}{d\theta} \\ &= \frac{dC_A}{d\theta} \end{aligned} \tag{5.6.2}$$

where C_A = molar concentration of A at time θ.

Equation (5.6.2) states that the reaction rate is equal to the rate of change in concentration of one of the components with respect to time. When rates are expressed in terms of concentration changes, the assumption of constant volume is implied. The units of the rate become lbmol/h-ft^3 and gmol/s-L in the engineering and metric systems, respectively. When the reaction is nonhomogeneous, the geometric factor is usually some parameter other than the reaction volume. For fluid–fluid surface reactions, for example, the geometric factor is usually the interfacial area, in which case the units of the reaction rate are lbmol/h-ft^2. For fluid–solid reaction systems, the factor is often the mass of the solid; for example, in gas-phase catalytic combustion reactions, the factor is the mass of the catalyst and the units of the rate are pound moles per hour pound (lbmol/h-lb).

Based on Eq. (5.6.2), the reaction rate is positive if species A is being formed since C_A increases with time, and negative if A is reacting since C_A decreases with time. The rate is zero if the system is at chemical equilibrium (see Section 5.5).

An equation expressing the rate in terms of measurable and/or desirable quantities may now be developed. Based on experimental evidence, the rate of reaction is a function of the concentration of the components present in the reaction mixture (this includes reacting and inert species), temperature, pressure, catalyst variables, or, in equation form,

$$R_A = R_A(C_i, P, T, \text{catalyst variables}) \tag{5.6.3}$$

where C_i = molar concentration of each component present in the reaction mixture. Equation (5.6.3) may be condensed into

$$R_A = \pm k_A f(C_i) \tag{5.6.4}$$

where k_A incorporates all the variables in Eq. (5.6.3) other than the concentration variable. The $\pm$ notation is included to indicate whether component A is being consumed ($-$) or produced ($+$). The k_A may be regarded as a constant of proportionality; it is termed the *specific reaction rate* or, more commonly, the *reaction velocity constant*. Although this "constant" is independent of concentration, it is a function of the other variables. This approach has, in a sense, isolated one of the variables. The reaction velocity constant, like the rate of reaction, must be based on one of the components of the reacting system, almost always the same component on which the reaction rate is based. This term is very definitely influenced by temperature and catalyst activity. For the present, however, it is assumed that k is solely a function of temperature. Thus,

$$R_A = \pm k_A(T) f(C_i) \tag{5.6.5}$$

The effect of temperature on k is generally represented by the Arrhenius equation:

$$k = Ae^{-E/RT} \tag{5.6.6}$$

where A = frequency factor, same units as k
E = activation energy, same units as RT
R = universal gas constant
T = absolute temperature

The values of A and E are usually obtained experimentally.

The functional relationship for the molar concentration, $f(C_i)$, can be obtained from the stoichiometric reaction equation, which assumes elementary or power law kinetics to apply. Consider again the reaction

$$aA + bB \rightarrow cC + dD \tag{5.6.7}$$

where the single arrow represents an *irreversible reaction*, that is, if stoichiometric amounts of A and B are initially present, the reaction will proceed to the products until all the A and B have reacted. The rate of this reaction is given by

$$R_A = -k_A C_A^a C_B^b \tag{5.6.8}$$

where the negative sign is introduced to account for the fact that A is being consumed. The product concentrations do not affect the rate. The term $f(C_i)$ is then the product of the molar concentrations of the reactants, each raised to the power of its stoichiometric coefficient.

The *order* of this reaction with *respect to a particular species* is given by the exponent of that concentration term appearing in the rate expression. The reaction of Eq. (5.6.8) is, therefore, of order a with respect to A, and of order b with respect to B. The overall order n, usually referred to as the *order of the reaction*, is the sum of the individual orders, that is,

$$n = a + b$$

This analysis has assumed elementary law kinetics to apply. This is frequently an idealization of more complex theory and should therefore be looked upon as a general and first introduction to the rate laws.

All real and naturally occurring reactions are reversible. A *reversible* reaction is one in which the products also react to form the reactants. Unlike irreversible reactions that proceed to the right until completion, reversible reactions achieve an equilibrium state after an infinite period of time. Reactants and products are still present in the system. At this equilibrium state the net reaction rate is zero. Consider the following reversible reaction:

$$a\text{A} + b\text{B} \leftrightarrow c\text{C} + d\text{D} \tag{5.6.9}$$

where the double-headed arrow is a reminder that the reaction is reversible; the notation = is employed if the reaction system is at equilibrium. The rate of this reaction is given by

$$R_A = \underbrace{k_A C_A^a C_B^b}_{\text{forward reaction}} = \underbrace{k'_A C_C^c C_D^d}_{\text{reverse reaction}} \tag{5.6.10}$$

where k_A = reaction velocity constant for the forward reaction
k'_A = reaction velocity constant for the reverse reaction

The orders of the forward and reverse reactions are $(a + b)$ and $(c + d)$, respectively. For equilibrium systems, the forward reaction rate is equal to the reverse rate and Eq. (5.6.10) becomes

$$R_A = 0 \tag{5.6.11}$$

so that

$$k_A C_A^a C_B^b = k'_A C_C^c C_D^d \tag{5.6.12}$$

If an equilibrium constant K_C is defined by the ratio of the forward to the reverse reaction velocity constant

$$K_C = \frac{k_A}{k'_A} \tag{5.6.13}$$

then

$$K_C = \frac{C_C^c C_D^d}{C_A^a C_B^b} \tag{5.6.14}$$

This is the same as Eq. (5.5.13). Thus, if k_A is given and K_C is known, then k'_A can be calculated. Note that K_C is approximately the true equilibrium constant K referred to in Section 5.5.

Gas-phase reactions often have rates expressed in a different form. For the nth-order irreversible reaction

$$a\text{A} + b\text{B} \rightarrow \text{products}$$

where $n = a + b$, it was previously stated that

$$R_A = \frac{dC_A}{d\theta} = -k_A C_A^a C_B^b \tag{5.6.15}$$

This equation may also be written in terms of the partial pressures of A and B. Assuming ideal gas conditions permits the following substitution:

$$C_i = \frac{P_i}{RT} \tag{5.6.16}$$

where P_i = partial pressure of component i
R = universal gas constant
T = absolute temperature

Substituting Eq. (5.6.16) for each reactant into the right-hand side of Eq. (5.6.15) gives

$$R_A = \frac{dC_A}{d\theta} = -k_{PA}P_A^aP_B^b \tag{5.6.17}$$

where k_{PA} = reaction velocity constant for A based on partial pressure. Further substitution of Eq. (5.6.16) into Eq. (5.6.17) yields

$$R_A^* = \frac{dP_A}{d\theta} = -k_{PA}^*P_A^aP_B^b \tag{5.6.18}$$

where $R_A^* = R_A(RT)$

k_{PA}^* = reaction velocity constant of A based on R_A^*

The three reaction velocity constants defined by Eqs. (5.6.15), (5.6.17), and (5.6.18) are related as follows:

$$\begin{aligned} k_{Pi} &= \frac{k_i}{(RT)^n} \\ k_{Pi}^* &= \frac{k_i}{(RT)^{n-1}} \\ k_{Pi} &= \frac{k_{Pi}^*}{RT} \end{aligned} \tag{5.6.19}$$

These results are now applied to different kinetic systems. Some of the more common and elementary chemical reactions are presented here for constant volume systems.

1. Zero-Order, Irreversible Reaction

$$A \rightarrow \text{products}$$

$$R_A = dC_A/d\theta = -k_A$$

This is an idealized reaction system. It has been shown that all real *zero-order* reactions can be represented by this rate equation only at high concentrations.

2. First-Order, Irreversible Reaction

$$A \rightarrow \text{products}$$

$$R_A = dC_A/d\theta = -k_AC_A$$

3. Second-Order, Irreversible Reaction

(a)

$$A + B \rightarrow \text{products}$$

$$R_A = dC_A/d\theta = -k_AC_AC_B$$

(b)

$$2A \rightarrow \text{products}$$

$$R_A = dC_A/d\theta = -k_A C_A^2$$

4. Third-Order, Irreversible Reaction

(a)

$$3_A \rightarrow \text{products}$$

$$R_A = dC_A/d\theta = -k_A C_A^3$$

(b)

$$2A + B \rightarrow \text{products}$$

$$R_A = dC_A/d\theta = -k_A C_A^2 C_B$$

(c)

$$A + B + C \rightarrow \text{products}$$

$$R_A = dC_A/d\theta = -k_A C_A C_B C_C$$

5. Fractional- or Higher-Order, Irreversible Reaction

$$nA \rightarrow \text{products}$$

$$R_A = dC_A/d\theta = -k_A C_A^n$$

For many complex reactions, the order is often a large integer or a fraction.

6. First-Order, Reversible Reaction

$$A \leftrightarrow B$$

$$R_A = dC_A/d\theta = -k_A C_A + k'_A C_B$$

$$= -k_A[C_A - (C_B/K_A)]$$

where $K_A = k_A/k'_A$.

7. Second-order, Reversible Reaction

(a)

$$A \leftrightarrow 2B$$

$$R_A = dC_A/d\theta = -k_A C_A + k'_A C_B^2$$

$$= -k_A[C_A - (C_B^2/K_A)]$$

(b)

$$A \leftrightarrow B + C$$

$$R_A = dC_A/d\theta = -k_A C_A + k'_A C_B C_C$$

$$= -k_A[C_A - (C_B C_C/K_A)]$$

(c)

$$A + B \leftrightarrow C$$

$$R_A = dC_A/d\theta = -k_A C_A C_B + k_A' C_C$$

$$= -k_A[C_A C_B - (C_C/K_A)]$$

(d)

$$2A \leftrightarrow B$$

$$R_A = dC_A/d\theta = -k_A C_A^2 + k_A' C_B$$

$$= -k_A[C_A^2 - (C_B/K_A)]$$

(e)

$$A + B \leftrightarrow C + D$$

$$R_A = dC_A/d\theta = -k_A C_A C_B + k_A' C_C C_D$$

$$= -k_A[C_A C_B - (C_C C_D/K_A)]$$

8. Similar equations can be written for third- and higher-order reversible reactions.
9. Simultaneous Irreversible Reactions

$$A \xrightarrow{k_{A1}} \text{products}$$

$$A + B \xrightarrow{k_{A2}} \text{products}$$

$$3A \xrightarrow{k_{A3}} \text{products}$$

$$R_A = dC_A/d\theta = -k_{A_1} C_A - k_{A_2} C_A C_B - k_{A_3} C_A^3$$

$$R_B = dC_B/d\theta = -k_{A_2} C_A C_B$$

10. Consecutive Irreversible Reactions (two first-order reactions)

$$A \xrightarrow{k_A} B$$

$$B \xrightarrow{k_B} C$$

$$R_A = dC_A/d\theta = -k_A C_A$$

$$R_B = dC_B/d\theta = k_A C_A - k_B C_B$$

$$R_C = dC_C/d\theta = k_B C_B$$

In recent times, investigators in the waste combustion field have taken the path of least resistance in analyzing waste combustion reactions and assumed the reactions to approach first-order behavior. Available experimental data are employed to generate *average* or *best-fit* values of k for first-order models. The rate equation

$$dC/d\theta = -kC \tag{5.6.20}$$

can then be integrated to provide a simple quantitative relationship between concentration and time, as shown in Eq. (5.6.21):

$$C = C_0 e^{-k\theta} \tag{5.6.21}$$

where C_0 = initial concentration. The extent of the reaction or the destruction efficiency (DRE) may then be calculated.

Experimentally determined reaction velocity constants for waste combustor reactors are based on *thermal destruction*, as opposed to *flame mode destruction.* When a waste passes through the flame front, some of it is destroyed by flame zone radicals at a rate much faster than the rate of waste destruction in the *thermal destruction* zone beyond the flame zone. Most of the waste is consumed in the thermal destruction zone. Rate constants for specific compounds and waste samples are obtained by a variety of experimental methods. Some of the most common are the TDAS (Thermal destruction analytical system) method of the University of Dayton Research Institute and thermogravimetric analysis.[4]

5.7 ILLUSTRATIVE EXAMPLES

Example 5.7.1 Waste material is fed into a waste incinerator at a rate of 10,000 lb/h in the presence of 20,000 lb/h of air. Due to the low heating value of the waste, 2000 lb/h of methane is added to assist in the combustion of the waste. At what rate (lb/h) do the product gases exit the incinerator?

SOLUTION Apply the conservation law for mass to the incinerator. Assume steady-state conditions to apply.

$$\text{Rate of mass in } (w_{\text{in}}) = \text{rate of mass out } (w_{\text{out}})$$

$$w_{\text{in}} = (10{,}000 + 20{,}000 + 2000)$$

$$= 32{,}000 \text{ lb/h}$$

Therefore, $$w_{\text{out}} = 32{,}000 \text{ lb/h}.$$

Example 5.7.2 C_6H_5Cl is fed into a thermal oxidizer at a rate of 5000 scfm (60°F, 1 atm) and is combusted in the presence of air fed at a rate of 3000 scfm (60°F, 1 atm). Both streams enter the oxidizer at 70°F. The products are cooled from 2000°F and exit a cooler at 180°F. At what rate (lb/h) do the products exit the cooler? The molecular weight of C_6H_5Cl is 112.5; the molecular weight of air is 29.

SOLUTION First convert scfm to acfm using Charles' law.

$$5000 \text{ scfm} \times \frac{460 + 70}{460 + 60} = 5096 \text{ acfm of } C_6H_5Cl$$

$$3000 \text{ scfm} \times \frac{460 + 70}{460 + 60} = 3058 \text{ acfm of air}$$

One pound-mole of any ideal gas occupies 387 ft^3 at 70°F and 1 atm. Therefore, the molar flow rate ($\dot{n}$) may be calculated by dividing these results by 387:

$$\dot{n}(C_6H_5Cl) = \frac{5096}{387}$$

$$= 13.17 \text{ lbmol/min}$$

$$= \dot{n}(\text{air}) = \frac{3058}{387}$$

$$= 7.90 \text{ lbmol/min}$$

The mass flow rate is obtained by multiplying these results by the molecular weight:

$$w(C_6H_5Cl) = (13.17)(112.5)(60)$$

$$= 88{,}898 \text{ lb/h}$$

$$w(\text{air}) = (7.90)(29)(60)$$

$$= 13{,}746 \text{ lb/h}$$

Note: One can calculate these results by simply applying the conservation of mass law. Since 1 lbmol of any ideal gas occupies 379 ft^3 at 60°F and 1 atm, $\dot{n}$ can be calculated directly, and w by multiplying $\dot{n}$ by the molecular weight:

$$w_{\text{in}}(C_6H_5Cl) = \frac{(5000)(112.5)(60)}{379}$$

$$= 88{,}898 \text{ lb/h}$$

$$w_{\text{in}}(\text{air}) = \frac{(3000)(29)(60)}{379}$$

$$= 13{,}746 \text{ lb/h}$$

Since mass is conserved, w_{in} is equal to w_{out}:

$$w_{in,total} = 88{,}898 + 13{,}746$$
$$= 102{,}644 \text{ lb/h}$$

Example 5.7.3 A flue gas from an incinerator enters a waste heat boiler at 1800°F at a rate of 50,000 lb/h. It transfers 2.3×10^7 Btu/h of heat adiabatically. Calculate the outlet temperature of the gas stream. Assume the average heat capacity of the flue gas to be 0.36 Btu/lb-°F.

SOLUTION Since average heat capacity information is given, an equation based on Eq. (5.2.8) may be employed:

$$q = w\overline{c_P}\,\Delta T$$

$$\Delta T = \frac{q}{w\overline{c_P}} = \frac{2.3 \times 10^7}{(50{,}000)(0.36)}$$

$$= 1278°\text{F} = (1800 - T_2)$$

The outlet temperature (T_2) is then 522°F.

Example 5.7.4 Propane is combusted in the presence of stoichiometric oxygen at 1500°F to form carbon dioxide and water as the combustion products. Balance the combustion reaction

$$C_3H_8 + O_2 \rightarrow CO_2 + H_2O$$

SOLUTION Since there are three carbon atoms on the left-hand side of the reaction, it is balanced with $3CO_2$ on the right-hand side:

$$C_3H_8 + O_2 \rightarrow 3CO_2 + H_2O$$

Since eight hydrogen atoms appear on the left-hand side, this is balanced by 4 mol of H_2O:

$$C_3H_8 + O_2 \rightarrow 3CO_2 + 4H_2O$$

There are 10 oxygen atoms on the right-hand side of the reaction. This is balanced by 5 oxygen molecules on the left-hand side to yield the final balanced equation for the combustion of propane:

$$C_3H_8 + 5O_2 \rightarrow 3CO_2 + 4H_2O$$

Note: This result could have been obtained directly from Table 5.3.1.

Example 5.7.5 The offensive odor of butanol (butyl alcohol) can be removed from stack gases by its complete combustion to CO_2 and H_2O. It is of interest that the incomplete combustion of butanol can actually result in a more serious environmental and odor pollution problem than the original one. Write the reactions showing the formation of the two intermediate malodorous products formed if butanol undergoes incomplete combustion. The three possible products of combustion are:

1. Butylaldehyde (C_4H_8O)
2. Butyric acid (C_3H_7COOH)
3. Carbon dioxide (CO_2) and water (H_2O)

SOLUTION For butylaldehyde:

$$C_4H_9OH + \tfrac{1}{2}O_2 \rightarrow C_4H_8O + H_2O$$

For butyric acid:

$$C_4H_9OH + O_2 \rightarrow C_3H_7COOH + H_2O$$

For "complete" combustion:

$$C_4H_9OH + 6O_2 \rightarrow 4CO_2 + 5H_2O$$

This is one of the main disadvantages of waste management control by incineration. As shown, it is possible to create a worse pollution problem if the combustion is incomplete. These products of incomplete combustion are defined as PICs. Incomplete combustion can arise because of low temperatures, insufficient oxygen, incomplete mixing, reduced retention time, or a combination of these.

Example 5.7.6 The combustion reaction for C_6H_5Cl is

$$C_6H_5Cl + O_2 \rightarrow CO_2 + H_2O + HCl$$

1. Balance the equation.
2. 2000 lb/h of C_6H_5Cl is fed into an incinerator with 50% excess air. What quantities of combustion products are produced (lb/h)? The molecular weight of C_6H_5Cl is 112.5. The molecular weight of air is 29.

SOLUTION

1. The balanced reaction is

$$C_6H_5Cl + 7O_2 \rightarrow 6CO_2 + 2H_2O + HCl$$

2. This reaction for air becomes

$$C_6H_5Cl + 7O_2 + 26.3N_2 \rightarrow 6CO_2 + 2H_2O + HCl + 26.3N_2$$

This appears as Eq. (5.3.7) in the text. For 50% excess air,

$$\frac{\text{lbmol of } O_2}{\text{lbmol } C_6H_5Cl} = 7 \times 1.5$$

$$\frac{\text{lbmol of } N_2}{\text{lbmol } C_6H_5Cl} = 26.3 \times 1.5$$

$$= 39.45$$

The balanced reaction for 50% excess air is

$$C_6H_5Cl + 10.5O_2 + 39.45N_2 \rightarrow 6CO_2 + 2H_2O + HCl + 3.5O_2 + 39.45N_2$$

Convert the mass flow rate of C_6H_5Cl to a molar flow rate:

$$\dot{n}(C_6H_5Cl) = \frac{2000}{112.5}$$

$$= 17.78 \text{ lbmol/h}$$

$$w(CO_2) = \dot{n}(C_6H_5Cl)(6)(MW) = (17.78)(6)(44)$$

$$= 4694 \text{ lb/h}$$

$$w(H_2O) = (17.78)(2)(18)$$

$$= 640 \text{ lb/h}$$

$$w(HCl) = (17.78)(1)(36.46)$$

$$= 648 \text{ lb/h}$$

$$w(O_2) = (17.78)(3.5)(32)$$

$$= 1991 \text{ lb/h}$$

$$w(N_2) = (17.78)(39.45)(28)$$

$$= 19640 \text{ lb/h}$$

The reader is left the exercise of calculating the mass fractions of the flue components and the total mass flow rate.

Note: On combusting C_6H_5Cl, the resulting equilibrium conditions will not permit all of the Cl to go to HCl. The amount of Cl_2 formed (which will be small) is a function of temperature, excess air, residence time, and the H/Cl ratio. This effect is treated in a later example.

Example 5.7.7 Verify that the enthalpy of combustion (Table 5.4.1) may be calculated from enthalpy of formation data (also in Table 5.4.1). Use *n*-hexane as an example.

SOLUTION From Table 5.4.1,

$$\Delta H^0_C(n\text{-hexane}) = -1{,}002{,}570 \text{ cal/gmol}$$

First, write the combustion reaction:

$$C_6H_{14} + 9.5O_2 \rightarrow 6CO_2 + 7H_2O(l)$$

From Table 5.4.1, one obtains

$$C_6H_{14}(g) \qquad \Delta H^0_f = -36{,}960 \text{ cal/gmol}$$

$$CO_2(g) \qquad \Delta H^0_f = -94{,}052 \text{ cal/gmol}$$

$$H_2O(l) \qquad \Delta H^0_f = -68{,}317 \text{ cal/gmol}$$

$$\Delta H^0_c = \sum \Delta H^0_{f,p} - \sum \Delta H^0_{f,r} = 6(-94{,}052) + 7(-68{,}317) - (-39{,}960)$$
$$= -1{,}002{,}571 \text{ cal/gmol}$$

Example 5.7.8 Using standard heat of combustion data, calculate the standard heat of formation of $C_{14}H_9Cl_5$. [The organic and generic names for this compound are 1,1,1-trichloro-2,2-bis(*p*-chlorophenyl)ethane and DDT, respectively. DDT is the abbreviation for dichlorodiphenyl-trichloroethene, another name sometimes used for this compound.]

SOLUTION The standard heat of combustion for this organic is obtained directly from Table 5.4.1, noting that the H_2O and HCl formed are in the liquid and gaseous states, respectively.

$$\Delta H^0_c = -1600 \text{ kcal/gmol}$$

First write a balanced stoichiometric equation for this combustion reaction:

$$C_{14}H_9Cl_5 + 15O_2 \rightarrow 14CO_2 + 2H_2O(l) + 5HCl(g)$$

For this reaction,

$$\Delta H^0_c = 14\Delta H^0_{f,CO_2} + 2\Delta H^0_{f,H_2O(l)} + 5\Delta H^0_{f,HCl(g)} - \Delta H^0_{f,C_{14}H_9Cl_5}$$

From Table 5.4.1,

$$\Delta H^0_{f,CO_2} = -94.052 \text{ kcal/gmol}$$

$$\Delta H^0_{f,H_2O(l)} = -68.317 \text{ kcal/gmol}$$

$$\Delta H^0_{f,HCl(g)} = -22.063 \text{ kcal/gmol}$$

Solving this equation for $\Delta H^0_{f,C_{14}H_9C_{15}}$ yields

$$\Delta H^0_{f,C_{14}H_9Cl_5} = 36.32 \text{ kcal/gmol}$$

The reader is again reminded that the heat of formation of elements in their standard states is zero. In addition, care should be exercised in using heat of combustion data for highly chlorinated organics.

Example 5.7.9 Calculate the theoretical adiabatic flame temperature of C_6H_5Cl. Assume that the heat capacity variation with temperature takes the form

$$C_P = a + bT + cT^{-2}$$

SOLUTION The theoretical adiabatic flame temperature for an organic compound has been previously defined as the maximum temperature the flue products of combustion will achieve if the reactants of combustion are at ambient conditions. This temperature is achieved with adiabatic (no heat lost to the surroundings) operation and with theoretical (stoichiometric or 0% excess) air.

The standard heat of combustion for chlorobenzene is obtained from the heats of formation data in Table 5.4.1. Since

$$C_6H_5Cl + 7O_2 \rightarrow 6CO_2 + 2H_2O(g) + HCl(g)$$

$$\begin{aligned}\Delta H^0_c &= (6)(-94{,}052) + (2)(-57{,}789) + (-22{,}063) - 12{,}390 \\ &= 714{,}361 \text{ cal/gmol}\end{aligned}$$

Note: This value is the net or lower heating value since the water is in the gaseous state.

This stoichiometric reaction is now written for combustion in air. We first note that there are 7.0(79/21) or 26.33 lbmol of nitrogen present in the theoretical combustion air.

$$C_6H_5Cl + 7O_2 + 26.33N_2 \rightarrow 6CO_2 + 2H_2O(g) + HCl(g) + 26.33N_2$$

The heat capacity for the flue gas products in the form

$$C_P = a + bT + cT^{-2}$$

are read from Table 5.2.4.

$$C_{P,\mathrm{CO_2}} = 10.57 + 2.10 \times 10^{-3}T - 2.06 \times 10^5 T^{-2}$$

$$C_{P,\mathrm{H_2O}} = 7.30 + 2.46 \times 10^{-3}T + 0.0 \times 10^5 T^{-2}$$

$$C_{P,\mathrm{HCl}} = 6.27 \times 1.24 \times 10^{-3}T + 0.3 \times 10^5 T^{-2}$$

$$C_{P,\mathrm{N_2}} = 6.83 + 0.90 \times 10^{-3}T - 0.12 \times 10^5 T^{-2}$$

However,

$$\Delta C_P = 6C_{P,\mathrm{CO_2}} + 2C_{P,\mathrm{H_2O}} + C_{P,\mathrm{HCl}} + 26.33C_{P,\mathrm{N_2}}$$

Thus,

$$\begin{aligned}\Delta C_P = {} & 6(10.57 + 2.10 \times 10^{-3}T - 2.06 \times 10^5 T^{-2}) \\ & + 2(7.30 + 2.46 \times 10^{-3}T - 0.0 \times 10^5 T^{-2}) \\ & + 1(6.27 + 1.24 \times 10^{-3}T + 0.30 \times 10^5 T^{-2}) \\ & + 26.33(6.83 + 0.90 \times 10^{-3}T - 0.12 \times 10^5 T^{-2})\end{aligned}$$

If this is expressed in the form $\Delta C_P = \Delta a + \Delta bT + \Delta cT^{-2}$, then

$$\Delta C_P = 264.12 + 0.0425T - 1.522 \times 10^6 T^{-2} \text{ cal/gmol-K or Btu/lbmol-°R}$$

Equation (5.4.14) applies in calculating the adiabatic flame temperature. The energy liberated on combustion appears as sensible energy in heating the flue (product) gas. The sum of these two effects is zero if the operation is conducted adiabatically. This is treated in more detail in Chapter 6.

$$\Delta H_c^0 + \Delta H_p = \Delta H = 0$$

Since 25°C = 298 K, the enthalpy change associated with heating the flue products is given by

$$\Delta H_p = \int_{298}^{T_2} \Delta C_P \, dT \qquad T_2 = \text{theoretical adiabatic temperature (K)}$$

Substituting ΔC_P obtained previously and integrating yields

$$\begin{aligned}\Delta H_p &= \Delta a(T_2 - 298) + \left(\frac{\Delta b}{2}\right)(T_2^2 - 298^2) - \Delta c\left(\frac{1}{T_2} - \frac{1}{298}\right)\\ &= 264.12(T_2 - 298) + \left(\frac{0.0425}{2}\right)(T_2^2 - 298^2) + 1.522 \times 10^6\left(\frac{1}{T_2} - \frac{1}{298}\right)\\ &= -\Delta H_c^0\\ &= 714{,}361 \text{ cal/gmol}\end{aligned}$$

This equation may now be rewritten in the form

$$800{,}063 = 264.12T_2 + 0.02125T_2^2 + (1.522 \times 10^6)/T_2$$

This is a nonlinear cubic equation. It may be solved by any one of several analytical or numerical methods. The final result can be obtained by a crude trial-and-error procedure. However, this trial-and-error process can be simplified by recognizing that on the right-hand side, the last term is an order of magnitude smaller than the first two terms. Therefore, an excellent first guess can be obtained by solving the equation

$$0.02125T_2^2 + 264.12T_2 - 800{,}063 = 0$$

$$T_2 = \frac{-264.12 \pm \sqrt{(264.12)^2 + (4)(0.02125)(800{,}063)}}{(2)(0.02125)}$$

$$T_2 = 2519 \text{ K} = 4534°\text{R} = 4074°\text{F}$$

Final theoretical adiabatic flame temperature is 4074°F.

Example 5.7.10 Calculate the theoretical adiabatic flame temperature for DDT. Perform the calculations using SI (cgs) units.

SOLUTION The standard enthalpy (heat) of combustion is again obtained from Table 5.4.1.

$$H_c^0 = -1600 \text{ kcal/gmol}$$

As noted earlier, these combustion values assume that the water and hydrogen chloride are in the liquid and gaseous state, respectively. The enthalpy of vaporization of water at 60°F is 10.60 kcal/gmol of H_2O vaporized. The balanced stoichiometric combustion equation is

$$C_{14}H_9Cl_5 + 15O_2 + 56.43N_2 \rightarrow 14CO_2 + 2H_2O + 5HCl + 56.43N_2$$

With water in the vapor state, the standard heat of combustion (the net heating value) is

$$\Delta H_c^0 = -1600 + 2(10.60)$$
$$= -1578.8 \approx -1580 \text{ kcal/gmol}$$

(A value of -1580 kcal/gmol will be used in this and subsequent problems for the net heating value of DDT at 60°F.)

The describing equation for the adiabatic flame temperature is

$$\Delta H_c^0 + \Delta H_p = \Delta H = 0$$

Heat capacity data of the form

$$C_P = a + bT + \frac{c}{T^2}$$

are again obtained from Table 5.2.4.

The term is ΔC_P is

$$\Delta C_P = 14C_{P,CO_2} + 2C_{P,H_2O} + 5C_{P,HCl} + 56.43C_{P,N_2}$$

Substituting values for $C_{P,i}$ yields

$$\Delta C_P = 579.35 + 9.13 \times 10^{-2}T - 3.41 \times 10^6 T^{-2}$$

The sensible enthalpy term is

$$\Delta H_p = \int_{298}^{T_2} C_P \, dT$$
$$= 579.35(T_2 - 298) + \frac{9.13 \times 10^{-2}}{2}(T_2^2 - 298^2) + 3.41 \times 10^6 \left(\frac{1}{T_2} - \frac{1}{298}\right)$$

Setting the right-hand side of this equation equal to the net heating value yields

$$1580 \times 10^3 = 579.35(T_2 - 298) + \frac{9.13 \times 10^{-2}}{2}(T_2^2 - 298^2)$$
$$+ 3.41 \times 10^6 \left(\frac{1}{T_2} - \frac{1}{298}\right)$$

A trial-and-error solution yields

$$T_2 = 2541\ \text{K} = 2268°\text{C} = 4574°\text{R} = 4114°\text{F}$$

The theoretical adiabatic flame temperature for DDT is therefore 2541 K.

The reader should note that this is a *theoretical* thermochemical calculation. For practical purposes, 3200°F is about the maximum flame temperature that a refractory-lined incinerator can withstand.

Example 5.7.11 Calculate the chemical reaction equilibrium constant K for reactions that have ΔG^0 values of $+20.0$ and -20.0 kcal/gmol at 70°F.

SOLUTION From Eq. (5.5.4)

$$\ln K = -\frac{\Delta G_T^0}{RT}$$

With $T = 70°\text{F} = 294$ K.

For $\Delta G^0 = 20.0$ kcal/gmol

$$\ln K - \frac{20{,}000}{(1.99)(294)}$$

$$K = 1.4251 \times 10^{-15}$$

For $\Delta G^0 = 20.0$ kcal/gmol

$$\ln K - \frac{20{,}000}{(1.99)(294)}$$

$$K = 7.017 \times 10^{14}$$

The reader should note the effect of ΔG^0 has on the numerical value of K.

Example 5.7.12 The equilibrium constant K_P for the reaction (at one atm)

$$2HCl(g) + 0.5O_2 \rightarrow Cl_2 + H_2O(g)$$

may be expressed in the form[5]

$$K_P = Ae^{B/T}$$

where $A = 0.229 \times 10^{-3}$

$B = 7340$

T = absolute temperature (K)

As part of a hazardous waste incineration project, Dr. Theodore is required to develop a more rigorous equation for K_P as a function of the absolute temperature T. Tied up by other nonessential departmental assignments, he has chosen to hire you as a subcontractor on the project to obtain this information.

SOLUTION First calculate ΔG^0_{298} for the reaction using Eq. (5.5.3):

$$\Delta G^0_{298} = 1\Delta G^0_{\mathrm{f,H_2O}} - 1\Delta G^0_{\mathrm{f,HCl}}$$

From Table 5.5.1, $\Delta G^0_{\mathrm{f,H_2O}}$ and $\Delta G^0_{\mathrm{f,HCl}}$ can be obtained:

$$\Delta \mathrm{G}^0_{\mathrm{f,H_2O}} = -54{,}635 \text{ cal/gmol}$$

$$\Delta G^0_{\mathrm{f,HCl}} = -22{,}778 \text{ cal/gmol}$$

$$\Delta G^0_{298} = (1)(-54{,}635) - (2)(-22{,}778)$$

$$= -9079 \text{ cal/gmol}$$

Similarly, ΔH^0_{298} is obtained from Table 5.4.1:

$$\Delta H^0_{298} = -13{,}672 \text{ cal/gmol}$$

According to Eq. (5.5.9),

$$\ln K = \frac{-\Delta H_0}{RT} + \frac{\Delta\alpha}{R}\ln T + \frac{\Delta\beta}{2R}T + \frac{\Delta\gamma}{6R}T^2 + I$$

or, its equivalent form for $C_P = a + bT + cT^{-2}$:

$$\ln K = \frac{-\Delta H_0}{RT} + \frac{\Delta\alpha}{R}\ln T + \frac{\Delta b}{2R}T + \frac{\Delta c}{2R}T^{-2} + I$$

Next, ΔH_0 and I must be determined. ΔH_0 is found by a procedure similar to that used in Section 5.4. From Eq. (5.4.10),

$$\Delta H^0_T = \Delta H^0_{298} + \int_{298}^{T} \Delta C_P \, dT \tag{5.4.10}$$

The combustion reaction is

$$2\mathrm{HCl(g)} + 0.5\mathrm{O_2(g)} \rightarrow \mathrm{H_2O(g)} + \mathrm{Cl_2(g)}$$

The heat capacities for the flue gas products in the form

$$C_P = a + bT + cT^{-2}$$

are read from Table 5.2.4:

$$\begin{aligned}
\Delta a &= (7.30 + 8.85) - [(2)(6.27) + (0.5)(7.16)] \\
&= 0.03 \\
\Delta b &= (2.46 \times 10^{-3} + 0.16 \times 10^{-3}) - [(2)(1.24 \times 10^{-3}) + (0.5)(1.0 \times 10^{-3})] \\
&= -3.6 \times 10^{-4} \\
\Delta c &= (0.0 - 0.68 \times 10^{5}) - [(2)(0.30 \times 10^{5}) + (0.5)(-0.4 \times 10^{5})] \\
&= -1.08 \times 10^{5}
\end{aligned}$$

Equation (5.4.10) becomes

$$\Delta H_T^0 = \Delta H_{298}^0 + \int_{298}^{T} [\Delta a + (\Delta b)T + (\Delta c)T^{-2}]\, dT$$

or

$$\Delta H_T^0 = \Delta H_{298}^0 + \Delta a(T - 298) + \tfrac{1}{2}\Delta b(T^2 - (298)^2) - \Delta c\left(\frac{1}{T} - \frac{1}{298}\right)$$

Combining the constant terms into ΔH_0 (as in Section 5.4) yields the following:

$$\Delta H_T^0 = \Delta H_0 + (\Delta a)T + \tfrac{1}{2}(\Delta b)T^2 - (\Delta c)T^{-1}$$

where

$$\begin{aligned}
\Delta H_0 &= \Delta H_{298}^0 - 298\Delta a - \tfrac{1}{2}(298)^2\Delta b + \frac{1}{298}\Delta c \\
&= 13{,}682 - 298(0.03) - \tfrac{1}{2}(298)^2(-3.6 \times 10^{-4}) + \frac{1}{298}(-1.08 \times 10^{5}) \\
&= 14{,}027 \text{ cal/gmol}
\end{aligned}$$

From Eq. (5.5.4),

$$\begin{aligned}
\ln K = \frac{-\Delta G_{298}^0}{RT} &= \frac{9079}{(1.99)(298)} \\
&= 15.31
\end{aligned}$$

Therefore,

$$15.31 = \frac{14027}{(1.99)(298)} + \frac{0.03}{1.99}(\ln 298) + \frac{-3.6 \times 10^{-4}}{(2)(1.99)}(298)$$
$$+ \frac{-1.08 \times 10^{5}}{(2)(1.99)}(298)^{-2} + I$$

Solving for I,

$$I = -8.09$$

The final form of the equation for K is

$$\ln K = \frac{7048.7}{T} + 0.0151 \ln T - 9.06 \times 10^{-5} T - 2.714 \times 10^{4} T^{-2} - 8.09$$

Example 5.7.13 Refer to the previous example. The chemical reaction equilibrium constant based on partial pressure (K_P) was obtained as a function of temperature for the reaction

$$2HCl + 0.5O_2 \rightarrow Cl_2 + H_2O$$

The final result took the form:

$$\ln K = \frac{7048.7}{T} + 0.0151 \ln T - 9.06 \times 10^{-5} T - 2.714 \times 10^{4} T^{-2} - 8.09$$

If the initial partial pressures of HCl, O_2, Cl_2, and H_2O are 0.146, 0.106, 0.0, and 0.0292, respectively, calculate the equilibrium partial pressure of Cl_2, if the operating conditions are 1.0 atm and 1250 K.

SOLUTION By definition,

$$K_P = \frac{P_{Cl_2} P_{H_2O}}{P_{HCl}^{2} P_{O_2}^{0.5}}$$

A rigorous procedure to calculate the equilibrium partial pressures is available elsewhere.[3] The following procedure is satisfactory for purposes of engineering calculations. At equilibrium,

$$P_{Cl_2} = P_{Cl_2}(\text{initial}) + x$$
$$= x$$

The term x represents the increase in the partial pressure of the chlorine due to this equilibrium reaction:

$$P_{H_2O} = P_{H_2O}(\text{initial}) + x = 0.0292 + x$$

$$P_{HCl} = P_{HCl}(\text{initial}) - 2x = 0.146 - 2x$$

$$P_{O_2} = P_{O_2}(\text{initial}) - 0.5x = 0.106 - 0.5x$$

K_P can then be expressed as

$$K_P = \frac{(x)(0.0292 + x)}{(0.146 - 2x)^2(0.106 - 0.5x)^{0.5}}$$

Now, calculate K_P at 1250 K using the result from the previous example:

$$\ln K_P = \frac{7048.7}{1250} + 0.015(\ln 1250) - 9.06 \times 10^{-5}(1250) - 2.714 \times 10^4(1250)^{-2} - 8.09$$

$$= -2.475$$

$$K_P = 0.0842$$

Therefore,

$$0.0842 = \frac{(x)(0.0292 + x)}{(0.146 - 2x)^2(0.106 - 0.5x)^{0.5}}$$

Solving for x, which is the equilibrium partial pressure of Cl_2, by trial-and-error calculation yields

$$P_{Cl_2} = x = 0.01050 \text{ atm}$$

Note: Approximately 1% of the discharge flue gas is chlorine—a rather sizable amount.

Example 5.7.14 Calculate the residence time required to achieve 99.99% destruction efficiency of benzene if an incinerator is operated at a temperature of 980°C. The literature provides values of the Arrhenius constant (A) and the activation energy (E) of 3.3×10^{10} s^{-1} and 35,900 cal/gmol, respectively.

SOLUTION As discussed in the previous section, many chemical reactions have been shown to be satisfactorily described by a first-order reaction, that is,

$$\frac{dC}{d\theta} = -kC$$

where C = concentration of the material undergoing reaction
k = reaction rate constant
θ = time

If the initial concentration is C_0, the solution to this differential equation becomes [see Eq. (5.6.20)]:

$$\ln\left(\frac{C}{C_0}\right) = -k\theta$$

The temperature dependence of k is given by the Arrhenius equation (5.6.6):

$$k = A \exp\left(\frac{-E}{RT}\right)$$

where A = frequency factor (units of time^{-1})
E = activation energy (cal/gmol or Btu/lbmol)
T = K or °R
R = 1.987 cal/gmol-K or Btu/lbmol-°R

For 99.99% destruction efficiency,

$$\left(\frac{C}{C_0}\right) = 0.0001$$

From the Arrhenius equation,

$$k = A \exp\left(\frac{-E}{RT}\right) = 3.3 \times 10^{10} \exp \frac{(-35{,}900)}{(1.987)(980 + 273)}$$
$$= 1.804 \times 10^4 \text{ s}^{-1}$$

The required residence time is, therefore,

$$\theta = -\frac{\ln(C/C_0)}{k} = -\frac{(\ln 0.0001)}{(1.804 \times 10^4)}$$
$$= 5.11 \times 10^{-4} \text{ s}$$

PROBLEMS

1. Along with 2000 lb/h of hazardous waste, 5000 lb/h of air is fed into an incinerator. To assist in the complete combustion of the waste, 3000 lb/h of propane is added to the incinerator. What is the amount of product in lb/h?

2. Carbon tetrachloride at 7000 scfm (60°F, 1 atm) is combusted at 1250°F in the presence of oxygen fed at the rate of 12,000 scfm. All feeds enter

the incinerator at 120°F. At what rate (lb/min) do the products exit from the incinerator? The molecular weight of CCl_4 is 153.82.

3. Complete combustion of carbon disulfide results in combustion products of CO_2 and SO_2 according to the reaction

$$CS_2 + O_2 \rightarrow CO_2 + SO_2$$

(a) Balance this reaction.
(b) If 500 lb of CS_2 is combusted with 225 lb of oxygen, which is the limiting reactant?
(c) How much of each product is formed (lb)? MW of CS_2 is 76.14; MW of SO_2 is 64.07; MW of CO_2 is 44.

4. Combustion products at 25,000 lb/h exit an incinerator at 1700°F and then transfer 6.00×10^6 Btu/h of heat to a waste heat boiler. If the waste heat boiler operates adiabatically, and the heat capacity of the flue products over the appropriate temperature range is 0.25 Btu/lb-°F, at what temperature will the gases leave the boiler?

5. Benzene is incinerated at 2100°F in the presence of 50% excess oxygen. Balance the combustion reaction

$$C_6H_6 + O_2 \rightarrow CO_2 + H_2O + O_2$$

6. Find the enthalpy of combustion of cyclohexane from enthalpy of formation data. Perform the calculation for both liquid and gaseous cyclohexane.

7. Calculate the gross heating value in Btu/lbmol for a combustible gas mixture of 75 mol% methane, 10 mol% propane, and 15 mol% *n*-butane.

8. Calculate the theoretical adiabatic flame temperature of benzene.

9. Refer to Example 5.7.13. Calculate the equilibrium partial pressure of chlorine at a temperature of 2225°F.

10. Repeat Problem 9 with a temperature of 2770°F.

11. The equilibrium constant (K_P) for the reaction (at 1 atm pressure)

$$SO_2 + \tfrac{1}{2}O_2 = SO_3$$

may be expressed in the form[5]

$$Y = Ae^{BX}$$

where $A = 0.148 \times 10^{-4}$ $\quad Y = K_P$
$B = 11{,}700$ $\quad X = 1/T$; T = kelvin (K)

Develop a more rigorous equation for K_P as a function of the absolute temperature (K).

12. Refer to Problem 11. The chemical reaction equilibrium constant based on partial pressures (K_P) was obtained as a function of temperature for the reaction

$$SO_2 + \tfrac{1}{2}O_2 = SO_3$$

The final result took the form

$$\ln K = \frac{11{,}996}{T} - 0.362 \ln T + 9.36 \times 10^{-4}T - 2.969 \times 10^{5}T^{-2} - 9.88$$

If the initial partial pressures of SO_2, O_2, and SO_3 are 0.000257, 0.102, and 0.0 atm, respectively, calculate the equilibrium partial pressure of the SO_3. The operating conditions are 1.0 atm and 1250 K.

13. Repeat Problem 12 with an operating temperature of 2770°F.

REFERENCES

1. R. C. Weast, ed., *CRC Handbook of Chemistry and Physics*, 80th ed., CRC Press, Boca Raton, FL, 1999.
2. J. M. Smith and H. C. Van Ness, *Introduction to Chemical Engineering Thermodynamics*, 4th ed., McGraw-Hill, New York, 1992.
3. T. F. McGowan, private communication, 1999.
4. W. A. Rubey, *Design Considerations for a Thermal Decomposition Analytical System*, EPA-600/2-80-098, Cincinnati, OH, 1980.
5. L. Theodore, personal notes, 1985.

6

THERMOCHEMICAL APPLICATIONS

6.1 INTRODUCTION

In Chapter 5, a number of scientific principles, mostly from the fields of chemistry and chemical engineering, were presented. The presentation was general and the principles studied pertain to almost any type of chemical reaction. In this chapter, these same concepts are applied to the waste combustion process. Mass conservation and stoichiometry (Sections 5.2 and 5.3) are employed in the stoichiometric calculations of Section 6.3 in which the prediction of the flue gas composition from the combustion of waste–fuel mixtures of known composition is discussed. Thermochemistry (Section 5.4) and chemical reaction equilibrium (Section 5.5) principles are applied in Section 6.4, thermochemical calculations, to predict the combustor operating temperature as a function of the waste–fuel mixture, net heating value, and excess air ratio. Chemical kinetics (Section 5.6) are applied in Section 6.5, applications of chemical reaction principles, to predict the all-important residence time requirement for the destruction of a particular waste.

This chapter begins with a discussion of fuel options (Section 6.2). In practice, the wastes to be combusted may have too low a heating value to sustain combustion. In these cases, auxiliary fuel must be burned with the waste. The types of fuel available and a discussion of selection criteria are the subjects of this section. This chapter concludes with illustrative examples and problems.

6.2 FUEL OPTIONS

The properties of the fuels most often used in a waste combustion facility are reviewed here. Fuels burned in a waste combustor may be gaseous, liquid, or solid. Some of the more common fuels are the following:

- *Gaseous fuels* are principally natural gas (80–95% methane, the balance ethane, propane, and small quantities of other gases). Light hydrocarbons obtained from petroleum or coal treatment may also be used.
- *Liquid fuels* are mainly hydrocarbons obtained by distilling crude oil (petroleum). The various grades of fuel oil, gasoline, shale oil, and various petroleum cuts and residues are considered in this category.
- *Solid fuels* consist principally of coal (a mixture of carbon, water, noncombustible ash, hydrocarbons, and sulfur). Coke, wood, and solid waste (garbage) may also be employed.

These classifications are not mutually exclusive and necessarily overlap in some areas. The stoichiometric and thermochemical analyses of problems involving the different classes of fuels are similar. Consequently, the actual physical form of the fuel is not important in any calculational study.

In the following sections, fuels are discussed primarily in terms of their usefulness and economic utilization in assisting the waste combustion process. Properties of fuels considered to be of prime importance are composition and heating value. The heating value refers to the quantity of heat released during combustion of a unit amount of fuel gas. Thus, the heating value of a fuel is one of its most important properties. As described in Chapter 5, the major products of the complete combustion of a fuel are CO_2 and H_2O. Because of the presence of varying amounts of water in different fuels, two methods for expressing heating value are in common use. Once again, the *gross* or *higher heating value (HHV)* is defined as the amount of heat evolved in the complete combustion of a fuel with any water present in the liquid state; the *net* or *lower heating value (LHV)* is defined as the amount of heat evolved in the combustion of a fuel when all the products, including the water, are in the gaseous state. Thus, the gross heating value is greater than the net heating value by the latent heat of vaporization of the total amount of water originally present and that formed by oxidation of the hydrogen in the fuel.

In dealing with the combustion of some waste fuels, it is occasionally convenient to express the composition in terms of a single hydrocarbon, even though the fuel is a mixture of many hydrocarbons. For example, gasoline is usually considered to be octane (C_8H_{18}) and diesel fuel is considered to be dodecane ($C_{12}H_{26}$). However, the composition of most fuels employed in waste combustion is given in terms of its componential (or ultimate) analysis.

Approximate componential analyses of the three types of fuels most often employed at waste combustion facilities are given in Table 6.2.1. Estimated values for the net heating value are also included.

Natural Gas

Natural gas is perhaps the closest approach to an ideal fuel because it is practically free from noncombustible gas or particulate residue. Of the many gaseous fuels, natural gas is the most important for use at waste combustion fa-

Table 6.2.1 Componential analysis for common fuels

Element	Weight Fraction (lb element/lb fuel)		
	Residual fuel oil (No. 6)	Distillate fuel oil (No. 2)	Natural gas
C	0.866	0.872	0.693
H	0.102	0.123	0.227
N	—	—	0.08
S	0.03	0.005	—
Net heating value	19,500 Btu/lb	18,600 Btu/lb	1,000 Btu/scf (60°F)

cilities. No fuel preparation is necessary at the site because gases are easily mixed with air, and the combustion reaction proceeds rapidly once the ignition temperature is reached.

Natural gas is found compressed in porous rock or cavities, which are sealed between strata under the earth's surface. When these gas-bearing pools are tapped by drilling wells, the gas is found to be under pressure, which may be as high as 2000 psig. As gas is withdrawn, this pressure gradually decreases until it eventually becomes so low that the field must be abandoned. Thus, the natural gas from a well will usually not remain constant in its elemental composition and/or heating value.

Natural gas consists mainly of methane, with smaller quantities of other hydrocarbons, particularly ethane, and trace amounts of propane. Carbon dioxide and nitrogen are usually present in small amounts and sometimes there can be appreciable amounts of hydrogen sulfide present. (This is generally removed at the field before transmission.)

The physical and chemical characteristics of natural gas are influenced by the underground conditions existing in the localities where it is found. The gas may contain heavy saturated hydrocarbons, which are liquid at ordinary pressures and temperatures. Dry natural gas contains <0.1 gal of gasoline vapor/1000 ft^3; > 0.1gal/1000 ft^3 is termed *wet.* Note that the terms *wet* and *dry,* when dealing with natural gas, refer to its gasoline content, not to its moisture content. Natural gases are also classified as either *sweet* or *sour.* A *sour* gas is one that contains some mercaptans and a high percentage of hydrogen sulfide, while the *sweet* gas is one in which these compounds are present in trace quantities. Most natural gases employed at a hazardous waste combustion facility are *dry* and *sweet.*

Environmental concerns with natural gas are minimal. Most gaseous fuels, with the possible exception of some waste gases, are considered to be clean fuels. Pipeline-grade natural gas is virtually free of sulfur and particulates. Its flue products do not pollute water. In addition, natural gas transportation and distribution facilities have a minimal adverse ecological impact. However, leakage of natural gas can pose a very serious explosion hazard.

The principal air contaminants from gaseous fuels, which are affected by the combustion system design and operation, are the oxidizable materials—carbon monoxide, carbon, and unburned hydrocarbons. Burner design operating temperature and excess air also affect the production of the oxides of nitrogen. In some instances, the impurity in a gaseous fuel may be hydrogen sulfide. Another important sulfur impurity is carbon disulfide. In most gases, however, the total organic sulfur content is relatively small. Other sulfur compounds that may be present in trace amounts are the thiophenes, carbon oxysulfide, mercaptans, thioesters, and organic sulfide.

Proper operation of a waste combustion facility employing natural gas requires that the fuel rate be controlled in relation to the demand. The air supply must be appropriate to the fuel supply. This is usually accomplished by automatic control. The incoming gas supply is regulated at a constant pressure upstream of the control valve. This valve can be used to control the gas flow. Combustion air regulation is achieved through manipulating dampers or by a special draft controller. Large installations are likely to use more elaborate systems where the fuel and air flows are metered with automatic adjustment to compensate for any changes or disturbances. Natural gas passes through one or more fixed orifices before entering the combustion chamber. Since flow through an orifice is proportional to the square root of the pressure drop across it, small fluctuations of the upstream pressure will not have a very significant effect on the gas flow rate. However, should it be necessary to reduce the firing rate to 50% of its normal value (2-to-1 turndown), for example, a fourfold decrease in gas pressure would be required, with the airflow rate adjusted accordingly. This factor can present a control problem. Before initiating firing with natural gas, the gas injection orifices should be inspected to verify that all passages are unobstructed. Filters and moisture traps should be clean, in place, and operating effectively to prevent any plugging of gas orifices. Proper location and orientation of diffusers should also be confirmed.

Liquid Fuels

Fuel oil may be defined as petroleum or any of its liquid residues remaining after the more volatile constituents have been removed. Thus, the term *fuel oil* may conveniently cover a wide range of petroleum products. It may be applied to crude petroleum, to a light petroleum fraction similar to kerosene or gas oil, or to a heavy residue left after distillation. The principal industrial liquid fuels are therefore the by-products of natural petroleum. These fuel oils are marketed in two principal classes: distillates and residuals. The principal industrial boiler fuel is residual oil, known as No. 6 or *Bunker C*. The residual oil, as the name implies, is left over after the more valuable products are distilled off. As an industrial and utility fuel, it competes with coal, although its price has fluctuated wildly over the years. It is specified mainly by viscosity. Grades No. 1 and 2 are sometimes designated as *light* and *medium* domestic fuel oils and are specified mainly by the temperature of the distillation range and specific gravity.

Despite the multiplicity of chemical compounds found in fuel oils, there are some important physical and chemical properties that deserve comment:

1. *Specific Gravity* For fuel oils, specific gravity is usually taken as the ratio of the density of the oil at 60°F to the density of water at 60°F. Gravity determinations are readily made by immersing a specific gravity meter into the oil and reading the scale at the point to which the instrument sinks. The specific gravity is then read directly. The American Petroleum Institute (API), the U.S. Bureau of Mines, and the U.S. Bureau of Standards agreed to recommend that only one scale be used in the petroleum industry and that it be known as the API scale. This relationship is given by Eq. (4.3.7).

2. *Viscosity* The Saybolt Universal viscosity is expressed as the amount of time (in seconds) that it takes to run 60 cm^3 of the oil through a standard size orifice at any desired temperature. Viscosity is commonly measured at 100, 150, and 210°F. Fuel oil (particularly No. 6) is very viscous, and it takes a long time to make a determination with the Saybolt Universal viscometer. For this reason, the viscosity of fuel oil may be measured with a Saybolt Furol viscometer, which is the same as the Saybolt Universal viscometer except that the orifice is larger. Interestingly, the viscosity of fuel oil decreases as the temperature rises and becomes nearly constant above about 250°F. Therefore, when fuel oil is heated to reduce the viscosity for good atomization, there is little gain in heating the oil beyond 250°F.

3. *Heating Value* This may be expressed in either British thermal units per gallon or British thermal units per pound (at 60°F). The heating value per gallon increases with specific gravity because there is more weight for a given volume; values range from ~140,000 to 150,000 Btu/gal. The heating value per pound of fuel oil ranges in value from 18,600 to 19,500 Btu/lb (see Table 6.2.1).

4. *Flash and Fire Point* The flash point of fuel oil is the lowest temperature at which sufficient vapor is given off to form a momentary flash when flame is brought near the surface of the oil in a small container called a Cleveland Cup. The fire point is the lowest temperature at which the oil gives off enough vapor to burn continuously. Stored fuel oils (as well as liquid wastes) should be maintained significantly below their flash points.

The rate of combustion of fuel oil is limited by vaporization. Light distillate oils readily vaporize; other (heavier) fuel oils, because of their heavier composition, require additional equipment to assure vaporization and complete combustion. In order to achieve complete combustion, oils are atomized into small droplets for rapid vaporization. The rate of evaporation is dependent on surface area, which increases as the atomized droplet size becomes smaller. The desired shape and droplet size are influenced adversely if the fuel viscosity is too high. At ambient temperature, No. 2 fuel oil may be atomized with little difficulty, but typically No. 6 fuel oil must be heated to around 210°F to assure proper atomization. Dirt and foreign matter suspended in the oil may cause wear in the oil pump and blockage of the atomizing nozzles. Strainers or replaceable filters are required in both the oil suction line and the discharge line. During combustion of a distillate fuel oil, the droplet becomes uniformly smaller as it vaporizes. By contrast, a residual oil droplet undergoes thermal and catalytic cracking, and its composition and size undergo various changes with time. Vapor bubbles may form, grow, and burst within a droplet in such a way as to shatter the droplet as it is heated in the combustion zone.

The physical and chemical properties of the oil and the characteristics of the waste combustion equipment influence the air pollution emissions from a waste facility. Some of these properties are

- *Sulfur Content* Sulfur is a very undesirable element in fuel oil because its products of combustion are acidic and cause corrosion in the ducting, valves, downstream equipment, and so on.
- *Solid Impurities* Fuel oil usually contains all the solid impurities originally present in the crude oil. If these solids contain a large proportion of salt, they are very fusible and can stick to the ducting, valves, waste heat boiler, and so on.
- *Vanadium and Sodium Contents* The vanadium content in fuel oil may be deposited in the ash on surfaces. These deposits act catalytically in converting SO_2 to SO_3, thereby creating acid dew point problems. Both sodium and vanadium from fuel oil may form sticky ash compounds having low melting temperatures. These compounds increase the deposition of ash and are corrosive.

Some advantages of using No. 2 fuel oil as the fuel additive at a combustion waste facility are (1) the fuel oil requires less storage space than coal, (2) is not subject to spontaneous combustion, (3) is the least difficult to ignite of the liquid fuels, (4) is not subject to deterioration, and (5) is free from expensive manual handling.

Coal

Although coal is not usually utilized as an auxiliary fuel, it may be used in fluidized-bed waste combustion applications.

There is no satisfactory definition of coal. It is a mixture of organic, chemical, and mineral materials produced by a natural process of growth and decay, accumulation of both vegetal and mineral debris, and accomplished by chemical, biological, bacteriological, and metamorphic action. The characteristics of coal vary considerably with location, and even within a given mine some variation in composition is usually encountered. Coals are classified according to rank, which refers to the degree of conversion from one form of coal to another (e.g., lignite to anthracite). The following ranking of coals is employed today: anthracite, bituminous, subbituminous, and lignite.

Solid fuels, including coal, consist of free carbon, moisture, hydrocarbons, oxygen (mostly in the form of oxygenated hydrocarbons), small amounts of sulfur and nitrogen, and nonvolatile noncombustible materials designated as ash. Two types of analyses are in common use for expressing the composition of these fuels: the ultimate analysis and the proximate analysis.

In an *ultimate* analysis, determination is made of the carbon, moisture (water), ash, nitrogen, sulfur, net hydrogen, and perhaps combined water in the fuel. Ash content is usually determined as a whole (one quantity), although a separate analysis can be made on the ash. Since the hydrogen content is always in excess of the amount needed to form water with all the oxygen present, it is assumed that the oxygen is completely united with the available hydrogen, and this combination may be reported as *combined water.* Note that these two elements do react in the combustion process to form water. The hydrogen in excess of that necessary to combine with the oxygen in the fuel is often termed *net hydrogen*, which usually reacts with the oxygen in the combustion air, halogens (to form acids), and so on. Another common approximate method for expressing the composition of a solid fuel is to report it as moisture, volatile combustible matter, fixed carbon, and ash. This is known as a *proximate analysis.* The *moisture* in the proximate analysis is the same as that in the ultimate analysis. The *volatile combustible matter* consists of a large amount of the carbon in the fuel, which is lost as volatile hydrocarbons, plus the hydrogen and combined water, which are given off when combusted. The *fixed carbon* is the carbon left in the fuel after the volatile combustible matter has been removed. *Ash* is the noncombustible residue after complete combustion of the coal. The weight of ash is usually slightly less than that of the mineral matter originally present before burning. However, for high calcium content coals, the ash can be higher than the mineral matter due to retention of some of the sulfur present in the coal. When coal is heated, it becomes soft and sticky, and as the temperature in a waste combustor continues to rise, it becomes fluid. Coals with ash that softens and fuses at comparatively low temperatures are likely to deposit the ash within the system and create problems. The sulfur is separately measured, and its amount is useful in judging the air pollution potential of the combustion of the coal. Combustion of sulfur forms oxides, which usually combine with water to form acids that may condense when the flue gas is cooled below its dew point temperature.

Use of coal as a fuel in a waste combustion system results in particulate (ash) carryover and sulfur oxides (from sulfur) emissions. Fly ash is not considered a major problem; equipment employed for its control is presented in Chapter 10. Sulfur emissions can constitute a major and additional problem since most of the so-called hazardous wastes do not contain sulfur. The sulfur in coal is found in both organic and inorganic iron pyrites and/or marcasite. There appears to be no evidence that sulfur occurs in coal in its elemental state. The amount of sulfur found in coals varies significantly. During combustion, about 80% of the sulfur in the coal appears in the flue gas in the form of the oxides of sulfur; most of this is SO_2; <5% is SO_3. Coal cleaning at the mine reduces the ash content and simultaneously reduces the sulfur content by removing some of the iron pyrites. Today, cleaning is accomplished by gravimetric separation, which is a successful method because pyrite is nearly five times denser than coal. Unfortunately, methods to reduce organic sulfur are not economically viable at this time.

In selecting coal as the fuel at a waste combustion site, its storage must be considered. Coal slowly deteriorates when exposed to weathering. Attention must be given to the manner in which the coal is stockpiled; large piles loosely formed can ignite spontaneously. This problem is most severe with smaller sizes of coal and high sulfur content. Where large amounts must be stored, such as at power stations, stockpiles are created by using heavy equipment to form piles several hundred feet wide, several thousand feet long, and ~20 ft high. When smaller quantities are employed and turnover is fairly rapid, conical piles with a 12-ft depth or less are used. Where open piles are not permitted, silos can be used for coal storage; these should be equipped with fugitive dust control for use during loading.

Selection

The two major classes of hazardous waste incinerators are liquid injection and rotary kiln; fluidized-bed incinerators and fixed hearths, among other types, are used to a lesser extent (see Chapter 7). No set rule applies to the selection of a given fuel for a particular type of incinerator. However, some general comments and guidelines can be offered. Natural gas and No. 2 distillate oil are the usual options available for a liquid injection incinerator.

Number 6 Bunker oil is often not used with smaller units because of its storage and heating requirements. Rotary kilns can use natural gas, No. 2 or No. 6 oil. Natural gas appears to be the preferred choice for most existing small liquid injection and rotary kiln units. Due to today's economics, natural gas is most often used in large units. Coal appears to be ideally suited for the fluidized-bed incinerator, but natural gas and fuel oils have been widely used.

If a significant amount of fuel is to be employed at a waste combustion facility, the design engineer may require a thorough survey of the various fuels available in the location of the plant. The results of this study will assist in the fuel selection process. The factors in choosing a fuel for a proposed facility are

1. Cost of fuel
2. Delivered cost of fuel as fired. Transportation should be included in this cost.
3. Storage, heating, and handling costs.
4. Added cost of ash handling and removal, particularly with coal.

5. Market price history and trends. These are needed to serve as a guide for future price relationships.
6. Resources or reserves. These should be extensive to assure future supplies.
7. Relative availability. This should be considered under both normal and unusual circumstances or demands.
8. Long-range government conservation. This may be applied to limited reserves, especially when these reserves are in demand for specific essential uses.

6.3 STOICHIOMETRIC CALCULATIONS

In waste combustion applications, one of the key engineering calculations is the prediction of the flue gas flow rate and composition following combustion of a waste–fuel mixture. Unfortunately, no simple generalized procedure is available to accomplish this task. The resulting flue gas flow rate and composition are obviously strong functions of the amount of air and the composition of both the waste and fuel. However, this flow rate and composition also depend, to a lesser extent, on the operating temperature in the combustor because of chemical reaction equilibrium effects.

One of the difficulties in developing a comprehensive, overall calculational procedure to determine flue gas flows and compositions is coping with the different methods that are in use to characterize wastes. As described in Chapter 1, the waste characterization often takes the form of an ultimate analysis, but there are other ways in which the waste may be described. The four categories of characterization include

1. Ultimate or elemental analysis—mass basis, fraction or percentage
2. Ultimate analysis—mole basis
3. Componential or compound analysis—mass basis
4. Componential analysis—mole basis

If an ultimate analysis on a mass basis is provided for the waste–fuel mixture, stoichiometric (or theoretical) air mass requirements per unit mass of mixture (m_{st}) may be calculated from

$$m_{st} = 11.5m_C + 34.5m_H - 4.29m_O - 0.97m_{Cl} + 4.29m_S \qquad (6.3.1)$$

where m_{st} = stoichiometric air requirement per unit mass of waste–fuel mixture (lb air/lb mixture)

m_C = mass fraction of carbon (lb C/lb mixture)

m_H, m_O, m_{Cl}, m_S = mass fractions of hydrogen, oxygen, chlorine, and sulfur, respectively

The composition of the flue gas produced on a mass basis (for stoichiometric combustion only) is then

$$
\begin{aligned}
m_{CO_2} &= 3.67 m_C \\
m_{H_2O} &= 9.0 m_H - 0.25 m_{Cl} + \{m_{H_2O,W}\} \\
m_{SO_2} &= 2.00 m_S \\
m_{N_2} &= 8.78 m_C + 26.3 m_H - 3.29 m_O - 0.74 m_{Cl} + 3.29 m_S + m_{N,W} \\
m_{HCl} &= 1.03 m_{Cl}
\end{aligned}
\tag{6.3.2}
$$

where m_{CO_2} = mass CO_2 in the flue gas (lb CO_2/lb mixture)
m_{H_2O}, m_{SO_2}, m_{N_2}, m_{HCl} = masses of each product compound per unit mass of the waste–fuel mixture
$m_{N,W}$ = mass fraction of nitrogen in the waste–fuel mixture (lb N/lb mixture)
$m_{H_2O,W}$ = mass fraction of water in the waste–fuel mixture (lb H_2O/lb mixture); used only when part of the weight fraction of the waste–fuel mixture is expressed as water or moisture

This set of equations assumes that there is enough hydrogen available in the mixture to completely convert the chlorine to hydrogen chloride. The bracketed water term accounts for any water present in the waste–fuel mixture if a mass fraction of water or moisture is included in the analysis. The $m_{N,W}$ term accounts for any nitrogen that may be present in either the fuel or waste. If the ultimate analysis is on a *mole* basis, a balanced stoichiometric equation of the following form may be used:

$$
C_zH_yO_xCl_wS_vN_u + \left[z + \phi + v - \tfrac{1}{2}x\right]O_2 + \tfrac{79}{21}\left(z + \phi + v - \tfrac{1}{2}x\right)N_2
$$

$$
\rightarrow zCO_2 + 2\phi H_2O + wHCl + vSO_2 + \left[\tfrac{1}{2}u + \tfrac{79}{21}\left(z + \phi + v - \tfrac{1}{2}x\right)\right]N_2 \tag{6.3.3}
$$

where z, y, x, w, v, u = number of moles (or mole fractions) of C, H, O, Cl, S, N present in the waste–fuel mixture, respectively, and

$$
\begin{aligned}
\phi &= \tfrac{1}{4}(y - w) \quad \text{when } y > w \\
&= 0 \quad \text{when } y \leq w
\end{aligned}
$$

Note that the parameter ϕ is included if its value is positive and ignored if it is zero or negative. For this approach, the mole fraction for carbon in the waste–fuel mixture (x_C) is given by

$$x_C = \frac{z}{z + y + x + w + v + u} \tag{6.3.4}$$

with similar equations applicable to hydrogen, oxygen, and so on. This assumes that the waste–fuel mixture consists only of C, H, O, Cl, S, and N. Finally, Eqs. (6.3.1) through (6.3.4) are valid provided:

- Cl_2, SO_3, and NO_x formation is neglected.
- Oxygen in the waste–fuel mixture is available for combustion and is therefore treated as a *credit* in calculating stoichiometric air requirements.
- The air essentially consists of 79% nitrogen and 21% oxygen (by mole or volume).
- The air essentially consists of 76.8% nitrogen and 23.3% oxygen (by weight or mass).
- Combustion is complete.

Whether the calculations are performed on a mass or a mole basis, the *excess* air usage requirements for a particular application may easily be included in this analysis. For example, for 100% excess air, twice the stoichiometric air requirement would be employed; half of this air would be used in stoichiometrically combusting the waste–fuel mixture and the remaining half would be carried through the process in the flue gas as the *extra* or excess air.

The preferred choice of the waste characterization method in performing these calculations is often dictated by the contents of the waste mixture. If the waste contains an agglomeration of organics, ultimate analysis would make the calculations easier; if the waste consists of one hydrocarbon (e.g., benzene) in water, the componential analysis is preferred. The choice as to whether the analysis is on a mass or mole basis may also be dictated by the calculational procedure employed.

The Environmental Protection Agency (EPA) has provided worksheets to simplify the calculation of the mass fraction and standard volumetric flow rate of the flue gas. The worksheets in their original form are given in Tables 6.3.1 and 6.3.2 for liquid injection and rotary kiln incinerators, respectively.[1] This calculational procedure is based on the availability of an ultimate analysis for both the waste and any fuel employed. If the analysis data are based on moles or on one individual species, the user must convert the data to an ultimate analysis based on mass. In addition, this procedure does not take into account chemical reaction equilibrium effects, for example, how much of the HCl is converted to Cl_2 (see Sections 5.5 and 5.7.) Additional calculations, not provided on the worksheet, must be performed if the componential analysis is required on a mole fraction basis or if the actual volumetric flow rate is needed.

Although a simple all-encompassing calculational scheme cannot be provided for the flue gas composition and flow rates, a suggested step-by-step

Table 6.3.1 EPA worksheet: Procedure to calculate stoichiometric air requirements, combustion gas flow, and composition (liquid injection incineration)[a]

1. Identify the elemental composition and moisture content of the waste or waste mixture:

Carbon, C_w	______	lb/lb waste
Hydrogen, H_w	______	lb/lb waste
Moisture, H_2O_w	______	lb/lb waste
Oxygen, O_w	______	lb/lb waste
Nitrogen, N_w	______	lb/lb waste
Chlorine, Cl_w	______	lb/lb waste
Fluorine, F_w	______	lb/lb waste
Bromine, Br_w	______	lb/lb waste
Iodine, I_w	______	lb/lb waste
Sulfur, S_w	______	lb/lb waste
Phosphorus, P_w	______	lb/lb waste

2. If auxiliary fuel is to be burned in conjunction with the waste, identify the fuel type and approximate proposed fuel-to-waste ratio from the permit application. (If auxiliary fuel is to be used only for startup, proceed to step 5.)

Fuel type	______	
Fuel–waste ratio, n_f	______	lb fuel/lb waste

3. Determine the approximate elemental composition of the fuel from the following table:

	lb component/lb fuel		
Component	Residual Fuel Oil (e.g., No. 6)	Distillate Fuel Oil (e.g., No. 2)	Natural Gas
C_f	0.866	0.872	0.693
H_f	0.102	0.123	0.227
N_f	—	—	0.08
S_f	0.03	0.005	—

4. Calculate the composition of the combined waste/auxiliary fuel feed:

C $\quad \dfrac{C_w + n_f C_f}{1 + n_f} =$ ______ lb/lb feed

H $\quad \dfrac{H_w + n_f H_f}{1 + n_f} =$ ______ lb/lb feed

H_2O $\quad \dfrac{H_2O_w}{1 + n_f} =$ ______ lb/lb feed

N $\quad \dfrac{N_w + n_f N_f}{1 + n_f} =$ ______ lb/lb feed

O $\quad \dfrac{O_w}{1 + n_f} =$ ______ lb/lb feed

Table 6.3.1 (*Continued*)

Cl	$\frac{Cl_w}{1 + n_f}$	= ______	lb/lb feed
F	$\frac{F_w}{1 + n_f}$	= ______	lb/lb feed
Br	$\frac{Br_w}{1 + n_f}$	= ______	lb/lb feed
I	$\frac{I_w}{1 + n_f}$	= ______	lb/lb feed
S	$\frac{S_w + n_f S_f}{1 + n_f}$	= ______	lb/lb feed
P	$\frac{P_w}{1 + n_f}$	= ______	lb/lb feed

5. Calculate the stoichiometric oxygen requirement:

$C \times 2.67 \frac{\text{lb } O_2}{\text{lb C}}$	= ______	lb O_2/lb feed
$\left(H - \frac{Cl}{35.5} - \frac{F}{19}\right) \times 8.0 \frac{\text{lb } O_2}{\text{lb H}}$	= ______	lb O_2/lb feed
$S \times 1.0 \frac{\text{lb } O_2}{\text{lb S}}$	= ______	lb O_2/lb feed
$P \times 1.29 \frac{\text{lb } O_2}{\text{lb P}}$	= ______	lb O_2/lb feed
−O (in feed)	= ______	lb O_2/lb feed[b]
$(O_2)_{\text{stoich}} = \Sigma =$	______	lb O_2/lb feed

6. Calculate the combustion gas mass flows, based on the stoichiometric oxygen requirement:

CO_2	$C \times 3.67 \frac{\text{lb } CO_2}{\text{lb C}}$	= ___	lb CO_2/lb feed
H_2O	$\left[\left(H - \frac{Cl}{35.5} - \frac{F}{19}\right) \times 9.0 \frac{\text{lb } H_2O}{\text{lb H}}\right] + H_2O \text{ (in feed)}$	= ___	lb H_2O/lb feed
N_2	$\left[(O_2)_{\text{stoich}} \times 3.29 \frac{\text{lb } N_2}{\text{lb } O_2} \text{ (in air)}\right] + N \text{ (in feed)}$[c]	= ___	lb N_2/lb feed
HCl	$Cl \times 1.03 \frac{\text{lb HCl}}{\text{lb Cl}}$	= ___	lb HCl/lb feed

(*continued*)

Table 6.3.1 (*Continued*)

HF	$F \times 1.05 \dfrac{\text{lb HF}}{\text{lb F}}$	= ___ lb HF/lb feed
Br_2	Br	= ___ lb Br_2/lb feed[d]
I_2	I	= ___ lb I_2/lb feed[e]
SO_2	$S \times 2.0 \dfrac{\text{lb SO}_2}{\text{lb S}}$	= ___ lb SO_2/lb feed
P_2O_5	$P \times 2.29 \dfrac{\text{lb P}_2\text{O}_5}{\text{lb P}}$	= ___ lb P_2O_5/lb feed

Combustion products = CP = Σ = ______________________ lb/lb feed

7. Identify the total excess air rate:

   EA = ______________________ %/100

8. Calculate the additional nitrogen and oxygen present in the combustion gases due to excess air feed:

   $(O_2)_{EA} = EA \times (O_2)_{stoich}$ = ______________________ lb O_2/lb waste

   $(N_2)_{EA} = 3.29 \dfrac{\text{lb N}_2}{\text{lb O}_2}$ (in air) $\times (O_2)_{EA}$ = ______________________ lb N_2/lb waste

9. Calculate the total combustion gas flow:

   Combustion gases = CG = CP + $(O_2)_{EA}$ + $(N_2)_{EA}$ = ______________________ lb/lb waste

10. Calculate the mass fraction of each combustion gas component:

CO_2	$\dfrac{CO_2}{CG}$	= ______________________ lb/lb gas
H_2O	$\dfrac{H_2O}{CG}$	= ______________________ lb/lb gas
N_2	$\dfrac{N_2 \text{ (from No. 6)} + (N_2)_{EA}}{CG}$	= ______________________ lb/lb gas
O_2	$\dfrac{(O_2)_{EA}}{CG}$	= ______________________ lb/lb gas
HCl	$\dfrac{HCl}{CG}$	= ______________________ lb/lb gas
HF	$\dfrac{HF}{CG}$	= ______________________ lb/lb gas
Br_2	$\dfrac{Br_2}{CG}$	= ______________________ lb/lb gas

Table 6.3.1 (*Continued*)

I_2	$\dfrac{I_2}{CG}$	= ______	lb/lb gas
SO_2	$\dfrac{SO_2}{CG}$	= ______	lb/lb gas
P_2O_5	$\dfrac{P_2O_5}{CG}$	= ______	lb/lb gas

11. Identify those components that constitute <1–2% of the combustion gas. These components can be eliminated from further consideration in heat and material balance calculations. In most cases, CO_2, H_2O, N_2, and O_2 will be the only combustion gas components that need to be considered. If halogenated compounds are burned, HCl, HBr, HF, and HI must also be considered.

12. Calculate the volumetric flow of the major combustion products at standard conditions of 68°F and 1 atm.[f]

CO_2	$\dfrac{CO_2}{CG} \times CG \div 0.114 \dfrac{\text{lb}}{\text{scf}}$	= ______	scf/lb feed
H_2O	$\dfrac{H_2O}{CG} \times CG \div 0.0467 \dfrac{\text{lb}}{\text{scf}}$	= ______	scf/lb feed
N_2	$\dfrac{N_2}{CG} \times CG \div 0.0727 \dfrac{\text{lb}}{\text{scf}}$	= ______	scf/lb feed
O_2	$\dfrac{O_2}{CG} \times CG \div 0.083 \dfrac{\text{lb}}{\text{scf}}$	= ______	scf/lb feed
Other	$\dfrac{\text{other}}{CG} \times CG \div (0.00259\ M) \dfrac{\text{lb}}{\text{scf}}$	= ______	scf/lb feed
	where M = molecular weight[g]		

Total flow, $q = \Sigma =$ ______ $\dfrac{\text{scf}}{\text{lb feed}}$

$q \times m_{\text{feed}}$ (lb/h) $\div$ 60 = ______ scfm

The following footnotes are comments prepared by L. Theodore and J. Reynolds.

[a] The notation of the original EPA worksheet has been retained, and often does not correspond to the notation used in this text. Any illustrative examples or problems based on the use of this table will also retain the EPA notation.

[b] The quantity O in the feed need not be multiplied by 0.5 since there is 1.0 lb O_2/1.0 lb O. This term is subtracted in calculating stoichiometric oxygen requirements since this is a credit, that is, there is oxygen already present in the feed that is used for combustion.

[c] The N (in feed) term need not be multiplied by 0.5 since there is 1.0 lb N_2/1.0 lb N.

[d] The Br term need not be multiplied by 0.5.

[e] The I term need not be multiplied by 0.5.

[f] The density terms—0.114, 0.0467, and so on, lb/ft^3—are based on densities at 68°F, 30 in. Hg (std. conditions) per Example 5.7.2.

[g] M is the average molecular weight (abbreviated MW elsewhere in text) of the other components.

Table 6.3.2 EPA worksheet: Procedure to calculate stoichiometric air requirements, approximate combustion gas flows, and approximate gas compositions (rotary kilns)[a]

1. Identify the elemental composition and moisture content of the waste fed to the kiln.

	1. Solids (kiln)	*2. Liquids (kiln)*	
Carbon, C_w	______	______	lb/lb waste
Fuel hydrogen, H_w	______	______	lb/lb waste
Moisture, H_2O_w	______	______	lb/lb waste
Oxygen, O_w	______	______	lb/lb waste
Nitrogen, N_w	______	______	lb/lb waste
Chlorine, Cl_w	______	______	lb/lb waste
Fluorine, F_w	______	______	lb/lb waste
Bromine, Br_w	______	______	lb/lb waste
Iodine, I_w	______	______	lb/lb waste
Sulfur, S_w	______	______	lb/lb waste
Phosphorus, P_w	______	______	lb/lb waste

2. Identify the approximate liquid and solid waste feed rates to the kiln, and calculate the liquid/solid feed fractions:

Liquid feed rate, m_1 = ______ lb/h
Solid feed rate, m_2 = ______ lb/h
Total feed, $m_{12} = m_1 + m_2$ = ______ lb/h
Liquid fraction, $n_1 = m_1/m_{12}$ = ______ lb liquid/lb waste
Solid fraction, $n_2 = 1 - n_1$ = ______ lb solid/lb waste

3. If auxiliary fuel is to be burned in conjunction with the wastes, identify the fuel type and approximate proposed fuel-to-waste ratio:

Fuel type: ______
Fuel–waste ratio in kiln; n_{fk} = ______ lb fuel/lb waste

4. Determine the approximate elemental composition of the fuel from the following table:

	lb component/lb fuel		
Component	Residual Fuel Oil (e.g., No. 6)	Distillate Fuel Oil (e.g., No. 2)	Natural Gas
C_f	0.866	0.872	0.693
H_f	0.102	0.123	0.227
N_f	—	—	0.08
S_f	0.03	0.005	—

5. Calculate the composition of the combined waste/auxiliary fuel feed to the kiln:

$$C_k \quad \frac{n_1C_1 + n_2C_2 + n_fC_f}{1 + n_f} = \text{\_\_\_\_\_\_ lb/lb feed}$$

Table 6.3.2 (*Continued*)

H_k	$\dfrac{n_1 H_1 + n_2 H_2 + n_f H_f}{1 + n_f}$	= ______	lb/lb feed
H_2O_k	$\dfrac{n_1 H_2O_1 + n_2 H_2O_2}{1 + n_f}$	= ______	lb/lb feed
N_k	$\dfrac{n_1 N_1 + n_2 N_2 + n_f N_f}{1 + n_f}$	= ______	lb/lb feed
O_k	$\dfrac{n_1 O_1 + n_2 O_2}{1 + n_f}$	= ______	lb/lb feed
Cl_k	$\dfrac{n_1 Cl_1 + n_2 Cl_2}{1 + n_f}$	= ______	lb/lb feed
F_k	$\dfrac{n_1 F_1 + n_2 F_2}{1 + n_f}$	= ______	lb/lb feed
Br_k	$\dfrac{n_1 Br_1 + n_2 Br_2}{1 + n_f}$	= ______	lb/lb feed
I_k	$\dfrac{n_1 I_1 + n_2 I_2}{1 + n_f}$	= ______	lb/lb feed
S_k	$\dfrac{n_1 S_1 + n_2 S_2 + n_f S_f}{1 + n_f}$	= ______	lb/lb feed
P_k	$\dfrac{n_1 P_1 + n_2 P_2}{1 + n_f}$	= ______	lb/lb feed

6. Calculate the stoichiometric oxygen requirement for the kiln:

$C_k \times 2.67 \dfrac{\text{lb } O_2}{\text{lb C}}$	= ______	lb O_2/lb feed
$H_k - \dfrac{Cl_k}{35.5} - \dfrac{F_k}{19} \times 8.0 \dfrac{\text{lb } O_2}{\text{lb H}}$	= ______	lb O_2/lb feed[b]
$S_k \times 1.0 \dfrac{\text{lb } O_2}{\text{lb S}}$	= ______	lb O_2/lb feed
$P_k \times 1.29 \dfrac{\text{lb } O_2}{\text{lb P}}$	= ______	lb O_2/lb feed
$\dfrac{-O_k \text{ (in feed)}}{(O_2)_{\text{stoich(k)}}}$	= ______	lb O_2/lb feed[c]

7. Calculate the combustion gas mass flows, based on the stoichiometric oxygen requirement (assume complete combustion is achieved for purposes of gas flow estimation):

CO_{2k}	$C_k \times 3.67 \dfrac{\text{lb } CO_2}{\text{lb C}}$	= ______	lb CO_2/lb feed

(continued)

Table 6.3.2 (*Continued*)

H_2O_k	$H_k - \frac{Cl_k}{35.5} - \frac{F_k}{19} \times 9.0 \frac{\text{lb } H_2O}{\text{lb H}} + H_2O_k$ (in feed)	=	______	lb H_2O/lb feed[d]
N_{2k}	$(O_2)_{stoich(k)} \times 3.29 \frac{\text{lb } N_2}{\text{lb } O_2}$ (in air) + N_k (in feed)	=	______	lb N_2 lb feed
HCl_k	$Cl_k \times 1.03 \frac{\text{lb HCl}}{\text{lb Cl}}$	=	______	lb HCl/lb feed
HF_k	$F_k \times 1.05 \frac{\text{lb HF}}{\text{lb F}}$	=	______	lb Hf/lb feed
Br_{2k}	Br_k	=	______	lb Br_2/lb feed[e]
I_{2k}	I_k	=	______	lb I_2/lb feed[f]
SO_{2k}	$S_k \times 2.0 \frac{\text{lb } SO_2}{\text{lb S}}$	=	______	lb SO_2/lb feed
P_2O_{5k}	$P_k \times 2.29 \frac{\text{lb } P_2O_5}{\text{lb P}}$	=	______	lb P_2O_5/lb feed
Kiln combustion products = CP_k		=	______	lb/lb feed

8. Identify the elemental composition and moisture content of the liquid wastes to be burned in the afterburner (if any):

C_3 ______ lb/lb waste
H_3 ______ lb/lb waste
H_2O_3 ______ lb/lb waste
O_3 ______ lb/lb waste
N_3 ______ lb/lb waste
Cl_3 ______ lb/lb waste
F_3 ______ lb/lb waste
Br_3 ______ lb/lb waste
I_3 ______ lb/lb waste
S_3 ______ lb/lb waste
P_3 ______ lb/lb waste

9. Identify the fuel type and approximate, proposed fuel-to-waste ratio for the afterburner:

Fuel type: ______
Fuel–waste ratio, n_{fa} = ______ lb fuel/lb waste

10. Determine the approximate elemental composition of the fuel from the table shown in step 4:

C_{fa} = ______ lb/lb fuel
H_{fa} = ______ lb/lb fuel
N_{fa} = ______ lb/lb fuel
S_{fa} = ______ lb/lb fuel

Table 6.3.2 (*Continued*)

11. Calculate the composition of the combined waste/auxiliary fuel feed to the afterburner:

$C_a \quad \dfrac{C_3 + n_{fa}C_{fa}}{1 + n_{fa}} =$ __________ lb/lb feed

$H_a \quad \dfrac{H_3 + n_{fa}H_{fa}}{1 - n_{fa}} =$ __________ lb/lb feed

$H_2O_a \quad \dfrac{H_2O_3}{1 + n_{fa}} =$ __________ lb/lb feed

$N_a \quad \dfrac{N_3 + n_{fa}N_{fa}}{1 + n_{fa}} =$ __________ lb/lb feed

$O_a \quad \dfrac{O_3}{1 + n_{fa}} =$ __________ lb/lb feed

$Cl_a \quad \dfrac{Cl_3}{1 + n_{fa}} =$ __________ lb/lb feed

$F_a \quad \dfrac{F_3}{1 + n_{fa}} =$ __________ lb/lb feed

$Br_a \quad \dfrac{Br_3}{1 + n_{fa}} =$ __________ lb/lb feed

$I_a \quad \dfrac{I_3}{1 + n_{fa}} =$ __________ lb/lb feed

$S_a \quad \dfrac{S_3 + n_{fa}S_{fa}}{1 + n_{fa}} =$ __________ lb/lb feed

$P_a \quad \dfrac{P_3}{1 + n_{fa}} =$ __________ lb/lb feed

12. Calculate the stoichiometric oxygen requirement for the afterburner feed:

$C_a \times 2.67 \dfrac{\text{lb } O_2}{\text{lb C}} =$ __________ lb O_2/lb feed

$\left(H_a - \dfrac{Cl_a}{35.5} - \dfrac{F_a}{19}\right) \times 8.0 \dfrac{\text{lb } O_2}{\text{lb H}} =$ __________ lb O_2/lb feed

$S_a \times 1.0 \dfrac{\text{lb } O_2}{\text{lb S}} =$ __________ lb O_2/lb feed

$P_a \times 1.29 \dfrac{\text{lb } O_2}{\text{lb P}} =$ __________ lb O_2/lb feed

$-O_a$ (in feed) = __________ lb O_2/lb feed[g]

$(O_2)_{stoich(a)} = \Sigma =$ __________ lb O_2/lb feed

(continued)

Table 6.3.2 (*Continued*)

13. Calculate the combustion gas mass flows, based on the stoichiometric oxygen requirement:

CO_{2_a}	$C_a \times 3.67 \dfrac{\text{lb } CO_2}{\text{lb C}}$	= ____	lb CO_2/lb feed
H_2O_a	$\left[\left(H_a - \dfrac{Cl_a}{35.5} - \dfrac{F_a}{19}\right)\left(9.0 \dfrac{\text{lb } H_2O}{\text{lb H}}\right)\right] + H_2O_a \text{ (in feed)}$	= ____	lb H_2O/lb feed
N_{2_a}	$\left[(O_2)_{\text{stoich(a)}} \times 3.29 \dfrac{\text{lb } N_2}{\text{lb } O_2} \text{ (in air)}\right] + N_a \text{ (in feed)}$	= ____	lb N_2/lb feed[h]
HCl_a	$Cl_a \times 1.03 \dfrac{\text{lb HCl}}{\text{lb Cl}}$	= ____	lb HCl/lb feed
HF_a	$F_a \times 1.05 \dfrac{\text{lb HF}}{\text{lb F}}$	= ____	lb HF/lb feed
Br_{2_a}	Br_a	= ____	lb Br_2/lb feed[i]
I_{2_a}	I_a	= ____	lb I_2/lb feed[j]
SO_{2_a}	$S_a \times 2.0 \dfrac{\text{lb } SO_2}{\text{lb S}}$	= ____	lb SO_2/lb feed
$P_2O_{5_a}$	$P_a \times 2.29 \dfrac{\text{lb } P_2O_5}{\text{lb P}}$	= ____	lb P_2O_5/lb feed

Afterburner combustion products = $CP_a = \Sigma$ = ________________ lb/lb feed

14. Calculate the ratio of total afterburner feed to total kiln feed.

Liquid waste to kiln: m_1	=	________________	lb/h
Solid waste to kiln: m_2	=	________________	lb/h
Auxiliary fuel to kiln: $(m_1 + m_2)n_{fk}$	=	________________	lb/h
Liquid waste to afterburner: m_3	=	________________	lb/h
Auxiliary fuel to afterburner: $m_3 n_{fa}$	=	________________	lb/h

$$n_{ak} = \frac{m_3(1 + n_{fa})}{(m_1 + m_2)(1 + n_{fk})} = \text{lb afterburner feed/lb kiln feed}$$

15. Calculate the total combustion gas mass flows, based on stoichiometric oxygen requirements.

CO_2	$\dfrac{CO_{2_k} + CO_{2_a} n_{ak}}{1 + n_{ak}}$	= ________________	lb/lb feed
H_2O	$\dfrac{H_2O_k + n_{ak} H_2O_a}{1 + n_{ak}}$	= ________________	lb/lb feed

Table 6.3.2 (*Continued*)

N_2	$\dfrac{N_{2_k} + n_{ak}N_{2_a}}{1 + n_{ak}}$	= ______	lb/lb feed
HCl	$\dfrac{HCl_k + n_{ak}HCl_a}{1 + n_{ak}}$	= ______	lb/lb feed
HF	$\dfrac{HF_k + n_{ak}HF_a}{1 + n_{ak}}$	= ______	lb/lb feed
Br_2	$\dfrac{Br_{2_k} + n_{ak}Br_a}{1 + n_{ak}}$	= ______	lb/lb feed
I_2	$\dfrac{I_{2_k} + n_{ak}I_{2_a}}{1 + n_{ak}}$	= ______	lb/lb feed
SO_2	$\dfrac{SO_{2_k} + n_{ak}SO_{2_a}}{1 + n_{ak}}$	= ______	lb/lb feed
P_2O_5	$\dfrac{P_2O_{5_k} + n_{ak}P_2O_{5_a}}{1 + n_{ak}}$	= ______	lb/lb feed

Combustion products = CP = Σ = ______ lb/lb feed

16. Identify the total excess air rate for the system (i.e., to be maintained in the afterburner):

    EA = ______ percent

17. Calculate the additional nitrogen and oxygen present in the combustion gases due to excess air feed:

$$(O_2)_{EA} = EA \times \frac{(O_2)_{stoich(k)} + n_{ak}(O_2)_{stoich(a)}}{1 + n_{ak}} = \text{______ lb } O_2\text{/lb waste}$$

$$(N_2)_{EA} = 3.29 \frac{\text{lb } N_2}{\text{lb } O_2} \text{ (in air)} \times (O_2)_{EA} = \text{______ lb } N_2\text{/lb waste}$$

18. Calculate the total combustion gas flow:

    Combustion gas flow = CG = CP + $(O_2)_{EA}$ + $(N_2)_{EA}$ = ______ lb/lb feed

19. Calculate the mass fraction of each combustion gas component:

CO_2	$\dfrac{CO_2}{CG}$	= ______	lb/lb gas
H_2O	$\dfrac{H_2O}{CG}$	= ______	lb/lb gas
N_2	$\dfrac{N_2 \text{ (from 15)} + (N_2)_{EA}}{CG}$	= ______	lb/lb gas

(continued)

Table 6.3.2 (*Continued*)

N_2	$\dfrac{N_2\text{ (from 15)} + (N_2)_{EA}}{CG}$	= ______________	lb/lb gas
O_2	$\dfrac{(O_2)_{EA}}{CG}$	= ______________	lb/lb gas
HCl	$\dfrac{HCl}{CG}$	= ______________	lb/lb gas
HF	$\dfrac{HF}{CG}$	= ______________	lb/lb gas
Br_2	$\dfrac{Br_2}{CG}$	= ______________	lb/lb gas
I_2	$\dfrac{I_2}{CG}$	= ______________	lb/lb gas
SO_2	$\dfrac{SO_2}{CG}$	= ______________	lb/lb gas
P_2O_5	$\dfrac{P_2O_5}{CG}$	= ______________	lb/lb gas

20. Identify those components that constitute <1–2% of the combustion gas. These components can be eliminated from further consideration in heat and material balance calculations. In most cases, CO_2, H_2O, N_2, and O_2 will be the only combustion gas components that need to be considered. If halogenated compounds are burned, HCl, HBr, HF, and HI must also be considered.

21. Calculate the volumetric flow of the major combustion products from the kiln at standard conditions of 68°F and 1 atm:[k]

CO_2	$\left(\dfrac{CO_2}{CG}\right) \times CG \div 0.114$ lb/scf	= ______________	scf/lb
H_2O	$\left(\dfrac{H_2O}{CG}\right) \times CG \div 0.0467$ lb/scf	= ______________	scf/lb
N_2	$\left(\dfrac{N_2}{CG}\right) \times CG \div 0.0727$ lb/scf	= ______________	scf/lb
O_2	$\left(\dfrac{O_2}{CG}\right) \times CG \div 0.083$ lb/scf	= ______________	scf/lb
Other	$\left(\dfrac{\text{other}}{CG}\right) \times CG \div (0.00259\ M)$ lb/scf	= ______________	scf/lb

where M = molecular weight[l]

Table 6.3.2 (*Continued*)

Total flow $q = \Sigma$	= ______	scf/lb feed
$q \times (m_1 + m_2)(1 + n_{fk})(1 + n_{ak}) \div 60$	= ______	scfm

The following footnotes are comments prepared by L. Theodore and J. Reynolds.

[a]The notation of the original EPA worksheet has been retained, and often does not correspond to the notation used in this text. Any illustrative examples or problems based on the use of this table will also retain the EPA notation.

[b]This term should read:
$[H_k - (Cl_k/35.5) - (F_k/19)]8.0$(lb O_2/lb H)

[c]This term is misleading and the alignment is awkward. It should read:
$-O_k$ (in feed) = ______ lb O_2 (in feed)/lb feed
stoich $(O_2)_k$ = Σ ______ lb O_2/lb feed

[d]The first three terms should be bracketed.

[e,f]These terms need not be multiplied by 0.5.

[g]This term is misleading and the alignment is awkward. It should read
$-O_a$ (in feed) = ______ lb O_2/lb feed
stoich $(O_2)_a$ = Σ ______ lb O_2/lb feed

[h]The N_a term need not be multiplied by 0.5.

[i,j]Both terms need not be multiplied by 0.5.

[k]The lb/scf terms are densities at standard conditions.

[l]M should read the average molecular weight)abbreviated MW elsewhere in text) of the other components.

outline for this engineering calculation is given here. It should assist the reader in obtaining these results for most engineering applications.

1. Select a basis for the waste–fuel mixture. This is a very critical step to which the engineer often does not give enough thought. A good choice of basis can often greatly simplify the calculations.
    a. If an ultimate analysis is given on a mole basis, select 1 mol (or 100 mol) as a basis.
    b. If a componential analysis is given on a mole basis, select 1 mol as a basis.
    c. If an ultimate analysis is given on a mass basis, select 100 units of mass.
    d. If a componential balance is given on mass basis, select 100 units of mass.

For (c) and (d), the data should be converted to an elemental or componential mole basis.

2. Convert the individual elements to their end products of stoichiometric combustion. As described earlier,

$$C \rightarrow CO_2$$

$$H \rightarrow H_2O$$

$$Cl \rightarrow HCl$$

$$O \rightarrow O_2$$

$$N \rightarrow N_2$$

$$S \rightarrow SO_2$$

and so on.

If the analysis is on a component basis, it may be preferable to write the balanced stoichiometric equation for the individual species participating in the combustion reaction [see Eq. (6.3.3)]. Also note that the quantity of water formed must reflect the hydrogen combined with the chlorine in HCl.

3. Based on the results of step 2, calculate the stoichiometric oxygen or air requirement on a mole basis, being careful to correct for any oxygen present in the waste–fuel mixture. This will appear as a credit in the oxygen balance. This calculation may be performed on an elemental basis or directly from a balanced stoichiometric equation. The reader may again choose to utilize Eq. (6.3.3) for assistance.

4. Calculate and then add the excess air required for operation of the waste combustor to the flue products of stoichiometric combustion. Once again, care should be exercised to adjust the quantity of nitrogen to reflect that contained in the waste–fuel mixture. This step will complete the *initial* or first determination for the flue gas at the outlet of the waste combustor.

5. Based on the initial calculations in step 4 for HCl, H_2O, O_2, CO_2, SO_2, and so on determine the amount of HCl and SO_2 converted to Cl_2 and SO_3, respectively. This will require a chemical reaction equilibrium calculation as described in Chapter 5. The combustor operating temperature must be specified since the reaction equilibrium constant used to calculate each of these conversions is a function of the temperature.

6. Calculate the *final* componential or compound balance of the flue gas at the outlet of the combustor. The results will be on a mole basis. Since this product is a gas, a volume basis is equivalent to a mole basis. These can then be converted to other units, including a mass basis (ppm, etc.) depending on the choice of the user.

7. The total molar flow rate can also be calculated by referring to the basis used for steps 1 through 6. The total (or componential) volumetric flow rate at actual or standard conditions can also be calculated via either the ideal gas law or use of the appropriate flue gas density.

Several illustrative examples detailing this calculational approach are presented in Section 6.6.

6.4 THERMOCHEMICAL CALCULATIONS

The operating temperature in a waste combustor is a function of many variables. For most waste combustors, the operating temperature is calculated by determining the flame temperature under adiabatic or near-adiabatic conditions. From a calculational point of view, the flame temperature is a function of several variables, with a particularly strong dependence on the excess air requirement and the heating value of the combined waste–fuel mixture. Both a rigorous and an approximate approach to this calculation are provided in this section.

Rigorous Approach

A rigorous flame temperature calculation is schematically depicted in Fig. 6.4.1. The overall enthalpy approach introduced to the reader in Chapter 5 is now applied to this combustor calculation. It should be kept in mind that any convenient calculation path can be selected in determining the enthalpy change between reactants (initial state) and products (final state). Since heat (enthalpy) of reaction and/or net (or gross) heating values are available at a reference temperature (T_0) only, the calculations proceed along a chosen path that includes the reaction at this reference temperature. Referring to Fig. 6.4.1, the enthalpy change (gain) associated with the cooling of the reactants (feed) is

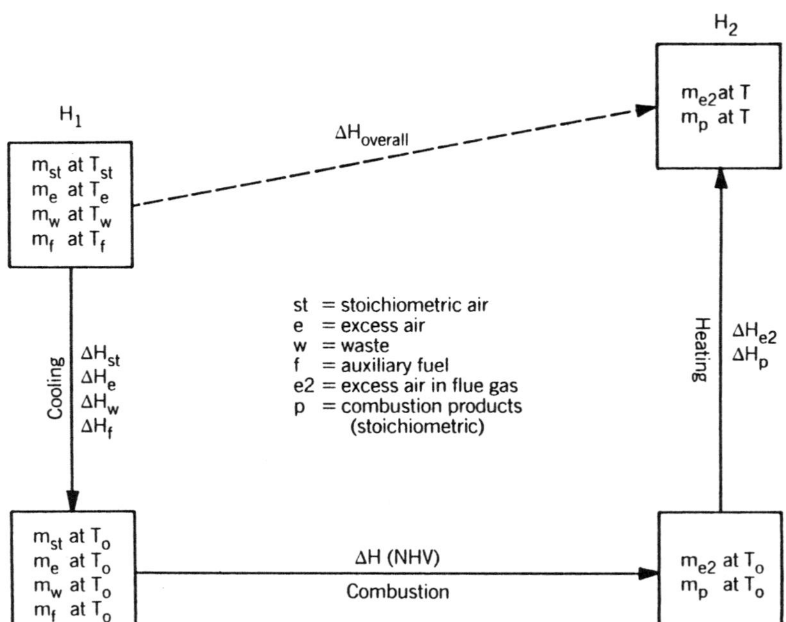

Figure 6.4.1 Flame temperature calculation schematic.

first obtained. This is followed by the enthalpy changes associated with the combustion reaction, the heating of the flue products, and the heat loss from the waste combustor. These thermochemical calculations have been developed on a mass basis as is customary in waste combustion applications. Heating values—both higher (gross) and lower (net)—are expressed as British thermal units per pound (Btu/lb).

Cooling Step

$$
\begin{aligned}
\Delta H_{\mathrm{st}} &= m_{\mathrm{st}} c_{P\mathrm{st}}(T_0 - T_{\mathrm{st}}) \\
\Delta H_{\mathrm{e}} &= m_{\mathrm{e}} c_{P\mathrm{e}}(T_0 - T_{\mathrm{e}}) \\
\Delta H_{\mathrm{w}} &= m_{\mathrm{w}} c_{P\mathrm{w}}(T_0 - T_{\mathrm{w}}) \\
\Delta H_{\mathrm{f}} &= m_{\mathrm{f}} c_{P\mathrm{f}}(T_0 - T_{\mathrm{f}})
\end{aligned}
\tag{6.4.1}
$$

where ΔH_{st} = sensible enthalpy change for cooling stoichiometric air (st) from its initial temperature T_{st} to reference temperature T_0
m_{st} = mass of st per unit mass of waste–fuel mixture (lb/lb mixture)
$c_{P\mathrm{st}}$ = average specific heat capacity of st over the temperature range T_{st}–T_0 (Btu/lb-°F)
T_{st} = initial temperature of st (°F)
T_0 = reference temperature (°F)

The subscripts st, e, w, f refer to the stoichiometric air, excess air, waste, and fuel, respectively.

Combustion Step

$$
\mathrm{NHV} = m_{\mathrm{w}}(\mathrm{NHV}_{\mathrm{w}}) + m_{\mathrm{f}}(\mathrm{NHV}_{\mathrm{f}}) \tag{6.4.2}
$$

where NHV = net heating value of waste–fuel mixture at T_0 (Btu/lb mixture)
$\mathrm{NHV}_{\mathrm{w}}$ = net heating value of waste at T_0 (Btu/lb waste)
$\mathrm{NHV}_{\mathrm{f}}$ = net heating value of fuel at T_0 (Btu/lb fuel)

Heating Step

$$
\begin{aligned}
\Delta H_{\mathrm{e2}} &= m_{\mathrm{e2}} c_{P\mathrm{e2}}(T - T_0) \\
\Delta H_{\mathrm{p}} &= \sum_{\mathrm{prod}} m_i c_{Pi}(T - T_0)
\end{aligned}
\tag{6.4.3}
$$

where ΔH_{e2} = sensible enthalpy change for heating excess air in the flue gas (e2) from reference temperature T_0 to the incinerator operating temperature T

ΔH_p = sensible enthalpy change for heating products of combustion (p) from reference temperature T_0 to the incinerator operating temperature T

m_{e2} = mass of e2 per unit mass of waste–fuel mixture (lb/lb mixture)

m_i = mass of i per unit mass of waste–fuel mixture (lb/lb mixture)

c_{Pi} = average specific heat capacity of i over the temperature range T_i to T

e2 = subscript indicating excess air in the flue gas

p = products of combustion (CO_2, H_2O, HCl, SO_2, N_2, etc.)

Overall Enthalpy Balance

$$H_2 - H_1 = \Delta H_{st} + \Delta H_e + \Delta H_w + \Delta H_f + (1 - R)\text{NHV} + \Delta H_{e2} + \Delta H_p = (R)\text{NHV} \tag{6.4.4}$$

The term R refers to the *fraction* (not percentage) of the net heating value associated with the combustion step that is lost by heat across the walls of the combustor; R is determined by the type and thickness of refractories and insulation used in the walls of the combustor. This value may vary from a low of 0.02 to a high of 0.15. This is also dependent upon the location, whether indoors or outdoors. If the system is assumed to be adiabatic, R is set equal to zero.

Equating these cooling and combustion enthalpy changes with the heating and overall enthalpy changes provides a rigorous equation relating the combustor operating temperature (T) with the system's variables. Application of this approach is given in Section 6.6.

Approximate Calculations Some reasonable assumptions can be made to simplify the rigorous approach provided previously since the calculations are somewhat cumbersome. These are detailed in this section. When compared to the rigorous approach, a simpler (and in many instances, a more informative) set of equations results that are valid for purposes of engineering calculation.

Assumptions

1. The sensible enthalpy change associated with the cooling step can in many instances be neglected compared to the net heating value (combustion step) of the combined waste–fuel mixture. For this condition,

$$\Delta H_{st} = \Delta H_e = \Delta H_w = \Delta H_f = 0 \tag{6.4.5}$$

2. Although the products of combustion consist of many components, the

major or primary components are nitrogen, carbon dioxide, and water (vapor). The average heat capacities of these components over the temperature range 60–2000°F (the latter being a typical incinerator operating temperature) are 0.27, 0.27, and 0.52 Btu/lb-°F, respectively. The arithmetic average of these three components is 0.35. However, since this product stream consists primarily of nitrogen, the average heat capacity of the combined mixture (not including the excess air) may be assigned a value of approximately 0.3 Btu/lb-°F. For this condition,

$$\Delta H_p = m_p(0.3)(T - T_0) \tag{6.4.6}$$

where m_p = mass of stoichiometric air, fuel, and waste entering the combustor per unit mass of waste–fuel mixture, or equivalently, mass of products less that of excess air per unit mass of waste–fuel mixture (lb/lb mixture).

Note: Energy required to raise the water resulting from combustion from the base temperature of 77°F as a liquid to boiling point at 1 atm of 212°F is equal to 1150 Btu/lb. This must be added as shown in Tables 6.4.5 and 6.4.6, step 5.

3. The average heat capacity of the (excess) air is ~0.27 Btu/lb-°F over the temperature range 60–2000°F. If this value is rounded to 0.3,

$$\Delta H_e = m_p(0.3)(T - T_0) \tag{6.4.7}$$

where m_e = mass of excess air per unit mass of waste–fuel mixture (lb/lb mixture).

4. Under truly adiabatic conditions the term R in Eq. (6.4.4) is zero. For this condition,

$$H_2 - H_1 = 0 \tag{6.4.8}$$

Under actual operating conditions, there will be some heat loss across the walls of the incinerator. However, if there is some preheat of the feed (usually the air), the effect of this assumption, to some extent, may balance the heat gained in the preheat (see assumption 1).

5. Since most of the heating value data are available at 60°F, the reference or standard temperature should be arbitrarily set to this condition, that is,

$$T_0 = 60°\text{F} \tag{6.4.9}$$

and NHVs should be obtained at approximately this temperature. It should be noted that in combusting liquids it is reasonable to assume 100% release of the heating value of the incoming liquid. However, it is not always valid to make the same assumption for incoming solids or for an incoming mixture of solids and liquids.

6. Perhaps the key assumption in this development is that associated with the stoichiometric air requirements for the combined waste–fuel mixture. Referring to Table 5.3.1, the stoichiometric air requirement, ν_{st} (ft^3 air/lb mixture), divided by NHV for any of the listed hydrocarbons is approximately 0.01 ft^3 air/Btu (or 100 Btu/ft^3 air). For example, the ratios for methane (M), benzene (B), and toluene (T) are

M $\quad \nu_{st}/\text{NHV} = 9.53/913 = 0.0104$ ft^3 air/Btu

B $\quad \nu_{st}/\text{NHV} = 35.73/3601 = 0.0099$ ft^3 air/Btu

T $\quad \nu_{st}/\text{NHV} = 42.88/4284 = 0.0100$ ft^3 air/Btu

Using the density of air at 60°F, this ratio can be converted to approximately 750 lb air/10^6 Btu or 7.5×10^{-4} lb air/Btu. Thus, for this condition the stoichiometric air requirement (m_{st}) is given by

$$m_{st} = 7.5 \times 10^{-4} \text{ NHV} \tag{6.4.10}$$

It should be noted that the validity and applicability of this assumption is limited to waste–fuel mixtures consisting of pure hydrocarbons, that is, organics with only hydrogen and carbon atoms, and to chlorinated organic mixtures where the mass fraction of the chlorine is low. However, for purposes of engineering calculations, this value for stoichiometric air will apply for almost all waste combustor applications.

To further simplify the development to follow, the describing equations are again based on 1.0 lb of the combined waste–fuel mixture, that is,

$$m_w + m_f = m_b = 1.0 \tag{6.4.11}$$

where m_w = mass of w (waste) per unit mass of waste–fuel mixture (lb/lb mixture)

w, f, b = subscripts indicating waste, fuel, and burnable mixture, respectively

Thus, m_w and m_f now become the mass fractions of the waste and fuel, respectively, in the burnable mixture (b). Applying the six assumptions listed here with this basis results in the following equation:

$$\text{NHV} = (1.0 + 7.5 \times 10^{-4} \text{ NHV})(0.3)(T - 60) + m_e(0.3)(T - 60) \tag{6.4.12}$$

or

$$\text{NHV} = (1 + m_e + 7.5 \times 10^{-4} \text{ NHV})(0.3)(T - 60) \tag{6.4.13}$$

It is more convenient to work with the mass of excess air in terms of a *fraction* or *percentage* rather than on the basis of a unit mass of the waste–fuel mixture (m_e). The fraction of excess air (EA) is defined by

$$\text{EA} = m_e/m_{st} \tag{6.4.14}$$

Since m_{st} may be approximated by

$$m_{st} = 7.5 \times 10^{-4}\ \text{NHV} \tag{6.4.15}$$

Eq. (6.4.15) for m_e becomes

$$m_e = (\text{EA})(7.5 \times 10^{-4})(\text{NHV}) \tag{6.4.16}$$

The approximate enthalpy balance is then

$$\text{NHV} = [1 + (\text{EA})(7.5 \times 10^{-4})(\text{NHV}) + (7.5 \times 10^{-4})(\text{NHV})1(0.3)(T - 60) \tag{6.4.17}$$

This may be further simplified to

$$\text{NHV} = [1 + (1 + \text{EA})(7.5 \times 10^{-4})(\text{NHV})](0.3)(T - 60) \tag{6.4.18}$$

Three parameters or variables appear in Eq. (6.4.18). Specifying any two of these constitutes a complete set for this equation. Thus, three equations may be generated expressing one variable (dependent) in terms of the other two (independent), that is,

$$\begin{aligned} T &= f_1(\text{EA}, \text{NHV}) \\ \text{EA} &= f_2(T, \text{NHV}) \\ \text{NHV} &= f_3(T, \text{EA}) \end{aligned} \tag{6.4.19}$$

The three resulting equations are

$$T = 60 + \frac{\text{NHV}}{(0.3)[1 + (1 + \text{EA})(7.5 \times 10^{-4})(\text{NHV})]} \tag{6.4.20}$$

$$\text{EA} = \frac{[\text{NHV}/0.3(T - 60)] - 1}{7.5 \times 10^{-4}\text{NHV}} - 1 \tag{6.4.21}$$

$$\text{NHV} = \frac{0.3(T - 60)}{1 - (1 + \text{EA})(7.5 + 10^{-4})(0.3)(T - 60)} \tag{6.4.22}$$

Note: The units of T and NHV are °F and Btu/lb, respectively; EA is a dimensionless fraction.

The reader should also note that these equations become sensitive to the value assigned to c_P (0.3 in this case). This is a function of both the temperature (T) and excess air fraction (EA), and also depends on the flue products since the heat capacities of air and CO_2 are about half that of H_2O. In addition, the 7.5×10^{-4} term "derived" earlier may vary slightly with the composition of the waste–fuel mixture combusted. The overall relationship between operating temperature and composition is therefore rather complex, and its prediction not necessarily as straightforward as shown here. These variabilities can be compensated for by treating the c_P term as a statistical parameter in Eqs. (6.4.20) through (6.4.22). Theodore and Reynolds[2] have generated values for c_P for excess air percentage (EA) and operating temperature (T) ranging from 0 to 100% and 1000–3500 K, respectively, for benzene, n-octane, and dichloromethane. These results are presented in Tables 6.4.1–6.4.3. In addition, regression coefficients α, β, and γ were obtained for the equation

$$c_P = \alpha + \beta T + \gamma(\text{EA}) \tag{6.4.23}$$

These may be found in Table 6.4.4.

Solutions to Eqs. (6.4.20)–(6.4.22) are graphically provided in Fig. 6.4.2. The equations and graph clearly demonstrate the interdependence of temperature, excess air, and heating value. Application of these equations to a waste combustor system can also be found in Section 6.6.

EPA Worksheets The EPA has also provided a procedure to estimate the excess air requirement for an incinerator. This procedure is shown in Tables 6.4.5 and 6.4.6 (with the original worksheet numbers and notation retained). Note that the stoichiometric oxygen requirement $(O_2)_{\text{stoich}}$ as well as the mass composition of the flue gas must be provided. In addition to the assumed heat capacities for the CO_2, N_2, and so on, the final form of the equation (see steps 4 and 7 in Table 6.4.5) is based on an assumption of 5% heat loss across the walls of the incinerator. Use of this table is also demonstrated later in Section 6.6.

Finally, the reader should note that the stoichiometric calculations presented in the previous section and the thermochemical calculations presented in this section were developed independently of one another. Strictly speaking, both of these calculations are interrelated since the products of combustion are a function of the operating temperature that impacts on the quantity of Cl_2, SO_3, and so on. The operating temperature, on the other hand, can affect the excess air requirement that, in turn, significantly affects the products of combustion. Thus, an actual industrial application may require the simultaneous solution of both the stoichiometric and thermochemical equations since they are interdependent. A number of operating conditions are often specified to eliminate this interdependence, enabling one to perform the calculations separately.

Table 6.4.1 Specific heat capacity of benzene (C_6H_6) as a function of temperature and excess air [results of a regression analysis using Eq. (6.4.23)][a]

Excess Air (EA) (%)	Temperature (K)										
	1000	1250	1500	1750	2000	2250	2500	2750	3000	3250	3500
0	0.34	0.35	0.37	0.38	0.38	0.39	0.40	0.40	0.41	0.41	0.41
10	0.32	0.33	0.35	0.35	0.36	0.37	0.37	0.38	0.38	0.38	0.39
20	0.31	0.32	0.33	0.34	0.35	0.36	0.36	0.36	0.37	0.37	0.37
30	0.30	0.32	0.33	0.33	0.34	0.35	0.35	0.35	0.36	0.36	0.36
46	0.30	0.31	0.32	0.33	0.33	0.34	0.34	0.35	0.35	0.35	0.35
50	0.29	0.31	0.31	0.32	0.33	0.34	0.34	0.34	0.35	0.35	0.35
60	0.29	0.30	0.31	0.32	0.33	0.33	0.34	0.34	0.34	0.34	0.34
70	0.29	0.30	0.31	0.32	0.32	0.33	0.33	0.34	0.34	0.34	0.34
80	0.29	0.30	0.31	0.31	0.32	0.33	0.33	0.33	0.34	0.34	0.34
90	0.29	0.30	0.31	0.31	0.32	0.32	0.33	0.33	0.33	0.34	0.34
100	0.28	0.29	0.30	0.31	0.32	0.32	0.33	0.33	0.33	0.33	0.34

[a] c_P = cal/g-°C or Btu/lb-°F.

Table 6.4.2 Specific heat capacity of octane (C_8H_{18}) as a function of temperature and excess air [results of a regression analysis using Eq. (6.4.23)][a]

Excess Air (EA) (%)	Temperature (K) 1000	1250	1500	1750	2000	2250	2500	2750	3000	3250	3500
0	0.38	0.39	0.41	0.42	0.43	0.44	0.45	0.46	0.46	0.47	0.47
10	0.35	0.36	0.38	0.39	0.40	0.41	0.41	0.42	0.42	0.42	0.43
20	0.33	0.35	0.36	0.37	0.38	0.38	0.39	0.39	0.40	0.40	0.40
30	0.32	0.33	0.34	0.35	0.36	0.37	0.38	0.38	0.38	0.39	0.39
40	0.31	0.33	0.34	0.35	0.35	0.36	0.37	0.37	0.37	0.37	0.38
50	0.31	0.32	0.33	0.34	0.35	0.35	0.36	0.36	0.36	0.37	0.37
60	0.30	0.31	0.32	0.33	0.34	0.35	0.35	0.36	0.36	0.36	0.36
70	0.30	0.31	0.32	0.33	0.34	0.34	0.35	0.35	0.35	0.36	0.36
80	0.30	0.31	0.32	0.33	0.33	0.34	0.34	0.35	0.35	0.35	0.35
90	0.29	0.31	0.31	0.32	0.33	0.34	0.34	0.34	0.35	0.35	0.35
100	0.29	0.30	0.31	0.32	0.33	0.33	0.34	0.34	0.34	0.35	0.35

[a] c_P = cal/g-°C or Btu/lb-°F.

Table 6.4.3 Specific heat capacity of dicloromethane (CH_2Cl_2) as a function of temperature and excess air [results of a regression analysis using Eq. (6.4.23)][a]

Excess Air (EA) (%)	Temperature (K)										
	1000	1250	1500	1750	2000	2250	2500	2750	3000	3250	3500
0	0.24	0.25	0.26	0.27	0.27	0.28	0.29	0.29	0.30	0.31	0.31
10	0.24	0.25	0.26	0.27	0.28	0.28	0.29	0.30	0.30	0.31	0.31
20	0.25	0.26	0.26	0.27	0.28	0.28	0.29	0.30	0.30	0.31	0.31
30	0.25	0.26	0.27	0.27	0.28	0.29	0.29	0.30	0.30	0.31	0.31
40	0.25	0.26	0.27	0.28	0.28	0.29	0.30	0.30	0.31	0.31	0.31
50	0.25	0.26	0.27	0.28	0.28	0.29	0.29	0.30	0.30	0.31	0.31
60	0.25	0.26	0.27	0.28	0.28	0.29	0.30	0.30	0.31	0.31	0.31
70	0.25	0.26	0.27	0.28	0.29	0.29	0.30	0.30	0.31	0.31	0.31
80	0.26	0.27	0.27	0.28	0.29	0.29	0.30	0.30	0.31	0.31	0.31
90	0.26	0.27	0.27	0.28	0.29	0.29	0.30	0.30	0.31	0.31	0.31
100	0.26	0.27	0.27	0.28	0.29	0.29	0.30	0.30	0.31	0.31	0.31

[a] c_P = cal/g-°C or Btu/lb-°F.

Table 6.4.4 Regression coefficients for use with Eq. (6.4.23)[a]

Compound	Formula	α	$\beta \times 10^5$	$\gamma \times 10^3$
Benzene	C_6H_6	0.336	2.230	−0.942
n-Octane	C_8H_{18}	0.372	2.434	−1.311
Dichloromethane	CH_2Cl_2	0.233	2.645	−0.102
Benzene and *n*-octane		0.354	2.332	−1.126
Overall		0.314	2.436	−0.785

[a]Temperature range: 1000–2750 K; excess air range: 0–100%.

6.5 APPLICATION OF CHEMICAL REACTION PRINCIPLES

There have been a host of companion terms used to describe incinerators; among these are *oxidizer, afterburner, combustor, combustion device, burner, chemical reactor*, and *organic reactor.* Whichever term is used, the overall process of waste combustion is best characterized by those phenomena occurring in a *chemical reactor.* In effect, an incinerator is one of a number of units that fits into the class of what industry describes as a *chemical reactor.* With this in mind, one may apply chemical reactor principles to either design and/or predict the performance of waste combustors.

Kinetic and design calculations for reactors require information on the rate of reaction and equations to describe the concentration and temperature within

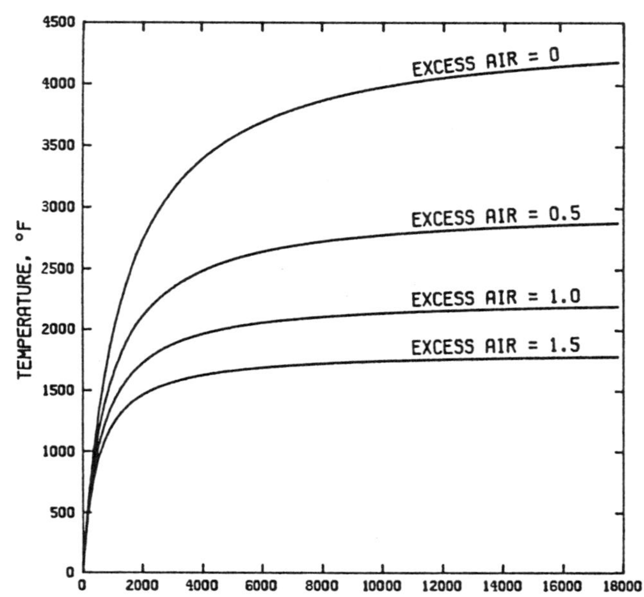

Figure 6.4.2 Operating temperature as a function of excess air and net heating value.

Table 6.4.5 EPA worksheet: Procedure to calculate excess air rate for a specified temperature and feed composition[a]

1. Identify the following input variables:
   From Table 6.3.1, step 5
   $(O_2)_{stoich}$ = ______________ lb/lb feed
   From Table 6.3.1, step 6
   CO_2 = ______________ lb/lb feed
   H_2O = ______________ lb/lb feed
   N_2 = ______________ lb/lb feed
   Other major component(s) = ______________ lb/lb feed
   From thermochemical data on the waste
   NHV_{waste} = ______________ Btu/lb waste
   From proposed operating conditions
   Operating temperature, T = ______________ °F
   Air preheat temperature, T_{air} = ______________ °F
   (if applicable)
   If auxiliary fuel is to be burned in conjunction with the waste, also identify the following from Table 6.3.1
   n_f = ______________ lb fuel/lb waste
   HV_f = ______________ Btu/lb fuel

2. If air preheating is employed, calculate the corresponding enthalpy input to the incinerator. If the combustion air is not to be preheated, proceed to step 3:
   $\Delta H'_1 = 1.12(T_{air} - 77)(O_2)_{stoich}$ = ______________ Btu/lb feed[b]
   $\Delta H_1 = \Delta H'_1(1 + \text{EA})$

3. Calculate the heat generated by combustion of the waste or waste/auxiliary fuel mix:
   $$\Delta H_2 = \frac{NHV_{waste} + n_f HV_f}{1 + n_f}$$ = ______________ Btu/lb feed

4. Calculate the heat loss through the walls of the incinerator, assuming 5% loss:
   $Q = 0.05\Delta H_2$ = ______________ Btu/lb feed

5. Calculate the enthalpy of the combustion products leaving the incinerator:[c]
   $0.26(CO_2 + N_2)(T - 77)$ = ______________ Btu/lb feed[d]
   $0.49(H_2O)(T_k - 212) + H_2O_k(1150)$ = ______________ Btu/lb feed
   other $\times \bar{C}_{P\,other}\,(T - 77)$ = ______________ Btu/lb feed

   ---
   $\Delta H_3 = \Sigma$ = ______________ Btu/lb feed

6. Calculate the enthalpy of excess air leaving the incinerator.
   $\Delta H'_1 = 1.1(T - 77)(O_2)_{stoich}$ = ______________ Btu/lb feed
   $\Delta H_4 = \Delta H'_4 \times \text{EA}$

Table 6.4.5 (*Continued*)

7. Calculate the excess air percentage as follows:

$$\text{EA} = 100\left(\frac{\Delta H_1' + \Delta H_2 - Q - \Delta H_3}{\Delta H_4' - \Delta H_1'}\right) = \underline{\hspace{6cm}} \%^e$$

The following footnotes are comments prepared by L. Theodore and J. Reynolds.

[a]The notation of the original EPA worksheet has been retained, and often does not correspond to the notation used in this text. Any illustrative examples or problems based on the use of this table will also retain the EPA notation.

[b]The term 1.12 has been derived in the following manner. The enthalpy change for air is

$$\Delta H_1' = (\text{air})_{\text{stoich}}(C_{P,\,\text{air}})(T_{\text{air}} - 77)$$

Every lbmol of air contains approximately 0.21 lbmol of O_2. Every lb of air therefore contains approximately (0.21)(32/29) lb O_2 where the 32 and 29 are the molecular weights of the oxygen and air, respectively. Thus,

$$(\text{air})_{\text{stoich}} = (O_2)_{\text{stoich}}/(0.21)(32/29)$$

If the heat capacity of air is roughly 0.25 Btu/lb-°F, this equation reduces to that given in step 2.

[c]Here 0.26 and 0.49 represent the heat capacity, in Btu/lb-°F, of CO_2 and H_2O, respectively.

[d]Care should be exercised with this term. The N_2 represents the nitrogen contribution from the *stoichiometric* air and the fuel–waste. The nitrogen in the *excess air* is treated separately in step 6. Another approach is to treat this stoichiometric nitrogen through step 6. There is approximately 0.77 lb N_2/lb air. In order to account for the N_2 in the flue gas associated with the stoichiometric air, this term would then read:

$$\Delta H_4 = \Delta H_4'(0.77 + \text{EA})$$

The contribution of fuel and waste nitrogen can usually be safely neglected.

[e]This equation has been derived by an enthalpy balance across the incinerator.

Energy in:

- Enthalpy of preheat air in $= \Delta H_1'(1 + \text{EA})$
- Enthalpy of fuel/waste combustion $= \Delta H_2$

Energy out:

- Heat loss across walls $= Q$
- Enthalpy of flue gas $= \Delta H_3$
- Enthalpy of excess air in flue gas $= (\Delta H_4')(\text{EA})$

Equating the "in" with the "out" terms leads to the final equation given in step 7. Care should be exercised with the ΔH_2 term. Combustion reactions are exothermic, that is, energy is liberated or given off. It is standard thermodynamic (engineering) practice to record these enthalpy of combustion terms as *negative* values. The approach here uses this term as a *positive* quantity.

the reactor. Chapter 5 developed equations to describe the rate of chemical reaction, and discussed energy (temperature) effects and chemical equilibrium. The contents of those sections can be combined and the results applied to the three major classifications of reactors: batch, tank flow, and tubular flow. These three classes of reactors are discussed here.

Table 6.4.6 Procedure to calculate the maximum achievable excess air rate for a rotary kiln operating at a specified temperature with a specified feed composition[a]

1. Identify the following input variables:
   From Table 6.3.2, step 6
   $(O_2)_{stoich(k)}$ = ________ lb O_2/lb kiln feed
   From Table 6.3.2, step 7
   $CO_{2(k)}$ = ________ lb/lb feed
   $H_2O_{(k)}$ = ________ lb/lb feed
   $N_{2(k)}$ = ________ lb/lb feed
   Other major combustion products(s) = ________ lb/lb feed
   From Table 6.3.2, step 2
   Liquid waste feed fraction, n_1 = ________ lb liquid/lb waste
   Solid waste feed fraction, n_2 = ________ lb solid/lb waste

   From thermochemical data on the waste
   Liquid waste heating value, NHV_1 = ________ Btu/lb
   Solid waste heating value, NHV_2 = ________ Btu/lb

   From proposed operating conditions
   Kiln operating temperature, T_k = ________ °F
   Air preheat temperature, T_{air} = ________ °F
   (if applicable)

   If auxiliary fuel is to be burned in the kiln along with the wastes during normal operation, identify the following from Table 6.3.2:
   n_{fk} = ________ lb fuel/lb waste
   HV_{fk} = ________ Btu/lb fuel

2. If air preheating is used, calculate the corresponding enthalpy input to the kiln. If the combustion air is not preheated, proceed to step 3:
   $\Delta H'_{1_{(k)}} = 1.12(T_{air} - 77)(O_2)_{stoich(k)}$ = ________ Btu/lb feed
   $\Delta H_{1_{(k)}} = \Delta H'_{1_{(k)}}(1 + EA_k)$

3. Calculate the maximum heat generated in the kiln by combustion of the wastes or waste/auxiliary fuel mix:
   $$\Delta H_2 = \frac{n_1 NHV_1 + n_2 NHV_2 + n_{fk} HV_{fk}}{1 + n_{fk}} =$$ ________ Btu/lb feed

4. Estimate the heat loss through the walls of the kiln, assuming 5% loss:
   $Q_{(k)} = 0.05\ \Delta H_{2(k)}$ = ________ Btu/lb feed

5. Calculate the enthalpy of the combustion products leaving the kiln:
   $0.26[CO_{2(k)} + N_{2(k)}](T_k - 77)$ = ________ Btu/lb feed
   $0.49(H_2O)(T_k - 212) + H_2O_k(1150)$ = ________ Btu/lb feed
   $\text{other}_k \left(\frac{\text{lb}}{\text{lb feed}}\right) \times \bar{C}_{P\,\text{other}}(T_k - 77)$ = ________ Btu/lb feed
   $\Delta H_{3(k)} = \Sigma$ = ________ Btu/lb feed

Table 6.4.6 (*Continued*)

6. Calculate the enthalpy of excess air leaving the kiln:
   $\Delta H'_{4_{(k)}} = 1.1(T_k - 77)(O_2)_{stoich(k)} =$ ________________ Btu/lb feed
   $\Delta H_{4_{(k)}} = \Delta H'_{4_{(k)}} EA_k$

7. Calculate the excess air percentage as follows:
   $$EA_k = 100\left(\frac{\Delta H'_{1_{(k)}} + \Delta H_{2_{(k)}} - Q_k - \Delta H_{3_{(k)}}}{\Delta H'_{4_{(k)}} - \Delta H'_{1_{(k)}}}\right) = \text{\_\_\_\_\_\_\_\_\_\_\_\_}\ \%^b$$

[a] The notation of the original EPA worksheet has been retained, and often does not correspond to the notation used in this text. Any illustrative examples or problems based on the use of this table will also retain the EPA notation.

[b] The reader should carefully review the note comments in the previous worksheet, that is, Table 6.4.5.

There are three basic modes of operation for any type of reactor. Reactors may be operated isothermally, nonisothermally, or adiabatically. In *isothermal operation*, the temperature is maintained constant during the course of the reaction. This condition can be approached in practice by providing sufficient heat exchanger facilities to account for enthalpy effects arising during reaction. This mode of operation finds its major application in laboratory kinetic studies. Most industrial reactors are described by *nonisothermal operation.* Some energy in the form of heat is added to or removed from the reactor, but isothermal conditions are not satisfied. For *adiabatic operation*, the reactor is insulated to minimize heat transfer between the reactor contents and the surroundings. Many industrial reactors, including waste combustors, attempt to operate in this manner.

Batch Reactors

A batch reactor is a vessel or container that may be open or closed. Reactants are usually added to the reactor simultaneously. The contents are then mixed (if necessary) to ensure no variations in the concentrations of the species present. The reaction then proceeds. During this period, there is no transfer of mass into or out of the reactor. Because the concentrations of reactants and products change with time, this is a *transient* or *unsteady-state* operation. The reaction is terminated when the desired chemical change has been achieved. The contents are then discharged and sent elsewhere for further processing. Line diagrams of three typical batch reactors are given in Fig. 6.5.1. A more detailed schematic is presented in Fig. 6.5.2.

The describing equation for this device is obtained by applying the conservation law for either mass or number of moles on a time rate basis to the contents of the reactor. It is preferable to work with moles rather than mass in this and the following subsections since the rate of reaction is most conven-

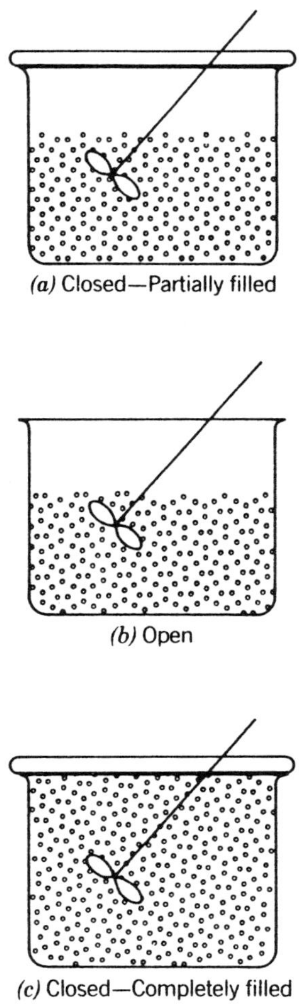

Figure 6.5.1 Different batch reactors.

iently described in terms of molar concentrations. The resulting equation takes the form

$$\theta = C_{A0} \int_0^{X_A} \frac{dX_A}{-R_A} \tag{6.5.1}$$

where R_A = rate of chemical reaction for A (lbmol/h-ft^3)
C_{A0} = initial molar concentration of reactant A (lbmol/ft^3)
X_A = fractional molar conversion of A in time θ (dimensionless)
θ = residence time in the reactor (h)

If the reaction is first order and irreversible with respect to A (see Chapter 5), this equation becomes

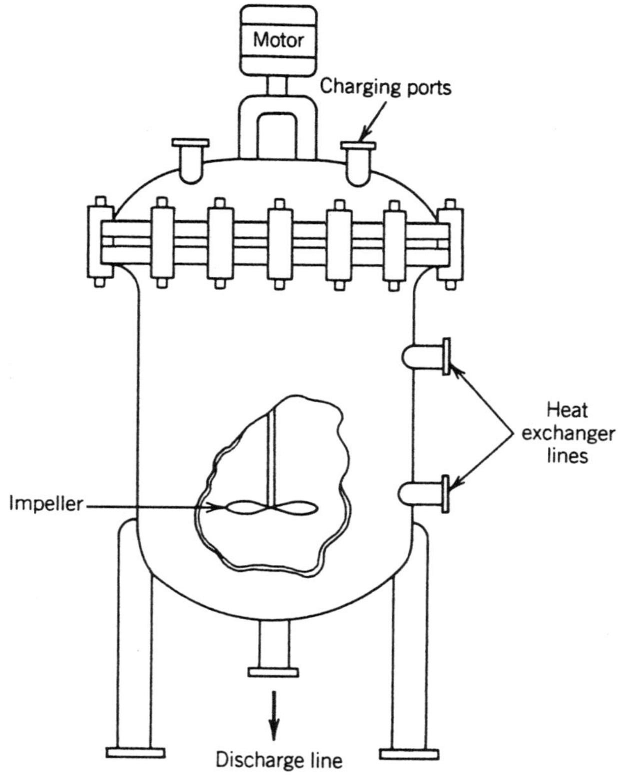

Figure 6.5.2 Typical batch reactor.

$$\theta = -\frac{1}{k}\ln\left(\frac{C_A}{C_{A0}}\right) \quad \text{or} \quad \theta = \frac{1}{k}\ln\left(\frac{C_{A0}}{C_A}\right) \tag{6.5.2}$$

where k = reaction velocity constant (h^{-1}).

Since this equation contains the ratio of the two concentrations, the concentration may be based on either mass or number of moles. For gas-phase reactions, a volume basis may also be used (e.g., ppm, ppb, or the equivalent). The choice of basis is often a matter of convenience or preference, the form of the data, and the available expression for the rate of reaction. However, Eq. (6.5.2) applies only to first-order reactions; molar concentrations should be used for other or more complex reactions.

Continuously Stirred Tank Reactor

Another reactor where mixing is important is the *tank flow* or *continuously stirred tank reactor* (CSTR). This type of reactor also consists of a tank or kettle equipped with an agitator (see Fig. 6.5.3). It may be operated under

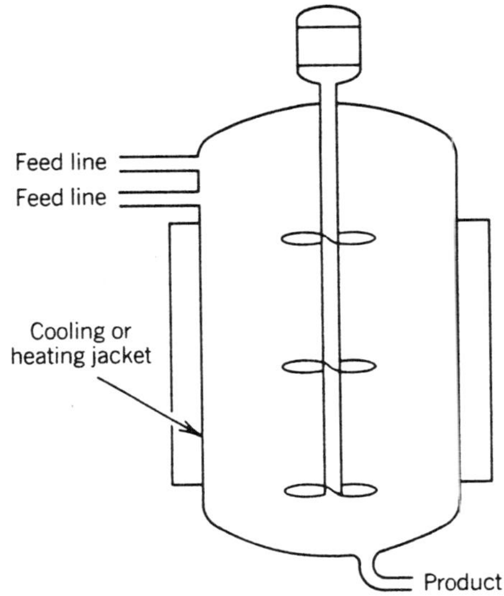

Figure 6.5.3 Typical tank flow reactor.

steady-state or transient conditions. Reactants are fed continuously, and the products withdrawn continuously. The reactants and products may be in the liquid, gas, or solid state, or a combination of these. If the contents are perfectly mixed, the reactor design problem is greatly simplified for steady-state conditions because the mixing results in uniform concentration, temperature, and so on, throughout the reactor. This means that a *fluidized-bed incinerator*, operating under nearly perfect mixing conditions, may be best described by a CSTR model (see Chapter 7). For perfect mixing, the rate of reaction is constant; therefore the describing equations are not differential and do not require integration. High reactant concentrations can be maintained with low flow rates so that conversions approaching 100% may be achieved. A disadvantage of this practice, however, is that very large reactors must be required to obtain high conversions. The describing equation for this type of reactor takes the form

$$\theta = \frac{C_{A1} - C_{A0}}{R_{A1}} \tag{6.5.3}$$

where θ = residence time in the reactor (h)
C_{A0} = inlet concentration of reactant A (lbmol/ft^3)
C_{A1} = outlet concentration of reactant A (lbmol/ft^3)
R_{A1} = rate of chemical reaction for A based on outlet conditions (lbmol/h-ft^3)

Note that the concentration of A in the reactor is equal to the outlet concentration since perfectly mixed conditions are assumed. For a first-order irreversible reaction in A, this equation becomes

$$\theta = \frac{C_{A1} - C_{A0}}{-kC_{A1}} \tag{6.5.4}$$

where k = *velocity* rate constant (h^{-1})

Since a ratio of concentrations appears again for this first-order reaction, the concentration may be based on either mass or number of motes.

Tubular Flow Reactors

The last and most important class of reactor to be examined is the *tubular flow reactor.* The most common type is the single-pass cylindrical tube (see Fig. 6.5.4); another important type is one that consists of a number of tubes in parallel (see Fig. 6.5.5). The reactor(s) may be vertical or horizontal. The feed is charged continuously at the inlet of the tube, and the products are continuously removed at the outlet. If heat exchange with the surroundings is required, the reactor tube is *jacketed.* If adiabatic conditions are required, the reactor is *covered* with insulation. Tubular flow reactors are usually operated under steady-state conditions so that physical and chemical properties do not vary with time. Unlike the batch and tank flow reactors, there is no mechanical mixing. Thus, the state of the reacting fluid will vary from point to point in the system, and this variation may be in both the radial r and axial z directions (see Fig. 6.5.6). The describing equations are then differential, with position as the independent variable. In terms of the final results, the reacting system is assumed to move through the reactor in plug flow (see Fig. 6.5.7*a*). It is further assumed that there is no mixing in the axial direction and *complete* mixing in the radial direction so that the concentration, temperature, and so on, do not vary through the cross section of the tube. Thus, the reacting fluid flows through the reactor in an undisturbed plug of mass. Note that the time for this hypothetical plug to flow through this type of reactor is the same as the contact or reaction time in a batch reactor. Under these conditions, the form of the describing equations for batch reactors will also apply to tubular flow reactors. From a qualitative point of view, as the length of the reactor approaches infinity,

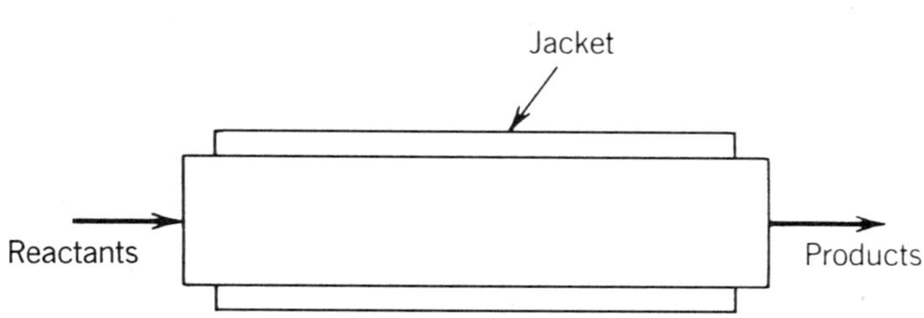

Figure 6.5.4 Tubular flow reactor, single pass.

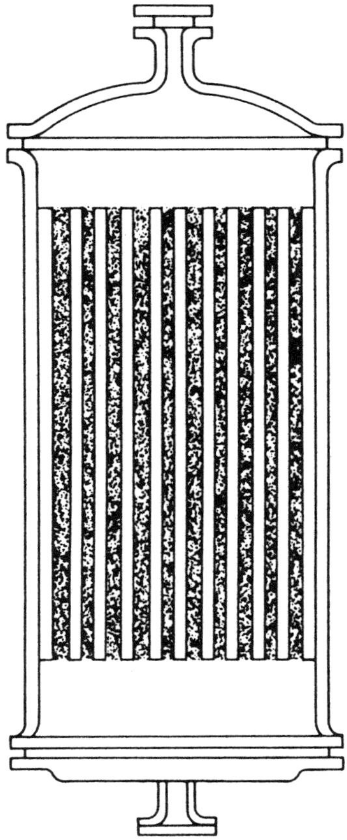

Figure 6.5.5 Multitube flow reactor.

the concentration of the limiting reactant approaches zero for irreversible reactions (except zero order) and the equilibrium concentration for reversible reactions. In actual practice, tubular flow reactors deviate from the plug flow model because of velocity variations in the radial direction (see Fig. 6.5.7*b*–*d*). For any of these conditions, the residence time for annular elements of fluid within the reactor will vary from some minimal value at a point where the velocity is a maximum to a maximum value near the wall where the velocity approaches zero. The concentration and temperature profiles, as well as the

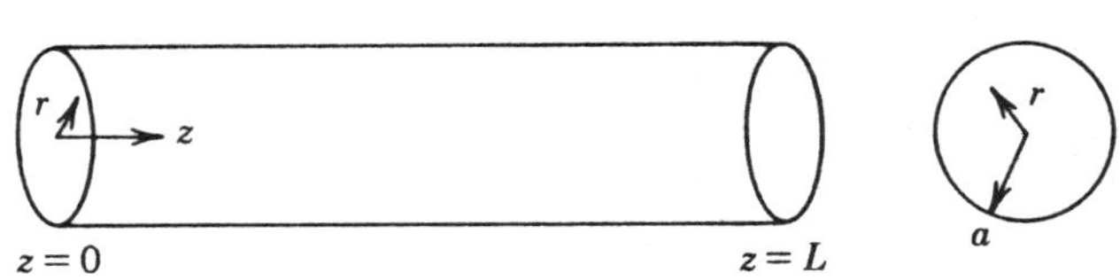

Figure 6.5.6 Reactor system, cylindrical coordinates.

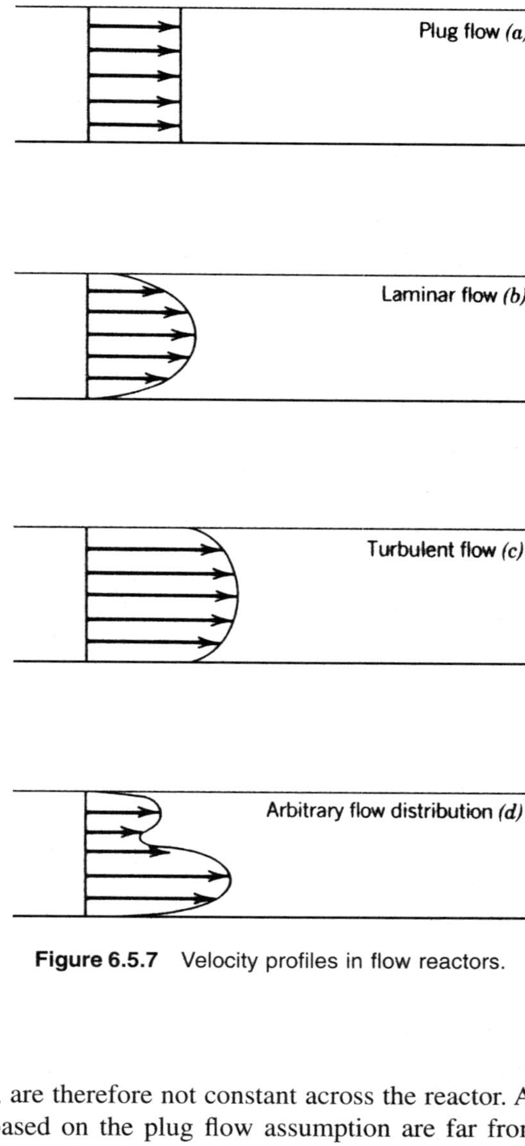

Figure 6.5.7 Velocity profiles in flow reactors.

velocity profile, are therefore not constant across the reactor. Although describing equations based on the plug flow assumption are far from accurate, most practitioners still employ the plug flow model. The describing equation for this unit is obtained by applying the conservation law for mass to a differential volume element in a flow reactor. The resulting equation is

$$V = \dot{n}_{A0} \int_0^{X_A} \frac{dX_A}{-R_A} \tag{6.5.5}$$

where V = volume of the reactor (ft^3)
$\dot{n}_{A0}$ = inlet molar feed rate of A (lbmol/h)

For first-order irreversible kinetics, Eq. (6.5.5) becomes

$$\theta = \frac{1}{k} \ln \left(\frac{C_A}{C_{A0}} \right) \tag{6.5.6}$$

where θ = average residence time in the reactor (h).

As mentioned previously, this equation is identical in form to that developed for a batch reactor. Finally, this particular class of reactor may be used to describe both the *liquid injection incinerator* and *rotary kiln incinerator* (see Chapter 7). These are the two most common types employed in hazardous waste incineration facilities.

As mentioned earlier, the chemical reaction sequence that takes place during the destruction of organics by thermal oxidation is a complicated process. Several intermediate products appear before the original material is completely oxidized into carbon dioxide, water, nitrogen oxides, and so on. However, most of the intermediate products exist for only a very short time, and for engineering purposes complete destruction of the organics is the principal concern. Lee et al.[3] have shown from available experimental data that the following equations fit the data better than the first-order model described previously:

$$\ln \left(\frac{C_A}{C_{A0}} \right) = -k(\theta - \theta_1), \qquad \theta > \theta_1 \tag{6.5.7}$$

$$= 0, \qquad \theta \leq \theta_1$$

$$\theta_1 = x_1 - x_2 T \tag{6.5.8}$$

where θ_1 = an *apparent delay time* (a lag prior to onset of the oxidation reaction)
x_1, x_2 = constants (h and h/°F, respectively)
T = temperature (°F)

The temperature dependence of the reaction rate constant k is correlated by the Arrhenius equation (see Chapter 5).

$$k = Ae^{-E/RT} \tag{6.5.9}$$

where E = *activation* energy (Btu/lbmol)
R = universal gas constant 1.987 Btu/lbmol-°R
T = absolute temperature (°R)

Regarding Eq. (6.5.7), the *apparent lag time* (θ_1) is defined for each temperature as the intercept of the line representing the data with the zero reaction line, that is, $\ln(C_A/C_{AO}) = 0$, on a graph of $\ln(C_A/C_{AO})$ versus θ. The lag time (θ_1) and the reaction temperature were found to be best correlated by a linear equation [Eq. (6.5.8)] using a least-square analysis. The experimental data were modeled by Eq. (6.5.7), which assumes that no reaction occurs until the residence time is longer than the lag time. (The Arrhenius constants, A and E, were estimated using a nonlinear least-squares estimation procedure.) The estimated parameters and some results are shown in Table 6.5.1 for a variety of organic compounds. The lower limiting temperature is indicated for each compound. Use of the parameters at temperatures lower than this value or at temperatures higher than about 1700–1800°F is not recommended. At temperatures >1700–1800°F, the reaction rates are believed to be much faster than that predicted by the parameters in this table because different reaction mechanisms with more free radical involvement take place at higher temperatures.

Equations to predict the time–temperature kinetic rate information for untested compounds applicable in the conventional thermal oxidation temperature ranges (1200–1800°F) were developed based on the data shown in Table 6.5.1. Temperatures required for three destruction levels (99.0, 99.9, and 99.99%) at four residence times (0.5, 1, 1.5, and 2 s) were calculated for the compounds listed in Table 6.5.1. These temperatures were correlated with the characteristic compound structures, the autoignition temperature, and the residence times. The temperatures required at these three destruction levels are designated as T_{99}, $T_{99.9}$, and $T_{99.99}$. The temperature models that appear in Table 6.5.2 were developed using multiple linear regression. A *student's t* value is given for each coefficient.

The usefulness of this *reactor* approach to describing incinerators should be obvious to the reader. As described in Chapter 3, the intent of a trial burn is to determine the incinerability of a particular waste stream in a particular incinerator under a particular set of operating conditions. The purpose of selecting the primary organic hazardous constituents (POHCs) for a trial burn from among the constituent compounds in the hazardous waste feed is to obtain those few compounds for analysis that are most difficult to destroy. Thus, if meaningful kinetic data are available at system and operating conditions, the destruction efficiency of these POHCs could be predicted a priori. This prediction would be generated without recourse to pilot- or field-test data.

The ability to realistically use this approach is based on the applicability of the kinetic model and reactor equation to the system in question. Note, however, that the conditions under which the data were obtained may not be practically achieved in a real incinerator operating at an industrial facility. Notwithstanding this fact, one should appreciate the assumption that the kinetics of destruction can be expressed as a psuedo-first-order reaction. This essentially assumes that the reaction is first order even though the overall destruction process may be an extremely complex function of a host of variables.

6.6 ILLUSTRATIVE EXAMPLES

Example 6.6.1 Assume perfect combustion of a contaminated fuel oil with stoichiometric air. The gravimetric weight percentage analysis of a sample of

Table 6.5.1 Thermal oxidation parameters

Compound	A	E (cal/gmol)[a]	x_1	x_2	Lower Limiting Temperature (°F)	Autoignition Temperature (°F)	Calculated Destruction Temperature from Eq. (6.5.7)	
							$T_{99.99}$ at 1 s (°F)	$T_{99.99}$ at 2 s (°F)
Acrolein	3.30×10^{10}	35,900	0	0	800	453	1020	975
Acrylonitrile	2.13×10^{12}	52,100	0.375	0.000250	1250	898	1345	1297
Allyl alcohol	1.75×10^{6}	21,400	1.971	0.00146	1050	713	1176	1077
Allyl chloride	3.89×10^{7}	29,100	0.525	0.00035	1150	905	1276	1200
Benzene	7.43×10^{21}	95,900	2.59	0.00178	1275	1044	1351	1322
1-Butene	3.74×10^{14}	58,200	2.12	0.00157	1150	723	1232	1195
Chlorobenzene	1.34×10^{17}	76,600	1.22	0.0008	1350	1180	1408	1372
1,2-Dichloroethane	4.82×10^{11}	45,600	0.948	0.00073	1050	775	1216	1173
Ethane	5.65×10^{14}	63,600	7.33	0.0052	1275	959	1368	1328
Ethanol	5.37×10^{11}	48,100	2.10	0.0015	1250	793	1307	1256
Ethyl acrylate	2.19×10^{12}	46,000	0	0	1000	721	1132	1092
Ethylene	1.37×10^{12}	50,800	0	0	1200	842	1328	1281
Ethyl formate	4.39×10^{11}	44,700	0.325	0.00024	1100	851	1191	1145
Ethyl mercaptan	5.20×10^{5}	14,700	1.87	0.0022	700	570	778	704
Methane	1.68×10^{11}	52,100	1.90	0.00117	1200	999	1545	1486
Methyl chloride	7.34×10^{8}	40,900	1.518	0.00084	1500	1170	1597	1514
Methyl ethyl ketone	1.45×10^{14}	58,400	1.92	0.00136	1200	960	1290	1247
Propane	5.25×10^{19}	85,200	1.02	0.000686	1200	871	1330	1300
Propylene	4.63×10^{8}	34,200	4.54	0.00323	1200	851	1318	1247
Toluene	2.28×10^{13}	56,500	1.35	0.000922	1275	997	1340	1295
Triethylamine	8.10×10^{11}	43,200	1.20	0.0010	950	450	1101	1058
Vinyl acetate	2.54×10^{9}	35,900	1.076	0.00078	1150	800	1223	1164
Vinyl chloride	3.57×10^{14}	63,300	0	0	1250	882	1371	1332

[a] To convert cal/gmol to Btu/lbmol, divide by 1.8.

Table 6.5.2 Predictive destruction temperature models[a]

$$T_{99.00} = 577 - \underset{(3.5)^a}{10.0V_1} + \underset{(6.3)}{110.2V_2} + \underset{(9.1)}{67.1V_3} + \underset{(7.3)}{72.6V_4} + \underset{(32.7)}{0.586V_5} - \underset{(5.9)}{23.4V_6} - \underset{(34.8)}{430.9V_7} + \underset{(15.5)}{85.2V_8} - \underset{(10.2)}{82.2V_9} + \underset{(5.8)}{65.5V_{10}} - \underset{(19.2)}{76.1V_{11}} \quad R^2 = 0.989,\ S = 19.8$$

$$T_{99.90} = 594 - \underset{(4.1)}{12.2V_1} + \underset{(6.3)}{117.0V_2} + \underset{(9.2)}{71.6V_3} + \underset{(7.7)}{80.2V_4} + \underset{(31.2)}{0.592V_5} - \underset{(4.8)}{20.2V_6} - \underset{(32.1)}{420.3V_7} + \underset{(15.0)}{87.1V_8} - \underset{(7.8)}{66.8V_9} + \underset{(5.2)}{62.8V_{10}} - \underset{(18.0)}{75.3V_{11}} \quad R^2 = 0.987,\ S = 20.9$$

$$T_{99.99} = 605 - \underset{(4.3)}{13.8V_1} + \underset{(6.3)}{122.5V_2} + \underset{(9.2)}{75.7V_3} + \underset{(7.8)}{85.6V_4} + \underset{(29.8)}{0.597V_5} - \underset{(4.0)}{17.9V_6} - \underset{(29.8)}{412.0V_7} + \underset{(14.5)}{89.0V_8} - \underset{(6.1)}{55.3V_9} + \underset{(4.8)}{60.7V_{10}} - \underset{(17.0)}{75.2V_{11}} \quad R^2 = 0.985,\ S = 22.1$$

Variables

$T_{99.00}$ = 99.00% destruction temperature (°F)
$T_{99.90}$ = 99.90% destruction temperature (°F)
$T_{99.99}$ = 99.99% destruction temperature (°F)
R^2 = Multiple correlation coefficient
S = Standard deviation (°F)

V_1 = number of carbons
V_2 = aromatic (0 = no, 1 = yes)
V_3 = carbon–carbon double bond (0 = no, 1 = yes)
V_4 = number of nitrogens
V_5 = autoignition temperature (°F)
V_6 = number of oxygens
V_7 = number of sulfurs
V_8 = hydrogen/carbon ratio
V_9 = allyl compound (0 = no, 1 = yes)
V_{10} = carbon double bond, chlorine interaction (0 = no, 1 = yes)
V_{11} = ln (time in s)

[a]Number in parenthesis is student's t value for the coefficient.

this fuel oil is

88.52% carbon	0.10% nitrogen
10.87% hydrogen	0.06% oxygen
0.40% sulfur	0.05% ash

Calculate

1. The gravimetric analysis of the flue gas
2. Total volume of flue gas (at 500°F and 1 atm) per pound of oil burned
3. Volume percent of CO_2 in the dry flue gas

SOLUTION

1. Convert the gravimetric percent of each component to lbmol selecting 100 lb of fuel as a basis.

C	88.52/12 = 7.38 lbmol
H	10.87/2 = 5.44 lbmol
S	0.40/32 = 0.0125 lbmol
O_2	0.06/32 = 0.00188 lbmol
N_2	0.10/28 = 0.00357 lbmol

Since the molecular weight (MW) of the ash is unknown and its content is small, it is neglected in the calculation.

Write the combustion reaction for each component in the fuel:

$$C + O_2 \rightarrow CO_2$$

$$H_2 + \tfrac{1}{2}O_2 \rightarrow H_2O$$

$$S + O_2 \rightarrow SO_2$$

Determine the amount of oxygen required for each component. Use the previous results to complete the first column:

Component	lbmol in Fuel	lbmol O_2 Required
C	7.38	7.38
H_2	5.44	2.72
S	0.0125	0.0125
O_2	0.00188	−0.00188
		Total = 10.1

Note: Oxygen in the fuel reduces air requirements. Therefore, the amount of oxygen in the fuel has been subtracted from the amount of O_2 required.

Calculate the amount of nitrogen from the oxygen requirement.

$$\text{Amount of } N_2 = (\text{amount of } O_2)\,\frac{0.79 \text{ lbmol } N_2}{0.21 \text{ lbmol } O_2}$$

Component	lbmol N_2 Required
C	27.76
H_2	0.23
S	0.047
O_2	0.0
	Total = 38.03

Determine the gravimetric composition (percentage) of the flue gas.

Component	lbmol	MW	lb	wt %
CO_2	7.38	44	324.7	21.8
H_2O	5.44	18	97.9	6.6
SO_2	0.00125	64	0.08	0.005
O_2	0.0	32	0.0	0.0
N_2	38.03	28	1064.3	71.6
	Total = 50.85	Total = 1487.0		

2. Calculate the total lbmol of flue gas produced per pound of oil burned:

$$\text{lbmol of flue gas} = \frac{\text{total lbmol from above}}{100}$$

$$= \frac{50.85}{100}$$

$$= 0.5085 \text{ lbmol gas/lb oil}$$

Remember that 100 lb of fuel was used as a basis.

Calculate the total volume of flue gas at 500°F and 1 atm using the ideal gas law:

$$V = \frac{nRT}{P} = \frac{(\text{lbmol flue gas/lb oil})(0.7302)(500 + 460)}{1.0}$$

$$= 356.1 \text{ ft}^3/\text{lb oil}$$

3. One can now complete the following table:

Component of Dry Flue Gas	lbmol
CO_2	7.38
SO_2	0.00125
N_2	38.01
	Total = 45.39

The volume percentage of CO_2 in the dry flue gas is

$$\%CO_2 = \frac{\text{lbmol } CO_2}{\text{total lbmol dry flue gas}}(100) = \frac{7.38}{45.39}(100)$$

$$= 16.26\%$$

Example 6.6.2 A waste mixture to be incinerated in a liquid injection system has the following composition on a lb/lb basis:

$$\text{Carbon } (C_w) = 0.732$$

$$\text{Hydrogen } (H_w) = 0.109$$

$$\text{Moisture } (H_2O_w) = 0.022$$

$$\text{Oxygen } (O_w) = 0.102$$

$$\text{Chlorine } (Cl_w) = 0.035$$

The incinerator process will be assisted with 0.5 lb of natural gas per pound of waste. Thirty percent excess air is to be employed. Calculate the required quantities in the worksheet provided in Table 6.3.1.

SOLUTION Skip steps 1 and 2 of Table 6.3.1. Identify the elemental composition of the natural gas from step 3:

Components	lb component/lb fuel
C_f	0.693
H_f	0.227
N_f	0.080

Calculate the combined composition of waste and fuel feed.

$$n_C = \frac{C_w + n_f C_f}{1 + n_f}$$

$$= 0.719 \text{ lb/lb feed}$$

$$n_{\mathrm{H}} = \frac{H_{\mathrm{w}} + n_{\mathrm{f}} H_{\mathrm{f}}}{1 + n_{\mathrm{f}}}$$

$$= 0.148 \text{ lb/lb feed}$$

$$n_{\mathrm{H_2O}} = \frac{\mathrm{H_2O_w}}{1 + n_{\mathrm{f}}}$$

$$= 0.147 \text{ lb/lb feed}$$

$$n_{\mathrm{O}} = \frac{\mathrm{O_w}}{1 + n_{\mathrm{f}}}$$

$$= 0.0680 \text{ lb/lb feed}$$

$$n_{\mathrm{Cl}} = \frac{\mathrm{Cl_w}}{1 + n_{\mathrm{f}}}$$

$$= 0.0233 \text{ lb/lb feed}$$

$$n_{\mathrm{N}} = \frac{n_{\mathrm{f}} \mathrm{N_f}}{1 + n_{\mathrm{f}}}$$

$$= 0.0267 \text{ lb/lb feed}$$

Remember that the fuel-to-waste ratio (n_{f}) for this problem is 0.5.

Calculate the stoichiometric oxygen requirements:

$$(\mathrm{O_2})_{\mathrm{stoich}} = (n_i) \frac{\text{lb } \mathrm{O_2}}{\text{lb of species } i} - n_{\mathrm{O}}$$

$$= (n_{\mathrm{C}}) \frac{2.67 \text{ lb } \mathrm{O_2}}{\text{lb C}} + \left(n_{\mathrm{H}} - \frac{n_{\mathrm{Cl}}}{35.5} \right) \frac{8.0 \text{ lb } \mathrm{O_2}}{\text{lb H}} - n_{\mathrm{O}}$$

$$= (0.719)(2.67) + \left(0.148 - \frac{0.0233}{35.3} \right) (8.0) - 0.068$$

$$= 3.030 \text{ lb } \mathrm{O_2}/\text{lb feed}$$

Note: Here 1 lbmol of C with 1 lbmol of O_2 forms 1 lbmol of CO_2 when combusted completely. Also 2.67 lb O_2/lb C is obtained by converting lbmol to lb. The same explanation applies to H. However, H also reacts to form HCl, and the amount of H that forms HCl should be subtracted from n_{H}, hence $n_{\mathrm{H}} - (n_{\mathrm{Cl}}/35.5)$.

Calculate the combustion gas mass fraction.

$$N_{\mathrm{CO_2}} = (n_{\mathrm{C}}) \frac{3.67 \text{ lb } \mathrm{CO_2}}{\text{lb C}}$$

$$= 2.639 \text{ lb } \mathrm{CO_2}/\text{lb feed}$$

$$N_{H_2O} = \left(n_H - \frac{n_{Cl}}{35.5}\right)\frac{9.0 \text{ lb } H_2O}{\text{lb H}} + n_{H_2O}$$

$$= 1.341 \text{ lb/}H_2O\text{/lb feed}$$

$$N_{N_2} = (O_2)_{\text{stoich}}\left(\frac{3.29 \text{ lb } N_2}{\text{lb } O_2}\right) + n_N$$

$$= 10.03 \text{ lb } N_2\text{/lb feed}$$

$$N_{HCl} = (n_{Cl})\frac{1.03 \text{ lb HCl}}{\text{lb Cl}}$$

$$= 0.0240 \text{ lb Hcl/lb feed}$$

Note: Capital N is used to distinguish combustion gas mass fractions from feed mass fractions (n).

Calculate the combustion products (CP) per pound of feed:

$$\text{CP} = \text{N}_{CO_2} + N_{H_2O} + N_{N_2} + N_{HCl}$$

$$= 14.03 \text{ lb/lb feed}$$

Calculate the additional nitrogen and oxygen present in the combustion gases due to the excess air feed:

$$(O_2)_{EA} = (\text{EA})(O_2)_{\text{stoich}} \qquad (N_2)_{EA} = \frac{3.29 \text{ lb } N_2}{\text{lb } O_2}(O_2)_{EA}$$

$$= (0.3)(O_2)_{\text{stoich}} \qquad = 3.009 \text{ lb } N_2\text{/lb feed}$$

$$= 0.9090$$

Calculate the total combustion gas products (CG) per pound of feed:

$$\text{CG} = \text{CP} + (O_2)_{EA} + (N_2)_{EA} = 14.03 + 0.9090 + 3.009$$

$$= 17.95 \text{ lb/lb feed}$$

Calculate the mass fraction of each combustion gas component:

$$CO_2 \qquad \frac{\text{N}_{CO_2}}{\text{CG}} = \frac{2.639}{17.95} = 0.1470 \text{ lb/lb gas}$$

$$H_2O \qquad \frac{N_{H_2O}}{\text{CG}} = \frac{1.314}{17.95} = 0.0732 \text{ lb/lb gas}$$

$$N_2 \qquad \frac{N_{N_2} + (N_2)_{EA}}{CG} = \frac{13.04}{17.95} = 0.7264 \text{ lb/lb gas}$$

$$O_2 \qquad \frac{(O_2)_{EA}}{CG} = \frac{0.9090}{17.95} = 0.0506 \text{ lb/lb gas}$$

$$HCl \qquad \frac{N_{HCl}}{CG} = \frac{0.0240}{17.95} = 0.00133 \text{ lb/lb gas}$$

Identify those components <1 to 2% of the combustion gas. These components can be eliminated from heat and material balance calculations. Hydrogen chloride is the only component that constitutes <1 to 2% of the combustion gas.

Calculate the volumetric flow of the combustion products at standard conditions of 68°F and 1 atm:

$$CO_2 \qquad \frac{N_{CO_2}}{0.114 \text{ lb/scf}} = 23.15 \text{ scf/lb feed}$$

$$H_2O \qquad \frac{N_{H_2O}}{0.0467 \text{ lb/scf}} = 28.76 \text{ scf/lb feed}$$

$$N_2 \qquad \frac{N_{N_2} + (N_2)_{EA}}{0.0727 \text{ lb/scf}} = 179.4 \text{ scf/lb feed}$$

$$O_2 \qquad \frac{(O_2)_{EA}}{0.083 \text{ lb/scf}} = 10.96 \text{ scf/lb feed}$$

$$HCl \qquad \frac{N_{HCl}}{0.0945 \text{ lb/scf}} = 0.25 \text{ scf/lb feed}$$

Total volumetric flow rate = 242.52 scf/lb feed

Note: The density terms can be obtained from tables or calculated from the ideal gas law.

Example 6.6.3 A waste mixture to be incinerated by a rotary kiln has the following composition:

	Solids (1)	Liquids (2)	
Carbon (C_w)	0.74	0.30	lb/lb waste
Hydrogen (H_w)	0.10	0.12	lb/lb waste
Moisture (H_2O_w)	0.02	0.40	lb/lb waste
Oxygen (O_w)	0.11	0.10	lb/lb waste
Sulfur (S_w)	0.03	0.08	lb/lb waste

The liquid feed rate is 1500 lb/h and the solid feed rate is 2000 lb/h. The incineration process will be assisted with 0.5 lb of natural gas/lb of waste. It is to employ 50% excess air. Calculate the required quantities in Table 6.3.2.

SOLUTION Identify the elemental composition of the natural gas. See Table 6.2.1.

Components	lb component/lb fuel
C_f	0.693
H_f	0.227
N_f	0.080

Calculate the solid waste fraction (n_1) and liquid waste fraction (n_2):

$$n_1 = \text{(solid waste rate)/(liquid waste rate + solid waste rate)}$$

$$= \frac{w_s}{w_l + w_s}$$

$$= \frac{2000}{1500 + 2000}$$

$$= 0.57 \text{ lb solid/lb waste}$$

$$n_2 = \frac{w_l}{w_l + w_s}$$

$$= 0.43 \text{ lb liquid/lb waste}$$

Calculate the composition of the combined waste/auxiliary fuel feed to the kiln:

$$n_C = \frac{n_1 C_1 + n_2 C_2 + n_f C_f}{1 + n_f} = 0.5982 \text{ lb/lb feed}$$

$$n_H = \frac{n_1 H_1 + n_2 H_2 + n_f H_f}{1 + n_f} = 0.1481 \text{ lb/lb feed}$$

$$n_{H_2O} = \frac{n_1 (H_2O)_1 + n_2 (H_2O)_2}{1 + n_f} = 0.1223 \text{ lb/lb feed}$$

$$n_N = \frac{n_f N_f}{1 + n_f} = 0.0267 \text{ lb/lb feed}$$

$$n_O = \frac{n_1 O_1 + n_2 O_2}{1 + n_f} = 0.0705 \text{ lb/lb feed}$$

$$n_s = \frac{n_1 S_1 + n_2 S_2}{1 + n_f} = 0.0343 \text{ lb/lb feed}$$

Note: Subscripts 1 and 2 designate the composition of a given component in the solid waste and liquid waste, respectively. Remember that the fuel–waste ratio (n_f) for this problem is 0.5.

Calculate the stoichiometric oxygen requirement for the kiln using the following equation:

$$(O_2)_{stoich} = \sum_{i=1}^{n} (n_i)(\text{lb } O_2/\text{lb of species } i) - n_O$$

$$= (n_C)(2.67 \text{ lb } O_2/\text{lb C}) + (n_H)(8.0 \text{ lb } O_2/\text{lb H}) + (n_S)(1.0 \text{ lb } O_2/\text{lb S}) - n_O$$

$$= 2.746 \text{ lb } O_2/\text{lb feed}$$

See the note in Example 6.6.2 explaining this equation.

Calculate the combustion gas mass fractions:

$$N_{CO_2} = (n_C)(3.67 \text{ lb } CO_2/\text{lb C}) = 2.1954 \text{ lb } CO_2/\text{lb feed}$$

$$N_{H_2O} = (n_H)(9.0 \text{ lb } H_2O/\text{lb H}) + n_{H_2O} = 1.4552 \text{ lb } H_2O/\text{lb feed}$$

$$N_{N_2} = (O_2)_{stoich}(3.29 \text{ lb } N_2/O_2) + n_N = 9.1160 \text{ lb } N_2/\text{lb feed}$$

$$N_{SO_2} = (n_S)(2.0 \text{ lb } SO_2/\text{lb S}) = 0.0686 \text{ lb } SO_2/\text{lb feed}$$

Calculate the amount of combustion products per pound of feed:

$$\text{CP} = N_{CO_2} + N_{H_2O} + N_{N_2} + N_{SO_2}$$

$$= 12.8352 \text{ lb/lb feed}$$

Calculate the additional nitrogen and oxygen present in the combustion gases due to the excess air feed:

$$(O_2)_{EA} = (\text{EA})(O_2)_{stoich}$$

$$= 1.373 \text{ lb } O_2/\text{lb feed}$$

$$(N_2)_{EA} = (3.29 \text{ lb } N_2/\text{lb } O_2)(O_2)_{EA}$$

$$= 4.517 \text{ lb } N_2/\text{lb feed}$$

Calculate the total combustion gas products per pound feed:

$$\text{CG} = \text{CP} + (O_2)_{EA} + (N_2)_{EA} = 12.83 + 1.373 + 4.517$$

$$= 18.72 \text{ lb/lb feed}$$

Calculate the mass fraction of each combustion gas component:

$$CO_2 \qquad \frac{N_{CO_2}}{CG} = 0.1171 \text{ lb/lb gas}$$

$$H_2O \qquad \frac{N_{H_2O}}{CG} = 0.0776 \text{ lb/lb gas}$$

$$N_2 \qquad \frac{N_{N_2} + (N_2)_{EA}}{CG} = 0.727 \text{ lb/lb gas}$$

$$O_2 \qquad \frac{(O_2)_{EA}}{CG} = 0.0732 \text{ lb/lb gas}$$

$$SO_2 \qquad \frac{N_{SO_2}}{CG} = 0.0037$$

Calculate the volumetric flow of the combustion products at standard conditions of 68°F and 1 atm:

$$CO_2 \qquad \frac{N_{CO_2}}{0.114 \text{ lb/scf}} = 19.26 \text{ scf/lb feed}$$

$$H_2O \qquad \frac{N_{H_2O}}{0.0467 \text{ lb/scf}} = 31.16 \text{ scf/lb feed}$$

$$N_2 \qquad \frac{N_{N_2} + (N_2)_{EA}}{0.0727 \text{ lb/scf}} = 187.5 \text{ scf/lb feed}$$

$$O_2 \qquad \frac{(O_2)_{EA}}{0.083 \text{ lb/scf}} = 16.54 \text{ scf/lb feed}$$

$$SO_2 \qquad \frac{N_{SO_2}}{0.166 \text{ lb/scf}} = 0.413 \text{ scf/lb feed}$$

Total volumertric flow rate = 254.8 scf/lb feed

Example 6.6.4 Manhattan Waste Generators plan to incinerate pure C_6H_5Cl in a liquid injection incinerator with 100% excess air. A consulting engineer from ABC Associates has assured the client that the operating temperature will exceed the permit requirement of 2100°F. Verify or dispute the consultant's claim.

SOLUTION This problem involves an extension of the material presented in Section 5.3. However, unlike the previous development, it will require an adiabatic flame temperature calculation with 100% excess air. For this condition, the stoichiometric oxygen and nitrogen (or air) requirement is increased by a factor of 2. These values and the stoichiometric equation are again given here:

$$O_2 = 14 \text{ lbmol } N_2 = 52.6 \text{ lbmol}$$

$$C_6H_5Cl + 14O_2 + 52.6N_2 \rightarrow 6CO_2 + 2H_2O + HCl + 7O_2 + 52.6N_2$$

The heat capacity term (ΔC_P) is now calculated using Eq. (5.2.9) and (5.4.11).

$$\Delta C_P = 6C_{P,CO_2} + 2C_{P,H_2O} + C_{P,HCl} + 7C_{P,O_2} + 52.6C_{P,N_2}$$

The heat capacities are obtained from Table 5.2.4:

$$C_{P,CO_2} = 10.57 + 2.10 \times 10^{-3}T - 2.06 \times 105T^{-5}$$

$$C_{P,H_2O} = 7.30 + 2.46 \times 10^{-3}T + 0.00 \times 105T^{-2}$$

$$C_{P,HCl} = 6.27 + 1.24 \times 10^{-3}T + 0.30 \times 105T^{-2}$$

$$C_{P,O_2} = 7.16 + 1.00 \times 10^{-3}T - 0.40 \times 105T^{-2}$$

$$C_{P,N_2} = 6.83 + 0.90 \times 10^{-3}T - 0.12 \times 105T^{-2}$$

Thus,

$$\Delta C_P = 493.67 + 0.0731T - 2.12 \times 10^6 T^{-2}$$

The sensible enthalpy change for the fuel gas products is

$$\Delta H_p = \int_{298}^{T_2} \Delta C_P \, dT$$

In addition,

$$(\Delta H_c^\circ) = (\Delta H_p)$$

where (ΔH_c°) was calculated to be $-714{,}361$ cal/gmol from Example 5.7.9. Therefore,

$$714{,}361 = 493.67(T_2 - 298) + \frac{1}{2}(0.0731)(T_2^2 - 298^2) + 2.12 \times 10^6 \left(\frac{1}{T_2} - \frac{1}{298}\right)$$

$$871{,}834 = 493.67T_2 + 0.036557T_2^2 + \frac{2.12 \times 10^6}{T_2}$$

Solving for T_2 (see Example 5.7.9 for a simplified trial-and-error procedure),

$$T_2 = 1579 \text{ K} = 2842°\text{R} = 2382°\text{F}$$

Therefore, the operating temperature does exceed the permit requirement of 2100°F.

Example 6.6.5 Calculate the operating temperature of an incinerator burning pure DDT ($C_{14}H_9C_{15}$) with 150% excess air. Assume the DDT and air enter the incinerator at ambient conditions (25°C, 1 atm).

SOLUTION This problem is very similar to Example 5.7.10. Here, instead of stoichiometric air, 150% excess air is used. The following values can be used:

$$_c(\Delta H) = -1580.56 \text{ kcal/gmol}$$

$$C_{14}H_9O_5 + 37.5O_2 + 141.08N_2 \rightarrow 14CO_2 + 2H_2O + 5HCl + 22.5O_2 + 141.08N_2$$

$$\Delta C_P = 14C_{P,CO_2} + 2C_{P,H_2O} + 5C_{P,HCl} + 22.5C_{P,O_2} + 141.08C_{P,N_2}$$

Substituting in the heat capacities from Table 5.2.4,

$$\Delta C_P = 1318.60 + 0.1899T - 5.327 \times 10^6 T^{-2}$$

Since

$$\Delta H_c = \Delta H_p = \int_{298}^{T} \Delta C_P \, dT$$

$$1580.56 \times 10^3 = 1318.60(T_2 - 298) + \frac{1}{2}(0.1899)(T_2^2 - 298^2)$$

$$+ 5.327 \times 10^6 \left(\frac{1}{T_2} - \frac{1}{298}\right)$$

$$1{,}999{,}815.0 = 1318.60T_2 + 0.095T_2^2 + \frac{5.327 \times 10^6}{T_2}$$

Solving for T_2 by a trial-and-error method,

$$\begin{aligned} T_2 &= 1377 \text{ K} \\ &= 1104°\text{C} \\ &= 2019°\text{F} \end{aligned}$$

Example 6.6.6 Thirty percent by mass of benzene in water is combusted in an incinerator with stoichiometric air. Estimate the operating temperature of the incinerator using the EPA worksheet provided in Table 6.4.5. Repeat the calculation for 25, 50, 75, 100, and 150% excess air.

SOLUTION First write down the combustion equation for 0% excess air (or stoichiometric air):

$$C_6H_6 + 7.5O_2 + 28.2N_2 \rightarrow 6CO_2 + 3H_2O(g) + 28.2N_2$$

Converting from lbmol to pounds of flue gases (per lb of benzene incinerated),

$$\text{lb } CO_2 = \frac{(6)(\text{MW of } CO_2)}{\text{MW of } C_6H_6} = \frac{(6)(44)}{(78)}$$
$$= 3.38$$
$$\text{lb } H_2O = \frac{(3)(18)}{(78)} = 0.69$$
$$\text{lb } N_2 = \frac{(28.2)(28)}{(78)}$$
$$= 10.12$$

Since no auxiliary fuel or air preheating is employed, go to step 3 of Table 6.4.5 directly:

$$\Delta H_2 = (0.3)(\text{NHV of benzene})$$

The NHV of benzene is 17,451 Btu/lb from Table 5.3.1.

$$\Delta H_2 = (0.3)(17{,}451)$$
$$= 5235.3 \text{ Btu/lb feed}$$

Calculate the heat loss through the walls of the incinerator, assuming 5% loss:

$$Q = (0.05)(\Delta H_2)$$
$$= 261.76 \text{ Btu/lb feed}$$

Calculate the enthalpy of the combustion products leaving the incinerator:

$$\begin{aligned}\Delta H_3 &= (0.26)(0.3 \times 3.38 + 0.3 \times 10.12)(T - 77) \\ &\quad + (0.49)[(0.3 \times 0.69) + 0.7](T - 212) + [(0.3 \times 0.69) + 0.7](1150) \\ &= 1.053(T - 77) + 0.907(T - 212) + (0.907)(1150)\text{Btu/lb feed} \\ &= 1.96T - 770 \text{ Btu/lb feed}\end{aligned}$$

Calculate the enthalpy of stoichiometric air:

$$\begin{aligned}\Delta H_4' &= (1.1)(T - 77)(O_2)_{\text{stoich}} = (1.1)(T - 77)(0.3)(7.5)(32)/(78) \\ &= 1.015(T - 77)\end{aligned}$$

For 0% EA, $\Delta H_4 = 0$. From step 7 of Table 6.4.5,

$$\begin{aligned}O &= (\Delta H_2 - Q - \Delta H_3)/\Delta H_4' \\ &= [5235.3 - 261.76 - (1.053)(T - 77) + 0.907(T - 212) \\ &\quad + 0.907(1150)]/(1.015T - 77)\end{aligned}$$

Solving for T,

$$T = 2900°\text{F}$$

For 25% excess air,

$$25 = \frac{(100)(\Delta H_2 - Q - \Delta H_3)}{\Delta H_4'}$$

$$0.25 = \frac{5235.3 - 261.76 - (1.053)(T - 77) - 0.907(T - 212) - 0.907(1150)}{(1.015)(T - 77)}$$

Therefore,

$$T(25\%\text{ EA}) = 2421°\text{F}$$

Similarly,

$$\begin{aligned}T(50\%\text{ EA}) &= 2117°\text{F} \\ T(75\%\text{ EA}) &= 1883°\text{F} \\ T(100\%\text{ EA}) &= 1697°\text{F} \\ T(150\%\text{ EA}) &= 1420°\text{F}\end{aligned}$$

Example 6.6.7 A fuel–waste mixture is to be combusted in an incinerator at an operating temperature of 2000°F. Calculate the minimum net heating value of the mixture in Btu/lb if 0, 20, 40, 60, 80, and 100% excess air is employed. Use Eq. (6.4.22) to perform the calculation.

SOLUTION Referring to Eq. (6.4.22),

$$\text{NHV} = \frac{(0.3)(T - 60)}{1 - (1 + \text{EA}(7.5 \times 10^{-4})(0.3)(T - 60)}$$

For 0% excess air,

$$\text{NHV (0\% excess air)} = \frac{(0.3)(2000 - 60)}{1 - (1)(7.5 \times 10^{-4})(0.3)(2000 - 60)}$$
$$= 1032.8 \text{ Btu/lb}$$

For 20% excess air,

$$\text{NHV (20\% excess air)} = \frac{(0.3)(2000 - 60)}{1 - (1.2)(7.5 \times 10^{-4})(0.3)(2000 - 60)}$$
$$= 1222.2 \text{ Btu/lb}$$

Similarly,

$$\text{NHV (40\% excess air)} = 1496.5 \text{ Btu/lb}$$
$$\text{NHV (60\% excess air)} = 1929.7 \text{ Btu/lb}$$
$$\text{NHV (80\% excess air)} = 2715.8 \text{ Btu/lb}$$
$$\text{NHV (100\% excess air)} = 4582.7 \text{ Btu/lb}$$

Example 6.6.8 A plant engineer has calculated the stoichiometric oxygen requirement and combustion gas mass flow rates using the approach in Example 6.6.6. Apply these results and calculate the excess air rate. Use Table 6.4.5 to perform the calculation:

$$(O_2)_{\text{stoich}} = 3.195 \text{ lb } O_2\text{/lb feed}$$

Combustion gas mass fractions:

$$N_{CO_2} = 2.61 \text{ lb } CO_2\text{/lb feed}$$
$$N_{H_2O} = 1.52 \text{ lb } H_2O\text{/lb feed}$$
$$N_{N_2} = 10.60 \text{ lb } N_2\text{/lb feed}$$

Additional Data

NHV of the feed = 7200 Btu/lb feed
Operating temperature, T = 1900°F
Air preheat temperature, T_{air} = 500°F
No auxiliary fuel is used.

SOLUTION Since no auxiliary fuel is used, proceed to step 2 of Table 6.4.5 directly.

Calculate the enthalpy input to the incinerator through air preheating:

$$\Delta H_1' = 1.115(T_{air} - 77)(O_2)_{stoich}$$
$$= 1513.7 \text{ Btu/lb feed}$$
$$\Delta H_1 = \Delta H_1'(1 + \text{EA})$$

Note: ΔH_1 is the enthalpy input with an excess air rate of EA. See footnote *b* of Table 6.4.5 for the derivation of this equation.

Determine the heat generated by the combustion of the waste:

$$\Delta H_2 = 7200 \text{ Btu/lb feed}$$

Note: Since no auxiliary fuel is used, the NHV of the feed is ΔH_2.

Calculate the heat transferred across the walls of the incinerator, assuming a 5% loss:

$$Q = 0.05\Delta H_2$$
$$= 360 \text{ Btu/lb feed}$$

Calculate the enthalpy of the stoichiometric combustion products leaving the incinerator:

$$\Delta H_3 = 0.26(N_{CO_2} + N_{N_2})(T - 77) + 0.49(N_{H_2O})(T - 212) + N_{H_2O}(1150)$$
$$= 8032 \text{ Btu/lb feed}$$

See footnote *d* of Table 6.4.5 for the derivation of the previous equation. Calculate the enthalpy of the excess air leaving the incinerator:

$$\Delta H_4' = 1.115(T - 77)(O_2)_{stoich}$$
$$= 6523.4 \text{ Btu/lb feed}$$
$$\Delta H_4' = (\Delta H_4')(\text{EA})$$

$\Delta H'_4$ is the enthalpy of the air leaving the incinerator if no excess air is used.
Calculate the excess air percentage by the following equation:

$$\begin{aligned} EA &= 100(\Delta H'_1 + \Delta H_2 - Q - \Delta H_3)/(\Delta H'_4 - \Delta H'_1) \\ &= 6.4\% \end{aligned}$$

Note: This equation is obtained by an energy balance around the incinerator.

Finally, the percentage of excess air employed in an incinerator may be estimated from the following equation:

$$EA = \frac{(95)O_2}{21 - O_2}$$

where O_2 is the percentage of oxygen in the discharge gas on a dry basis.

Example 6.6.9 A liquid waste consisting of 5% benzene in an aqueous solvent (with zero heating value) is to be incinerated. The combustion process will be supplemented with 0.4 lb natural gas (NHV = 19,500 Btu/lb) per lb waste. Ninety percent excess air at 300°F is to be employed. Calculate the operating temperature of the incinerator if 5.0% of the total heat released is lost across the walls of the unit. Use EPAs worksheet as presented in Table 6.4.5.

SOLUTION First refer back to Example 6.6.2. Since the waste consists of 5% benzene,

$$\begin{aligned} C_w &= \frac{(0.05)(6)(\text{MW of C})}{\text{MW of } C_6H_6} = \frac{(0.05)(6)(12)}{(78)} \\ &= 0.046 \text{ lb/lb waste} \\ H_w &= \frac{(0.05)(6)(1)}{78} = 0.004 \text{ lb/lb waste} \\ H_2O_w &= 0.95 \text{ lb/lb waste} \end{aligned}$$

Identify the elemental composition of the natural gas (see Table 6.3.1):

$$\begin{aligned} C_f &= 0.693 \text{ lb/lb fuel} \\ H_f &= 0.227 \text{ lb/lb fuel} \\ N_f &= 0.08 \text{ lb/lb fuel} \end{aligned}$$

Calculate the combined composition of waste and fuel feed:

$$n_C = \frac{C_w + n_f C_f}{1 + n_f} = \frac{0.046 + (0.04)(0.693)}{1 + 0.4}$$

$$= 0.23 \text{ lb/lb feed}$$

$$n_H = \frac{H_w + n_f H_f}{1 + n_f} = \frac{0.004 + (0.04)(0.227)}{1.4}$$

$$= 0.0677 \text{ lb/lb feed}$$

$$n_N = \frac{n_f N_f}{1 + n_f} = \frac{(0.04)(0.08)}{1.4}$$

$$= 0.023 \text{ lb/lb feed}$$

$$n_{H_2O} = \frac{H_2O_w}{1 + n_f} = \frac{0.95}{1.4}$$

$$= 0.679 \text{ lb/lb feed}$$

Calculate the stoichiometric oxygen requirement:

$$(O_2)_{stoich} = (n_C)(2.67) + (n_H)(8.0) = (0.23)(2.67) + (0.0677)(8.0)$$

$$= 1.16 \text{ lb/lb feed}$$

See footnote *c* of Table 6.3.1 for the details of this equation. Calculate the combustion gas mass fraction:

$$N_{CO_2} = (n_C)(3.67 \text{ lb } CO_2/\text{lb C}) = (0.23)(3.67)$$

$$= 0.844 \text{ lb } CO_2/\text{lb feed}$$

$$N_{H_2O} = (n_H)(9.0 \text{ lb } H_2O/\text{lb H}) = n_{H_2O} = (0.0677)(9) + (0.679)$$

$$= 1.289 \text{ lb } H_2O/\text{lb feed}$$

$$N_{N_2} = (O_2)_{stoich}(3.29 \text{ lb } N_2/\text{lb } O_2) + n_N = (1.16)(3.29) + 0.023$$

$$= 3.8394 \text{ lb/}N_2\text{/lb feed}$$

Now refer to Table 6.4.5.

Calculate the enthalpy input by the preheated air:

$$\Delta H'_1 = (1.115)(T_{air} - 77)(O_2)_{stoich} = (1.12)(300 - 77)(1.16)$$

$$= 289.7 \text{ Btu/lb feed}$$

Calculate the heat generated by combustion of the waste–fuel mixture:

$$\Delta H_2 = \frac{\text{NHV}_{\text{waste}} + n_f \text{NHV}_f}{1 + n_f} - n_{H_2O}(\Delta H_v)$$

$$= \frac{0.05 \times 17{,}451 + 0.4 \times 19{,}500}{1.4} - (0.679)(1050)$$

$$= 5481.7 \text{ Btu/lb feed}$$

Note: The heat required to vaporize the solvent is subtracted from (ΔH_2).

The heat capacity of the solvent is assumed to be 1.0 Btu/lb and the heat of vaporization is 1050 Btu/lb.

Calculate the heat loss through the walls of the incinerator:

$$Q = (0.05)(\Delta H_2)$$

$$= 274 \text{ Btu/lb feed}$$

Calculate the enthalpy of the combustion products leaving the incinerator:

$$(\Delta H_3) = (0.26)(N_{CO_2} + N_{N_2})(T - 77) + (0.49)(N_{H_2O})(T - 212) + N_{H_2O}(1150)$$

$$= (0.26)(0.844 + 3.8394)(T - 77) + (0.49)(1.289)(T - 212) + (1.289)(1150)$$

$$= 1.217(T - 77) + 0.63(T - 212) + 1482 \text{ Btu/lb feed}$$

$$= 1.847T + 1255 \text{ Btu/lb feed}$$

Calculate the enthalpy of the stoichiometric air:

$$(\Delta H_4) = (1.1)(T - 77)(O_2)_{\text{stoich}} = (1.1)(T - 77)(1.16)$$

$$= 1.276(T - 77)$$

Use the equation in step 7 of Table 6.4.5 to calculate the incinerator temperature:

$$\text{EA} = \frac{100(\Delta H_1' + \Delta H_2 - Q - \Delta H_3)}{\Delta H_4' - \Delta H_1'}$$

$$50 = \frac{100[289.7 + 5481.7 - 274 - (1.251)(T - 77) + .6316(T - 212) + 1482}{(1.276)(T - 77) - 289.7}$$

Solving for T, $T = 1784°\text{F}$.

Example 6.6.10 In order to meet recently updated pollution regulations for discharging hydrocarbons to the atmosphere, a gas stream must be reduced by 99.5% of its present hydrocarbon concentration. Due to economic considerations, it is proposed to meet this requirement by combusting the hydrocarbons in a thermal oxidizer operating at 1500°F. The gas and methane (fuel) are to be fed to the reactor at 80°F and 1 atm. Design the proposed oxidizer using kinetic principles.

Data

Flue gas flow rate (from fuel combustion) = 2500 scfm (80°F, 1 atm)

Process gas flow rate = 7200 scfm

Hydrocarbon: Essentially toluene

Reaction rate constant, $k = 7.80\ \mathrm{s}^{-1}$ at 1500°F

Average velocity = 20 ft/s; $C_f/C_0 = 0.005/1.0$

C_f and C_0 are the final and initial concentrations, respectively.

SOLUTION Based on the actual operating conditions of the oxidizer, the volumetric flow rate is

$$Q_a = Q_s\left(\frac{T_a}{T_s}\right) = (2500 + 7200)\left(\frac{460 + 1500}{460 + 80}\right)$$
$$= 35{,}207 \text{ acfm}$$

Calculate the cross-section area of the oxidizer:

$$A = \frac{Q_a(\text{acfm})}{V(\text{fpm})} = \frac{35{,}207}{(20)(60)}$$
$$= 29.33 \text{ ft}^2$$

Calculate the diameter of the oxidizer:

$$D = \left(\frac{4A}{\pi}\right)^{0.5} = \left(\frac{4 \times 29.33}{3.1416}\right)^{0.5}$$
$$= 6.11 \text{ ft}$$

Calculate the residence time required assuming a plug flow oxidizer. Use Eq. (6.5.6):

$$\theta = -\left(\frac{1}{k}\right) \ln \left(\frac{C_f}{C_0}\right) = -\frac{1}{7.8} \ln \left(\frac{0.005}{1.0}\right)$$

$$= 0.68 \text{ s}$$

Calculate the oxidizer volume required to achieve the residence time calculated previously:

$$V = \theta Q_a = (0.68) \frac{35{,}207}{60}$$

$$= 399 \text{ ft}^3$$

Calculate the length of the oxidizer:

$$L = \frac{V}{A} = \frac{399}{29.33}$$

$$= 13.6 \text{ ft}$$

Example 6.6.11 A fluidized-bed incinerator is to be designed to destroy 99.99% of a unique hazardous waste. Based on laboratory and pilot plant studies, researchers have described the waste reaction by a first-order reversible mechanism. Their preliminary findings are given here.

$$A \underset{k'}{\overset{k}{\longleftrightarrow}} B$$

$$k = 1.0 \exp(-10{,}000/T) \qquad T = \text{Rankine}$$

$$k' = 9.89 \exp(-35{,}000/T) \qquad T = \text{Rankine}$$

Calculate a fluidized-bed incineration operating temperature that will minimize the volume of the incinerator and achieve the desired degree of waste conversion (destruction). The operating temperature must be in the 1400–1700°F range.[4]

SOLUTION A fluidized-bed incinerator is best described by a CSTR. According to Eq. (6.5.3),

$$\theta = \frac{V}{Q_a} = \frac{C_{A0} - C_{A1}}{-R_A} = \frac{C_{A0} - C_{A1}}{kC_{A1} - k'C_{B1}}$$

For 99.99% destruction of the waste A, the conversion variable X_A becomes

$$X_A = 0 - 9999$$

So that

$$C_{A1} = 0.0001C_{AO} \qquad C_{B1} = 0.9999C_{AO}$$

Therefore,

$$\frac{V}{Q_a} = \frac{C_{AO} - 0.0001C_{AO}}{(k)(0.0001C_{AO}) - (k')(0.9999C_{AO})} = \frac{0.9999}{0.0001k - 0.9999k'} \tag{6.6.1}$$

To minimize the incinerator volume, dV/dT is set to zero. Since Q_a is assumed to be constant, $d(V/Q_a)/dT$ can also be set to 0:

$$\frac{d(V/Q_a)}{dT} = \frac{-9999[(dk/dT) - 9999(dk'/dT)]}{(k - 9999k')^2} = 0$$

Thus,

$$\frac{dk}{dT} = \frac{10^4 dk'}{dT} \tag{6.6.2}$$

Substituting in the values of the reaction velocity constants,

$$k = 1.0 \exp\left(\frac{-10{,}000}{T}\right)$$

$$\frac{dk}{dT} = \left(\frac{-10{,}000}{T^2}\right) \exp\left(\frac{-10{,}000}{T}\right)$$

$$k' = 9.89 \exp\left(\frac{-35{,}000}{T}\right)$$

$$\frac{dk'}{dT} = \left(\frac{3.4615 \times 10^5}{T^2}\right) \exp\left(\frac{-35{,}000}{T}\right)$$

Therefore,

$$(10{,}000)\exp\left(\frac{-10{,}000}{T}\right) = (3.4165 \times 10^9)\exp\left(\frac{-35{,}000}{T}\right)$$

Solving for T,

$$T = 1960°\text{R} = 1500°\text{F}$$

There are simpler approaches that could have been used to solve this problem:

1. Rather than maximizing the derivative of Eq. (6.6.1), the equation could have been inverted and the derivative of the resulting equation would immediately lead to Eq. (6.6.2).
2. The rate constant, which is essentially the derivative of Eq. (6.6.1), could be maximized to obtain the minimum volume. This again would immediately lead to Eq. (6.6.2).
3. The right side of Eq. (6.6.1) could be plotted against temperature (upon which the k's depend) to determine the minimum of the function.

Example 6.6.12 It is proposed to decompose pure diethyl peroxide (A) at 225°C in a bench-scale incinerator. This pollutant will be entering the incinerator at a flow rate of 12.1 L/s. It is desired to decompose 99.995% of the diethyl peroxide. The following data are available:

$$R_A = -k_A C_A \text{ gmol/L-s} \qquad k_A = 38.3 \text{ s}^{-1} \text{ at } 225°\text{C}$$

The inside diameter of the incinerator is 8.0 cm. What should the length of the incinerator be?

SOLUTION Assume a plug flow reactor and use Eq. (6.5.6) since the reactor is first order and irreversible:

$$\theta = -\frac{l}{k} \ln\left(\frac{C_A}{C_{A0}}\right)$$

For 99.995% destruction,

$$C_A = 5 \times 10^{-5} C_{A0}$$

$$\theta = -\left(\frac{1}{38.3}\right) \ln(5 \times 10^{-5})$$

$$= 0.259 \text{ s}$$

Calculate the incinerator volume:

$$V = (\theta)(Q_a) = (0.259)(12.1)$$

$$= 3.13 \text{ L} = 3130 \text{ cm}^3$$

Calculate the length of the reactor:

$$L = \frac{V}{\pi D^2/4} = \frac{3130}{(\pi)(8^2)/(4)}$$

$$= 62.3 \text{ cm}$$

PROBLEMS

1. Consider a gaseous fuel with the following volume fraction composition:

$$N_2 = 0.05$$
$$CH_4 = 0.81$$
$$C_2H_6 = 0.10$$
$$C_3H_8 = 0.04$$

Calculate the standard volumetric flow rate of air required for complete combustion of 1.0 scfm of this fuel with 100% theoretical (or stoichiometric) air. Also calculate the standard volumetric flow rates of the combustion products.

2. A waste mixture to be incinerated in a liquid injection system has the following composition on a lb/lb waste basis:

$$\text{Carbon } (C_w) = 0.732$$
$$\text{Hydrogen } (H_w) = 0.109$$
$$\text{Moisture } (H_2O_w) = 0.022$$
$$\text{Oxygen } (O_w) = 0.102$$
$$\text{Chlorine } (Cl_w) = 0.035$$

The incineration process will be assisted with 1.0 lb of natural gas/lb of waste.

One hundred percent excess air is to be employed. Calculate the required quantities in the worksheet provided in Table 6.3.1.

3. A waste mixture to be incinerated by a rotary kiln has the following composition:

	Solids (1)	Liquids (2)	
Carbon (C_w)	0.74	0.30	lb/lb waste
Hydrogen (H_w)	0.10	0.12	lb/lb waste
Moisture (H_2O_w)	0.02	0.40	lb/lb waste
Oxygen (O_w)	0.11	0.10	lb/lb waste
Sulfur (S_w)	0.03	0.08	lb/lb waste

The liquid feed rate is 2000 lb/h and the solid feed rate is 1500 lb/h. The incineration process will be assisted with 1.0 lb of natural gas/lb of waste. Employ 120% excess air. Use Table 6.3.2 in performing the calculations.

4. Calculate the adiabatic operating temperature of DDT ($C_{14}H_9C_{15}$) when it is incinerated with 50% excess air.

5. Estimate the incinerator temperature for the DDT in the previous problem using the simplified equation developed in Section 6.4 [Eq. (6.4.20)].

6. Repeat the calculations of Problems 4 and 5 if the combustion is accomplished with 100% excess air.

7. Pure chlorobenzene is to be incinerated in a fluid-bed unit with 150% excess air. Verify that the operating temperature of the incinerator will be in the 1500–2000°F range. Assume the feed (chlorobenzene–air mixture) is at 25°C.

8. It is planned to conduct a trial burn for a rotary kiln incinerator using pure DDT ($C_{14}H_9C_{15}$) with 200% excess air. Assuming negligible heat losses from the incinerator, calculate the operating temperature of the kiln. The feed (DDT–air mixture) is at ambient conditions.

9. Calculate the adiabatic operating temperature of C_6H_5Cl when it is incinerated with 50% excess air.

10. Estimate the adiabatic flame temperature of the C_6H_5Cl in Problem 9 using the simplified equation presented in Section 6.4 [Eq. (6.4.20)].

11. Forty percent by mass of xylene in an aqueous solvent is combusted in an incinerator with stoichiometric air. Estimate the operating temperature of the incinerator using the EPA worksheet provided in Table 6.4.5. Repeat the calculation for 50 and 100% excess air.

12. Perform the calculations required in Problem 11 using the simplified approach given in Eq. (6.4.20).

13. A waste–fuel mixture is to be burned in an incinerator at an operating temperature of 1900°F. Calculate the minimum net heating value of the mixture in Btu/lb if 0, 20, 40, 60, 80, and 100% excess air is employed. Use Eq. (6.4.22) to perform the calculations.

14. A liquid mixture consisting of 3% waste (NHV = 20,650 Btu/lb) in an aqueous solvent (with zero heating value) is to be incinerated. The combustion process will be supplemented with 0.25 lb natural gas (NHV = 20,000 Btu/lb)/lb waste. Sixty percent excess air at 450°F is to be employed. Calculate the operating temperature of the incinerator if 5% of the total heat released is lost across the walls of the unit. Use the EPAs worksheet as presented in Table 6.4.5.

15. Repeat Problem 14 using the simplified approach given in Eq. (6.4.20).

16. A fluidized-bed incinerator is to be designed to destroy 99.99% of a hazardous waste. The waste reaction is described by a first-order reversible mechanism where

$$A \underset{k'}{\overset{k}{\longleftrightarrow}} B$$

$$k_1 = Ae^{-E/T} \qquad T = \text{Rankine}$$

$$k_1' = A'e^{-E'/T} \qquad T = \text{Rankine}$$

$$A = 1.0 \text{ s}^{-1} \qquad E = 10{,}000$$

$$A' = 111 \text{ s}^{-1} \qquad E' = 40{,}000$$

Calculate a fluidized-bed incinerator operating temperature that will minimize the volume of the incinerator and achieve the desired degree of waste conversion.

17. Design a proposed incinerator using kinetic principles for the destruction of toluene vapor at operating conditions of 1800°F and 1 atm. The gas and methane fuel are fed to the incinerator at 60°F and 1 atm.

DATA Flue gas flow rate (from fuel combustion) = 2800 scfm (60°F, 1 atm)
Process gas flow rate = 8600 scfm
Reaction rate constant, $k = 8.2 \text{ s}^{-1}$ at 1800°F
Average velocity = 20 ft/s
$C_f/C_0 = 0.006/1.0$ (99.4% DRE)

REFERENCES

1. U.S. EPA, *Engineering Handbook for Hazardous Waste Incineration*, Monsanto Research Corporation, EPA Contract No. 68-03-3025, Sept., 1981.
2. L. Theodore and J. Reynolds, personal notes, 1987.
3. K. C. Lee, N. Morgan, J. L. Hanson, and G. Whipple, "Revised Predictive Model for Thermal Destruction of Dilute Organic Vapors and Some Theoretical Explanation," Paper No. 82-5.3, Air Pollution Control Association Annual Meeting, New Orleans, 1982.
4. L. Theodore, *Chemical Reaction Kinetics*, A Theodore Tutorial, Air and Waste Management Association (Bookstore), Pittsburgh, 1994.

III

EQUIPMENT

Part II provided the reader with some basic scientific and design principles needed to understand the waste combustion process. The purpose of Part III is to show how these principles are applied in practice by discussing the various pieces of equipment that constitute a hazardous waste combustion facility.

The lead chapter, Chapter 7, describes the major types of incinerators and waste combustors in use today and discusses the types of wastes that each one is best suited to handle; new and emerging waste combustion technologies are also described. The remaining four chapters cover equipment that are always or often part of a waste combustion system. The hot gases exiting the combustion chamber contain gaseous and particulate pollutants that cannot be permitted to escape into the atmosphere. Their removal from the gas stream requires the use of air pollution control devices (see Chapter 10). Because of materials limitations (e.g., the fabric materials used in filter baghouses), most types of air pollution control equipment cannot handle gas temperatures typical of incinerator exhaust streams. Before reaching the air pollution control equipment, therefore, the hot gases must be cooled. This is usually accomplished either through the use of a waste heat boiler (Chapter 8), in which a good portion of the high enthalpy of the combustion chamber exhaust gases is recovered, or by utilizing a quenching technique (Chapter 9). The choice of method to cool the gases is dictated by the economics of the particular installation. In Chapter 11, equipment necessary for receipt and storage of waste feeds, transport systems, ash and residuals removal, and ancillary equipment used with incinerator, waste combustor and air pollution control system operation (e.g., pumps, fans, compressors, ducts, and stacks) are covered.

7

HAZARDOUS WASTE COMBUSTORS

Contributing authors: Donald L. Corwin, Thomas F. McGowan, and John F. Mullen

7.1 INTRODUCTION

Waste combustion and incineration is not a new technology and has been commonly used for treating organic hazardous waste for many years in Europe and the United States. Incineration is a mature and thoroughly proven technology that has been successfully used for the treatment of hazardous (chemical) wastes since the 1930s. The first rotary kiln incinerator for the treatment of hazardous wastes was installed in 1948. The major benefit of waste incineration is that the process actually destroys most of the waste rather than disposing of or storing it.

As of January 1997, there were 156 permitted commercial and industrial hazardous waste incinerators in the United States at 110 locations. There were 27 permitted commercial hazardous waste incinerators in the United States at 17 locations. The Environmental Protection Agency's (EPA's) Land Disposal Restriction regulations (40 CFR 268) require that many hazardous wastes can only be treated with either incineration or combustion in a boiler or industrial furnace. It can be used for a variety of specific wastes and is reasonably competitive in cost compared to other disposal methods.

Hazardous waste incineration involves the application of combustion processes under controlled conditions to convert wastes containing hazardous material to inert mineral residues and gases. The required three *T*'s of combustion must be present along with sufficient oxygen for the reaction:

- The most important T, *temperature*, of the combustion zone, must be maintained; exothermic reactions of combustion of the waste and auxiliary fuel (if required) must provide enough heat to raise the burning mixture (air, wastes, and fuel) to a sufficient temperature to destroy all organic components.
- *Turbulence*, the second most important T, the constant mixing of waste and oxygen, must exist.
- Elapsed time of exposure to combustion temperatures (*residence time*) must be adequately long in duration to ensure that even the slowest combustion reaction has gone to completion. In other words, transport of the burning mixture through the high-temperature region must occur over a sufficient period of *time*.
- Adequate free *oxygen* must always be available in the combustion zone.

Thus, four parameters influence the mechanisms of incineration: temperature, turbulence, residence time, and oxygen.

This chapter deals with the types of incinerators and waste combustors used to destroy hazardous waste. The liquid injection incinerator is the most common type used for destroying a wide range of liquids generated in the process industries. The liquid burners are also adapted to the solid waste incinerators, the most common being the rotary kiln incinerator. Liquid injection types are also applied to boilers, cement kilns, and lightweight aggregate kilns burning hazardous wastes as a source of energy for the industrial furnace or boiler.

7.2 LIQUID INJECTION INCINERATION

Overview of the Technology and Its Applications

In the field of hazardous waste incineration, there is more experience with liquid injection (LI) incinerators than all other types combined. In a survey conducted in 1981, approximately 64% of the total number of incinerators in service at the time were of the LI type.[1] Various types of incinerators are used to handle liquid waste as well as wastes in other forms, that is, solids, sludges, slurries, and fumes. In this section, the discussion is confined to the liquid injection type of incinerator, which, as the name implies, is applicable almost exclusively to pumpable liquid waste. The waste is burned directly in a burner (combustor) or injected into the flame zone or combustion zone of the incinerator chamber (furnace) via nozzles. The heating value of the waste is the primary determining factor for nozzle location.

Waste liquids in the process industries are numerous in kind and therefore defy easy definition. Disposal of these waste liquids has become a serious problem for the plant operator. Regulations today cover wastes once considered wastewater streams as well as the organic streams known to be made up of hazardous compounds. Liquid injection incinerators are available today to handle the various liquid streams that are being generated by the process industries. These units may be on site (at the location of the generator) or off site (at a sister facility or at a commercial disposal operation). This section will attempt to explore the various designs available and the basic components necessary to provide an adequate liquid injection system.

Liquid Injection Incinerators

Liquid injection incinerators are usually refractory-lined chambers (horizontal, vertical with orientation up or down), generally cylindrical and equipped with a primary combustor (waste and auxiliary fuel fired) and often secondary combustors or injection nozzles for low-calorific-value materials (aqueous wastes containing either organic or inorganic compounds, or both).

These units operate at temperature levels from 950°C (1742°F) to 1700°C (3092°F). Residence time in the chamber may vary from milliseconds to 2.5 s. The viscosity determines whether the material being incinerated is considered a liquid, slurry, or sludge. Liquid units today are capable of burning high-viscosity materials of 4500 Saybolt Seconds Universal (SSU) or less. Critical to the operation of the unit would be the atomizing nozzle used to convert the liquid stream into finely atomized droplets.

When considering the liquid injection incinerator, one must be concerned about the ability of the system; that is, the design, components, controls, and air pollution control system, to meet the requirements of the federal and state regulations. All units handling hazardous compounds (as listed in Sec. VIII Par. 261.30) have had to meet the basic requirements of 40 CFR Subpart O—Incinerators. As noted in Chapter 3, the MACT regulations of Sept. 30, 1999 set the emission requirements of existing as well as new incinerators.

Waste Liquid Data

The physical, chemical, and thermodynamic properties of the waste must be considered in the basic design requirements of the total incinerator system. This includes storage tanks, mixers, pumps, control valves, piping, atomizers, combustors, refractory, heat recovery, quench system, and air pollution equipment. The types of data needed by the designer to properly engineer the total system and the individual components are listed in Table 7.2.1.

Liquid Incinerator System Design

Storage In order to design the storage system to permit a feed of uniform mixture and heating value, proper mixing in the storage tank is required. Layering often occurs in large tanks, especially where wastes are introduced from various sources.[2] Mixing may be accomplished by internal mixers, external recycling pumps, or air or steam spargers. Venting of the tank is also necessary to prevent pressure buildup. Vent gases should be exhausted to the incinerator

Table 7.2.1 Waste liquid fuel data

Chemical composition (including halides, sulfur, nitrogen, etc.)
Specific gravity
Heat of combustion
Viscosity
Corrosivity
Ignitability
Reactivity
Polymerization
Solids content (type, percentage, physical data)
Metals
Slagging properties (temperature, eutectic data)

chamber. A more detailed discussion of storage systems is covered in Chapter 11.

Materials of construction of all components in the storage and feed system —tanks, pumps, mixers, piping, valves, and nozzles—are determined by chemical composition, viscosity, corrosivity, and solids content of the waste stream(s). If solids are known to be inert materials, filters should be used to prevent unnecessary carryover of these materials into the incinerator and the air pollution control system. If, however, these solids are organic, they must be included as part of the waste stream and the design should provide for moving this material directly to the incinerator.

Transport High-viscosity materials (slurries and sludges) and liquids containing solids are a serious concern in the design of the feed system to the incinerator. Various pump types are available, and their selection and use is predicated on the properties of the waste stream. Progressive-cavity pumps have been used successfully, provided the materials of the rotor and stator are selected properly. Gear pumps have been the workhorse for most lower-viscosity materials (equivalent to No. 6 fuel oil). However, in this case, the combination of viscosity and speed are critical for long-life, low-maintenance operation. In systems where a variety of waste streams must be pumped into the incinerator, the operator must be instructed about the proper feed system to use with each waste. In some cases, pumps may not be practical because of the high maintenance costs and shutdowns that result. Pressurizing with an inert gas (nitrogen) may be the safest and most economical method. Diaphragm pumps are in use today at many installations where other pumps have failed.

The atomizer design usually dictates the pressure requirements of the transport system (see below). When a hydraulic (mechanical) nozzle is used, requiring high pressures, a pump design needing close clearances between the impeller and the housing is also usually indicated. High-viscosity materials and liquids containing solids will score the surfaces of the pump, causing excessive wear and rapid deterioration of pump pressure. This results in a reduction of flow as well as poor atomization.

Materials transport is critical in the proper design of an incinerator. Certain liquid wastes and sludges must be maintained at temperature levels sufficiently high to permit pumping. If cooled, the material will solidify in the pipeline. However, if heating is not controlled, the material may polymerize if heated to too high a temperature. Once polymerized, it is impossible to liquefy the material. As a result, the piping with the polymer presents a solid waste disposal problem.[3]

Atomizers The method of injection of the liquid into the burner or the waste combustor is one of the most important features of a well-designed system. The main purposes for injecting the liquid as a spray are

- To break up the liquid into fine droplets
- To place the liquid droplets in a specific zone with a specific pattern with sufficient kinetic energy and penetration into the burning zone
- To control the rate of flow of the liquid discharged to the nozzle

Liquid waste atomization can be achieved by any of the following means:

- Rotary cup atomization
- Single-fluid pressure atomization
- Two-fluid, low-pressure air atomization (internal mix)
- Two-fluid, high-pressure air or steam atomization (internal mix)
- Two-fluid, high-pressure air or steam atomization (external mix)

The *rotary cup* consists of an open cup mounted on a hollow shaft. The cup is spun rapidly and liquid is admitted through the hollow shaft. A thin film of the liquid to be atomized is torn from the lip of the cup and surface tension reforms it into little droplets. To achieve conical-shaped flames an annular high-velocity jet of air (primary air) must be directed axially around the cup. If too little primary air is admitted, the fuel will impinge on the sides of the incinerator. If too much primary air is admitted, the flame will not be stable and will be blown off the cup. For fixed-firing rates, the proper adjustment can be found and the unit operated for long periods of time without cleaning. This requires little pressurization of the liquid and is ideal for atomizing liquids with relatively high solids content. Burner turndown is about 5:1 and capacities from 1 to 250 gal/h, are available.

In *single-fluid pressure* atomizing nozzle burners, the liquid is given a swirl as it passes through an orifice with internal tangential guide slots. Moderate liquid pressures of 100–150 psig provide good atomization with low-to-moderate liquid viscosity. In the simplest form, the waste is fed directly to the nozzle but turndown is limited from 2.5 to 3:1 since the degree of atomization drops rapidly with decrease in pressure. In a modified form involving a bypass return of liquid, turndown up to 10:1 can be achieved. When this type of atomization is used, secondary combustion air is generally introduced around the conical spray of droplets. The flame tends to be short, bushy, and of low velocity. Combustion tends to be slower as only secondary air is supplied and a larger combustion chamber is required. Typical burner capacities are in the range of 10–100 gal/h. Disadvantages of single-fluid pressure atomization are erosion of the burner orifice and a tendency toward pluggage with solid or liquid pyrolysis products, particularly in smaller sizes.

In air or steam atomizing burners, atomization can be accomplished internally by impinging the gas and liquid streams inside the nozzle before spraying (see Fig. 7.2.1*a*), externally by impinging jets of gas and liquid outside the nozzle (see Figure 7.2.1*b*), or by sonic means. Sonic atomizers use compressed gas to produce high-frequency sound waves that are directed onto the liquid stream. The liquid nozzle diameter is relatively large, and little waste pressurization is required. Some slurries and liquids with relatively large particulates can be handled by externally atomized and sonic atomizers without plugging problems.

Two-fluid atomizing nozzles may be either the low-pressure or high-pressure variety, with the latter being more common with high-viscosity materials. In

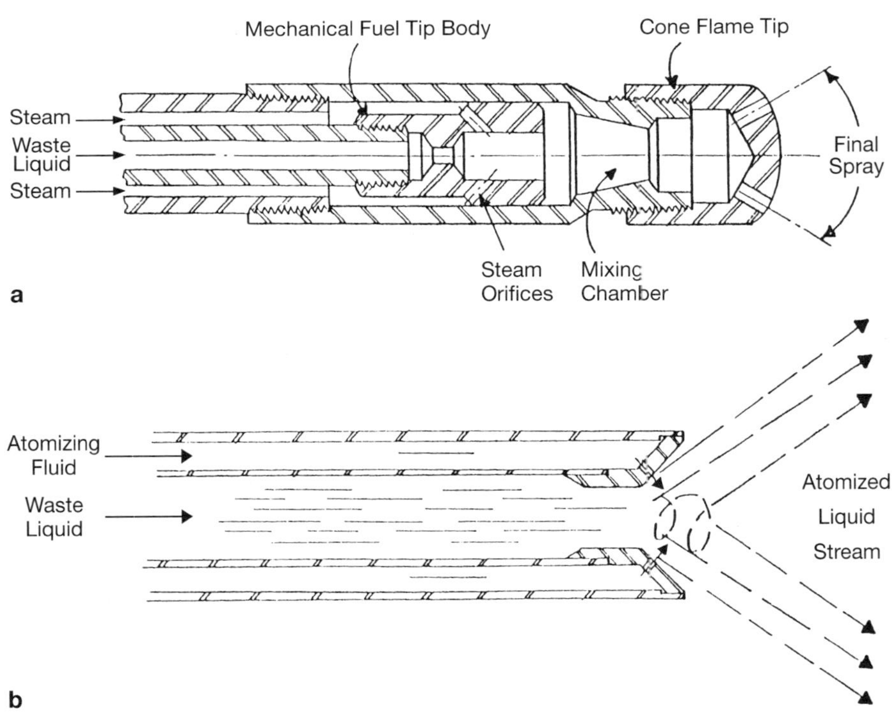

Figure 7.2.1 (*a*) Internal mix nozzle. (*b*) External atomizer tip.

low-pressure atomizers, air from blowers at pressures from 0.5 to 5.0 psig is used to aid atomization of the liquid. A viscous tar, heated to a viscosity of 75–90 SSU, requires air at a pressure of somewhat more than 1.5 psig, while a low-viscosity or aqueous waste can be atomized with 0.5-psig air. The waste liquid is supplied at a pressure of 4.5–17.5 psig. Burner turndown ranges from 3:1 to 6:1. Atomization air required varies from 350 to 1000 ft^3/gal (1.5 lb air/lb liquid to 4.5 lb/lb) of waste liquid. Less air is required as atomizing pressure is increased. The flame is relatively short as up to 40% of the stoichiometric air may be mixed with the liquid in atomization. For aqueous waste atomizers, compressed air at 40–100 psig is used. Air required is in the 0.1- to 0.5-lb/lb range.

High-pressure two-fluid burners require compressed air or steam at pressures from 30 to 150 psig. Air consumption is from 80 to 200 ft^3/gal waste, and steam requirements may be 2–4 lb/gal waste with careful control of the operation. Turndown is relatively poor (3:1 or 4:1) and considerable energy is employed for atomization. The major advantage of such burners is the ability to burn barely pumpable liquids without further viscosity reduction. Steam atomization also tends to reduce soot formation with wastes that would normally burn with a smoky flame. Compressed air atomizers provide oxygen immediately to mix with the waste or fuel and increases overall combustion efficiency by reducing the energy required to raise the steam to combustion temperature.

Table 7.2.2 identifies typical kinematic viscosity and solids-handling limitations for the various atomization techniques. These data are based on a survey of 14 burner manufacturers.[3] In evaluating a specific incinerator design, however, the viscosity and solids content of the wastes should be compared with the manufacturer specifications for the particular burner employed.

Table 7.2.2 Kinematic viscosity and solids handling limitations of various atomization techniques[a]

Atomization Type	Maximum Kinematic Viscosity (SSU)	Maximum Solids	
		Mesh Size	Concentration (%)
Rotary cup	130–175	35–100	20
Single-fluid pressure	150		0
Internal low-pressure air[b]	100		0
External low-pressure air[b]	200–1500	200[d]	30[d]
External high-pressure air[c]	150–5000	100–200	70
External high-pressure steam[c]	150–5000	100–200	70

[a]From T. Bonner et al., *Hazardous Waste Incineration Engineering*, Noyes Data Corp., Park Ridge, NJ, 1981.[2]

[b]Lower than 30 psi.

[c]Greater than 30 psi.

[d]Depends on nozzle i.d.

Organic and aqueous liquids pass through three phases before actual oxidation is initiated. Hydrocarbons ignite at temperatures as low as 230°C (446°F) and as high as 650°C (1022°F). They are heated, vaporized, and superheated to the ignition temperature (see Fig. 7.2.2). These droplets must be exposed to high-temperature surroundings so as to receive heat by radiation and convection as rapidly as is practical. At the same time, they must also be in intimate contact with oxygen. If the droplet diameter is large, fewer droplets will be produced and the total surface available for heat transfer will be minimized.

An example is shown below:

$$\text{Ratio of number of droplets} \quad \frac{N_{125}\mu\text{m}}{N_{300}\mu\text{m}} = 13.8$$

$$\text{Ratio of surface areas} \quad \frac{A_{125}\mu\text{m}}{A_{300}\mu\text{m}} = 0.1736$$

For an equal flow rate of liquids, the total surface available for heat transfer is as follows:

$$\begin{aligned} \frac{A_{t125}}{A_{t300}} &= \frac{N_{125}}{N_{300}} \times \frac{A_{125}}{A_{300}} \\ &= 13.8 \times 0.1736 \\ &= 2.396 \end{aligned}$$

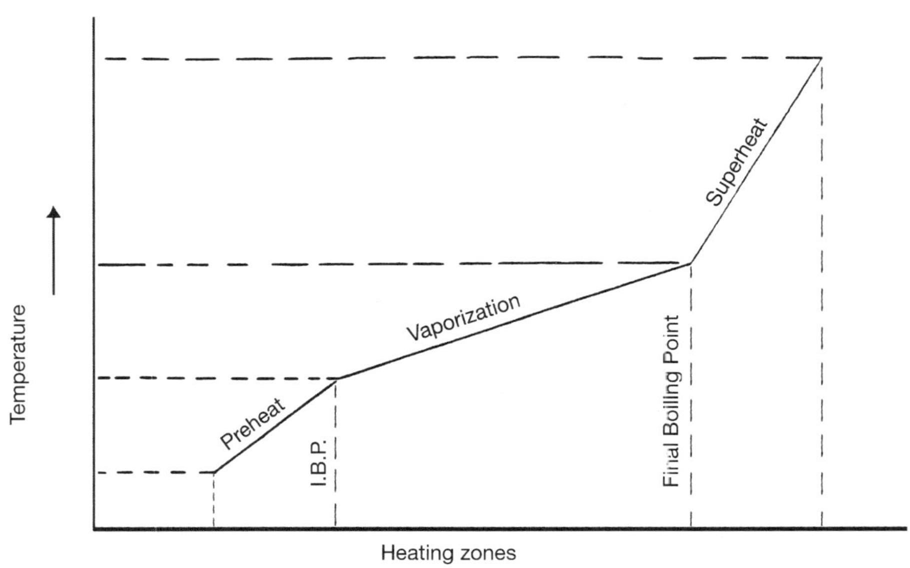

Figure 7.2.2 Liquid temperature vs. heating zone.

where A_t = total surface area
N = number of droplets
A = surface area of droplet
125 = 125-μm droplet; 300 = 300-μm droplet

Therefore, the rate of heat transfer to the 125-μm droplets will be at least 2.4 times as great as that for the 300-μm droplets. Because of their smaller diameter and volume, the 125-μm droplets will be vaporized at an even higher rate. The fact that there are almost 14 times more droplets means turbulence will be generated with flames initiated at their surfaces. (Note in Fig. 7.2.3 that time is 150 ms versus 30 ms, which is 5 times more.)

Combustion Air System Proper mixing of the combustion air with the liquid droplets is necessary. As the liquid is vaporized and superheated to ignition temperature, oxygen reacts with the hydrocarbon vapor to allow release of the energy. As this occurs, there is a sudden rise in temperature. This increases the velocity of the gases in the zones surrounding the droplet, causes increased mixing, and finalizes the oxidation reaction. With low-boiling-point temperature liquids, the rapid rise to reaction temperatures (950–1700°C or 1742–3092°F) will provide the necessary heat sink to the incoming colder liquid droplets.

As the viscosity of the liquid increases, droplet sizes tend to get larger. In most cases, the boiling point of the hydrocarbon is at a much higher level. To completely vaporize and superheat the droplet, more time is needed. Increased turbulence resulting from high-intensity mixing permits this reaction to be achieved rapidly. Energy is imparted to the combustion air of the combustor

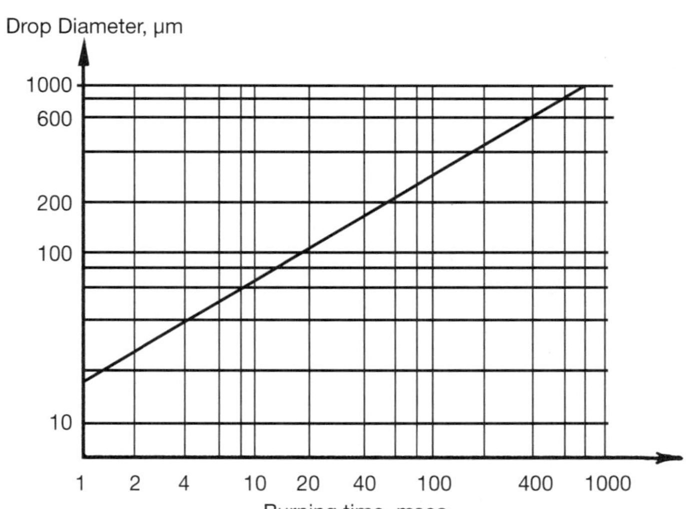

Figure 7.2.3 Droplet burning time vs. droplet diameter.

and this, in turn, permits more rapid mixing (turbulence) of the air with the fuel droplet. In many burners, this turbulence permits an internal recycling of the hot products of combustion, which transfer heat to the atomized droplets and raise them to the ignition point.

As the superheated liquid reaches ignition temperature, oxygen must be available to complete the oxidation reaction. Many problems result if the vaporized hydrocarbon contacts a low-oxygen-level stream. Pyrolysis results with a cracking of the hydrocarbon into carbon, carbon monoxide, and hydrogen, and time for completion of combustion is increased.

A secondary reaction will occur when these combustion gases eventually reach the proper oxygen supply. In a true thermal oxidizer, the most rapid oxidation occurs when single-stage combustion is designed into the system.

If the vaporized liquid contains solids, the incinerator design must allow the particles to be carried into the gas stream without agglomeration. A high-swirl or a cyclonic design may cause the solids to be reagglomerated into larger particles that become more difficult to burn. Proper design of the air-mixing device and the nozzle location are therefore important. Proper oxygen concentration at the surface of these solids is needed to ensure that gradual oxidation will occur. Since they are solid, the particles will burn at the surface and heat will be transferred inward to the core of the liquid droplet by conduction. Sufficient time should be provided to permit complete burnout of the solids in suspension.

If inert materials are carried with the liquid as it is vaporized, these tend to be carried into the gas stream as particulates or ash. Depending on the type of atomizer, the composition of the solid, and the temperature of the oxidizer, a percentage will become submicron in size and be carried with the gas stream. Since the heavier particulates will become molten slag and agglomerate, the combustor should be designed to collect the slagged ash without plugging the flow passages of the system. A slanted horizontal design or a vertical orientation is most often used for these types of waste.

Combustors Primary and secondary combustion units are utilized in the liquid injection incinerator systems. Primary units are utilized for those wastes with sufficient heating value to supply the necessary heat input to the system without need for auxiliary fuel. With nozzle-type burners, wastes with heating values from approximately 2500 kcal/kg (4500 Btu/lb) and above can be burned satisfactorily. The type of burner design, air mixing and turbulence, determine the minimum heating value for burning without auxiliary fuel.

Figure 7.2.4 provides a guide to the ability of a combustor to burn a particular waste. Low-intensity or laminar flame burners (air pressure drop varies from 50 to 150 mm H_2O or from 2 to 6 in. H_2O) utilize high excess air levels (25–60%) to provide the mixing of air with the waste fuel. These usually demonstrate low turbulence. High-intensity burners operate with combustion air pressures of 200–500 mm H_2O (8 to 20 in. H_2O). As a result, the increased turbulence resulting from the energy imparted by the air allows operation at

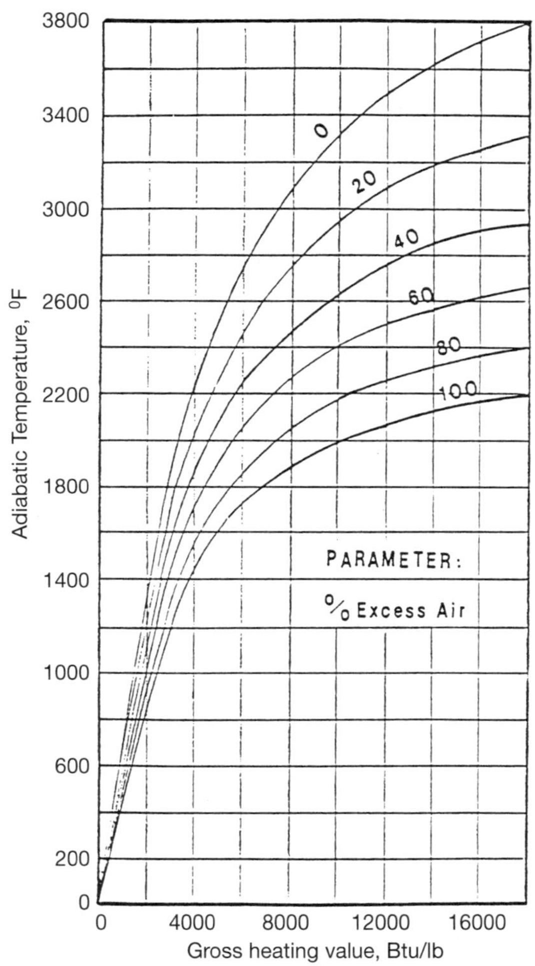

Figure 7.2.4 Adiabatic temperature vs. heating value and excess air.

much lower excess air levels (0–20%). The ideal minimum combustion temperature for burning a fuel is approximately 1200–1315°C (2200–2400°F). A review of the curve indicates that a high-intensity burner would be able to oxidize a 2500-kcal/kg (4500-Btu/lb) waste material at 1200°C and 10% excess air. A low-intensity burner operating at 40% excess air and 1200°C (2200°F) would be limited to a minimum heating value of 3220 kcal/kg (5800 Btu/lb). This burner would require auxiliary fuel to properly burn a 2500-kcal/kg (4500-Btu/lb) material.

Consider the following. If the waste to be burned had a heating value of 2770 kcal/kg (5000 Btu/lb) and 454 kg/h (1000 lb/h) had to be disposed of, a high-intensity burner rated at 5×10^6 Btu/h would be required. The burner would be designed to ignite with auxiliary fuel at approximately 2×10^6 Btu/h for initial warmup. Once the chamber temperature reached a level of 1093°C (2000°F) or higher, the waste could be introduced.

The low-intensity burner requires a minimum heating value of 3220 kcal/kg (5800 Btu/lb) to operate. In order to burn the 2770-kcal/kg material, auxiliary fuel will be required to elevate the heating value of the fuel mixture to 3220 kcal/kg. Assuming auxiliary fuel is available with a heating value of 10,000 kcal/kg (18,000 Btu/lb), the amount of fuel required and the size of burner needed can be determined as follows:

$$\text{Waste LHV}(x) + \text{aux fuel LHV}\ (1 - x) = \text{min. burner LHV}\ (1)$$

$$5000(x) + 18000(1 - x) = 5800\ (1)$$

where LHV = lower heating value
x = fraction of low heat content waste
$1 - x$ = fraction of high heat content waste (auxiliary fuel)

Solving for x,

$$x = 0.93846 \quad \text{and} \quad 1 - x = 0.06154$$

$$\text{Total fuel flow} = 1000/0.93846 = 1065 \text{ lb/h}$$

$$1065-1000 = 65 \text{ lb/h of aux. fuel}$$

$$\text{Total heat input} = (5000)\ (1000) + 18000(65) = 6.179 \times 10^6 \text{ Btu/h}$$

$$\text{Additional heat input required by auxiliary fuel} = 18000(65)$$

$$= 1.179 \times 10^6 \text{ Btu/hr}$$

Therefore, burner must be rated at 6.18×10^6 Btu/h versus the 5.0×10^6 Btu/h for the high-intensity burner.

Each burner, regardless of type, is usually mounted in a refractory block or ignition tile (see Fig. 7.2.5*a* and *b*). This is necessary to confine the primary combustion air introduced through the burner, to ensure proper air–fuel–waste mixing and to maintain ignition. Some burners and tiles are arranged to aspirate hot combustion gases back into the tile, which aids in vaporizing the liquid and increasing flame temperature more rapidly. The dimensions of the burner block, or ignition tile, vary depending on the burner design. Manufacturers each have their own geometric specifications, which have been developed through past experience. Therefore, it is not possible to specify a single burner block geometry for design evaluation purposes. However, this aspect can be checked to eliminate systems that do not provide for any flame retention.

The location of each burner in the incinerator and its firing angle relative to the combustion chamber is another important consideration. In axial or side-fired nonswirling units, the burner is mounted either on the end, firing down

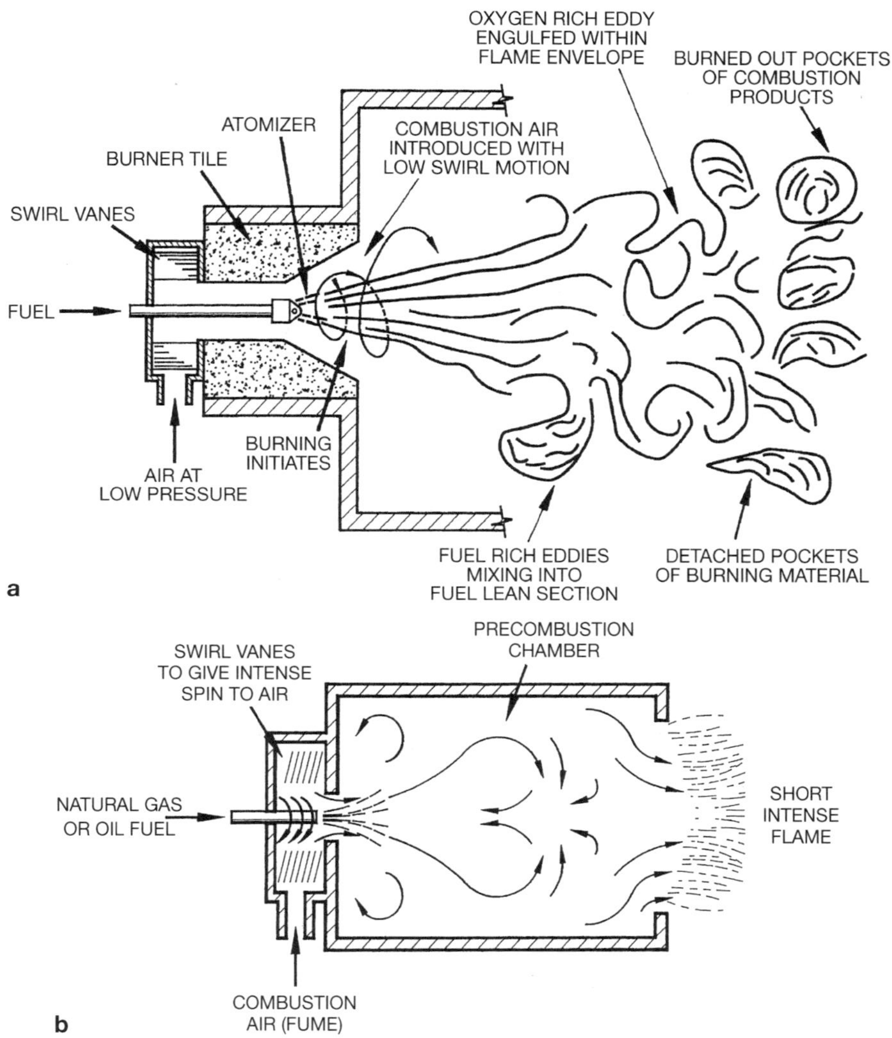

Figure 7.2.5 (*a*) Long flame burner. (*b*) Vortex-type burner.

the length of the chamber, or in a side-wall, firing along a radius. Such designs, while simple and easy to construct, are relatively inefficient in their use of combustion volume. Improved utilization of combustion space and higher release rates can be achieved with the utilization of swirl or vortex burners or designs involving tangential entry. Regardless of the burner location and/or gas flow pattern, the burner is placed so that the flame does not impinge on refractory walls. Impingement results in flame quenching and can lead to smoke formation or otherwise incomplete combustion. In multiple burner systems, each burner should be aligned so that its flame does not impact on other burners. This will also affect the flame scanning systems.

In most liquid injection incinerator designs, the desired temperature at the chamber outlet is preset by the operator and secondary air is fed to the system at a constant rate or by use of an oxygen monitor and control system. Fluctuations in temperature are controlled by increasing or reducing the waste or auxiliary fuel feed rate to the burner within the designed turndown ratio. This turndown ratio is fixed, in part, by the limited range of liquid waste injection velocities required to prevent flame liftoff or flashback. If waste is injected through the burner nozzle at too high a velocity, the flame will separate from the burner and be extinguished. If the injection velocity is too low, the waste will burn in the nozzle and damage it. The range of injection velocities needed to prevent these occurrences is determined by the flame propagation rate for the wastes and the flame retention characteristics of the burner.

Waste Characteristics Determine Incinerator Design

In the process industries, waste liquids are generated that are numerous in kind and defy easy definition. The wastes can be categorized as organic, aqueous organic, and aqueous organic-inorganic. Each of these can again be divided into those that are considered hazardous and those considered nonhazardous. Appendix VIII to Title 40 of the Code of Federal Regulations lists those compounds and waste mixtures that are considered hazardous by the EPA and must be treated accordingly. These organic wastes listed in Table 7.2.3 will in most cases sustain combustion and would be injected into the primary combustor. Many are utilized as a source of fuel in boilers, process heaters, process kilns, incinerator systems, and the like. Note that the list also contains halogenated wastes.

Halogenated Hydrocarbons With the increased production in the plastics industry, the generation of by-products from vinyl chloride monomer (VCM) and polyvinyl chloride (PVC) processes has expanded rapidly. In many of the

Table 7.2.3 Organic waste liquors

Organic (nonaqueous)	Aqueous with Ash	Aqueous without Ash
Solvents	Sodium glutamate	Phthalic anhydride
Waste oils, sludges	Molasses ferment	Phenol
Halogenated hydrocarbons	Synthetic phenols	Oil/water
Organic acids	Polyesters	Maleic anhydride
Organometal components	Caprolactum	
Aromatics	Agricultural process wastes	

larger facilities where the majority of product is produced, an increase in toxic gaseous vents and viscous liquid by-product material results. Since many of these materials are listed as Appendix VIII compounds, it is necessary to recycle or dispose of them in a safe and proper manner to protect the health and the environment of local communities.

Several schemes have been used for the on-site recycle or disposal of materials generated from these plants. They have included carbon adsorption, chemical fixation, solvent absorption, high-temperature chemical reaction, and incineration. Industry has gained a considerable amount of expertise in these areas over the past 30 years. Incineration has resulted in the ultimate answer to the disposal of these halogenated waste materials. There are many incineration systems operating today in which chlorinated waste materials are being disposed of properly.

There are also high-temperature reactor systems complete with staged absorption to produce anhydrous HCl that is recycled to the primary process that generated the by-product materials.

The objective in incinerating residues containing chlorine is to convert as much as possible of the chlorine content to hydrogen chloride. HCl can be absorbed in water from the reaction gas mixture. In this way, reusable hydrochloric acid is recovered in an environmentally safe way. The quantity of chlorine contained in the waste is often the measure of the heating value of the waste material. High chlorine percentages are typical of wastes produced in the VCM process. Heating values in the range of 1944–4444 kcal/kg (3500–8000 Btu/lb) are typical. Chlorinated hydrocarbon fuels tend to be slower burning and soot forming. As a result many systems have operated with high excess air to promote mixing of the oxygen with the hydrocarbons to minimize soot formation. This has required an addition of auxiliary fuel to overcome the drop in flame temperature with high excess air. This also increases the operating horsepower and the size of downstream pollution control equipment. High excess air also promotes the formation of a much higher level of chlorine.[4] These may not appear as problems in the incinerator since the results will indicate proper destruction of the principal organic hazardous constituents (POHCs); however, problems will occur in the waste heat recovery equipment as well as in the downstream scrubbers.

Free chlorine is much more aggressive than hydrogen chloride in attacking metals. Free chlorine requires caustic in the final absorber tower for proper pollution control when scrubbing, while HCl is freely absorbed in water. Free chlorine combined with ash has been found to be the source of dioxin reformation in the waste heat boiler and air pollution control (APC) systems. This occurs when the cooler zones downstream of the combustor are in a temperature range of 200–400°C (392–752°F).[5] It is also important that the combustor be designed for thorough mixing of air and wastes, thereby achieving maximum combustion efficiency. This promotes maximum temperature levels achievable under equilibrium conditions. This will also require proper selection of refractories so that these will withstand the temperatures reached in high-efficiency combustors. Water and steam cooling are often used for the purpose of safe temperature control. The incineration section should be designed with the following goals:

- Minimize auxiliary fuel requirements
- Operate at minimum excess air
- Minimize free chlorine
- Provide a highly turbulent combustion section to ensure maximum flame temperatures in a minimum chamber volume
- Minimize soot formation

Systems have been designed for operation at very low heat release rates (e.g., in the range of 1×10^6 to 2×10^6 Btu/h) where heat recovery is not critical. Systems have been designed for heat release rates as high as 70×10^6 to 80×10^6 Btu/h. These systems include not only waste heat recovery in the form of a steam boiler but also recovery of acid by-product. In most plants where VCM is manufactured, hydrochloric acid is required as a raw material for production. This aids the plant operator in investing in a waste disposal system that will provide a by-product and also is economical in terms of the return on investment (see Fig. 7.2.6).

Aqueous Waste Systems with Ash In reviewing the aqueous wastes in Table 7.2.3, one should note the distinction between those wastes with and without ash. Non-ash-bearing wastes are easily handled in either a horizontal or vertical (up-oriented) incinerator system. The waste is usually injected downstream of the burner combustion chamber into the flame of the main burner. Care must be taken to design for proper sizing, location, number, and direction of the nozzles for injecting hazardous waste.

Since a highly aqueous waste stream is not viscous, it may be atomized by hydraulic pressure alone in a mechanical atomizing nozzle or a dual fluid nozzle (steam or compressed air) may be utilized. Use of compressed air for these nozzles has been very successful. The air serves as a portion of the combustion air and reduces the heat load on the system that would be required with steam. The design and controls should provide the proper ratio to ensure adequate spray pattern and droplet size through the entire range of flow.

Nozzles should also be protected in the event of loss of waste flow by maintaining the flow of the atomizing fluid. In many units, these nozzles are designed to be physically removed from the incinerator chamber to prevent burnout of the nozzle tip.

Provision should also be made to provide sufficient air with the atomized liquid at the injection point to oxidize the organics contained in the waste stream. Incinerators have been known to produce soot as a result of a pyrolysis condition at the entry of the aqueous waste.

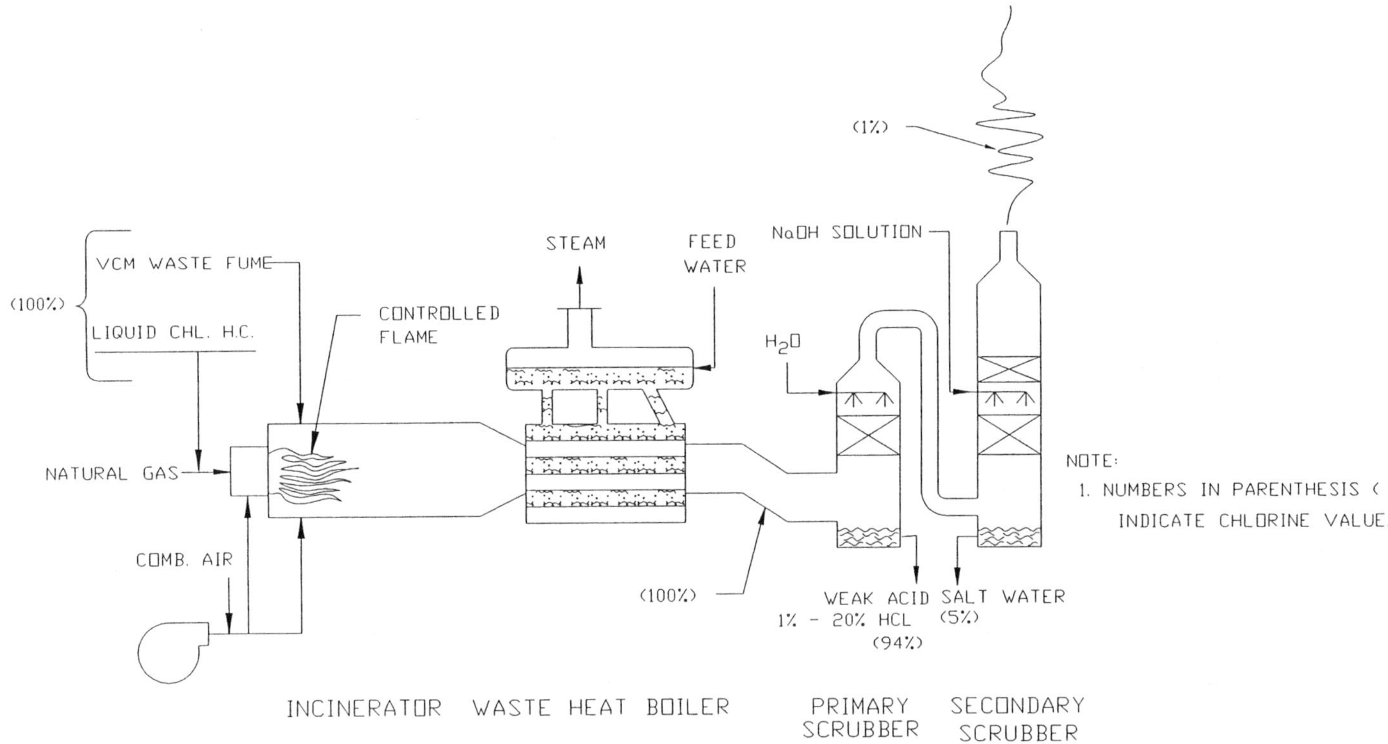

Figure 7.2.6 Typical incinerator–heat recovery–scrubber system for vinyl-chloride monomer process waste distribution of chlorine from wastes.

Aqueous wastes containing organics and inorganic material have been produced by the agricultural process industries in the manufacture of herbicides, pesticides, and other products. The pharmaceutical industry also produces a high water content waste containing organics with phosphorus, chlorine, and sulfur-bearing compounds. Aqueous solutions are not self-sustaining from a combustion standpoint when the water content is greater than 75%. If the wastewater contains ash (salts), consideration must be given to the maximum operating temperature to minimize refractory problems. These wastes are normally injected into the incinerator downstream of the primary combustor in a lower temperature zone rather than directly into the high-temperature refractory zone.

Incinerator designs for the oxidation of those wastes with high ash content are vertically down oriented (see Fig. 7.2.7). This allows gravity discharge of the molten ash material into the quench zone [quench pot with sprays, submerged exhaust (sub-X), or hot tap port]. Critical to the design and operation of incinerators for this waste type is the selection and design of the refractory. A reaction occurs at these elevated temperatures between the metallic oxides resulting from oxidation and the compounds that make up the refractory lining (typically alumina and silica). Severe spalling and excessive maintenance and down time have resulted from improper selection of refractories. Testing of various refractory types with the slag that will result from the incineration process at the operating temperatures expected is extremely important.[6]

Environmental Effects The amount of effort that has been directed toward the proper design of air pollution control and wastewater control systems associated with liquid injection incinerators is probably greater than that for all other areas of incineration. Since most liquid injection incinerators do not generate solids or ash, the air pollution control problem is minimized. In most cases, it is a matter of providing the proper scrubber for absorption of the acid gases generated, that is, HCl, SO_x, NO_x, and others. The technology for the air pollution control systems in this area is well established. See Chapter 10 for details.

In systems burning liquid wastes that contain solids or inert materials, particulates that form may be submicron in size because of the high temperatures generated in the combustion process. It is very important that pilot tests be conducted to evaluate the particulate loading and size distribution. This will affect the selection of an air pollution control system. Wastes containing phosphorus have used venturi scrubbers with required pressure drops at levels of 100 in. H_2O (186.8 mm Hg). An alternative would be either an ionizing wet scrubber, an electrostatic precipitator (ESP), a collision scrubber, or a baghouse. These systems are covered in greater detail in Chapter 10.

Scrubber Water Treatment The wastewater discharge is usually dependent on the acid gases formed in the incinerator, the ash composition and size, and the scrubber system employed. Many of the larger systems today are incor-

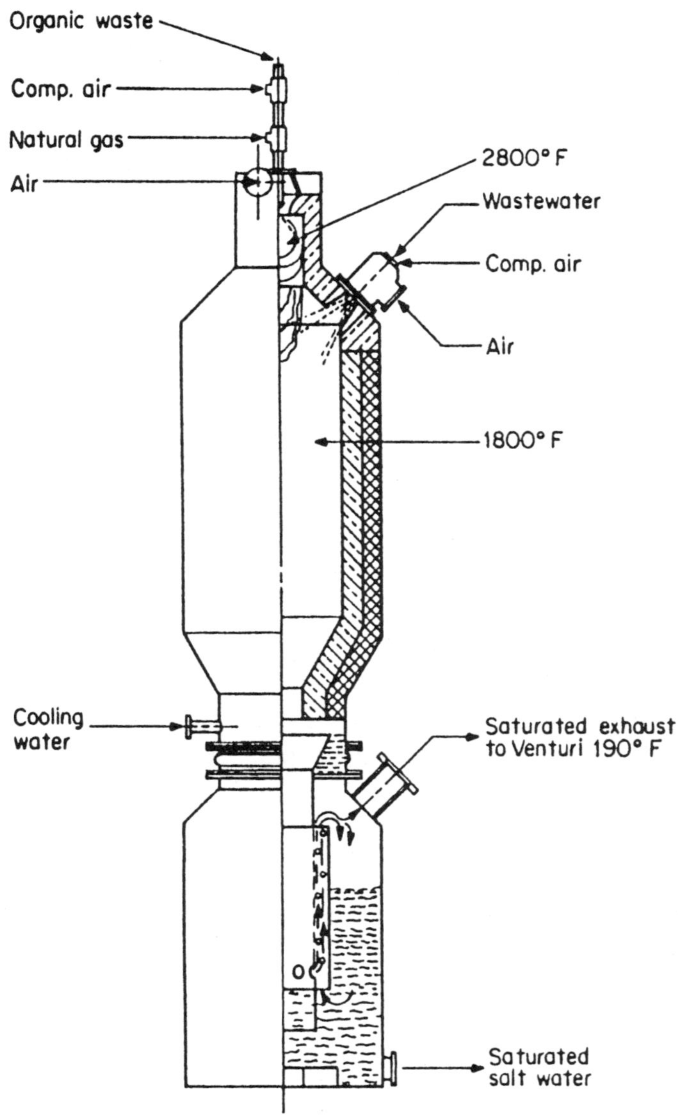

Figure 7.2.7 Sub-X incinerator (T-Thermal).

porating spray dryers for acid gas removal followed by ESPs or baghouses. Where packed-tower scrubbers or venturi scrubbers are used in HCl scrubbing, a lean acid solution is generated. This is then delivered to a lagoon where it is neutralized with lime (much less expensive than caustic) and then discharged to the plant's wastewater treatment system.

7.3 ROTARY KILN INCINERATION

Technology Overview

Rotary kiln (RK) process incinerators were originally designed for lime processing. In 1885, Frederick Ransom was the first to patent a rotary kiln. Peray provides an excellent summary of the early kiln.[7] Recently the rotary kiln has been utilized to process hazardous waste because it has the unique capability to stabilize the combustion process for many different materials that may be fed simultaneously. Kilns have demonstrated the ability to handle a wide variety of materials with minimal impact on its performance. This results in minimal preprocessing of the wastes and is one of the main reasons for the RK popularity. The processing of raw materials has been a historic application of the rotary kiln. The basic building materials of a community—cement, lime, coal, iron ore, aggregates, and other bulk materials—require heating to high temperatures. These materials historically have been produced in a rotary kiln. This high-temperature treatment and the continual mixing and blending of the raw materials provide the basis for the processing of hazardous waste. Just as the raw materials for cement are constantly exposed to the heat of the kiln, so too are wastes that are processed by the RK. Lime, lightweight aggregate, cement, and other process kilns may operate at temperatures up to 1925°C (3500°F). Hazardous waste incinerators will operate in the range of 700–1315°C (1300–2400°F) depending on the waste types and shapes fed.[8]

The rotary kiln is a cylindrical refractory-lined shell that is mounted at a slight incline from the horizontal plane to facilitate mixing the waste materials and exposing the surface to the auxiliary burner flame and the waste fuel flames and flames generated over the burning surface of solids (see Fig. 7.3.1). The kiln accepts all types of solid and liquid waste materials with heating values between 555 and 8333 kcal/kg (1000 and 15,000 Btu/lb), and even higher. Solid wastes and drummed wastes are usually fed by a pack-and-drum feed system, which may consist of a bucket elevator for loose solids and a conveyor system for drummed wastes. Pumpable sludges and slurries are injected into the kiln through nozzles. Temperatures for burning vary from 700 to 1315°C (1300 to 2400°F) dependent upon the waste types and shape.

Rotary kilns are classed as slagging [>1000°C (1832°F)] or nonslagging [<1000°C (1832°F)]. The nonslagging rotary kiln for hazardous waste destruction is not a total incineration system. The kiln does not attain high destruction and removal efficiencies (DRE) required of an incinerator due to the basic manner that it operates. It will volatilize the organic content of the materials placed into it. The volatilized organics are carried in the combustion products exit gas stream. A secondary combustion chamber (SCC) is located immediately downstream of the kiln to perform the final combustion (destruction) process on the kiln exit combustion products as is shown in Figure 7.3.2. The nonvolatile part of the material fed into the kiln such as metals, stones, sand, and other inorganics exit as ash. The ash and residue exiting the kiln should

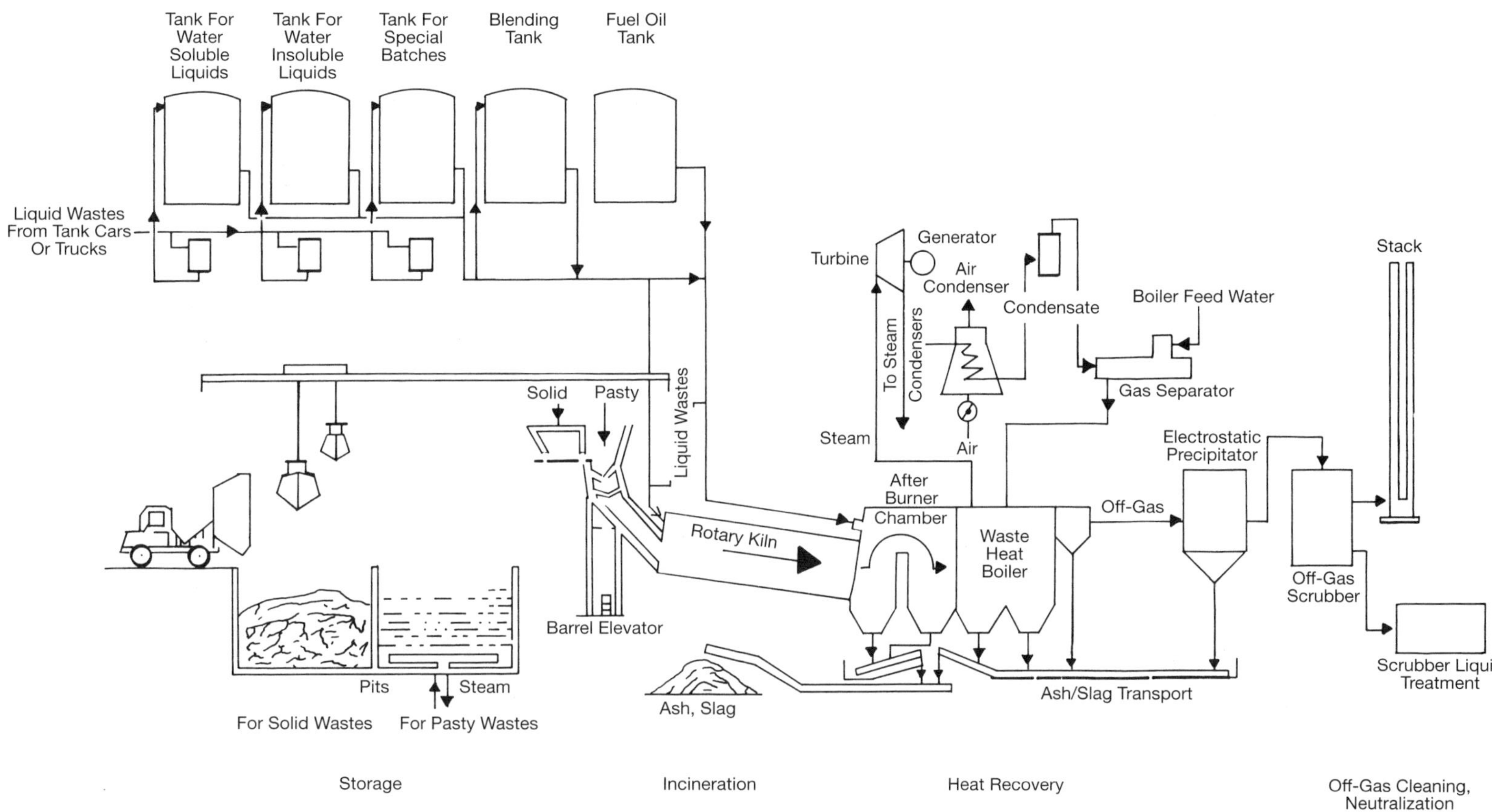

Figure 7.3.1 Rotary kiln incinerator.

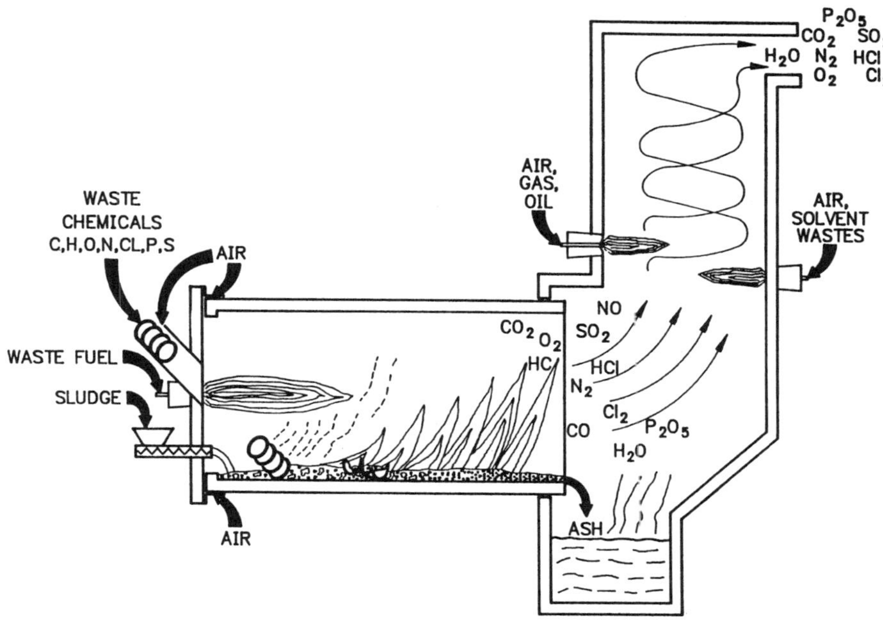

Figure 7.3.2 Rotary kiln—afterburner

contain minimum to zero organics and may be treated and disposed of in a hazardous waste landfill. Some metals and other elements, due to the operating temperature will vaporize within the kiln and be carried in the exhaust gas stream to the downstream equipment.

The rotary kiln combines three very important aspects of waste material processing into one piece of equipment: mixing, temperature, and time. The mixing process is well known in the production of cement where lime, iron, and sand or clay are mixed while being heated. Good mixing is required to obtain a quality product. Hazardous waste feed streams are generally not of uniform size, shape, or composition. Therefore, the processing equipment must allow for the wide variations in the material while achieving a uniform bottom ash that meets the regulatory requirements. The waste material that is placed into the kiln must have sufficient time for its organic components to volatilize. With proper time, the heat source within the kiln can raise the waste material to the desired temperature. The rotary kiln provides excellent mixing through the tumbling action to distribute the heat evenly to all of the material being processed and to expose all of the material to the heat source. The original kilns used for waste incineration were called "tumble burners."

Process Considerations

The first step in the design of a hazardous waste rotary kiln incinerator is the characterization of the waste stream as noted in Section 7.2. While chemical analysis of the waste stream is necessary prior to processing the waste, it may not always be sufficient to establish the process. The physical variations in the material dictate the pretreatment processes necessary such as mixing, shredding, or dilution before it can be placed into the kiln.

Prior to processing of waste, it is important to determine its heating value. This is a measure of the energy in the waste expressed as kilocalories per gram or Btu per pound. Materials with high heat content will range from 8333 to 11,100 kcal/kg (15,000 to 20,000 Btu/lb). This compares with fuel oil that has a heat content of approximately 10,555 kcal/kg (19,000 Btu/lb). Low heat content values range down to 0 Btu/lb.

The heat content is important since it is a direct measure of the temperature and the oxygen requirements that the specific waste will place on the system. Materials with very high heat content must be handled differently from materials with low heat content, just as oils must be handled differently than water. Materials with high heat content will burn within the kiln, releasing heat to the kiln. Water may be added to absorb the heat and lower the temperature to a reasonable value. Low heat content materials, on the other hand, generally require supplemental fuel to control and maintain the proper temperature.

Secondary characterizations of the waste are the elemental composition, moisture, specific heat, ash content, and physical form. Material size and physical characteristics are important in the handling and the processing of the material before it enters the kiln, while it is traveling through the kiln and in the bottom ash processing equipment. The elemental composition is important in defining the quantity of the waste that will enter the gas stream in vapor and particle form. Some metals such as lead, cadmium, and mercury will vaporize in the kiln and flow through to the air pollution control system. These high volatile materials will react with excess oxygen or a halogen to form a solid particle, for example, mercuric chloride. Other metals such as chromium and barium will be discharged with the bottom ash and will not require further processing. Wastes may also form into submicron particles in the kiln. If so, the APC system must be designed to remove them from the exhaust gas stream. Metals that vaporize in the kiln will form into submicron particles when the gas is cooled in the APC equipment. These particles must also be removed.

Elemental analysis is important when defining the characteristics of the ash that exit the kiln. Many of the materials processed will reduce to elemental form in the kiln. These will then combine with the other materials being processed forming new compounds. Classic examples are the alkali (sodium and potassium) and halide (chlorine and fluorine) groups. Many of the waste materials contain chlorine or fluorine. When processed with other materials containing sodium or potassium, salts will form. A typical salt, sodium chloride (NaCl) in its pure form has a melting point of 800°C (1472°F). When this material is formed in the kiln through the chemical combination of sodium and chlorine, it first becomes sticky and attaches itself to the other materials in the kiln. Large boulders or "logs" may form as the sticky mass attaches itself to more material. The material may not have been fully processed before it is

Table 7.3.1 Physical properties of selected materials

Material	Melting Temperature °C	Melting Temperature °F
$CaCl_2$	722	1421
$MgCl_2$	712	1313
$MgSO_4$	1185	2164
$Na2CO_3$	850	1563
$Na2SO_4$	884	1623
$CaCO_3$	1339	2442
KCl	790	1453
$CaCO_3$	1339	2442
NaCl	800	1472
NaBr	752	1390

enveloped in the sticky mass. Globs of materials the size of compact cars have been observed exiting kilns. These fall into the ash handling system and plug up the ash removal machinery. They have to be removed with water cannons or jack hammers. As the temperature of the material increases, the NaCl may become molten and has the potential of penetrating refractory or just flowing out of the kiln (see Table 7.3.1).

Several of the materials have melting points based on the composite materials. The combination of sodium chloride and sodium sulfate salts have a melting point 166°C (300°F) below the melting point of the pure materials when combined at approximately a 60/40 ratio.

If a combination of waste materials does not form a "log," it will have a tendency to stick to the walls of the kiln. The continual buildup of the material cools as it flows to the exit of the kiln. There it forms a "dam" within the kiln. Under this condition, material upstream of the dam becomes hotter and less viscous. When the slagged material reaches the crest, it cools rapidly, causing a continued buildup of the height and length of the dam. A kiln with 14-ft internal diameter can be reduced to approximately 6-ft internal diameter as a result of the slag cooling build up mechanism. Explosives and mobile jack hammers are required to remove the material, often causing significant refractory damage.[9]

The exit end of the kiln must be designed to handle the physical characteristics of the material leaving the kiln. Systems that cause slagging must maintain the slag heated until it exits the kiln. If there is cooling on the exit end, the slag may freeze and plug the end.

Process Design

Kilns are designed to have from 5 to 15% of their internal volume occupied by solids bed. This value allows for reasonable heat transfer to all of the materials within the bed. Material having a high heat content, must have a lower fill percentage to provide reasonable exposure time, and temperature control at the inlet section. Low heat content materials may be loaded up to 15% of the cross-sectional area provided there is sufficient length in the kiln to provide the residence time needed for the solids. If there are large variations in the material consistency, the percent loading must be adjusted to accommodate the material.

The waste ash and initial physical characteristics will determine the effectiveness of the heating. A majority of the heating may be performed by contact transfer from the heated refractory surface beneath the bed. Good heat transfer is obtained when the rotation of the kiln lifts the material up the side and the bed rolls down over the surface and is exposed to the combustion gases and flame (Fig. 7.3.3). This process exposes new material to the refractory surface and to the high-temperature gases.

Materials with high heat content will ignite and burn when placed in the kiln. The material will ignite from the surface of the material already in the kiln and upon contact with incoming combustion air to the kiln will start burn-

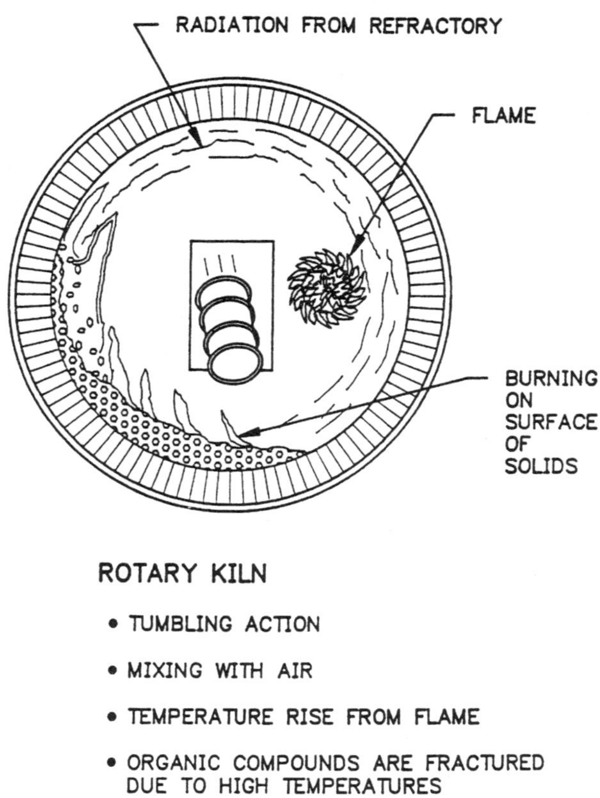

Figure 7.3.3 Cross section through kiln.

ing immediately on the surface. This surface burning will generate very high temperatures on the surface of the bed. Volatile organics close to the heat source will volatilize and provide additional energy for the burning action. The resulting combustion will raise the kiln wall temperature and evaporate water contained in the material. The heated walls, which are rotating under the bed, will raise the temperature of the material in contact with the wall.

If the bed is very deep or the bed material slides down the side walls, heat transfer to the internal portion of the bed is by conduction only. Heat transfer to the internal bed material is then poor, resulting in unburned material in the center of the bed. If this internal material does not reach the minimum temperature required to properly treat the material, the bottom ash will contain unburned organics. This is a problem with deep beds and short kilns where residence time is low for the solids.

The brick used to line the kiln must be selected to meet abrasion and chemical attack. The physical characteristics of the waste stream will have an important impact on the brick. If metal drums are to be directly fed into the kiln, the inner surface of the refractory must withstand the full impact and shock loading of the drum in addition to the heat loading of the instantaneous rupturing of the drum.

The chemical attack of sodium and chlorine must also be factored into the refractory selection process. After the kiln is at its designated operating temperature, the refractory will be at a temperature slightly below the gas exit temperature. However, there will be zones of higher temperature dependent upon the flame length and the location of the maximum solids burning. This may be observed with a radiamatic pyrometer sensor at the exit of the kiln. Outside surface temperatures of the kiln also provide a means to determine the points of highest internal temperatures of the refractory.

Kiln Selection Criteria

Two types of rotary kilns are currently being manufactured in the United States today: co-current (burner at the front end with waste feed) and countercurrent (burner at the back end). For a waste that easily sustains combustion, the positioning of the burner is arbitrary from an incineration standpoint since both types will destroy a waste. However, for a waste having low combustibility (such as a high-water-volume sludge), the countercurrent design offers the advantage of controlling temperature at both ends, which all but eliminates problems such as overheating the refractory lining. The countercurrent flow technique has been reputed to carry excessive ash over into the APC system due to the associated higher velocities involved. However, this condition also increases the turbulence during combustion, which is generally a desirable factor.[8]

Another feature of the rotary kiln is that it can operate under substoichiometric (oxygen-deficient) conditions to pyrolyze the wastes. The combustible off-gases can then be incinerated with liquid wastes in the secondary combustion chamber. This mode of operation reduces the total mass flow of combustion gases from the secondary combustion chamber into the APC system. It will also reduce the particulate matter carryover in the kiln gases. The rotating action allows better volatilization of the nonpumpable wastes than stationary or fixed-hearth incinerators. Care must be exercised to prevent air in-leakage into the kiln or into the interconnecting duct between the kiln and SCC. This will create an explosive mixture with a rich fuel, ignition temperatures, and oxygen availability.

Disadvantages of the rotary kiln include high capital and operating costs, highly trained personnel required to ensure proper operation, frequent replacement of the refractory lining if very abrasive or corrosive conditions exist in the kiln, and the generation of fine particulates (which become entrained in the exhaust gases) due to the cascading action of the burning waste. Unlike liquid injection incinerators, which have no moving parts, rotary kiln designs incorporate high-temperature seals between the stationary end plates and rotating section. These seals are inherently difficult to maintain airtight, which creates the potential for release of unburned wastes. Rotary kilns burning hazardous wastes are almost always operated at a negative pressure to circumvent this problem; however, difficulties can still arise when batches of waste are fed semicontinuously. When drums containing relatively volatile wastes are fed to the kiln, for example, extremely rapid gas expansion occurs. This results in a positive pressure surge at the feed end of the kiln (even though the discharge end may still be under negative pressure), which forces unburned waste out through the end plate seals. This phenomenon is known as *puffing* and can pose a major problem if extremely toxic or otherwise hazardous materials are being burned. Fugitive emissions can also exit the kiln through the feed chute if improperly designed. Therefore, the design of the solid waste feed system is an extremely important consideration in evaluating rotary kiln incinerators.

Rotary kiln systems usually have an SCC after the kiln to ensure complete combustion of the wastes. Airtight seals close off the front end of the kiln; the exit (lower) end is connected to the SCC or mixing chamber. In some cases, liquid waste is injected into the SCC. The kiln acts as the primary combustion chamber (PCC) to volatilize and oxidize combustibles in the wastes. Inert ash (bottom ash) is then removed from the exit end of the kiln. The volatilized combustibles exit the kiln and enter the SCC where additional oxygen is available and ignitable liquid wastes or fuel can be introduced. Complete combustion of the waste and fuel occurs in the SCC.

Both the SCC and RK are usually equipped with an auxiliary fuel firing system to bring the units up to the desired operating temperatures. The auxiliary fuel system may consist of separate burners for auxiliary fuel, dual-liquid burners designed for combined waste–fuel firing, or single-fuel burners equipped with a premix system, whereby fuel flow is turned down and liquid waste flow is increased after the desired operating temperature is attained.

If liquid wastes are to be burned in the RK and/or SCC, additional considerations are

- Flame retention characteristics of the burners
- Burner alignment to avoid flame impingement on refractory walls
- Burner alignment to avoid interference with the operation of other burners (for multiple burner systems)

Feed Systems

The type and method of handling of the waste stream must be reviewed in detail before designing or specifying the kiln. The feed mechanism is an integral part of the incineration process. The form that the material is in when it is placed into the kiln determines many of the combustion characteristics of that material. As was stated above, the kiln generally processes solid materials. The solid material can be either loose or packaged.

Containerized solids are fed through a batch injection system. This system feeds the material into the kiln with either a chute or a ram system. Air locks are provided to permit control of entrained air. Without air control locks on the feed system, door openings may cause either a rapid increase of the combustion process with high Btu materials and high air inflow or a rapid cool down with low Btu materials. Kiln puffing occurs when high Btu materials volatilize and burn, rapidly increasing the kiln temperature and volume of combustion products. This volumetric increase often creates an increase in the kiln pressure that may not be overcome by the available draft from the induced draft fan. Puffing causes a release of hot gas and fumes through the seals of the kiln.

The method and control of the waste feed must be closely monitored. When a container is added to the kiln, it may release all of its thermal energy in the very short period of time that it takes to degrade the container. This is dependent upon the volatility of the material in the container, the heat content in Btu per pound, and the percentage of the total heat release that the container represents. This often results in an instantaneous heat release rate that will exceed the system capabilities and generate a puff. These puffs have been studied.[8,9] A simulation of the heat release history of a container system is shown in Figure 7.3.4. It is extremely important that the characteristics of the waste material be known and the feed rate closely monitored to assure that the instantaneous heat release rate does not exceed the capability of the kiln. The combustion airflow rate to the kiln is almost constant due to the operation and control of the induced-draft fan. The oxygen available for combustion is also constant. Most kilns operate with excess air levels in the range of 50–100%. This establishes the range of heat release rate of the kiln. As higher Btu rates are fed to the kiln, the oxygen will be depleted and the temperature will rise. A batch-fed system will have a continuous rise and fall of the oxygen at the exit. As oxygen falls, the temperature will rise. As burnout occurs, the oxygen will increase since air is entering at a fixed rate. This is followed by a drop in temperature. As a result, feed heating value and feed rate are critical to the

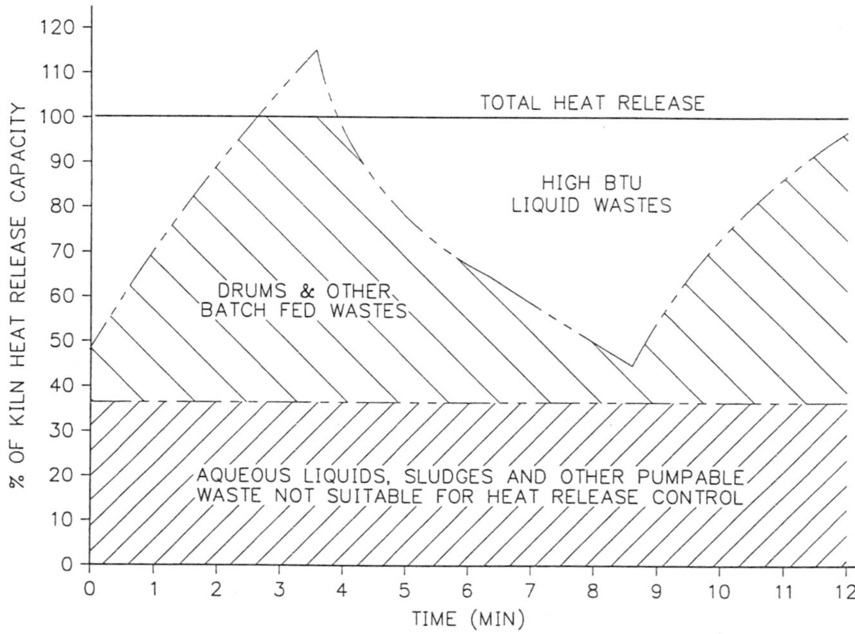

Figure 7.3.4 Typical kiln heat release.

proper operation and control of the kiln. Refractory damage often occurs due to high temperatures resulting from high heat input and stoichiometric oxygen levels.

***Introduction to Conveyors and Related Equipment*[8,10]** Feed systems include hoppers, live bottom bins, single- and double-screw auger, ram, apron, weigh belts, slinger belts, and chutes. Of these, the simplest is the steeply inclined (>60° from horizontal) chute. While cheap, rugged, and dependable, it does not offer a good seal for preventing air entry to the rotary kiln. Addition of a simple "flop gate" to the chute can reduce air leakage.

Loose solids from a belt feeder or apron feeder drop onto a weigh belt. The weigh belt is a constant-speed belt conveyor that has a load cell on one or more sets of idler rollers. By calibrating the electronic readout with preweighed amounts of solids, the weigh belt can be used to monitor the instantaneous feed rate to the rotary kiln and can also produce integrated hourly totals of solids processed. An alternate to a weigh belt is a volumetric feeder, such as a variable speed auger. These are less accurate than a weigh belt, particularly when moisture contents and bulk density vary. Systems handling certain types of solids, slurry, and sludge use a weigh box, which discharges a predetermined weight into a feed chute or ram feeder. However, these are not recommended for wet or sticky materials.

From the weigh belt, the solids pass into the final feed mechanism at the feed end of the kiln. Four types predominate:

- Chute
- Screw feeders, single or double auger
- Slinger belt
- Ram

Each type has disadvantages, arising more from the misapplication of a particular feeder than from any technical failing.

Screw feeders work well with sandy, gravel-type and nonsticky solids. Dual, counterrotating screws extend the range to mildly sticky solids. Very wet and sticky solids tend to wrap around the screw, causing feed jams.

The chute is a simple, low-cost feed device that can handle dry through mildly sticky solids. With close operator attention at the chute, it can handle medium-sticky solids by manually breaking up an occasional lump.

Slinger belts are high surface speed belt conveyors that throw the solids into the kiln. They must be retractable in case of a shutdown, as the belt will burn up if left in place. The slinger belt, and to a lesser degree the chute, allow a large amount of air to leak into the rotary kiln. When burning waste with high heat content, this may not be a problem, as combustion air must be injected into the kiln anyway. On low heat content materials and with higher kiln drafts, the air leakage can drive up fuel costs.

The ram feeder, and a weigh box using a cyclically loaded container at the kiln face, are frequently used for solids packages, drums, and the like. The feed is charged in batches. They are relatively good at keeping air leakage to a minimum; however, they are higher cost feeders, and the multiplicity of parts and limit switches leads to high maintenance.

Conveyors The two most popular types are screw augers and belt conveyors.

Screw Augers Screw augers can be used with dry, friable solids. They are a volumetric device, moving a certain volume per revolution of the screw shaft. They are normally run at 30% of maximum capacity and are prone to jamming if run at 100%. When run at angles above horizontal, capacity suffers further. Screw conveyors with standard U-shaped troughs are not recommended for inclines above 25°, and their capacity at this angle is but 40% of that in a horizontal position. Several systems use auger feed systems. These feed systems have physical size limitations not found with the belt system. Augers extend into the kiln and are exposed to kiln temperatures. Waste materials with low boiling points may volatilize within the auger housing creating an unsafe condition. Some materials may cake up and plug the auger if exposed to high temperatures. This results in high maintenance with augers dependent upon the waste type.

Troughs frequently have a sacrificial "wear liner" tack welded in place, and for abrasive materials, the edge of the screw can be wear hardened with abrasive-resistant stellite weld metal. Auger revolutions per minute (rpm) is normally 50 for large-diameter augers and 70 for small ones. These shaft speeds can be doubled if the auger is found to be short of capacity. For augers handling a wide range of particle size, it is a good idea to install a manual reverse switch near the auger. It allows operators to reverse and jog the screw to clear jams quickly.

Belt Conveyors Belt conveyors are another method of mechanical conveying, commonly used for feed solids and processed waste solids handling applications. Belt conveyors are typically used where the conveying rate or distance involved exceeds that of screw conveyors. For smaller particles and dusty materials, a totally enclosed conveyor is required with venting back to the storage area. Additionally, smaller particles require lower belt speeds to prevent dusting of the material. Like screw conveyors, belt conveyors suffer a decrease in capacity as the conveyor incline increases. For most standard belts, the maximum incline is limited to 20°; above this, special (and more costly) ribbed belts are required.

With loose material, the feed mechanism must be designed to control airflow; generally high flow rates are attained with a belt-type system. The bulk material is metered onto a standard belt that is operated at high speed. The material is "ejected from the belt into the kiln." The belt does not enter the kiln and hence is not exposed to the high kiln temperatures. This system requires a baffle arrangement to minimize air infiltration. Many of the thermal desorbers, described later in this chapter, use a minimum size opening and chain curtains to limit the air entrained with the bulk material.

Liquids

The liquids can enter either through a liquid injector, burner, or in containers. When injecting liquids, the heat content will determine how it is injected. High heat content materials can be used to supply heat to the process. These can be fired into the burner as a fuel reducing the natural gas or fuel oil that would be required to provide the energy.

When using waste as a fuel, the flame characteristics must be reviewed with respect to the standard types of bulk solid materials to be added. If the bulk material heat content is high or will volatilize rapidly, a long flame burner would be appropriate. If the material is of lower heat content or requires considerable energy to volatilize the material, a short flame burner may be preferable.

The low heat content liquid may also be added on the kiln front wall. The low heating value liquid may be used to cool the temperature in the kiln if it is high. The injection characteristics of the low heat content liquid injection stream must be matched to the kiln length and the heat characteristics of the material. This liquid stream is generally not injected on the exhaust end of the kiln.

Design Criteria

The flow of material through a rotary kiln is determined by the kiln slope, rotational speed, and the characteristics of the material at the kiln operating temperatures. The kiln is installed with a slight slope to allow the bed of solids to advance through the kiln by gravity. As the kiln is rotated, the solid material is carried up on the kiln walls and eventually tumbles or slides down the wall. Gravity forces the material to fall down at the point that it is released by the wall. The forward movement is directly related to the slope of the kiln.

The nominal residence time is controlled by the rotation of the kiln. Residence times for commercial hazardous waste incinerators range between 15 min up to maximum of 2 h for the countercurrent kilns. The kiln revolution rate ranges from 0.15 up to 2.5 rpm. The lengths normally range from 15 up to 60 ft for the co-current kilns. The countercurrent kilns will range up to 120 ft to obtain the additional residence times.

The slope of the kiln is normally between 0.01 and 0.07 vertical foot per horizontal foot. The slope is generally on the lower side on the higher rpm kilns to allow sufficient time to completely volatilize the organics in the material. If the slope is too steep, the material will exit the kiln before it is fully processed. The size and shape of the material to be processed will also define the slope. The larger materials require the lower slope to assure that they do not exit the kiln too quickly.

The combustion products remain in the kiln for a very short time. Due to the short times and low gas velocities, there is very little gas mixing. Air that enters the system may not mix with all of the organics emitted by the waste. Hot and cold layers will exist within the kiln. The SCC must process all of the kiln exit gases to properly oxidize all of the organics flowing in the gas stream.

Due to the large variations in the waste being added to the kiln, the burner must be designed to assure the ignition of the material added to the kiln. The burner must also be stable to provide a positive indication to the flame scanner to assure that there is always a source of ignition.

The method of air addition will impact the combustion characteristics within the kiln. If all of the air is added through the burner, there will be an oxygen deficiency on the solid surface. If the oxygen component of the air does not mix with the combustible material, all of the available heat will not be released in the kiln.

There are several methods of adding air to the kiln that do not benefit the combustion process. Due to the mechanical limitations of the kiln seals, air will enter the kiln through both the front and back seals. Oxygen measurements for systems having poor back seals have given false security that there is sufficient oxygen within the kiln to properly release all of the heat available. The container feed door has also been a source of uncontrolled air addition. The air enters with the container giving a very high volume for a short period.

Well-designed kiln combustion systems have air ports on the kiln front wall. These air ports are connected to a combustion air blower to provide forced airflow into the bed and promote the combustion process. This type of system is appropriate for the high heat content materials.

Solids Residence Time

A rough estimate of the solids retention time in a particular incinerator can be determined from the following equation:

$$\theta = 0.19L/NDs \qquad (7.3.1)$$

where θ = retention time (min)
L = kiln length (ft)
N = kiln rotational velocity (rpm)
D = kiln diameter (ft)
s = kiln slope (ft/ft)

Although liquid wastes are frequently incinerated in rotary kilns, kilns are primarily designed for combustion of solid wastes. They are exceedingly versatile in this regard, capable of handling slurries, sludges, bulk solids of varying size, and containerized wastes. The only wastes that create problems in rotary kilns are aqueous organic sludges that become sticky on drying and form a ring around the kiln's inner periphery and packaged solids (e.g., drums) that tend to roll down the kiln and are not retained as long as the bulk solids. To reduce this problem, drums and other cylindrical containers are usually not introduced to the kiln when it is empty. Other solids in the kiln help to impede the rolling action.

Rotary Kiln Temperatures

In rotary kilns, temperature is controlled within a specific range by automatically varying the liquid waste or auxiliary fuel firing rate within the design turndown ratio and/or manually or automatically controlling the solid waste feed. Regardless of which technique is employed, provisions should be included for the following:

- *Termination of Liquid Waste Feed on Loss of Ignition in the Burner* If more than one liquid waste burner is employed, feed need only be terminated in the burner where flameout occurs.
- *Termination of Solid Waste Feed to the Kiln When Low Temperatures Are Sensed at the Kiln Outlet* If the feed to the kiln is automatic or semi-automatic, then the low-temperature cutoff system should also be auto-

matic. If manual feeding is employed, an alarm system is needed to warn the operator. The low-temperature cutoff point should be such that solid waste burnout can be maintained, but at lower than the normal operating temperature to avoid shutdown due to routine temperature fluctuations. Engineering judgment must be used to determine acceptable minimum temperature.

Secondary Combustion Chamber Temperatures

The outlet temperature of the SCC can be controlled by varying the liquid or auxiliary fuel feed or by varying the secondary airflow rate. Regardless of which technique is employed, provisions should be included for at least the following:

- *Termination of Liquid Waste or Auxiliary Feed on Loss of Ignition* This cutoff is necessary to prevent the release of unburned waste contaminants and to prevent potential explosion on release of unburned fuel. However, it also eliminates the function of the SCC. Therefore, solid waste feed to the kiln should also be terminated on loss of ignition in the afterburner. To minimize the occurrence of flameout in the SCC, only *clean,* homogeneous liquid wastes (or fuel) should be burned.
- *Termination of Solid Waste Feed to the Kiln If Low Temperatures Are Sensed at the SCC Outlet* The SCC feed should be maintained, however, to minimize potential release of unburned contaminants. As previously stated, the low-temperature cutoff point should be such that combustion is maintained, but at lower than normal operating temperatures to avoid shutdown due to routine fluctuations.
- *Termination of Solid Waste Feed to the Kiln If High Temperatures Are Sensed at the SCC Outlet* This is necessary to prevent damage to the refractory lining and the downstream air pollution control devices. The high-temperature cutoff point should be well above the normal operating temperatures, but low enough to avoid damage to the system. In the event of this cutoff, some liquid waste or fuel feed to the SCC should be maintained to complete combustion of off-gases from solid wastes remaining in the kiln.

In addition to these criteria, all liquid waste burners in the kiln and SCC should be equipped with manufacturer-specified primary air control systems so that air–fuel stoichiometry is maintained on turndown. Numerous hazardous wastes that previously were disposed of by potentially harmful methods (ocean dumping, landfilling, and deep-well injection) are currently being safely and economically destroyed using RK/SCC systems combined with proper flue gas handling. Included in this list of primarily toxic wastes are polyvinyl chloride (PVC) wastes, polychlorinated biphenyl (PCB) wastes from capacitors, obsolete munitions, and obsolete chemical warfare agents such as GB, VX, and mustard gas. Beyond these specific wastes, the RK/SCC systems are generally applicable to the destruction and ultimate disposal of any form of hazardous waste material that is combustible at all. Unlikely candidates are noncombustibles such as heavy metals, high moisture content wastes, inert materials, inorganic salts, and the general group of materials having a high inorganic content.[11] Most commercial incinerator sites in the United States presently operate with at least one RK/SCC system. These have served as the workhorse of the industry due to the ability to burn a variety of waste streams, physical forms, and heat contents.

7.4 OTHER METHODS

Fluidized-Bed Combustion

Introduction Applications of fluidized beds for the combustion and destruction by incineration of municipal sludge and industrial waste began in the early 1960s. Since then, more than 100 commercial units have been installed in the United States and over 300 worldwide. These range in heat capacity from 1 to 230 million Btu/h and in size from 4 to 50 ft in diameter. The primary application has been to municipal sewage sludge, but they have also been used for industrial and hazardous waste treatment. Among the industrial wastes burned are chemical process wastes, petroleum tank bottoms, sludge from pharmaceutical, pulp and paper, and nylon manufacturing operations, waste plastics, carbon black, spent activated carbon, precoat filter cake, and waste oils and solvents. In the 1980s, industry began to apply fluid beds to the destruction of hazardous wastes.

Fluid-bed technology offers such advantages as efficient combustion, ease of control, the ability to handle variable feeds, the ability to operate intermittently, and reasonable-to-low capital and operating costs. Additional advantages relative to hazardous wastes include fuel savings and lower emissions of nitrogen oxides (NO_x) and metals.[12]

System Description A fluid-bed combustor, shown schematically in Figure 7.4.1 is a vertical, cylindrical, refractory-lined carbon steel vessel with four major components:

- A plenum, or windbox, into which the fluidizing/combustion air is received
- An air distribution plate, which transmits the air from the windbox to fluidize the bed and prevents the bed material from sifting into the windbox
- The fluid bed of inert material, usually sand or product ash, into which the feed and/or fuel is injected
- The freeboard area above the bed

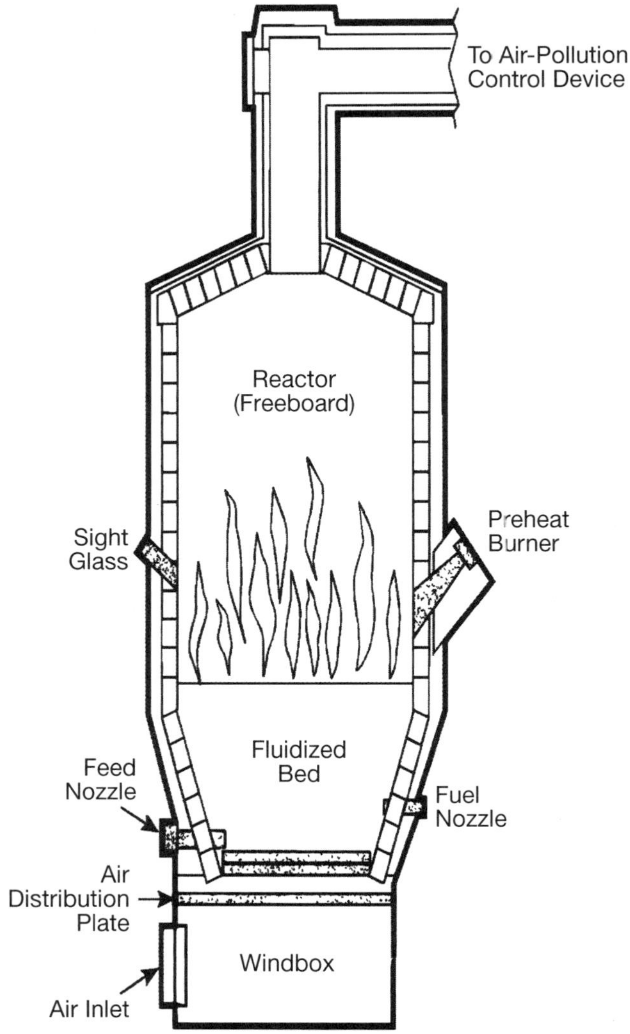

Figure 7.4.1 Fluid-Bed Incinerator.

The inert bed is levitated by the upcoming air. The solids–gas mixture takes on many of the physical and heat-transfer characteristics of a liquid (hence the name *fluid bed*).

The suspended-solids/gas mixture has a vigorous boiling action and an extremely large heat transfer area. These result in rapid and thorough mixing of the waste feeds, oxygen, and hot bed medium, and provide a near-ideal environment for combustion. At any given time, only a small portion (usually less than 1%) of the bed is combustibles, and the inert material acts as a large thermal flywheel, smoothing process variations. The flue gas proceeds from the bed through the freeboard, where combustion (particularly of volatiles) is completed and the larger solid particles entrained in the flue gas are disengaged and fall back into the bed. Due to attrition or burnout, ash is usually reduced to a sufficiently small size that it exits the combustor with the flue gas. Under certain conditions (such as waste feeds with high ash content or deliberate pelletization), bed material will accumulate. It is typically removed from the bed by an overflow drain, thus taking advantage of the bed's fluid characteristics.

In most instances, the system is of the forced-air type with a single fluidizing blower at the front end. When downstream heat recovery or air pollution control devices that cannot be subjected to significant pressure are included in the system, an induced-draft fan may be added to provide a balanced draft, with the null point (i.e., the point of atmospheric, or zero gage, pressure) set in the freeboard. Balanced draft systems are also utilized to provide a negative pressure in the combustion zone so that difficult-to-move materials can be fed via gravity through a chute-and-airlock arrangement. The choice of the air-pollution control device is not constrained by the use of a fluid bed, as there are existing fluid-bed combustors with virtually all types of APC equipment. As with other waste combustors, the choice of control device is a function of exhaust gas composition, ash content, the method of ultimate ash disposal, and, most significantly, local air emission regulations.

Limitations Of Fluid Beds Before a detailed discussion of the advantages of fluid beds as waste combustors or incinerators, it is important to note some of their limitations. The primary constraints are the physical form of the feed and the constituents of the waste.

Feed Physical Form The material fed into a fluid-bed waste combustor must be in such a form that it can be fluidized; that is, it will be suspended in the rising airstream. Large or bulky solids will sink to the bottom of the bed, smolder, and interfere with the fluidization process. Large is a relative term and maximum permissible size will depend upon specific gravity, fluidizing velocity, and other design parameters. Wet wood, for instance, can be fed with a maximum dimension of 3–4 in., while coal should be kept below 1.75 in., and particularly difficult-to-burn material, such as pure carbon, should be constrained to 0.75 in. or smaller.

The shape of the resultant ash also affects fluidization characteristics. The ideal form is as granular solids, which is the normal form that the ash will take.

A third limitation on feed physical form is economic rather than technical. Gases and atomizable liquids can and are being combusted in fluid-bed incineration units. However, if they are the only feeds to be combusted, it will normally ***be*** more cost-effective to burn them in a simple fume or liquid injection incinerator, or if they have low moisture and high heat contents [4170 kcal/kg (7500 Btu/lb) and higher] in a boiler.

Fluid beds find their primary application to feeds that are nonatomizable liquids, slurries, sludges, tars, and granular solids, with the potential for combusting liquids or gases in combination with these. Thus, in terms of feed form, fluid beds fall in an intermediate range between the simple liquid injection incinerator and the rotary kiln.

Feed Melting Point Fluid beds cannot be operated at temperatures above the melting point of the constituents in the bed. The designer must be concerned not only with the melting point of the individual components fed but also with the potential lower melting points resulting from the mixture of two or more materials (called eutectics). This problem is similar to that described in the rotary kiln section above. If ash material melts in the bed, the solid particles will agglomerate and the bed will eventually be defluidized. Two methods have been developed for dealing with potential melting in fluid beds—operating the incinerator as a "chemically active bed" and altering the melting point.

In a chemically active bed, the unit is operated close to but below the melting point of the ash constituents or of eutectics that may be formed, such that a controlled, limited amount of melting occurs. Such operation is possible in fluid beds because of their potential for close temperature control. Seed material of relatively small size (−8 mesh to +65 mesh) may be added to the bed to provide a base for material agglomeration. The bed tends to increase in depth due to the agglomeration, so a bed ash removal system is normally required. Since the bed acts like a fluid, such material removal can be accomplished relatively simply, for example, through an overflow valve. Chemically active beds of this type are used for spent liquor incineration in paper mills as well as other applications.[12]

Due to ash melting points, the chemically active fluid bed has an upper limit on the operating temperature of about 730–790°C (1350–1450°F) [although it can be as low as 620°C (1150°F)]. Higher temperature operation is often desirable, either to effectively destroy particular hazardous wastes or to improve the efficiency of heat recovery. In such cases, the recommended approach is to alter the melting point, when possible.

Waste streams, particularly hazardous streams, frequently contain alkali metal salts, which, in combination with silica either in the feed or as bed sand, form sticky alkali metal silicates or other compounds with eutectic melting points in the range of 635–815°C (1175–1500°F). However, chemical additives of relatively low cost, such as calcium hydroxide or kaolin clay, may be injected into the bed with the feed. These have the effect of tying up the alkali metals "in flight" and forming, instead of the eutectic, a refractory-like material with a melting point in the 1065–1288°C (1950–2350°F) range. In addition to alkali metal salt control, additives have also been shown to be effective for melting point control of wastes containing phosphorus.[12]

Thus, the maximum operating temperature of a fluid bed is normally determined by the feed constituent or eutectic melting point or modified melting point. As a practical matter, this normally limits operating temperatures to below 1038°C (1900°F). In most applications, such high temperature is neither required nor desirable, and most fluid-bed incineration systems operate in the 760–900°C (1400–1650°F) temperature range.

General Advantages of Fluid-Bed Incineration As mentioned previously, fluid-bed incineration offers a number of advantages. Several of these are common to fluid-bed systems in general, regardless of the type of waste being burned.

Efficient Combustion The intimate mixing in the bed with its resulting high turbulence and the large surface area of the inert bed material available for heat transfer provide virtually complete combustion at relatively low excess air levels (25–50% in typical incinerator applications) with minimal temperature variation throughout the bed. The constant collision of particles in the bed reduces the ash layer that may be formed on organic particles, thereby exposing additional combustible surface. Combustible materials entering the bed are subject to the full combustion temperature for gas residence times on the order of 5–8 s or more, since the combustor freeboard acts as an integral SCC. Solids, on the other hand, are subjected to significantly longer residence times, which depend on the solid's size and specific gravity. Small, light material will, as a limiting case, travel close to the gas velocity and have essentially the same residence time as the gas. Larger solids may be carried out of the bed by larger gas bubbles. Once in the freeboard, solids are acted on by two forces—the gas force trying to carry it out as particulate and the force of gravity pulling it back into the bed. The larger and denser the particle, the more likely it will return to the bed. For such particles, the residence time can be in minutes.

The long residence time, coupled with high turbulence and resultant oxygen availability, ensures efficient destruction of organic materials and produces an ash that is normally very low in organic content. Ash organic content will vary as a function of feed input and operating temperature, but for most cases it will be less than 3% and frequently less than 1%.

Ease of Control Uniform temperature in a single reaction chamber provides for simple, accurate measurement of process conditions, resulting in ease of process control and simplified automation of the process. It is not uncommon for fluid beds to operate at steady state with less than 8°C (15°F) temperature variation throughout the bed.

Variable Feed Acceptability Any incinerator is designed such that its total input fuel value is autogenous; that is, it has just enough heat content to sustain combustion at the design operating temperature. It is unusual, however, to have a waste feed that in itself is exactly autogenous. Furthermore, waste feeds will vary over time in their composition and resultant fuel value.

Fluid-bed combustors can be designed to burn subautogenous or overautogenous waste or waste that shifts between the two. A subautogenous waste is a waste with a heat content too low to sustain combustion and requiring supplemental fuel. Conversely, an overautogenous waste is a waste with a heat content greater than that required to sustain combustion and requiring some type of temperature control.

In the simplest system for burning overautogenous waste, direct water injection into the bed may be used for reaction temperature control, with the excess heat primarily absorbed by the water as latent heat of vaporization. This, however, prevents significant heat recovery in downstream devices such as waste heat boilers. If heat recovery is desirable, increased excess air may be utilized. However, this has the disadvantage of requiring a larger system because most components are sized based on the gas flow. For highly overautogenous feeds, heat exchange surface can be installed in the bed to take advantage of the higher heat transfer coefficients available on the bed side due to the fluid characteristics of the gas–solid mixture. (A fluid-bed boiler is an extension of a fluid-bed incinerator with bed coils.)

In combusting subautogenous feeds, the incoming combustion air can be preheated up to 675°C (1250°F) in an external heat exchanger by the sensible heat of the flue gas, thus reducing or eliminating the need for auxiliary fuel. To maintain structural integrity, a refractory arch is used in place of a metal distributor plate (as in the cold windbox design).

When feed variability is such that it alternately varies above and below autogenous, auxiliary fuel or cooling water can be injected directly into the bed, as required, by an automatic system activated by bed temperature changes.

Feed variations, particularly in waste incineration, can occur instantaneously or over short periods of time. The large inventory of hot inert bed material acts as a "thermal flywheel" so that major variations in the fuel or water content of the waste feed are stabilized by the large heat reservoir to prevent sudden temperature changes. Coupled with a long residence time, this tends to minimize variations in the exhaust gas composition, or puffing, which occurs in rotary kiln operations.

Cyclic Operation Capability Combustion reactions and water evaporation are virtually instantaneous and, therefore, there is a very low inventory of fuel or feed anywhere in the system. This permits startups and shutdowns in a matter of minutes when the combustor is near its operating temperature. This characteristic, coupled with the slow rate of bed cooling in a combustor that has been shut down (9°F/h), permits intermittent operation if desirable (e.g., one shift operation or weekend shutdown).

Capital and Operating Costs A major maintenance cost for most incinerators is repair and replacement of refractory. In fluid-bed units the large inventory of bed material acts as a thermal flywheel, eliminating thermal shock and resultant damage to the refractory, provided reasonable care is taken during startups and shutdowns. A number of fluid-bed incinerators have been in operation for over 20 years with no refractory replacement and only minor refractory repairs. This factor, combined with the absence of moving parts in the hot zone, results in relatively low maintenance.

Operating costs of fluid-bed and rotary kiln incinerators are approximately equal. Although the fluid bed will have higher power consumption, it tends to require less operating labor.

For systems where the waste feed is suitable for direct injection into a fluid bed, that is, sludges and small solids, the capital cost for a complete system of equivalent thermal capacity, from feeders through the APC device, will normally be on the order of 60–85% of the capital cost for a rotary kiln.

Application to Hazardous Wastes As noted in Chapter 3, the basic performance criterion for hazardous waste incinerator design is achieving a *99.99%* DRE of the primary organic hazardous constituents in the waste. The two former performance criteria defined by the Resource Conservation and Recovery Act (RCRA) were particulate emissions and hydrogen chloride emissions. The MACT rules issued on September 30, 1999, have imposed specific emission requirements that must be met by all hazardous waste combustors. See Chapter 3.

The efficiency of the destruction reaction is both time and temperature dependent. Important points in considering fluid-bed combustors are that they have relatively uniform temperature profiles in the bed and inherently long gas residence times (usually 5–8 s) at the operating temperature.

The range of applicability of fluid-bed incinerators to hazardous wastes can be illustrated by the following test program results. The first unit,[13] tested in 1975, is a fluid-bed combustor with a freeboard inner diameter of 25 ft and a heat input of about 50 million Btu/h; it has as its APC device a wet venturi scrubber/tray tower. In one test, phenol was destroyed to greater than 99.999% DRE at a bed temperature of 720–758°C (1330–1395°F) with a total gas residence time of 12–14 s. In another test, the same DRE was achieved for methyl methacrylate at 770–788°C (1420–1450°F) and a 12-s gas residence time. CO emissions were 6–26 ppm (instantaneous values) during steady-state operation.

More recently, trial burns have been conducted on fluid-bed incineration units applying for permits under the RCRA Part B program. At one such facility,[14] the fluid bed is 7 ft in diameter; auxiliaries include a downstream cyclone, a combustion air preheater, a dry venturi for calcium injection, and a baghouse. Heat input is about 3 million Btu/h. At bed temperatures of 834–848°C (1533–1558°F) and a residence time of about 5 s, the DREs for carbon tetrachloride and chloroform were 99.996 and 99.9975%, respectively. DREs for the remaining POHCs (1,1,1-trichloroethane, trichloroethylene, phenol, and perchloroethylene) were all in excess of 99.999%. Although no dioxins were detected in the input feed, the stack gas was tested for this constituent due to the presence of chlorinated hydrocarbons as potential precursors, and none was detected at a level of 0.5 parts per trillion. CO emissions were 2.0–8.0 ppm, averaged over each test run; peak instantaneous CO emissions were less than 50 ppm.

There is a common misconception that extremely high temperature is required to provide adequate destruction of hazardous waste. A unit at one petroleum refinery is a 28-ft freeboard inner diameter fluid bed with a combustion air heat exchanger, cyclone, and wet venturi scrubber/tray tower. Initial testing at 753°C (1387°F) bed temperature and 786°C (1447°F) freeboard temperature with an 8-s residence time produced DREs of 99.9996 and 99.9997% for toluene and 1,1,1-trichloroethane, respectively. A second test at lower temperatures—665°C (1229°F) in the bed and 711°C (1312°F) in the freeboard section—produced only slightly lower DREs of 99.9994 and 99.9992%. A consequence of such low-temperature operation, however, is increased CO release due to reaction kinetics—68 and 320 parts per million by volume corrected averages for the two tests above.

A fluid-bed incinerator with an 18-ft freeboard, an air preheater, and a wet venturi scrubber/tray tower underwent its trial burn in early 1991.[15] DRE for chlorobenzene was 99.9997%, at freeboard operating temperatures of 848°C (1560°F). Average CO emissions ranged from 6 to 70 ppmv during three tests, with a peak instantaneous value of 80 ppmv. Total hydrocarbon emissions averaged approximately 2 ppmv.

Advantages for Hazardous Waste Incineration Based on test data above, one can draw two primary conclusions about fluid beds burning hazardous waste:

- Because of residence times in the freeboard, no SCC is required to achieve acceptable DRE values.
- Due to superior turbulence and the near-isothermal operation of the bed, fluid beds achieve equivalent destruction efficiencies at temperatures 110–166°C (200–300°F) lower (on average) than rotary kilns. These features result in several operational advantages:
  - *Fuel Savings* Lower operating temperature and reduced excess air requirements due to the absence of an SCC result in lower auxiliary fuel requirements when burning subautogenous wastes.
  - *Lower NO_x Emissions* At the lower temperatures at which fluid beds operate for hazardous waste incineration [870°C (1600°F) and below], essentially no thermal NO_x, is formed.[16] There are also indications[16,17] that NO_x produced from fuel-bound nitrogen decreases with decreasing temperature. Should the level of NO_x from fuel-bound nitrogen be unacceptable based upon local air permitting regulations, it can be significantly reduced by operating the fluid-bed incinerator in a staged-combustion mode.[18,19] In this mode, the bed is operated slightly substoichiometrically (making oxygen unavailable to form NO_x) and combustion is completed by introducing additional secondary air in the freeboard (which operates at temperatures where thermal NO_x formation is minimal).

- *Lower Metal Emissions* The U.S. EPA has reported[16,20] that of the primary air emissions from an incinerator [POHCs, products of incomplete combustion (PICs), and metals], metal emissions represent the greatest health threat and account for almost all of the potential cancer risk. Increasing incineration temperature causes the average flue gas particulate metal distribution to shift significantly to particles of less than 10 μm, apparently due to vaporization followed by heterogeneous deposition and, more importantly, homogeneous nucleation.[21,22] As the incineration temperature increases, the emission of trace metals after the pollution control device increases, due to their concentration on smaller particulates; for lead and cadmium, in some instances, the increase is exponential. Because of increasing metal emissions with increased temperature, EPA has chosen to limit the maximum temperature for sewage sludge incinerators to 900°C (1650°F).[23]

Basic System Design Considerations Sizing of the combustor and downstream components is basically a function of the stack exhaust volume, reducing this volume will usually result in lower capital costs. The exhaust gas volume can be minimized by using a minimum of excess air and maintaining the minimum combustor temperature.

For overautogenous feeds where incineration only is desired, minimum capital cost is achieved by using direct bed water addition to control combustor temperature. On the other hand, if energy recovery as steam is desired, water addition is the least efficient method because of the unrecoverable heat of vaporization. In this case, excess air addition is more efficient.

The choice of the downstream APC system will be primarily a function of the recently enacted MACT rules. Wet venturi scrubbers with tray towers, for instance, will not have the high particulate removal efficiency of other devices but will have the lowest capital cost. Their total operating costs will generally be higher, both because they produce a wet ash (with its consequent increased disposal costs) and because their pressure drop and resultant power demand is higher. On the other hand, a wet venturi scrubber/tray tower is very efficient for acid gas control and a combined spray dryer–baghouse followed by a low pressure drop wet scrubber may be the most cost efficient in such cases.

Feed systems are generally chosen on the basis of the solids content of the waste stream. Liquids with little or no solids are fed by centrifugal pumps. Wastes containing less than about *20%* dry solids are fed by progressive cavity pumps, while wastes with more than about *20%* solids utilize screw pumps. Most recently, piston pumps have been gaining popularity because they can handle from a few percent to *30–40%* dry solids. Although their initial capital cost and power demand are high, they have a reputation for high reliability and low maintenance cost. Essentially dry material may be conveyed pneumatically, and difficult-to-handle material, as noted earlier, may be fed by gravity into a negative pressure incinerator via a conveyor/chute/air lock system.

The choice of components for the total system is often the result of a series of trade-offs between capital and life-cycle cost, with relative reliability and maintainability also considerations.

Fixed-Hearth Incinerators

Fixed-hearth incinerators, used extensively for medical waste incineration, are less commonly used to incinerate hazardous wastes. There are several units employed at commercial sites as well as plant sites for hazardous waste destruction. Fixed hearths can handle bulk solids and liquids; however, this design does not have the versatility of a rotary kiln.[24]

Process Description Fixed-hearth incinerators typically contain two furnace chambers: a primary combustion chamber (PCC) and a secondary combustion chamber (SCC). A typical fixed-hearth system is shown in Figure 7.4.2. Solid and liquid wastes may be charged into the PCC. Small units are normally batch-fed while larger units may be continuously fed with a screw feeder or moving grate or semicontinuously fed with a ram feeder. In some designs, there may be two or three step hearths on which the ash and waste are pushed with rams through the system. A controlled flow of "underfire" combustion air is introduced, usually up through the hearth on which the waste sits. In some designs, combustion air may also be provided from the front wall over the waste bed.

In many fixed-hearth incinerators, known as "controlled-air" or "starved-air" incinerators, 70–80% of the stoichiometric air required is provided in the PCC, thus the PCC is operated in a reducing, substoichiometric or starved-air mode. In this situation, the waste is pyrolyzed and partially combusted. In some cases, steam injection into the PCC is used to enhance waste fixed carbon burnout (fixed carbon is nonvolatile carbon that will burn only when exposed to combustion air).

Temperature in the PCC must be high enough to destroy hazardous organics in the waste 760°C (1400°F) but low enough to reduce the potential for slagging and refractory damage 1000°C (1832°F). Waste slagging may clog the underfire air ports. Bottom ash is removed using a continuous or semicontinuous method, depending upon the unit design. The ash is usually dumped into a water bath, which provides a seal between the PCC and the atmosphere.

For controlled air units, due to insufficient supply of air, the PCC flue gas will contain unburned hydrocarbons and high levels of CO and H_2. These volatiles are burned out in the SCC. In the SCC, 140–200% excess air is provided, as well as sufficient residence time at temperature for complete volatiles burnout. Again, the three T's of combustion and oxygen are required for high combustion efficiency. Liquid wastes may be injected in the PCC and/or the SCC. Supplemental fuel may be provided in both chambers to maintain temperature. SCC temperature may be as high as 1100°C (2012°F). Tempera-

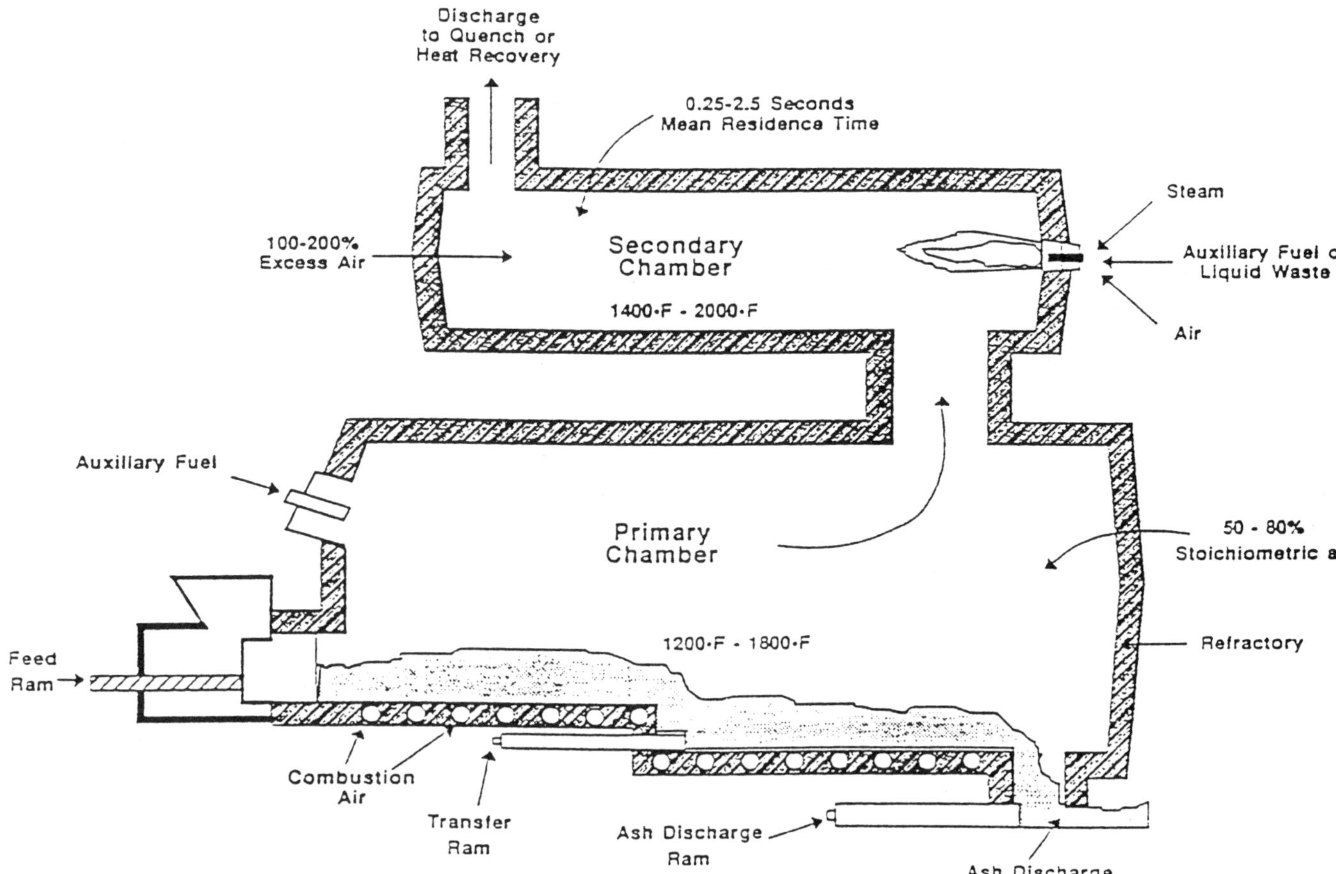

Figure 7.4.2 Typical fixed-hearth incinerator.

tures above this are avoided to prevent refractory damage, decreased residence time, and excess usage of auxiliary fuel.

Operating Parameters Critical fixed-hearth operating parameters included PCC and SCC temperatures. PCC temperature is controlled by the underfire air supply with a fixed waste feed rate. SCC temperature is controlled by auxiliary fuel firing into the SCC and PCC waste feed rate adjustments. The capacity of the SCC limits the PCC burning rate. The SCC must have adequate volume to accept and oxidize all volatile gases generated in the PCC. Additionally, the SCC must operate with excess air to ensure complete burnout of the volatiles.

Multiple-Hearth Incineration

The multiple-hearth incinerator (commonly called the Herreshoff furnace) is a flexible unit that has been utilized to dispose of sewage sludges, tars, solids, gases, and liquid combustible wastes. This type of unit was initially designed to incinerate sewage plant sludges in 1934. In 1968, there were over 125 installations in operation with a total capacity of 17,000 tons/day (wet basis) for this application alone.[25]

A typical multiple-hearth furnace includes a refractory-lined steel shell, a central shaft that rotates, a series of solid flat hearths, a series of rabble arms with teeth for each hearth, a combustion air blower, fuel burners mounted on the walls, an ash removal system, and a waste feed system. Side ports for tar injection, liquid waste burners, and an SCC may also be included. An illustration of a multiple-hearth incinerator and a typical flow scheme is shown in Figure 7.4.3. Sludge and/or granulated solid combustible waste is fed through the furnace roof by a screw feeder or belt and flapgate. The rotating air-cooled central shaft with air-cooled rabble arms and teeth distributes the waste material across the top hearth to drop holes. The waste falls to the next hearth and then the next until discharged as ash at the bottom. The waste is agitated as it moves across the hearths to make sure fresh surface is exposed to hot gases.

Units range from 6 to 25 ft in diameter and from 12 to 75 ft in height. The diameter and number of hearths are dependent on the waste feed, the required processing time, and the type of thermal processing employed. Generally, the uppermost hearth is used as the SCC. Normal incineration usually requires a minimum of six hearths, while pyrolysis applications require a greater number.

The rabble arms and teeth located on the central shaft all rotate in the same direction; additional agitation of the waste (back rabbling) is accomplished by reversing the angles of the rabble teeth. Waste retention time is controlled by the design of the rabble tooth pattern and the rotational speed of the central shaft. Liquid and/or gaseous combustible wastes may be injected into the unit through auxiliary burner nozzles; this utilization of liquid and gaseous waste represents an economic advantage because it reduces auxiliary fuel requirements, thus lowering operating costs.

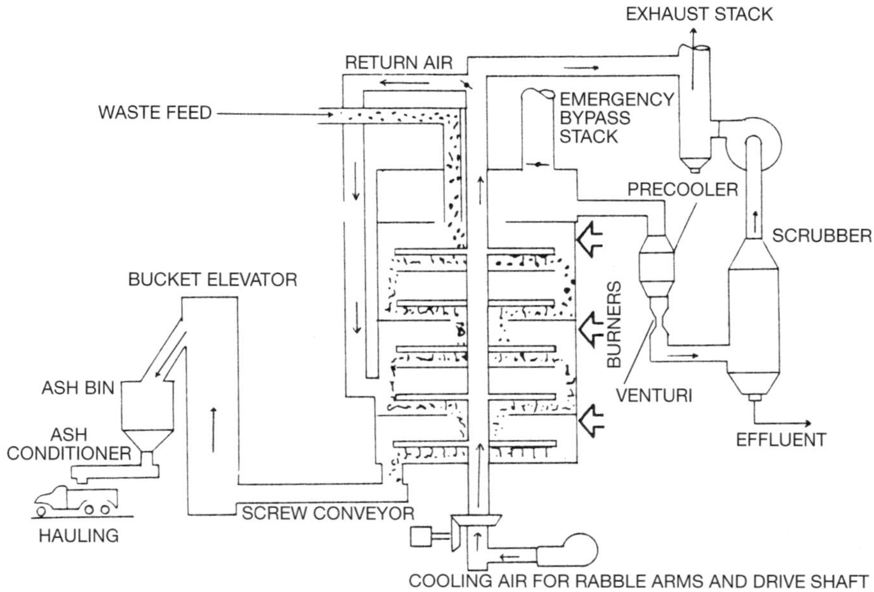

Figure 7.4.3 Multiple-hearth incinerator.

There are three temperature zones in a multiple-hearth furnace. The top hearths operate at temperatures between 300 and 550°C (570 and 1020°F) to dry the waste material. Incineration takes place in the middle hearths at temperatures between 750 and 1000°C (1380 and 1832°F), and the incinerated wastes are cooled by incoming combustion air in the bottom hearths at temperatures between 200 and 300°C (392 and 572°F). Exit gases have good potential for heat recovery, being around 300–590°C (570–1100°F). Due to the Clean Air Act Amendments (CAAA) of 1990, most multiple-hearth units have had to operate the top hearths at higher temperatures [700–870°C (1400–1600°F)] serving as SCCs for reduction of CO and THC.

Advantages of multiple hearth incinerators are

- The retention or residence time in multiple-hearth incinerators is usually higher for hazardous materials having low volatility than in other incinerator configurations.
- Large quantities of water can be evaporated.
- A wide variety of wastes with different chemical and physical properties can be handled.
- Multiple hearth incinerators are able to utilize many fuels including natural gas, reformer gas, propane, butane, oil, coal dust, waste oils, and solvents.
- Because of its multizone configuration, fuel efficiency is high and typically improves with the number of hearths used.

- Fuel burners can be added to any of the hearths to maintain a desired temperature profile.
- Multiple-hearth incinerators are capable of a turndown ratio of 35%.
- High fuel efficiency is allowed by the multizone configuration.

Disadvantages of multiple hearth incinerators are

- Due to the longer residence times of the waste materials, temperature response throughout the incinerator when the burners are adjusted is usually very slow.
- It is difficult to control the firing of supplemental fuels as a result of this slow response.
- Maintenance costs are high because of the moving parts (rabble arms, main shaft, etc.), which are subjected to high temperatures and corrosive conditions.
- Multiple-hearth incinerators are susceptible to thermal shock resulting from frequent feed interruptions and excessive amounts of water in the feed. These conditions can lead to early refractory and hearth failures.
- If used to dispose of hazardous wastes, an SCC will be required and higher operating temperatures will be necessary.
- These devices are not well suited for wastes containing fusible ash, wastes that require extremely high temperature for destruction, or irregular bulky solids.

Wastes that have been incinerated in multiple-hearth installations include sewage primarily, but also solid residues from the manufacture of aromatic amines, reactor bottoms from PVC manufacture, chemical sludge, oil refinery sludge, still bottoms, and pharmaceutical wastes. Examples of wastes not compatible with this process include heavy metals, inert materials, inorganic salts, and materials with a high inorganic content.[11]

Due to the more stringent requirements of RCRA and the CAAA of 1990, many multiple hearths are being phased out and replaced with fluid beds. Fluid-bed incinerators operate at much lower excess air rates resulting in higher combustion efficiency, lower CO emissions, and lower volumes of exhaust gas products. This reduces the size requirements for the APC system and horsepower for the induced-draft fans.[26]

7.5 CO-INCINERATION

Co-incineration refers to the combination of hazardous waste combustion with an industrial combustion process to provide additional energy or as a supplemental source of fuel. This is not a unique technology; any existing incineration process can be used for this special case of mixing waste streams to obtain better combustion of particularly intractable waste material. Co-incineration most often occurs in processes such as utility and industrial boilers, process kilns (e.g., cement kilns, lightweight aggregate kilns, and lime kilns), and refuse or sludge incinerators. The advantages of co-incineration include utilization of the waste's heat content to save fuel with usually little or no increase in emissions. The major disadvantage is the limited number of acceptable waste types based on the particular combustion process and the product manufactured.

Hazardous Waste Boilers

Hazardous Waste Boilers By definition, a boiler is a closed vessel in which water under pressure is transformed into steam by the application of heat. More specifically, EPA has defined the following characteristics of a boiler for regulatory purposes under the RCRA: (1) the combustion chamber and primary energy recovery section must be of integral design; (2) thermal recovery efficiency must be at least 60%; and (3) at least 75% of the recovered energy must be "exported" (i.e., not for internal boiler uses). Boilers come in a variety of sizes, configurations, and designs. The three common boiler design categories that burn hazardous waste are the fire-tube boiler, the water-tube boiler, and stoker-fired boiler.[24]

Fire-Tube Boilers The fire-tube boiler can be described in simple terms as a water-filled cylinder with tubes running through it that provide the escape path for the combustion gases or flue gas (as shown in Fig. 7.5.1). As the flue gas passes through the tubes, the hot gases heat the tubes which then heat the water to produce steam. The fire-tube boiler is primarily used in industrial applications.

Fire-tube boilers are compact, low in initial cost, and easy to modularize based on plant requirements. However, they are also slow to respond to changes in demand for steam (load) compared to water-tube boilers, and circulation is slower. Also, stresses are greater in fire-tube boilers because of their rigid design and subsequent inability to expand and contract easily. Fire-tube boilers usually range in size from less than 2 to 50 million Btu per hour. Most industrial fire-tube boilers are either horizontal return tube (HRT), scotch marine, or firebox.

Fire-tube boilers may be directly fired with waste fuels, either liquid or gaseous. They also serve to recover heat from incinerators fired on waste fuels.

Water-Tube Boiler The basic design of a water-tube boiler circulates the water through tubes, with the hot combustion gases passing over the outside surfaces of the tubes. Generally, the boiler can be physically divided into two sections, the furnace and the convection pass, as shown in Figure 7.5.2. Furnaces (fireboxes, combustion chambers) will vary in configuration and size, but their function is to contain the flaming combustion gases and transfer the heat energy to the water-cooled walls. The convection pass contains the superheat-

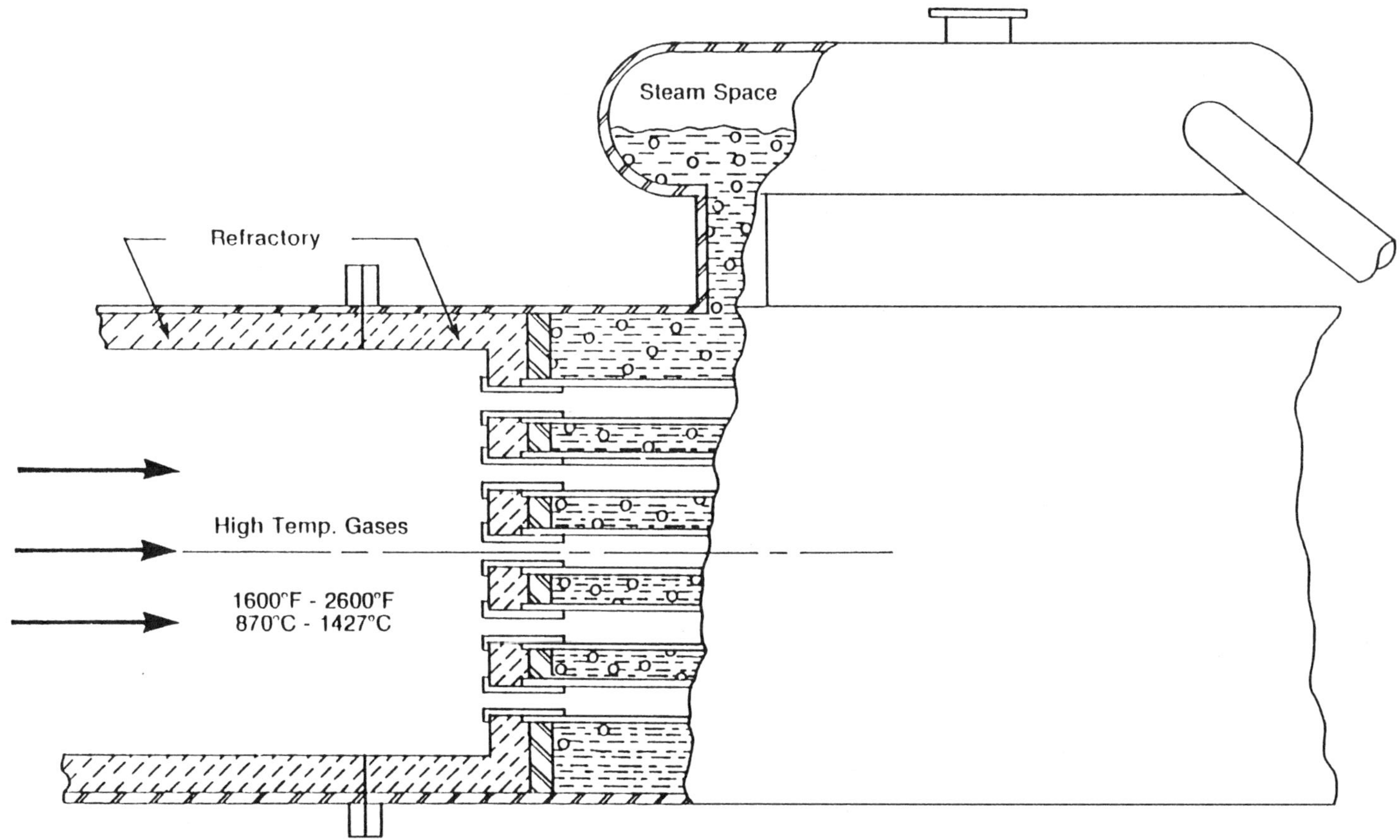

Figure 7.5.1 Fire-tube waste heat boiler.

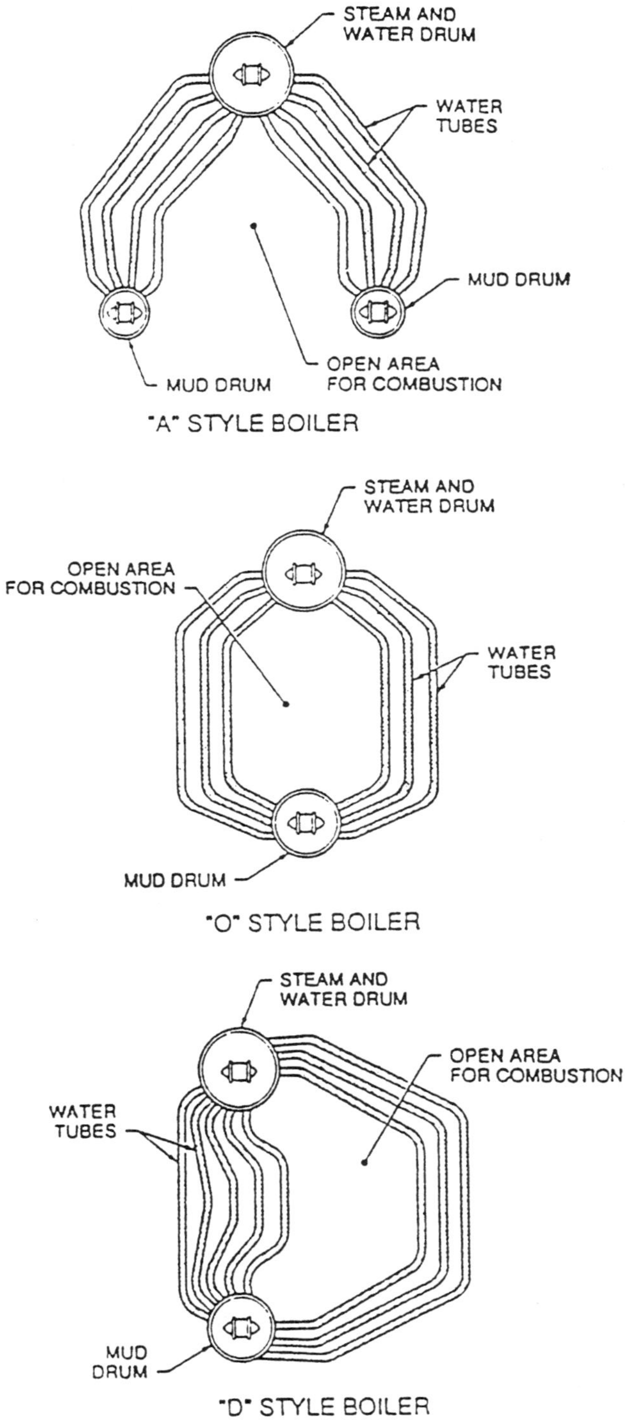

Figure 7.5.2 Water-tube boiler configurations.

ers, reheater, economizer, and air preheater heat exchangers, where the heat of the combustion flue gases is used to increase the temperature of the steam, water, and combustion air.

The *superheaters* and *reheaters* are designed to increase the temperature of the steam generated within the tubes of the furnace walls. Steam flows inside the tubes and flue gas passes along the outside surface of the tubes.

The *economizer* is a counterflow heat exchanger designed to recover energy from the flue gas after the superheater and the reheater. The boiler economizer is a tube bank type, hot-gas-to-water heat exchanger. It increases the temperature of the water entering the steam drum.

The *air heater* is not a portion of the steam–water circuit, but serves a key role in the steam generator system to provide heat transfer and efficiency. In many cases, especially in a high pressure boiler, the temperature of the flue gas leaving the economizer is still quite high. The air heater recovers much of this energy and adds it to the combustion air. Heating the combustion air prior to its entrance to the furnace reduces fuel usage.

Water-tube boilers are rapid steamers and respond quickly to changes in demand for steam due to improved water circulation. They can withstand much higher operating pressures and temperatures than fire-tube boilers. In addition, the water-tube boiler design is safer. They can also burn a wide variety of fuels as well as wastes and have the ability to expand and contract more easily than fire-tube boilers. The major drawback is that water-tube boilers are more expensive to install. They also require more complicated furnaces and repair techniques.

Overfire Air Systems The overfire air system on boilers has three functions: (1) to provide oxygen and turbulence to mix the fuel and oxygen in the lower high-temperature zone of the furnace for complete burnout, (2) to distribute fuel and assist in distribution when the design dictates the use of air for this purpose, and (3) to provide cooling air for mechanical fuel feeders. The air can come from a common fan that would have static pressure capabilities of 25–30 in. H_2O to provide the necessary energy for turbulence or fuel distribution.

Overfire air nozzles are placed in rows in the front and rear walls of the furnace. The design should place the air nozzles at elevations that will ensure mixing of the furnace gases in the high-temperature zones, so that the combustion processes can be completed in the furnace. Air nozzle spacing in each row should be on centers 10–12 in. for complete coverage of the furnace.

The size of the air nozzles in the various rows is a function of the depth of the penetration desired and the amount of overfire air required. Bituminous-coal-fired boilers are designed with 15–25% overfire air. Lignite-fired boilers are designed with 20–25% overfire air. The overfire air temperature may be either ambient or the temperature of the undergrate air. The choice of temperature is generally a function of air heater and boiler system design. Waste solids fired on grates will utilize overfire air to add the necessary turbulence and oxygen to the particles burning in suspension above the grate and carried by the rising combustion gases.

Cement Kiln Incineration

Cement kiln incineration is a proven technology for treating hazardous wastes. This process is used commercially since reliable data from research, demonstration, and operating units have been provided to overcome public opposition.[27]

Refer to Figure 7.5.3 for a cement kiln process schematic. It has been demonstrated that existing cement kilns, when properly operated, can destroy most toxic organic chemical wastes, including PCBs, which can be converted to less noxious forms during normal cement kiln operation.

Since many chemical wastes have significant heat content of 5555 kcal/kg (10,000 Btu/lb) or more, cement plants can save large quantities of energy from burning such liquid chemical wastes as alternate or synthetic fuels.The burning of halogenated PCB wastes at the St. Lawrence Cement Company, Mississauga, Ontario, Canada, was a technical success. During the 2-year test period, PCB-containing wastes of up to 10,000 ppm of PCB were burned successfully and completely destroyed without any adverse environmental effects or chemical accidents. *The general public was, however, never informed about these practices until after the tests.* When the tests became public knowledge, the reaction of the neighboring communities was severe, despite the fact that in most cases chemical wastes injected into the cement kilns were destroyed to below analytical limits of detection. Public emotion and fear had a part in suspending further testing pending extensive environmental public hearings.[28]

Cement Kiln Process Description This section presents an overview of the cement kiln process and the types of kilns used for cement making. Process operating parameters important to hazardous waste treatment and a characterization of cement kilns currently burning hazardous wastes are also presented.[24]

Process Components Cement is made from a carefully proportioned mixture of raw materials containing calcium (typically limestone), silica and alumina (typically clay, shale, slate, and/or sand), and iron (typically steel mill scale or iron ore). These materials are ground to a fine powder (80% passing 200 mesh), homogenized, and heated to a very high temperature to produce a cement "clinker" product. The raw feed material, known as "meal," is heated in a kiln that is a large, inclined, rotating cylindrical steel furnace lined with refractory materials. The kilns are operated in a *countercurrent* configuration; the gases and solids flow in opposite directions through the kiln, providing for more efficient heat transfer compared with *co-current* operations typical of rotary kilns used for hazardous waste incineration.. The raw meal is fed at the upper, "cold" end; the slope (3°–6°) and rotation (0.83–1.20 rpm) cause the meal to move toward the "hot" lower end. The kiln is fired at the hot end,

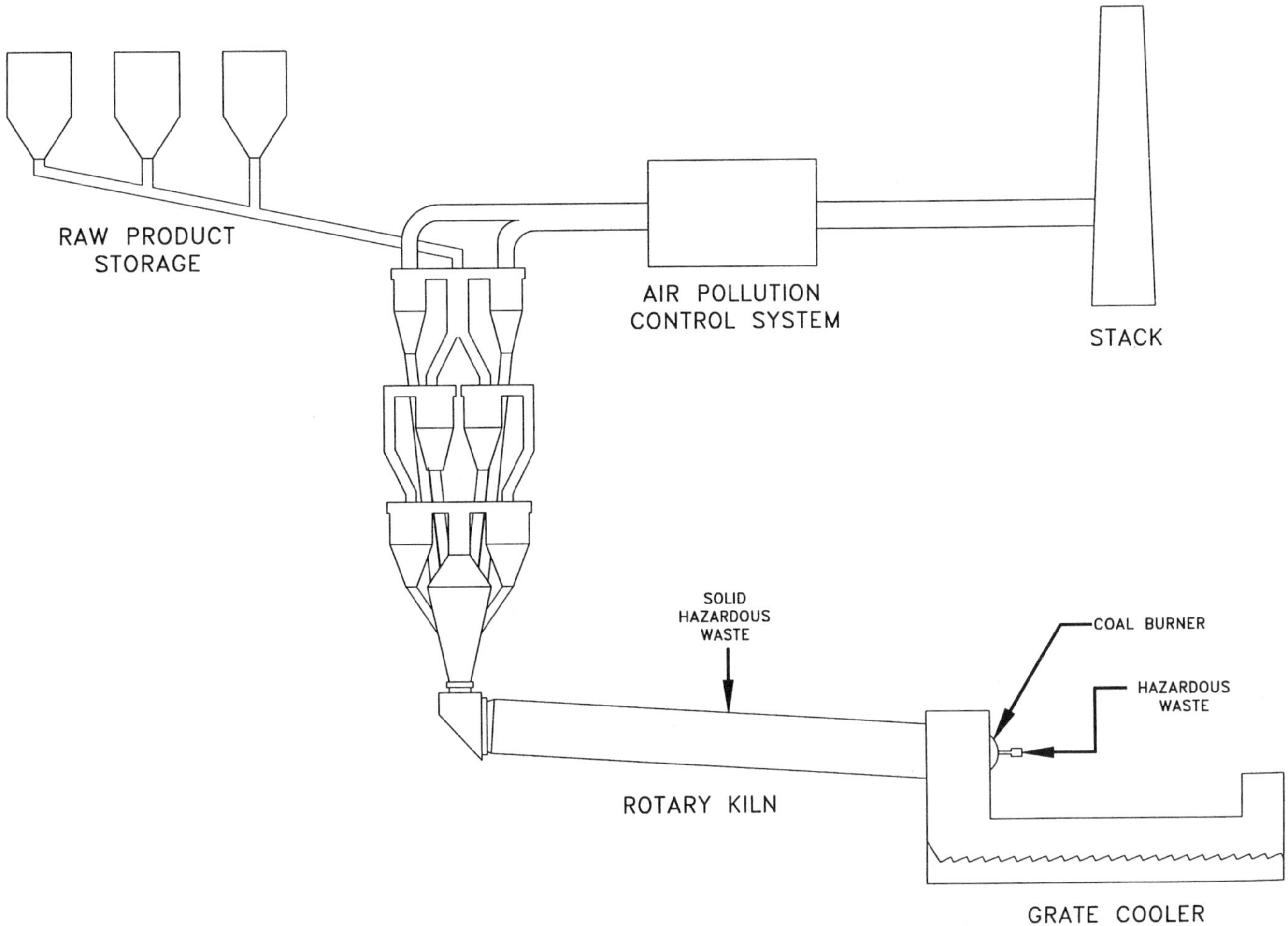

Figure 7.5.3 Cement kiln system.

usually with coal or petroleum coke as the primary fuel; natural gas or fuel oil may also be used as a supplemental fuel. As the meal moves through the kiln and is heated, it undergoes drying and pyro-processing reactions to form the clinker. The reactions can be categorized into three major stages:

- *Drying and Preheating Zone* Residual water is evaporated from the raw meal feed, and clay materials begin to decompose and are dehydrated (removal of bound water) in a temperature range of 20–600°C (70–1112°F).
- *Calcining Zone* Material is "calcined"; that is, calcium carbonate in the limestone is dissociated producing calcium oxide ("burnt lime") and carbon dioxide in the temperature range of 600–900°C (1112–1652°F).
- *Burning Zone* In the "burning" zone, also known as the "clinkering" or "sintering" zone, calcium oxide reacts with silicates, iron, and aluminum to form the clinker. The clinker is a chemically complex mixture of calcium silicates, aluminates, and alumino-ferrites. A minimum meal temperature of 1480°C (2696°F) is necessary in the burning zone of the kiln to produce the clinker.

The clinker is removed from the kiln at the hot end. After it passes through the burning zone and by the kiln flame, it enters a short cooling area where the clinker melt begins to solidify; the cooling rate from the burning zone to the kiln exit is important since it determines the microstructure of the clinker. The clinker leaves the kiln at about 1100°C (2012°F) and falls into a clinker cooler. The cooler is typically a moving grate onto which the clinker sits. Cooling air is blown through the clinker bed. The cooled clinker consists of gray-colored nodules of variable diameters, typically up to 2 in. The clinker is blended with gypsum and ground in a ball mill to produce the final product, cement. Hot exhaust air produced from the clinker cooler is either directed to the kiln where it is used as combustion air or used to predry the raw feed material in the case of dry process kilns.

Kiln exhaust flue gases contain significant amounts of entrained particulate matter due to the turbulence in the kiln from the rotary action and from the use of finely ground feed material. The entrained particulate matter, known as "cement kiln dust" (CKD), is removed from the flue gas by some type of APC device. Many plants return a portion of the CKD to the raw feed materials. However, in most cases, some CKD must be removed from the kiln system entirely to lower the buildup of alkali salts. CKD can be used in other industries as neutralizers or additives; however, the excess CKD is usually land disposed.

Cement Kiln Types Cement can be produced in three different types of arrangements—wet, dry, or semidry processes—as described below.

Wet Process Kilns A schematic of the wet cement making process is shown in Figure 7.5.4. Ground raw materials are mixed with water (about 30% by

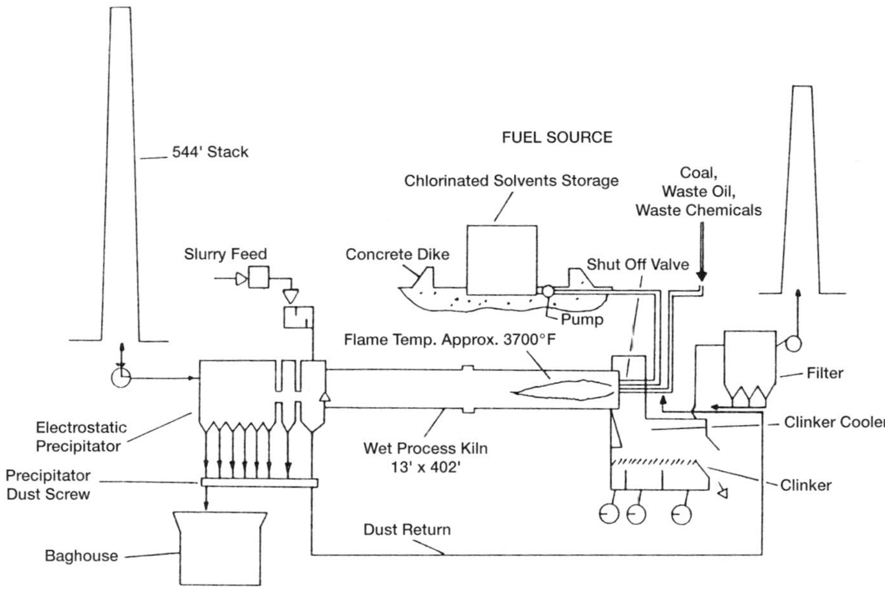

Figure 7.5.4 Wet process kiln.

weight) to form a slurried meal. The slurry is fed to the kiln through a flow metered pump. Solids in the slurry typically occupy no more than 15–20% of the internal kiln volume. Wet kilns typically have a length-to-diameter ratio of about 30:1 to 40:1 and rotate from 1.2 to 1.33 rpm. Kiln rotational speed is adjusted to maintain clinker quality and heat transfer. The wet cement making process is the older process, characterized in part by handling, mixing, and blending of the raw materials in the slurry form and lower emissions of kiln dust. However, because all water must be evaporated out of the slurry mixture, wet process kilns require greater energy input than other types of cement kilns —typically, from 5 to 7 million Btu/ton of clinker product is required.

To improve the energy efficiency of the process, steel chains are typically attached to the inside of the kiln shell in the drying zone. The chains act as a heat transfer mechanism. As the kiln rotates, the chains hang, by gravity and pick up heat from the kiln flue gases. When the chains settle back, the heat is transferred to the raw meal.

A typical solids/gas temperature profile of a wet kiln is shown in Figure 7.5.5. Note that the feed does not begin to heat up until all of the water in the raw meal slurry is evaporated. The gas temperature in the burning zone approaches 2200°C (4000°F), while the clinker reaches a maximum temperature of approximately 1480°C (2700°F), which is required for the formation of clinker. The kiln gas residence time is over 10 s. Typical solids residence time in the kiln is 2–3 h.

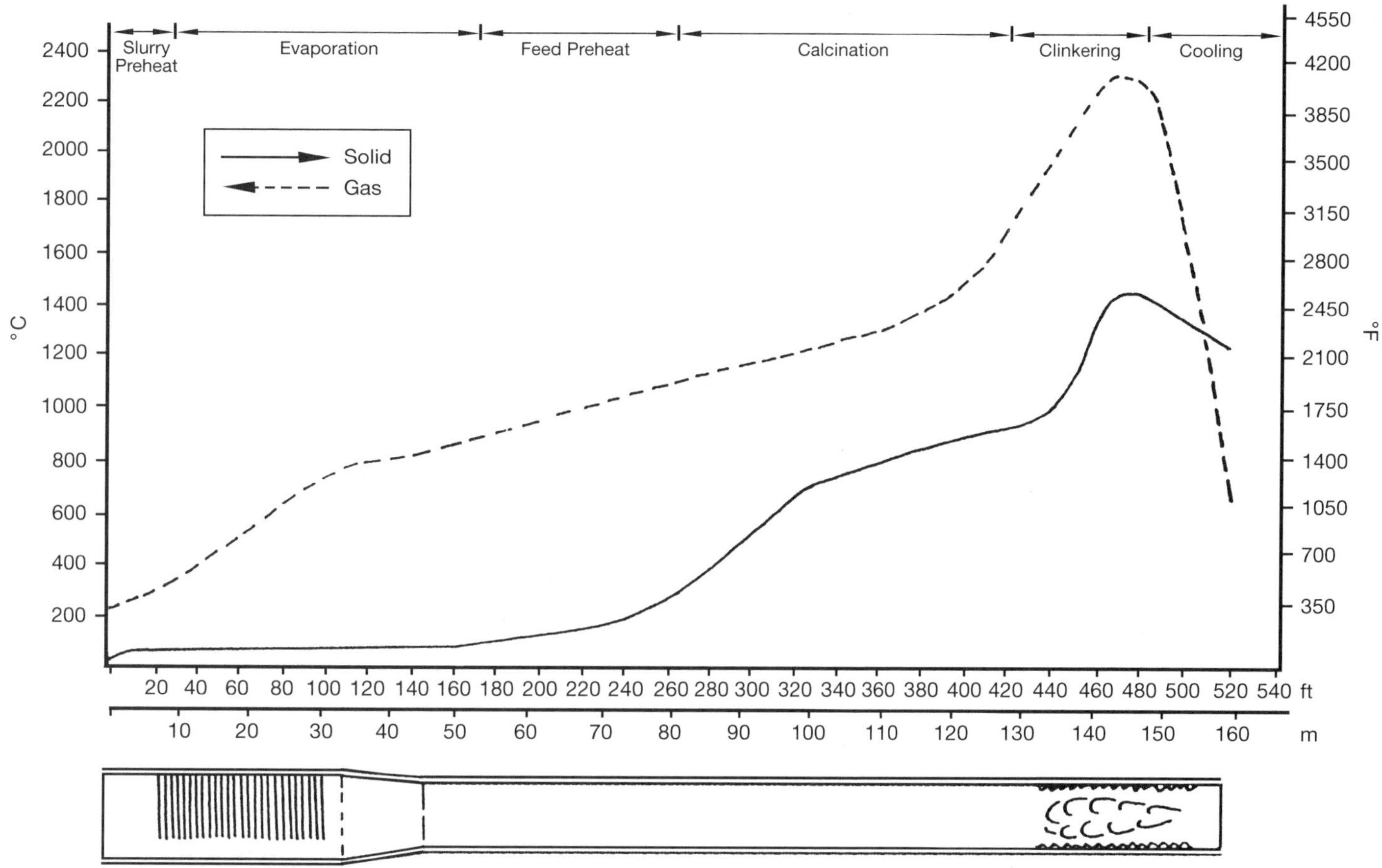

Figure 7.5.5 Wet-process kiln material and gas temperatures. Dellinger, H., Pershing, D., and Sarofim, A., "Evaluation of the Origin, Emissions and Control of Organic and Metal Compounds from Cement Kilns Co-fired With Hazardous Wastes," Scientific Advisory Board on Cement Kiln Recycling, June 8, 1993, p. 20.

Dry Process Kilns Due to higher energy efficiency compared with wet kilns (3.4–4.5 MM Btu/ton clinker), dry processes have become popular. In the dry process, the ground raw material is dried and pneumatically transported to the kiln or to a preheater if used. Kiln exhaust flue gases or hot clinker cooler air are typically used for drying the raw materials. There are three different dry process configurations: long, preheater, or preheater/precalciner.

A typical solid/gas temperature profile of a kiln with a preheater and precalciner is shown in Figure 7.5.6. It is similar to that of the preheater kiln; however, the time at which the gas temperature is above 1650°F is slightly less in a precalciner-type kiln.

Semi-dry Process Kilns In the semi-dry process, the ground feed material is pelletized with 12–14% water. The pellets are put on a moving Lepol grate on which they are dried and partially calcined by hot kiln exhaust gases before being fed to the rotary kiln.

Hazardous Waste Feeding The method of feeding hazardous waste into the kiln is dependent on temperature of the zone of injection and the resulting residence time at temperature. This is a primary factor in determining the destruction of the waste. The method of introducing the waste also depends on its physical state, as discussed in the following sections.

Liquids Liquid wastes are either blended directly with conventional fuels provided at the hot end of the kiln or they are injected separately through a separate burner/atomizer into the primary kiln flame.

Solids In the past, solid wastes were fed at the cold end of the kiln with the raw meal. However, the addition of wastes at the cold end is no longer practiced because as the wastes move down the kiln and get hotter, volatile components are driven off at temperatures below those required for complete oxidation and destruction. Thus, significant emissions of unburned volatiles may have occurred without the use of an afterburner, SCC, or other add-on volatile emissions control techniques. Solid wastes are not charged directly into the hot end burning zone since they can create local reducing conditions that adversely affect clinker quality such as strength, stability, setup time, and color. Oxidizing conditions must be maintained in the burning zone. If insufficient oxygen is present, a key component of cement clinker, tetra calcium alumino-ferrite, does not form; instead, Fe_2O_3 is reduced to FeO. This leads to a clinker product that produces a quick setting cement with decreased final strength. Additionally, the presence of unburned carbon in the burning region produces a clinker with an undesirable brown color. To overcome these feeding location limitations, a variety of techniques are used for the introduction of solid wastes into cement kilns. The choice depends on the solid waste type (composition and physical attributes) and kiln configuration (dry long, dry with preheater, wet, etc.). Three techniques rely on the introduction of the solid waste into the calcining zone.

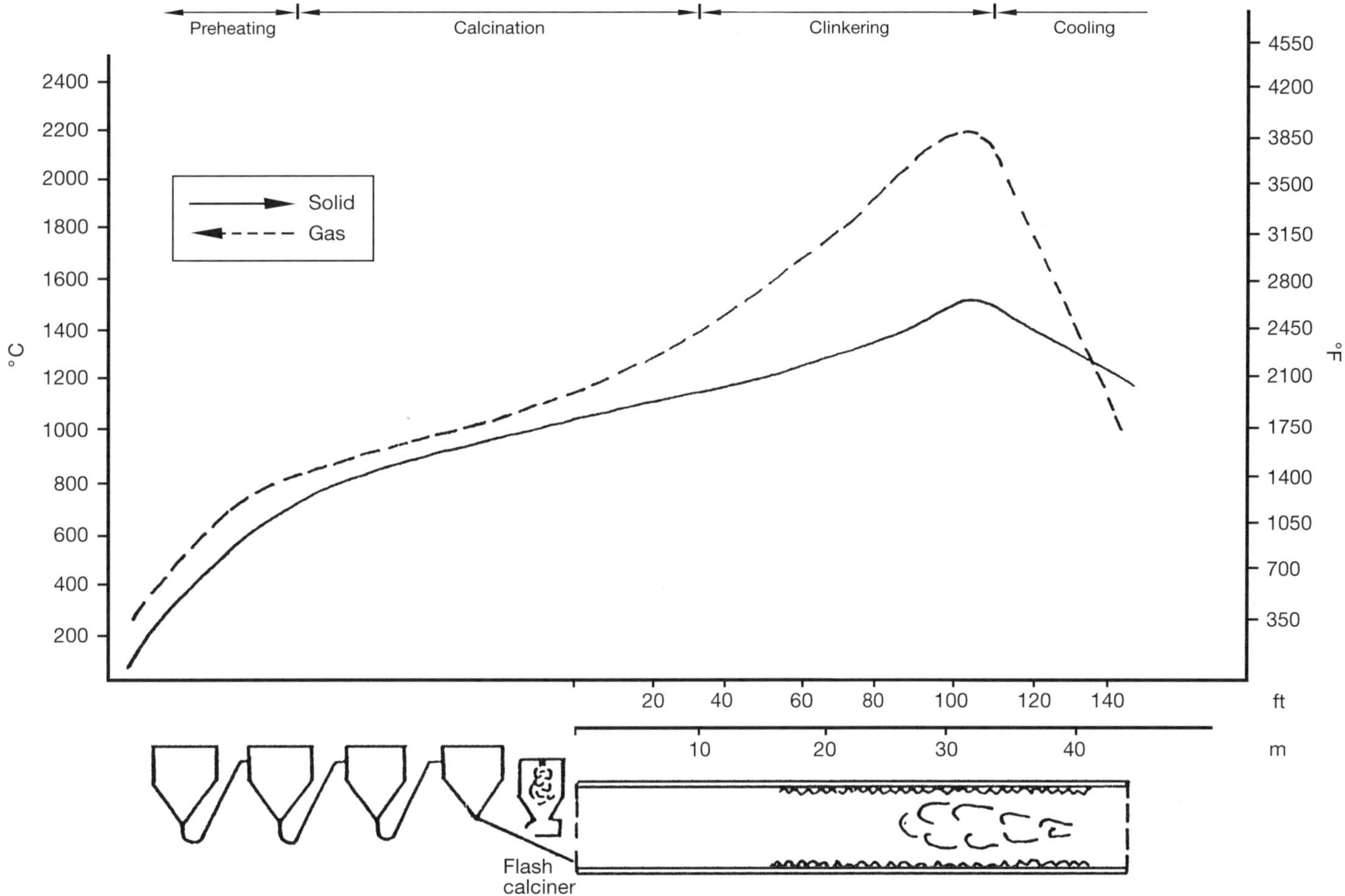

Figure 7.5.6 Preheater and precalciner kiln material and gas temperatures.[24]

Note that the 1991 boiler and industrial furnace (BIF) rules require that if hazardous waste is fed at any location other than the hot end, the combustion gas temperature must exceed 1000°C (1832°F) at the point of introduction and sufficient oxygen must be present to ensure complete combustion of organic constituents in the waste. For long kilns, this means that the waste is introduced midkiln, while for preheater/precalciner kilns, it is introduced onto the feed shelf. At this location, the solids have sufficient residence time for the organics to be driven off so that the presence of waste combustibles do not reach the burning zone and create reducing conditions that are detrimental to clinker quality, but yet far enough away from the cold end to provide sufficient temperature and residence time for complete destruction of volatile gaseous organics. These three methods are:

- For long-type wet and dry kilns and dry kilns with preheaters, a method has been patented by Benoit et al. in 1989[29] for charging solid wastes directly through a hatch on the rotating kiln wall at an intermediate location within the calcining zone. At each rotation, the hatch is opened, and containerized waste solids are fed down a drop tube that is inserted through the hatch and into the rotating kiln. The drop tube prevents hot mineral material from escaping through the port or contacting the enclosure. It is important that the volume of the volatile components do not exceed the capacity for their complete combustion in the gas stream. Thus, wastes are containerized to minimize the potential for overloading the combustion capacity (create local reducing conditions). See Figure 7.5.3.
- For preheater or precalcining kilns, the solid waste may be injected directly into the precalciner vessel or preheater inlet.
- Containerized solid wastes may also be injected at the hot end of the kiln at a high enough velocity so that they are projected into the calcining zone. An "air cannon," which is mounted to the kiln hood, is used to propel the waste containers. Lone Star Industries (Greencastle, IN) and Continental Cement Co. (Hannibal, MO) use this method for waste solids feeding.

If the waste is in powdered form, it may be injected directly into the primary burning zone coal flame of the cement kiln, as is done at Heartland Cement Co. (Independence, MO). In another arrangement, solid waste is burned in a separate kiln or combustor and the flue gas is routed to the hot end of the cement kiln, which is used as a SCC and for acid gas control. This is done at Giant Cement Co. (Harleyville, SC), which has four "resource recovery kilns" in addition to cement kilns.

Lightweight Aggregate Kiln Process Description

Lightweight aggregates include a wide variety of raw materials that when combined with cement form concrete products. Lightweight aggregate concrete is produced either for structural purposes or for thermal insulation purposes. Estimates indicate that the rotary kiln method is used in the production of 80% of all structural lightweight aggregates.[30] This section presents an overview of the lightweight aggregate kiln process and a characterization of the lightweight aggregate kilns currently burning hazardous waste.

Process Components A lightweight aggregate plant is composed of a quarry, a raw material preparation area, a kiln, a cooler, and a product storage area. The material flows from the quarry to the raw material preparation area. From there, the material is fed into the rotary kiln. Figure 7.5.7 provides a schematic of a lightweight aggregate kiln. The rotary kiln consists of a long, steel cylinder, lined internally with refractory bricks, which is capable of rotating about its axis and is inclined at an angle of about 5° to the horizontal. The length of the kiln depends in part upon the composition of the raw material to be processed but is usually 90–180 ft. The prepared raw material is fed into the kiln at the higher end, while firing takes place at the lower end. The dry raw material fed into the kiln is initially preheated by hot combustion gases. Once the material is preheated, it passes into a second furnace zone where it melts to a semiplastic state and begins to generate gases that serve as the bloating or expanding agent. In this zone, specific compounds begin to decompose and form gases such as CO_2, SO_x, and O_2 that eventually trigger the desired bloating action within the material. As temperatures reach their maximum at approximately 1150°C (2100°F), the semiplastic raw material becomes viscous and entraps the expanding gases. This bloating action produces small, unconnected gas cells, which remain in the material after it cools and solidifies. The product exits the kiln and enters the cooler section of the process where it is cooled with cold air and then conveyed to the discharge.

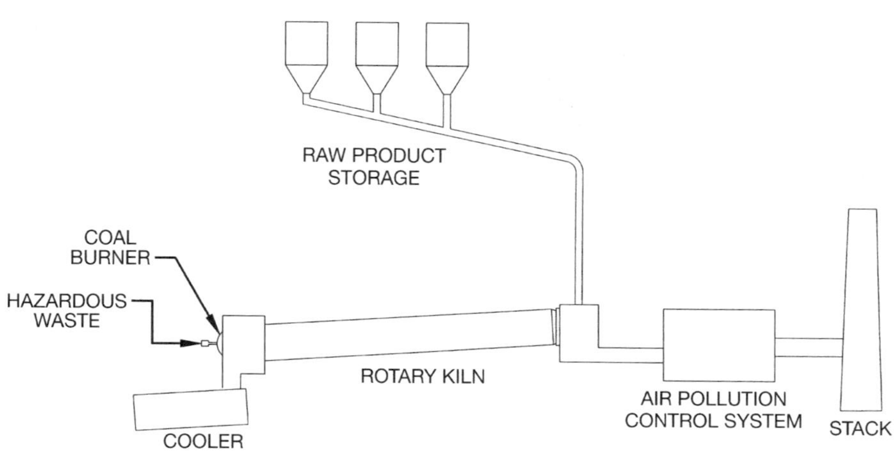

Figure 7.5.7 Lightweight aggregate system.

Operating Parameters Kiln operating parameters such as flame temperature, excess air, feed size, material flow, and speed of rotation vary from plant to plant and are determined by the characteristics of the raw material. Raw materials include clay, shale, or slate, which have varying properties linked to their geological formation. Maximum temperature in the rotary kiln varies from 1120 to 1260°C (2048 to 2300°F), again depending on the type of raw material being processed and its moisture content. Exit temperatures may range from 150 to 650°C (300 to 1200°F), again depending on the raw material and on the kiln's internal design. Approximately 80–100% excess air is forced into the kiln to aid in expanding the raw material.

A typical lightweight aggregate rotary kiln has a combustion gas residence time of 4 s based on maximum temperature and a temperature range of 370–1150°C (700–2100°F). A typical rotary kiln processes approximately 450 Mg (500 tons) of lightweight aggregate per day. This kiln has an average furnace volume of 1000 m^3 (35,343 ft^3) and an average volumetric flow rate of 1920 Nm^3/min (73,170 scfm). Lightweight aggregate kilns that bum hazardous waste typically bum 100% liquid hazardous waste (i.e., hazardous waste is the sole fuel).[24]

7.6 THERMAL TREATMENT OF SUPERFUND SOIL

Introduction

In the late 1980s, mobile and transportable high-temperature incinerators burst upon the scene for clean up of organic waste on Superfund sites. High cost and EPA's negative attitude toward incineration caused a transition to use of low-temperature thermal desorber systems. This started in 1989 with a pesticide project in Marianna, Florida. The low-temperature systems were able to work on the vast majority of Superfund waste at about half the cost of the high-temperature systems. As of the late 1990s, virtually all the thermal treatment at Superfund and similar non-Superfund sites were performed by low-temperature systems.[10,31] Most systems are of the rotary desorber type. Williams Environmental Services, Stone Mountain, GA, has been successful in using countercurrent rotary desorbers for processing of volatiles at up to 40 tons per hour (tph) throughputs. It has also processed lightly contaminated, high-boiling-point organics, including pesticides, coal tar creosote, and low-level PCB waste. It uses countercurrent rotary desorbers with return of contaminated fines from the baghouse to the rotary desorber via a "Dobson collar" injector. This forces the fines through the hot zone near the burner end of the drum to remove condensed organics prior to the fines leaving the system. Williams has used a variety of back-end treatments, including SCCs, scrubbers, and activated carbon.

Other firms have used co-current systems, rated up to 90 tph, for processing volatile organic compounds (VOCs) and high-boiling-point organics. Recent developments include the use of indirect desorbers, including the processing of high oil wastes such as drilling muds (up to 30% oil by weight range) in Mexico and Malaysia. The drilling muds, and many petrochemical industry wastes, are being processed with indirectly fired retorts equipped with condensers or indirectly heated screws with condensers. Maxymillian, Pittsfield, MA, has treated PCB waste with its indirectly fired retort system.

SRS, a subsidiary of American ECO, Irvine, CA, is one of the firms involved in high oil waste recovery. It has about 25 SAREX units in operation, using steam or hot oil indirectly heated equipment similar to paddle dryers and hollow flight screws for VOC contaminated waste. For higher boiling point material, a second (electrically heated) stage is added. The first-stage average capacity is 3–8 tph per unit with a 115–230°C (240–450°F) product. The second stage, with a 510°C (950°F) product temperature 593°C (1100°F) process temperature, has about 3 tph capacity. High-boiling-point materials, including coal tar creosote, polyaromatic hydrocarbon (PAH) compounds, PCBs, and pesticides have been removed successfully. The recovery system is a three-stage direct-contact quench with demister. Noncondensables go to vapor phase carbon adsorbers or are burned in a boiler to supply steam or hot oil for the process. Typical removal efficiency is about 98% for organics from the contaminated media. Oil recovery is about the same range. The process can handle up to 70% oil content or moisture. Feedstock physical properties can vary from soil to pumpable sludge, including such feeds as bio solids, American Petroleum Institute (API) separator sludge, tank bottoms, and still bottoms. Centrifuging is used to reduce water and oil content prior to thermal processing.

Units are being used in the midwestern states where live steam is directly injected into the soil to strip and heat the soil in a batch mode. Normal cycle time is 3–4 h, with system capacity of 20 tph. Off-gas is condensed and treated with activated carbon downstream. These work well with soil with open pore structure (e.g., with at least some sand content) and VOC contamination. They can handle some light VOC contamination. Thermal Remediation, Inc. has units that can handle up to 2000 ppm VOCs at full throughput and have gone as high as 24,000 ppm or organic contaminants by using much longer cycle times. As this is a desorber-condenser system, it tends to meet with greater regulatory and public acceptance.

McLaren Hart, Charlotte, NC, has used its high-vacuum low-temperature infrared heated batch system equipped with condenser for oil recovery. TerraTherm, Houston, TX, has used its heated blanket system to desorb soils in place. The blankets are placed on the top of the soil, which is heated from the top down to drive off organic vapors. This system is not a good match for sites that have high groundwater levels.

Several fixed sites exist, treating imported nonhazardous petroleum-contaminated soils and other nonhazardous waste. A few are permitted for MGP (manufactured gas plant) coal tars that are not an EPA hazardous waste.

Oxygen has been used on occasion to extend throughput and cut costs. This is particularly useful when high temperatures are required to achieve correspondingly high DREs in the SCC.

Heat exchangers have been tried with mixed success for cooling the gases leaving the SCC on co-current rotary desorbers prior to the gases entering the baghouse. At 870°C (1600°F) maximum input temperature and with low fines content, they may survive. With high fines content (from clays and silts), higher temperatures and chlorine content, they become a high-maintenance item and often fail. If the preheated air is used for combustion, lower fuel use results but NO_x levels rise. The best designs are arranged so the inlet of the heat exchanger does not see the hot refractory in the SCC. This prevents heat damage from sudden shutdowns (e.g., a power outage) where loss of cooling air to the heat exchanger causes distortion and cracking of the leading edge facing the hot refractory. Most designs have been horizontal, leading to silting and fouling of the heat exchanger. A vertical downflow arrangement with dropout chamber for dust at the bottom would fix this problem.

Incorrectly specified equipment has led to several baghouse fires. Manufacturers limit use of countercurrent/counterflow equipment to low-boiling-point organics such as gasoline, diesel fuel, and No. 2 fuel oil at low concentrations. The problem with higher boiling point organics is that the organics that desorb at the hot (burner) end of the primary treatment unit (PTU) drum condense at the cold end and enter the baghouse as liquid droplets or oil condensed on particulates. This fouls the bag filters and provides fuel for baghouse fires. Some units have a duct from the soil cooler to the baghouse and this can be a source of ignition. Other potential sources are baghouse fans (not a "nonsparking" design) incorrectly placed upstream of the baghouse, sparklers (usually cellulosic material in the feedstock) carried with the airstream, and compounds, such as sulfur, which have low autoignition temperatures. The best fix for these problems is to use co-current/parallel-flow equipment for high-boiling-point organics. In this design, the desorbed organics are kept in a hot, gaseous state between the PTU and the SCC, thereby eliminating fouling of the bags and associated baghouse fires.

Many desorbers use direct-fired burners without an external furnace. The lack of a furnace can cause quenching of the flame by cold soil for co-current units and increases the chance of creating sparklers that can cause baghouse fires for countercurrent units. Adding an external furnace allows the combustion to be completed prior to hot gases entering the drum, thereby reducing or preventing these problems.

Infrared Belt This is a mesh belt-type furnace using electrically heated "glow bars" over the soil on the belt to drive off organic vapors. The furnace can be equipped with an SCC or oxidizer and an air pollution control system. Alternately, the organic vapor can be condensed and adsorbed. The desorber/oxidizers do well on soil <4% organic by weight (about 800 Btu/lb) if suitably equipped per National Fire Protection Association (NFPA) requirements. The desorber/condensers may have an economic edge for soils of higher organic concentration. Good candidates are soils with stable organics that are nonreactive and do not decompose at the temperatures encountered in the processing system. The desorber/condensers are the obvious choice for sites where EPA has ruled out use of direct-fired desorbers.

Commercial Low-Temperature Systems Examples of commercially available equipment in use are Williams Environmental Services' 25-tph TPU-01 desorber/scrubber plus activated carbon adsorber and its 20-tph TPU-02 and 30-tph TPU-03 desorber/oxidizers; ENSCI Environmental's 40-tph co-current rotary desorber/oxidizer system; CWM's 5-tph X*TRAX desorber/condenser and 35-tph LT*X desorber/oxidizer; Weston's 7.5-tph LT^3 desorber/condenser; and Westinghouse/Haztech 8-tph infrared desorber/condenser. Many other systems exist for Underground Storage Tanks (UST) (petroleum-contaminated soils) that may also be used for MGP soils. However, most are countercurrent designs and will likely have problems with condensation of coal tars in their baghouses.[31]

Cost for Thermal Treatment Costs for thermal treatment of soil may be categorized by the level of contamination present and use of on-site or off-site facilities. Table 7.6.1 shows some typical "chute-to-chute" soil decontamination costs for a project of greater than 30,000 tons. When excavation and engineering are added, costs may double. Excavation and solids handling have been problematic in many soil cleanup projects. These problems can be overcome with experienced remedial design engineers and the proper selection of equipment.

The selection of the proper thermal equipment is predicated more on the concentration and boiling point of the contaminant than its toxicity. The largest transportable desorber/oxidizer systems have capacities of up to 45 tph and are used on jobs of 10,000 tons and above. Off-site treatment is available from a few firms that process contaminated soil at their own sites. Two examples are Southeastern Soil Recovery, Inc., Charleston, SC, and ENSCI Environmental (subsidiary of American ECO, Highpoint, SC).

7.7 EMERGING TECHNOLOGIES

Plasma Systems

In plasma systems, the extremely high temperature of plasma [approaching 10,000°C (18,032°F)] is employed for the destruction of toxic waste material.

Table 7.6.1 Typical cost of thermal treatment for contaminated soil[31]

Type of Treatment	Type of Waste	Cost $/ton
Off-site treatment low-temp desorber/oxidizer	<4% organic	$60
On-site treatment, low-temp desorber/oxidizer	<4% organic	$75
On-site treatment, high-temp incinerator	<4% organic	$150
On-site desorber/condenser	<2000 Btu/lb	$200+
Off-site incineration	<1200 Btu/lb	$1,000

A plasma temperature incinerator such as the one manufactured by Applied Energetics, Inc. (see Fig. 7.7.1), burns the toxic wastes, including PCBs, in a pressurized stream of preheated oxygen. The operation can achieve a temperature of 2760°C (5000°F) and maintain these temperatures at elevated levels for periods up to 1 s. The major components of a stand-alone incinerator that is designed for a feed rate of 67 ft^3/h are shown schematically in Figure 7.7.2. This incineration system consists of a preheater, combustion chamber, reaction residence chamber, quench chamber, scrubber tower, and exhaust stack. The preheater is a refractory-lined, water-cooled chamber in which a clean fuel is burned to preheat oxygen to about 1000°C (1832°F). The preheater consumes fuel at the rate of 265 lb/h to preheat 10,000 lb/h of oxygen. The preheated oxygen enters the combustion chamber through a disperser plate. The disperser plate contains a number of fuel nozzles and oxygen ports and is designed to achieve optimal mixing of the two reactants. The temperature of the preheated oxygen is sufficient to autoignite any fuel when the heat of combustion is at least 3333 kcal/kg (6000 Btu/lb). This arrangement minimizes the risk of flameout and subsequent contamination of the area with partially burned waste.

Advantages of the plasma temperature incinerator are

- In the plasma temperature operating range, many molecules are highly disassociated and have free electrons that greatly increase their reactivity.
- The combustion reaction rates have been shown to be much greater at these temperatures; in some studies a 10-fold increase has been achieved. In similar systems, solid fuels were completely consumed in 35–50 ms.

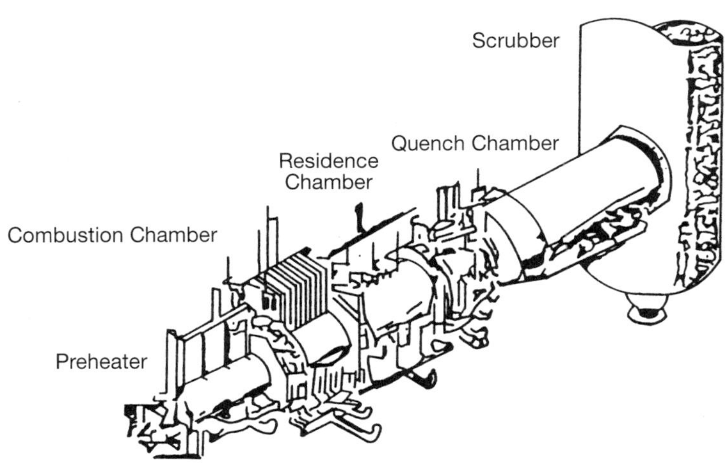

Figure 7.7.1 Plasma temperature incinerator.

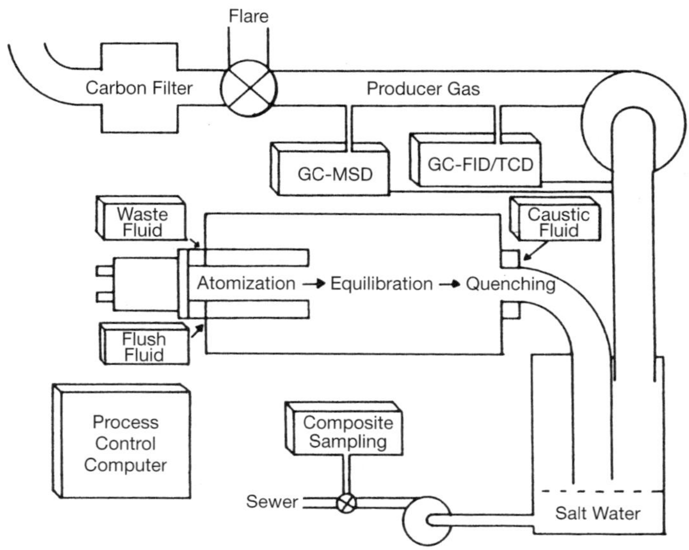

Figure 7.7.2 Pyroplasma process schematic (GC-MS, gas chromatography–mass spectroscopy; GC-FID/TCD, gas chromatography–flame ionization detection/thermal conductivity detection).

- The higher combustion efficiencies achieved by this design yield a more compact unit that may substantially reduce the cost of incineration. This unit can be built as a stand-alone incinerator or can be combined with heat recovery systems to conserve the energy that is released.

Disadvantages are

- With temperatures at about 10,000°C (18,032°F), the durability of the arc and refractory materials could be a potential problem.
- Operation of the system requires highly trained professionals.

Pyrolysis Systems, Inc. uses a plasma arc device in a mobile incineration system that has been demonstrated in Kingston, Ontario. In this device, a colinear electrode arrangement creates an electric arc that is stabilized by a field coil medium through which an electric current is passed. As air passes the arc, the electrical energy is converted to thermal energy by adsorption by the air molecules that are activated into ionized atomic states, losing electrons in the process. When the molecules or atoms relax from their highly activated states to lower energy levels, ultraviolet (UV) radiation is emitted. The resulting gas is in the plasma state and consists of charged and neutral particles with an overall charge near zero and with electron temperatures up to 27,780°C (50,000°F). As the activated components of the plasma decay, their energy is transferred to the waste material exposed to the plasma. The wastes are ultimately decayed and destroyed. A schematic of this process is shown in Figure 7.7.2. A pilot-sized plant has been used to gather data on the destruction of chlorinated organics, primarily PCBs. Destruction efficiencies in excess of 99.999% were reported, indicating that even greater destruction can be expected among the less stable chlorinated substances. The unit is sized to destroy approximately 500 lb/h of sludge, yet it is designed to fit on a semitrailer along with the necessary auxiliary equipment. Commercial viability must be established by testing actual waste streams over an extended period on a continuous operating basis.[32]

Under the Superfund Innovative Technology Evaluation Program, the U.S. EPA sponsored a plasma arc vitrification demonstration in July 1991 at a Department of Energy research facility in Butte, Montana. The plasma hardware was supplied by Retech, Inc. of Ukiah, California. During the demonstration, the furnace processed approximately 4000 lb of waste.[33]

In 1992, Mason and Hanger National, Inc. designed a waste processing system for installation at the Kaiser-Permanente Zion Medical Center in San Diego, California. This was reported in May 1992 at the International Incineration Conference.[34]

7.8 ILLUSTRATIVE EXAMPLES

Example 7.8.1 A vertical liquid injection incinerator is operating at 1204°C (2200°F) with a combustion gas flow rate of 6000 scfm 15°C (60°F). The incinerator is rectangular in shape and is 10 ft wide (across the front), 11 ft deep, and 28 ft high. The minimum residence time required for incineration is 2 s. Calculate (1) the actual combustion gas flow rate, (2) the volume required for minimum residence time, and (3) the maximum residence time.

SOLUTION

1. The actual gas flow rate is calculated by Charles' law:

$$Q_a = Q_s \left(\frac{T_a}{T_s}\right)$$

where T_a and T_s are in absolute units.

$$Q_a = 6000 \text{ scfm} \times \frac{2200 + 460}{60 + 460}$$

$$= 30{,}692 \text{ acfm}$$

2. The volume required for minimum residence time is

$$V_m = Q_a\theta = (30{,}692 \text{ acfm})(2 \text{ s}) \frac{\text{min}}{60 \text{ s}}$$
$$= 1023 \text{ ft}^3$$

3. The maximum residence time takes into account the actual incinerator volume V_a in the equation:

$$\theta_{max} = \frac{V_a}{Q_a}$$

Here

$$V_a = (10)(11)(28) = 3080 \text{ ft}^3$$
$$\theta_{max} = \frac{3080}{30{,}692}$$
$$= 0.1 \text{ min} = 6 \text{ s}$$

The minimum residence time requirement is satisfied. Note that the gas residence time, calculated by dividing the volume of the incinerator chamber by the combustion gas flow rate, is, however, an approximate value. Fully developed flow does not exist in a rectangular design. This does not include flow rate variations that arise due to chemical reaction and temperature changes through the unit. Thus a residence time distribution exists with real systems.

Example 7.8.2 An engineer involved in the analysis of a rotary kiln unit suspects that the kiln is operating below the desired residence time of 1 h. The available operating data are

Kiln length = 25 ft

Kiln diameter = 10 ft

Slope of kiln = 0.01 ft/ft of length

The present kiln rotation velocity is 0.9 rpm.

1. Is this velocity providing the necessary residence time?
2. What minimum kiln rotation velocity would meet the desired residence time?

SOLUTION

1. Equation (7.3.1) applies:

$$\theta = \frac{0.19L}{NDS}$$

where θ = retention time (min)
L = kiln length (ft)
N = kiln rotational velocity (rpm)
D = kiln diameter (ft)
S = kiln slope (ft/ft of length)

Calculate the residence time

$$\begin{aligned}\theta &= \frac{(0.19)(25\text{ ft})}{(0.9\text{ rpm})(10\text{ ft})(0.01\text{ ft/ft length})}\\ &= 53\text{ min}\end{aligned}$$

Therefore, the necessary residence time of 60 min is not being met.

2. Set the residence time of this equation to 60 min to find the minimum kiln rotation rate. Rearrange this equation and solve for N:

$$60\text{ min} = \frac{(0.19)(25\text{ ft})}{(N)(10\text{ ft})(0.01\text{ ft/ft length})}$$

$$\begin{aligned}N &= \frac{(0.19)(25\text{ ft})}{(60\text{ min})(10\text{ ft})(0.01\text{ ft/ft length})}\\ &= 0.8\text{ rpm}\end{aligned}$$

Therefore, a minimum rotation velocity of 0.8 rpm is required to achieve the desired residence time of 1 h in this rotary kiln.

The lead coefficient of 0.19 in this equation has been estimated from limited experimental data. Other values appear in the literature. Therefore, the residence time calculated through the use of this coefficient is, at the very best, a rough estimate.

Example 7.8.3 A 1000 scfm (60°F) effluent at 150°F needs to be treated to control a very low concentration of hazardous pollutant at the *parts-per-million* level. This is to be accomplished with a somewhat outdated thermal incinerator at 1200°F for at least 0.3 s.* The intake combustion air is at 60°F. Assuming a length-to-diameter ratio of 2.0 and a throughput velocity of 12 ft/s, calculate (1) the natural gas flow rate (NHV = 1059 Btu/scf) required for preheating the contaminated effluent to 1200°F using all fresh combustion air intake (primary air); (2) the afterburner throat diameter to give 20-ft/s throat velocity for good mixing (approximately 11.0 scf of combustion gases are produced from 1.0 scf of natural gas combusted); and (3) the length of the afterburner.

The available heat of a natural gas with an NHV of 1059 Btu/scf may be estimated from the equation[34]

$$\mathrm{HA}_T = -0.237T + 981$$

where HA_T is in Btu/scf and T is in °F.

SOLUTION

1. First, the waste effluent mass flow rate is calculated. The gas is assumed to have the physical properties of air:

$$w = Q_{\mathrm{a,w}}\rho = (1000\ \mathrm{scfm})(0.0766\ \mathrm{lb/scf})$$
$$= 76.6\ \mathrm{lb/min} = 4596\ \mathrm{lb/h}$$

The enthalpy of air at 1200 and 150°F is obtained from Table 5.2.2:

Enthalpy at 1200°F $\quad h_1 = 294.2$ Btu/lb

Enthalpy at 150°F $\quad h_2 = 28.4$ Btu/lb

The heat required to increase the effluent waste stream temperature from 150 to 1200°F, allowing for a 10% loss, is calculated as follows:

$$q = (1.\ 10)w(h_1 - h_2) = (1.10)(4596)(294.2 - 28.4)$$
$$= 1.34 \times 10^6\ \mathrm{Btu/h}$$

The available heat of the natural gas at 1200°F is next calculated:

$$\mathrm{HA}_T = -0.237T + 981 = 697\ \mathrm{Btu/scf}$$

It should be kept in mind that HA_T is the amount of heat remaining after

*The reader should note that some outdated incinerators may operate legally at 1200°F only because they are protected by a grandfather clause. With recent permitting requirements in most states, this temperature level has been revised for most installations. In modern fume incinerators, 1500 to 1600°F at a minimum of one second residence time is required for proper CO oxidation and 1800°F for halogenated hydrocarbon fumes.

the combustion products from 1 scf of gas are raised to the afterburner temperature (T). This heat is then available for heating the waste effluent to the afterburner temperature.

The natural gas (ng) flow rate required is determined as follows:

$$Q_{\mathrm{s,ng}} = \frac{q}{\mathrm{HA}_T} = \frac{1.34 \times 10^6}{697}$$

$$= 1.929 \times 10^3 \text{ scf/h}$$

2. The volumetric flow rate of combustion products at 1200°F is given by

$$Q_{\mathrm{a,c}} = Q_{\mathrm{a,ng}} \frac{11.0 \text{ scf of combustion gas}}{\text{scf ng}} \frac{460 + 1200}{460 + 60}$$

$$= \frac{1.929 \times 10^3 \text{ scf}}{\text{h}} \frac{11.0 \text{ scf combustion gas}}{\text{scf ng}} \quad (3.19)$$

$$= 6.774 \times 10^4 \text{ ft}^3/\text{h} = 18.8 \text{ ft}^3/\text{s}$$

The volumetric flow rate of waste effluent at 1200°F is calculated using Charles' law:

$$Q_{\mathrm{a,w,1200°F}} = (1000 \text{ scfm}) \frac{460 + 1200}{460 + 60}$$

$$= 3192 \text{ ft}^3/\text{min} = 53.2 \text{ ft}^3/\text{s}$$

The total volumetric flow rate of gases to the afterburner is given by

$$Q_{\mathrm{a,t}} = Q_{\mathrm{a,c}} + Q_{\mathrm{a,w,1200°F}} = 18.8 + 53.2$$

$$= 72.0 \text{ ft}^3/\text{s}$$

The afterburner throat diameter is calculated as follows:

$$A = \frac{\pi D_t^2}{4}$$

In addition,

$$A = \frac{Q_{\mathrm{a,t}}}{v_{\mathrm{t}}}$$

Therefore,

$$D_t = \left(\frac{4Q_{a,t}}{v_t \pi}\right)^{0.5} = \left[\frac{(4)(72.0)}{20\pi}\right]^{0.5}$$

$$= 2.14 \text{ ft}$$

The chamber diameter is given by

$$D = \left[\frac{(4)(72)}{12\pi}\right]^{0.5}$$

$$= 2.76 \text{ ft}$$

3. The length of the afterburner chamber is

$$L = 2D = (2)(2.76)$$

$$= 5.5 \text{ ft}$$

The residence time is now calculated to determine if it is greater than the required minimum residence time of 0.3 s.

$$\theta = \frac{L}{v_{\text{throughput}}} = \frac{5.5}{12}$$

$$= 0.46 \text{ s}$$

Since the residence time of 0.46 s is greater than the minimum required residence time of 0.3 s, this design is satisfactory. Because the hazardous waste concentration in the gas stream is very low (in the ppm range), there is no contribution to the heating value associated with the combustion process.

Note The effluent gas stream has been assumed to have the same physical and chemical properties as air. In addition, combustion of the fuel was accomplished with fresh (primary) air intake. Energy savings could be realized by employing secondary air (air from the waste effluent stream) for combustion. This would reduce fuel requirements to a considerable degree.

Example 7.8.4 An industrial organization proposes to build a hazardous waste liquid injection incinerator. The incinerator is 13.68 ft in diameter and 32 ft high. Approximately 100×10^6 Btu/h will be liberated during the combustion process. Determine if the incinerator volume is adequate. The local regulatory agency has suggested that a heat release rate of 25,000 Btu/h-ft^3 be employed in the calculation(s).

SOLUTION The furnace volume is calculated using the dimensions given:

$$V = \text{area} \times \text{height} = [(3.1417)(13.68^2)/4] \times (32)$$
$$= 4704 \text{ ft}^3$$

The energy release rate per cubic foot of the incinerator is

$$q_H = \frac{100 \times 10^6}{V} = \frac{100 \times 10^6}{4704}$$
$$= 2.13 \times 10^4 \text{ Btu/h-ft}^3$$

Since the calculated energy release rate is lower than the recommended heat release rate of 25,000 Btu/h-ft^3 for this liquid injection incinerator, the incinerator size is adequate.

Example 7.8.5 For preliminary design purposes, determine the size of an incinerator to operate with an average energy release rate (q_H) of 15,000 Btu/h-ft^3 of incinerator volume. It is estimated that during operation, 5000 lb/h of waste with an approximate heating value of 8000 Btu/lb is to be combusted. Assume the incinerator to be a rotary kiln with a *L*/*D* ratio of 3.5.

SOLUTION The amount of heat q released by the waste during operation is

$$q = (5000 \text{ lb/h})(8000 \text{ Btu/lb}) = 4 \times 10^7 \text{ Btu/h}$$

This result is used to calculate the furnace volume required:

$$V = \frac{q}{q_H} = \frac{4 \times 10^7}{15{,}000}$$
$$= 2666.67 \text{ ft}^3$$

The length and diameter of the kiln with a *L*/*D* ratio of 3.5 are calculated as follows:

$$V = \frac{\pi D^2}{4} L = \frac{\pi D^2}{4} (3.5D)$$
$$D = \left(\frac{V}{0.875\pi}\right)^{1/3} = \left(\frac{2666.67}{0.875\pi}\right)^{1/3}$$
$$= 9.897$$
$$L = (D)(3.5) = (9.897)(3.5)$$
$$= 34.64 \text{ ft}$$

Note The heat release rate of 15,000 Btu/h-ft^3 in this example is an average.

Typical kiln release rates vary from 5000 to 25,000 Btu/h-ft^3 dependent upon the wastes burned. For liquids only, the value is at the upper end. For dense and high surface area solids, the value is at the lower end.

PROBLEMS

1. A proposed design for a liquid injection incinerator is to be evaluated. The incinerator is 11 ft wide (across the front), 14 ft deep, and 29 ft high. Approximately 1.0×10^8 Btu/h will be liberated during the combustion process. Is the furnace volume adequate for the recommended heat release rate of 25,000 Btu/h-ft^3 for a liquid injection incinerator?

2. A rotary kiln with a *L/D* ratio of 2.45 is operating at temperatures high enough to achieve slagging. A residence time of 1 h is desired. The slope of the kiln should not exceed 0.01 ft/ft of length. (a) Determine the kiln rotation in rpm at the maximum allowable slope. (b) If the rotation rate in (a) was increased by 0.2 rpm, what would be the percentage change in the residence time?

3. A rotary kiln incinerator is operating with an average energy release rate (q_H) of 28,000 Btu/h-ft^3 of furnace volume. During operation, 4500 lb/h of waste with an approximate heat content of 8000 Btu/lb is to be combusted. Assume the *L/D* ratio of the rotary kiln to be 3.5. (a) Calculate the furnace volume required. (b) What are the dimensions of the kiln?

4. An exhaust process gas stream at 23,000 scfm (60°F) is to be heated from 200 to 1800°F. Assume that there are no heat losses and that the natural gas fuel has an available heat (HA) of 950 Btu/scf at 1800°F. The average heat capacity of air over the temperature range may be assumed to be equal to 7.4 Btu/lbmol-°F. (a) Determine the volumetric flow rate (in scfm) of natural gas. (b) Find the actual total products gas flow rate (in acfm) at the maximum final temperature.

REFERENCES

1. Telecon with U.S. EPA Office of Solid Wastes, Washington, DC, 1981.
2. Resource Conservation and Recovery Act, "Standards for Owners and Operators of Waste Facilities: Incinerators" 40 CFR 264, RCRA 3004, Jan. 23, 1981, Rev'd. June 24, 1982; April 1983; Jan. 4, 1985; Nov. 24, 1985; Feb. 21, 1991.
3. J. J. Santoleri, "Design and Operating Problems of Hazardous Waste Incinerators," *Environmental Progress*, **4**(4), 246–251, Nov. (1985).
4. K. Mitachi and H. Tsuruta, "Treatment of Liquid Wastes Derived from Chlorinated Hydrocarbon Plants," *Chemical Engineering*, 3, Oct. (1978).

5. J. J. Santoleri, "Dioxin Emissions—Effect of Chlorine/Time/Temperature at 300°C," 14th International Incineration Conference, Bellevue, WA, May 8–12, 1995.

6. J. S. Caprio and H. E. Wolfe, "Refractories for Hazardous Waste Incinerators—An Overview," ASME Solid Waste Processing Division Conference, May, 1982.

7. K. E. Peray, *The Rotary Cement Kiln*, Chemical Publishing Co., New York, 1986.

8. J. J. Santoleri, "Hazardous Waste Incineration Systems, Permit Procedures, Design and Operation," AWMA Continuing Education Course, WST-258 (Section 6), 84th Annual Meeting, Vancouver, BC, Canada, June, 1991.

9. Telecon and personal notes, J. J. Santoleri and D. L. Corwin, 1999.

10. Telecon and personal notes, J. Santoleri and T. F. McGowan, 1999.

11. T. Bonner, B. Desai, J. Fullenkamp, T. Hughes, E. Kennedy, R. McCormick, J. Peters, and D. Zanders, *Hazardous Waste Incineration Engineering*, Noyes Data Corp., Park Ridge, NJ, 1981.

12. J. F. Mullen, "Consider Fluid Bed Incineration for Hazardous Waste Destruction," *Chemical Engineering Progress*, American Institute of Chemical Engineers, New York, June, 1992.

13. D. G. Ackerman, J. F. Clausen, R. J. Johnson, and C. A. Zee. (TRW Defense and Systems Group), "Destroying Chemical Wastes in Commercial Scale Incinerators, Facility Report No. 3—Systems Technology," Prepared for U.S. EPA Office of Solid Waste Management, Washington, DC, NTIS No. PB-265-540, Apr. 1977.

14. "Part 373, Hazardous Waste Permit Application [Revision], General Electric Co., Selkirk, NY," New York State Dept. of Environmental Conservation, Albany, NY, May, 1987.

15. "Texas Eastman Trial Burn Program Emission Test Results, Final Report," prepared by Alliance Technologies Corp. for Texas Eastman Co., Longview, TX, Apr. 5, 1991.

16. E. T. Oppelt, "Air Emissions from the Incineration of Hazardous Waste," *Toxicology and Industrial Health*, **6**(5) (1990).

17. M. Horio, S. Mori, I. Furusawa, and S. Tamanuki, "Research and Development on NO, Emission Abatement in a Fluidized Bed Coal Combustor in Japan," Proceedings of the Sixth International Conference on Fluid Bed Combustion, sponsored by the U.S. Dept. of Energy, Assistant Secretary on Fossil Energy, Office of Coal Utilization, Washington, DC, Vol. 2, pp. 968–978, Apr. 9–11, 1980.

18. J. Tatebayashi, Y. Okada, K. Yano, and S. Ikeda, "Simultaneous NO_x and S02 Emission Reduction with Fluidized Bed Combustion," Proceedings of the Sixth International Conference on Fluid Bed Combustion, sponsored by the U.S. Dept. of Energy, Assistant Secretary on Fossil Energy, Office of Coal Utilization, Washington, DC, Vol. 2, pp. 986–995, Apr. 9–11, 1980.

19. G. Brem and J. J. S. Brouwers, "Mathematical Modeling of Staged Combustion in an AFBC," Proceedings of the 11th International Conference on Fluid Bed Combustion, American Society of Mechanical Engineers, pp. 449–458, Apr. 21–24, 1991.

20. E. T. Oppelt, "Incineration of Hazardous Waste—A Critical Review," *Journal of the Air Pollution Control Association*, **37**(5), 558–586, May (1987).

21. T. C. Ho, C. Chen, J. R. Hopper, and D. A. Oberacker, "Metal Capture During Fluidized Bed Incineration of Wastes Contaminated with Lead Chloride," Second International Conference on Toxic Combustion By-Products, Univ. of Utah, Mar. 26–29, 1990.
22. D. J. Fournier Jr., L. R. Waterland, and J. W. Lee, "The Behavior of Trace Metals in Rotary Kiln Incineration: Results of Incineration Research Facility Studies," 17th Annual Risk Reduction Engineering Laboratory (RREL) Hazardous Waste Research Symposium, U.S. Environmental Protection Agency, Cincinnati, OH, pp. 172–189, Apr. 9–11, 1991.
23. "Standards for the Disposal of Sewage Sludges, Proposed Rule," *Federal Register*, **54**(23), 5745–5902, Feb. 6, 1991.
24. U.S. EPA, Combustion Emissions Technical Resource Document, EPA 530-R-94-014, (CETRED) May, 1994.
25. C. Brunner, *Design of Sewage Sludge Incineration Systems,* Noyes Data Corp., Park Ridge, NJ, 1980.
26. L. A. Lundberg, B. D. Meckel, and N. J. Marchese, "Conversion of an Existing Multiple Hearth Furnace to a Fluidized Bed," AWMA—88th Annual Meeting, San Antonio, TX, June 1995, Paper No. 95-WP106.03.
27. T. Shen, U. Choi, and L. Theodore, *Hazardous Waste Incineration Manual, U.S. EPA Air Pollution Training Institute*, Research Triangle Park, NC.
28. L. P. MacDonald, D. J. Skinner, F. J. Hopton, and G. H. Thomas, "Burning Hazardous Waste Chlorinated Hydrocarbons in a Cement Kiln," Canadian Environmental Protection Service Report No. EAP-WP-77-22, March, 1977.
29. M. Benoit, E. Hansen, and T. Reese, U.S. Patent No. 4,850,290, "Method for Energy Recovery from Solid Hazardous Wastes," July 25, 1989.
30. Expanded Shale Clay and Slate Institute, *Lightweight Concrete—History, Applications and Economics*, Washington, DC, 1971.
31. T. F. McGowan, J. J. Santoleri, and B. M. Spratt, "A Comparison of Low and High Temperature Thermal Desorption for Superfund Sites," HMCRI Superfund XV, Washington, DC, Dec., 1994.
32. *The Superfund Innovative Technology Evaluation: Technology Profiles*, 4th ed., EPA/5405-91/008, Nov., 1991.
33. M. D. Springer and W. C. Burns, "Destruction of Medical Waste Using Plasma Energy," 1992 International Incineration Conference, Albuquerque, NM, May, 1992.
34. L. Theodore, personal notes, 1996.

8

WASTE HEAT BOILERS

8.1 INTRODUCTION

An obvious by-product of the waste incineration process is thermal energy—in most cases, a large amount of thermal energy. The total heat load generated by a typical waste incinerator is in the range of 10–150 million Btu/h. While waste heat boilers are capable of recovering 60–70% of this energy, the effort may or may not be justified economically. In assessing the feasibility of recovery, a number of factors must be taken into account; among these are the amount of heat wasted, the fraction of that heat that is realistically recoverable, the irregularity in availability of heat, the cost of equipment to salvage the heat, and the cost of energy. The last factor is particularly critical and may be the most important consideration in a decision whether or not to harness the energy generated by a particular waste incinerator. Other important considerations are the incinerator capacity and nature of the waste being handled. Generally, heat recovery on incinerators <2–7 million Btu/h may not be economical because of capital cost considerations. Larger capacity incinerators may also be poor candidates for heat recovery if steam is not needed at the plant site or if the combustion gases are highly corrosive; in the latter case, the maintenance cost of the heat recovery equipment may be prohibitive.

Because waste heat boilers are basically heat exchangers, the first part of this chapter covers some of the fundamentals of heat transfer. The remaining sections deal with the operation and design of waste heat boilers.

8.2 FUNDAMENTALS OF HEAT TRANSFER

A difference in temperature between two bodies in close proximity or between two parts of the same body results in a heat flow from the higher temperature to the lower temperature. There are three different mechanisms by which this heat transfer can occur: conduction, convection, and radiation. When the heat transfer is the result of molecular motion (e.g., the vibrational energy of molecules in a solid being passed along from molecule to molecule), the mechanism of transfer is *conduction.* When the heat transfer results from macroscopic motion, such as currents in a fluid, the mechanism is that of *convection.* When heat is transferred by electromagnetic waves, *radiation* is the mechanism.

In most industrial processes, more than one mechanism is usually involved in the transmission of heat. However, since each mechanism is governed by its own set of physical laws, it is beneficial to discuss them independently of each other.

Conduction As the temperature of a solid increases, the molecules that make up the solid experience an increase in vibrational kinetic energy. Since every molecule is bonded in some way to neighboring molecules, this energy can be passed through the solid. Thus, heating a wire at one end eventually results in raising the temperature at the other. This type of heat transfer is called conduction and is the principle mechanism by which solids transfer heat. Fluids are capable of transporting heat in a similar fashion. Conduction in a stagnant liquid, for example, occurs by the movement not only of vibrational kinetic energy, but also of translational kinetic energy as the molecules move throughout the body of the liquid. The ability of a fluid to flow, mix, and form internal currents on a macroscopic level (as opposed to the molecular mixing just described) allows fluids to carry heat energy by convection as well.

The rate of heat flow by conduction is given by Fourier's law:

$$q = -kA \frac{dT}{dx} \tag{8.2.1}$$

where q = heat flow rate (Btu/h)
x = direction of heat flow (ft)
k = thermal conductivity (Btu/h-ft-°F)
A = heat transfer area, a plane perpendicular to the x direction (ft^2)
T = temperature (°F)

The negative sign reflects the fact that heat flow is from high to low temperature, and therefore the sign of the derivative is opposite that of the heat flow. If the k can be considered constant over a limited temperature range (ΔT), Eq. (8.2.1) may be integrated to give

$$q = \frac{kA(T_h - T_c)}{x_h - x_c} = \frac{kA\,\Delta T}{L} \tag{8.2.2}$$

where T_h = higher temperature at point x_h (°F)
T_c = lower temperature at point x_c (°F)
$\Delta T = T_h - T_c$
L = distance between points x_h and x_c (ft)

It should be pointed out that the thermal conductivity (k) is a property of the material through which the heat is passing and, as such, does vary somewhat with temperature. Equation (8.2.2) should therefore be employed only for small values of ΔT with an average value of the conductivity used for k. Equation (8.2.1) is the basis for the definition of thermal conductivity as given in Chapter 4, that is, the amount of heat (Btu) that flows in a unit of time (1 h) through a unit area of surface (1 ft^2) of unit thickness (1 ft) by virtue of a unit difference in temperature (1°F). Values of k for some insulating solids are given in Table 8.2.1.

Equation (8.2.2) may be written in the form of the general rate equation:

$$\text{Rate} = \frac{\text{driving force}}{\text{resistance}} \tag{8.2.3}$$

Since q in Eq. (8.2.2) is the heat flow rate and ΔT the driving force, the L/kA term may be considered to be the resistance to heat flow. This approach is useful when heat is flowing by conduction through different materials in sequence. Consider, for example, a flat incinerator wall made up of three different layers, an inside insulating layer, a; a steel plate, b; and an outside insulating layer, c. The total resistance to heat flow through the incinerator wall is the sum of the three individual resistances:

$$R = R_a + R_b + R_c \tag{8.2.4}$$

Table 8.2.1 Thermal conductivities of some common insulating materials[a]

Material	k
Asbestos–cement boards	0.43
Sheets	0.096
Asbestos cement	1.202
Celotex	0.028
Corkboard, 10 lb/ft^3	0.025
Diatomaceous earth (Sil-o-cel)	0.035
Fiber, insulating board	0.028
Glass wool, 1.5 lb/ft^3	0.022
Magnesia, 85%	0.039
Rock wool, 10 lb/ft^3	0.023

[a] k = Btu/h-ft-°F.

At steady state, the rate of heat flow through the wall is, therefore, given by

$$q = \frac{T_a - T_d}{(L_a/k_aA_a) + (L_b/k_b/A_b) + (L_c/k_bA_c)} \tag{8.2.5}$$

where k_a, k_b, k_c = thermal conductivity of each section (Btu/h-ft-°F)
A_a, A_b, A_c = area of heat transfer of each section (ft) (these are equal for constant cross section of heat conduction)
L_a, L_b, L_c = thickness of each layer (ft)
T_a = temperature at inside surface of insulating wall a (°F)
T_d = temperature at outside surface of insulating wall d (°F)

In this example, the heat is flowing through a slab of constant cross section. In many cases of industrial importance, however, this is not the case. For example, in heat flow through the walls of a cylindrical vessel such as a rotary kiln incinerator, the heat transfer area increases with distance from the center of the cylinder. The heat flow in that case is given by

$$q = \frac{kA_{lm}\Delta T}{L} \tag{8.2.6}$$

In this equation A_{lm} represents the average heat transfer area, or more accurately, the log-mean average heat transfer area. The log-mean average can be calculated by

$$A_{lm} = \frac{A_2 - A_1}{\ln(A_2/A_1)} \tag{8.2.7}$$

where A_2 = outer surface area of cylinder (ft^2)
A_1 = inner surface area of cylinder (ft^2)

Convection When a pot of water is heated on a stove, the portion of water adjacent to the bottom of the pot is the first to experience a temperature rise. Eventually, the water at the top will also become hotter. Although some of the heat transfer from bottom to top is explainable by conduction through the water, most of the heat transfer is due to the second mechanism of heat transfer, convection. As the water at the bottom is heated, its density becomes lower. This results in convection currents as gravity causes the low-density water to move upwards to be replaced by the higher density cooler water from above. This macroscopic mixing is a far more effective mechanism for moving heat energy through fluids than conduction. This process is called *natural* convection because no external forces, other than gravity, need be applied to move the heat energy. In most industrial applications, however, it is more economical to speed up the mixing action by artificially generating a current by the use of a pump, agitator, or some other mechanical device. This is referred to as *forced* convection.

The flow of heat from a hot fluid to a cooler fluid through a solid wall is a situation often encountered in engineering equipment; examples of such equipment are heat exchangers, condensers, evaporators, boilers, and economizers. The heat absorbed by the cool fluid or given up by the hot fluid may be *sensible* heat, causing a temperature change in the fluid, or it may be *latent* heat, causing a phase change such as vaporization or condensation. In a typical waste heat boiler, for example, hot flue gas gives up heat to water through thin metal tube walls separating the two fluids. As the flue gas loses heat, its temperature drops. As the water gains heat, its temperature quickly reaches the boiling point where it continues to absorb heat with no further temperature rise as it changes into steam. The rate of heat transfer between the two streams, assuming no heat loss to the surroundings, may be calculated by the enthalpy change of either fluid:

$$q = w_h(h_{h1} - h_{h2}) = w_c(h_{c2} - h_{c1}) \tag{8.2.8}$$

where q = rate of heat flow (Btu/h)
w_h = mass flow rate of hot fluid (lb/h)
w_c = mass flow rate of cold fluid (lb/h)
h_{h1} = enthalpy of entering hot fluid (Btu/lb)
h_{h2} = enthalpy of exiting hot fluid (Btu/lb)
h_{c1} = enthalpy of entering cold fluid (Btu/lb)
h_{c2} = enthalpy of exiting cold fluid (Btu/lb)

Equation (8.2.8) is applicable to the heat exchange between two fluids whether a phase change is involved or not. In the waste heat boiler example, the enthalpy change of the flue gas is calculated from its temperature drop:

$$q = w_h(h_{h1} - h_{h2}) = w_h C_{Ph}(T_{h1} - T_{h2}) \tag{8.2.9}$$

where C_{Ph} = specific heat of hot fluid (Btu/lb-°F)
T_{h1} = temperature of entering hot fluid (°F)
T_{h2} = temperature of exiting hot fluid (°F)

The enthalpy change of the water involves a small amount of sensible heat to bring the water to its boiling point plus a considerable amount of latent heat to vaporize the water. Assuming all of the water is vaporized and no superheating of the steam occurs, the enthalpy change is

$$q = w_c(h_{c2} - h_{c1}) = w_c c_{Pc}(T_{c2} - T_{c1}) + w_c \lambda_c \tag{8.2.10}$$

where C_{Pc} = specific heat capacity of cold fluid (Btu/lb-°F)
T_{c1} = temperature of entering cold fluid (°F)
T_{c2} = boiling point of cold fluid (°F)
λ_c = heat of vaporization of cold fluid (Btu/lb)

In order to design a piece of heat transfer equipment, it is not enough to know the heat transfer rate calculated by the enthalpy balances described previously. The rate at which heat can travel from the hot fluid, through the tube walls, into the cold fluid, must also be considered in the calculation of the contact area, that is, the area of the tube walls is a direct function of this rate. The slower this rate is, for given hot and cold fluid flow rates, the more contact area is required. The rate of heat transfer through a unit of contact area is referred to as the *heat flux density* and, at any point along the tube length, is given by

$$\frac{dq}{dA} = U(T_h - T_c) \tag{8.2.11}$$

where dq/dA = local heat flux density (Btu/h-ft^2)
U = local overall heat transfer coefficient (Btu/h-ft^2-°F)

The use of the overall heat transfer coefficient (U) is a simple yet powerful concept. In most applications it combines both conduction and convection effects, although heat transfer by radiation can also be included. Methods for calculating the overall heat transfer coefficient are presented later in this section. In actual practice it is not uncommon for the vendors to provide a numerical value for U. A typical value for U for estimating heat losses from an incinerator is approximately 0.1 Btu/h-ft^2-°F.

With reference to Eq. (8.2.11) the temperatures T_h and T_c in Eq. (8.2.11) are actually averages. When a fluid is being heated or cooled, the temperature will vary throughout the cross section of the stream. If the fluid is being heated, its temperature will be highest at the tube wall and decrease with increasing distance from the tube wall. The average temperature across the stream cross section is, therefore, T_c, that is, the temperature that would be achieved if the fluid at this cross section was suddenly mixed to a uniform temperature. If the fluid is being cooled, on the other hand, its temperature will be lowest at the tube wall and increase with increasing distance from the wall.

In order to apply Eq. (8.2.11) to an entire heat exchanger, the equation must be integrated. This cannot be done unless the geometry of the exchanger is first defined. For simplicity, one of the simplest geometries will be assumed here—the double-pipe heat exchanger. This device consists of two concentric pipes. The outer surface of the outer pipe is well insulated so that no heat exchange with the surroundings may be assumed. One of the fluids flows through the center pipe, the other flows through the annular channel between the pipes. The fluid flows may be either co-current, where the two fluids flow in the same direction, or countercurrent, where the flows are in opposite directions. The countercurrent arrangement is more efficient and is more commonly used. For this heat exchanger, integration of Eq. (8.2.11) along the exchanger length, applying several simplifying assumptions, yields

$$q = UA\,\Delta T \tag{8.2.12}$$

The simplifying assumptions are that U and ΔT do not vary along the length. Since this is not actually the case, both U and ΔT must be regarded as averages of some type. A more careful integration of Eq. (8.2.11), assuming that only U is constant, would show that the appropriate average for ΔT in Eq. (8.2.12) is the log-mean average, that is,

$$\Delta T_{\text{lm}} = \frac{\Delta T_2 - \Delta T_1}{\ln(\Delta T_2/\Delta T_1)} \tag{8.2.13}$$

where ΔT_1 and ΔT_2 are the temperature differences between the two fluids at the ends of the exchanger. The area term (A) in Eq. (8.2.12) is the cylindrical contact area between the fluids. However, since a pipe of finite thickness separates the fluids, the cylindrical area (A) is not definite. Any one of the infinite number of areas between and including the inside and outside surface areas of the pipe may be arbitrarily chosen for this purpose. The usual practice is to use either the inside (A_i) or outside (A_o) surface areas; the outside area is the more commonly used of the two. Since the value of the overall heat transfer coefficient depends on the area chosen, it should be subscripted to match the area on which it is based. Equation (8.2.12) now becomes

$$q = U_o A_o \Delta T_{\text{lm}} \tag{8.2.14}$$

Comparing Eq. (8.2.14) to the general rate equation (8.2.3), it can be seen that $(U_o, A_o)^{-1}$ may be regarded as the resistance to heat transfer between the two fluids, that is,

$$R_t = \frac{1}{U_o A_o} \tag{8.2.15}$$

In practice, the flow of the fluids in a heat exchanger is turbulent, which means that the main bulk of the stream is well mixed and therefore of fairly uniform temperature. Immediately adjacent to the pipe wall, however, is a thin layer of fluid (referred to as the *boundary layer*) that is in laminar flow and is therefore not mixed. In the absence of macroscopic mixing, heat must be transferred through the film by conduction alone, and hence the laminar film presents a considerable contribution to the total resistance (R_t). The total resistance may therefore be divided into three contributions: the inside film, the tube wall, and the outside film:

$$R_t = R_i + R_w + R_o \tag{8.2.16}$$

or

$$\frac{1}{U_o A_o} = \frac{1}{h_i A_i} + \frac{x}{k A_{lm}} + \frac{1}{h_o A_o} \tag{8.2.17}$$

where h_i = inside film coefficient (Btu/h-ft^2 -°F)
h_o = outside film coefficient (Btu/h-ft^2 -°F)
A_o = outside surface area of pipe (ft^2)
A_i = inside surface area of pipe (ft^2)
A_{lm} = log-mean average surface area of pipe (ft^2)
x = pipe thickness (ft)
k = thermal conductivity of pipe (Btu/h-ft-°F)

Film coefficients are usually determined experimentally. Many empirical correlations can be found in the literature for a wide variety of types of fluids and exchanger geometries. Typical ranges of film coefficients are given in Table 8.2.2. For fully developed turbulent flow in smooth tubes, the following relationship is recommended by Dittus and Boelter[1] for determining the inside film coefficient, h_i:

$$\text{Nu} = 0.023\ \text{Re}^{0.8}\ \text{Pr}^{n} \tag{8.2.18}$$

where Nu = Nusselt number = $h_i D_i / k$ (dimensionless)
Re = Reynolds number = $D_i G/\mu$ (dimensionless)
Pr = Prandit number = $c_P \mu / k$ (dimensionless)

Table 8.2.2 Typical film coefficients[a]

	Inside Pipes	Outside Pipes[b,c]
Gases	10–50	1–3 (n), 5–20 (f)
Water (liquid)	200–2000	20–200 (n), 100–1000 (f)
boiling water	500–5000	300–9000
condensing steam		1000–10000
Nonviscous liquids	50–500	50–200 (f)
boiling liquid		200–2000
condensing vapor		200–400
Viscous liquids	10–100	20–50 (n), 10–100 (f)
condensing vapor		50–100

[a] h = Btu/h-ft^2-°F.
[b] (n) = natural convection.
[c] (f) = forced convection.

and

D_i = inside tube diameter (ft)
k = thermal conductivity of the fluid (Btu/h-ft-°F)
c_P = specific heat capacity of fluid (Btu/lb-°F)
μ = viscosity of fluid (lb/ft-h)
$G = w/A_{cs}$ = *fluid* mass flux (lb/h-ft^2)
n = constant (see below)
w = fluid mass flow rate (lb/h)
A_{cs} = inside cross-sectional area for fluid flow (ft^2)

The properties in Eq. (8.2.16) are evaluated at the fluid bulk temperature, and the exponent n has the following values:

$$n = 0.4 \text{ for heating}$$

$$n = 0.3 \text{ for cooling}$$

Heat transfer by natural convection is another mechanism that comes into play in calculating heat losses from incinerators. If the incinerator can be considered a stationary vertical cylinder, the heat transfer coefficient may be approximated by

$$h = 0.29\left(\frac{T_w - T_a}{L}\right)^{0.25}$$

where h = natural convection film coefficient (Btu/h-ft^2-°F)
T_w = temperature at incinerator wall (°F)
T_a = ambient air temperature (°F)
L = incinerator length (ft)

For stationary horizontal cylinders,

$$h = 0.27\left(\frac{T_w - T_a}{D}\right)^{0.25}$$

where D = incinerator diameter (ft).

For rotating horizontal cylinders (as with a rotary kiln), the film coefficient may be estimated by

$$h = 0.18(T_w - T_a)^{0.333}$$

After a period of service, thin films of foreign materials such as dirt, scale, or products of corrosion build up on the tube wall surfaces. As shown in Eqs.

(8.2.19) and (8.2.20), these films introduce added resistances to heat flow, R_{fi} and R_{fo} and reduce the overall heat transfer coefficient:

$$R_t = R_i + R_{fi} + R_w + R_{fo} + R_o \tag{8.2.19}$$

For this condition,

$$\frac{1}{U_o A_o} = \frac{1}{h_i A_i} + \frac{1}{h_{di} A_i} + \frac{x}{k A_{lm}} + \frac{1}{h_{do} A_o} + \frac{1}{h_o A_o} \tag{8.2.20}$$

where h_{di} = inside fouling factor (Btu/h-ft^2-°F)
h_{do} = outside fouling factor (Btu/h-ft^2-°F)

Typical values of fouling factors are shown in Table 8.2.3.

Radiation In the heat transfer mechanisms of conduction and convection, the movement of heat energy takes place through a material medium—a fluid in the case of conduction. This transfer medium is not required for the third mechanism, where the energy is carried by electromagnetic radiation. If a piece of steel plate is heated in a furnace until it is glowing red and then placed several inches away from a cold piece of steel plate, the temperature of the cold steel will rise, even if the process takes place in an evacuated container.

Radiation becomes important as a heat transfer mechanism only when the temperature of the source is very high. The driving force for conduction and convection is the temperature difference between the source and the receptor; the actual temperatures have only a minor influence. For these two mechanisms, it does not make too much difference whether the temperatures are 100 and 50°F or 500 and 450°F. Radiation on the other hand is strongly influenced by the temperature level; as the temperature level increases, the effectiveness of radiation as a heat transfer mechanism increases rapidly. It follows that, at very low temperatures, conduction and convection are the major contributors to the total heat transfer; at very high temperatures, radiation is the controlling factor.

Table 8.2.3 Typical fouling factors

	h_d[a]
Steam	1500–3000
Clean water	500–2000
Dirty water	150–500
Petroleum vapors and condensates	200–2000
Petroleum residuals and crudes	100–200

[a] h_d = Btu/h-ft^2-°F.

At temperatures in between, the fraction contributed by radiation depends on such factors as the convection film coefficient and the nature of the radiating surface. To cite two extreme examples,[2] for large pipes losing heat by natural convection, the temperature at which radiation accounts for roughly one half of the total heat transmission is around room temperature; for fine wires, this temperature is above a red heat. In an industrial boiler the radiation heat transfer from the burning gases to the solid surfaces inside the combustion chamber is considerable; in waste heat boilers, however, the contribution of radiation as a heat transfer mechanism is usually minor.

A perfect or ideal radiator, referred to as a *black body,* emits energy at a rate proportional to the fourth power of the absolute temperature of the body. When two bodies exchange heat by radiation, the net heat exchange is then proportional to the difference between each T^4. This is shown by Eq. (8.2.21), which is based on the Stefan–Boltzmann law of thermal radiation:

$$q = \sigma FA(T_h^4 - T_c^4) \tag{8.2.21}$$

where $\sigma = 0.1714 \times 10^{-8}$ Btu/h-ft^2-R^4 (Stefan–Boltzmann constant)
A = area of either surface (chosen arbitrarily) (ft^2)
F = view factor (dimensionless)
T_h = absolute temperature of the hotter body (°R)
T_c = absolute temperature of the colder body (°R)

The view factor (F) depends on the geometry of the system, that is, the surface geometries of the two bodies plus the spatial relationship between them, and on the surface chosen for A. Values of F are available in the literature for many geometries. It is important to note that Eq. (8.2.21) applies only to black bodies and is valid only for thermal radiation, that is, that type of electromagnetic radiation resulting from a temperature difference.

Other types of surfaces besides black bodies are not as capable of radiating energy, although the T^4 law is generally obeyed. The ratio of energy radiating from one of these *gray* bodies to that radiating from the black body under the same conditions is defined as the *emissivity* (ε). For gray bodies, Eq. (8.2.21) becomes

$$q = \sigma FF_\varepsilon A(T_h^4 - T_c^4) \tag{8.2.22}$$

where F_ε = emissivity function dependent on the emissivity of each body and the geometry of the system (dimensionless).

Relationships for F_ε for various geometries are available in the literature. In many applications, the system may be approximated by two long concentric cylinders where the energy exchange is between the outside surface of the inner cylinder (A_h) and the inside surface of the outside cylinder (A_c). An example would be heat radiation from the outside surface of a hot pipe to the surrounding air. In this case, the view factor (F) (based on either area) is 1.0 and

$$F_\varepsilon = 1 \bigg/ \left[\frac{1}{\varepsilon_h} + \frac{A_h}{A_c}\left(\frac{1}{\varepsilon_c} - 1\right)\right] \tag{8.2.23}$$

For radiation to the surrounding air, (A_h/A_c) approaches zero and Eq. (8.2.23) reduces to

$$F_\varepsilon = \varepsilon_h \tag{8.2.24}$$

Substitution for F and F_ε in Eq. (8.2.22) gives

$$q = \sigma\varepsilon_h A(T_h^4 - T_c^4) \tag{8.2.25}$$

In the development of convection heat transfer, the film heat transfer coefficient (h_c) was defined by

$$q_c = h_c A(T_h - T_c) \tag{8.2.26}$$

where T_h and T_c are the temperatures of the two bodies exchanging heat by convection. In the case of the pipe surface convecting energy to the surrounding air, T_h would be the outside pipe wall temperature and T_c the average (bulk) temperature for the air. Since radiation heat transfer often accompanies convection and the total heat transfer is the sum of both contributions, it is worthwhile to put both processes on a common basis by defining a radiation heat transfer coefficient (h_r) from

$$q_r = h_r A(T_h - T_c) \tag{8.2.27}$$

where q_r = the heat transfer rate by radiation[3] (Btu/h).

The total heat transfer is then the sum of the convection and radiation contributions.

$$q = (h_c + h_r)A(T_h - Tc) \tag{8.2.28}$$

For the example of the hot pipe radiating and convecting to the surrounding air, h_r can be evaluated by solving Eqs. (8.2.25) and (8.2.27) simultaneously to obtain

$$h_r = \sigma\varepsilon_h \frac{(T_h^4 - T_c^4)}{T_h - T_c} \tag{8.2.29}$$

Note: T_h and T_c are absolute temperatures in degrees Rankine (°R).

In most instances, the convection heat transfer coefficient is not strongly dependent on temperature; it should be apparent from Eq. (8.2.29) that this is not the case for the radiation heat transfer coefficient. At the gas temperatures normally employed in waste heat boilers the radiation coefficient is relatively small in comparison to the convection coefficient.

8.3 OPERATION OF WASTE HEAT BOILERS

The main purpose of a boiler is to convert a liquid, usually water, into a vapor. In most industrial boilers, the energy required to vaporize the liquid is provided by the direct firing of a fuel in the combustion chamber. The energy is transferred from the burning fuel by convection and radiation to the metal wall separating the liquid from the combustion chamber. Then, conduction through the metal wall and conduction–convection into the body of the vaporizing liquid takes place. In a waste heat boiler, no combustion occurs in the boiler itself; the energy for vaporizing the liquid is provided by the sensible heat of hot gases, which are usually product (flue) gases generated by a combustion process occurring elsewhere in the system. The waste heat boilers found at many incinerator facilities make use of the incineration flue gases for this purpose.

In a typical waste heat boiler installation, the water enters the unit after it has passed through a water treatment plant or the equivalent. This boiler feedwater is passed through heaters/economizers and then into a steam drum. Steam is generated in the boiler by indirectly contacting the water with the hot combustion (flue) gases. These hot gases are typically around 2000°F. The steam is separated from the water in the steam drum, may pass through a superheater, and is then available for internal use or export. The required steam rate for the process or facility and the steam temperature and pressure are the key design and operating variables on the water side. The inlet and outlet flue gas temperatures also play a role, but it is the chemical properties of the flue gas that can significantly impact on boiler performance. For example, acid gases arise due to the presence of chlorine in the waste. As described earlier, the principal combustion product of chlorine is hydrogen chloride, which is extremely corrosive to most metal heat transfer surfaces. This problem is particularly aggravated if the temperature of the flue gas is below the dew point temperature of HCl (i.e., the temperature at which the HCl will condense). This usually occurs at temperatures of about 300°F. In addition to acid gases, problems may also arise from the ash of hazardous waste incineration. Some wastes contain fairly high concentrations of alkali metal salts that have melting points below 1500°F. These lower melting point salts can slag and ultimately foul (and in some cases corrode) boiler tubes and/or heat transfer surfaces.

Boilers may be either fire tube or water tube (water-wall). Both are commonly used in hazardous waste incineration; the fire-tube variety is generally employed for smaller applications ($<50 \times 10^6$ Btu/h). In the fire-tube waste heat boiler, the hot gases from the incinerator are passed through the boiler tubes. The bundle of tubes are immersed in the water to be vaporized; the vaporizing water and tube bundle are encased in a large insulated container called a shell. The steam generated is stored in a surge drum usually located above the shell and connected to the shell through vertical tubes called *risers*. Because of construction constraints, steam pressure in fire-tube boilers is usually limited to around 1000 psia.

In the water-tube waste heat boiler, the water is contained inside the tubes and the hot flue gases flow through the tube bundle usually in a direction perpendicular to the tubes (cross flow). Because of the increased turbulence that accompanies cross flow, the overall heat transfer coefficient for water-tube boilers is higher than that for fire-tube boilers. This advantage is somewhat offset, however, because it is more difficult to clean the outside surfaces of the tubes than the inside surfaces. As a result, heat transfer losses and maintenance problems due to flue gas fouling tend to be greater in water-tube boilers.

8.4 DESIGN OF WASTE HEAT BOILERS

The design of waste heat boilers involves calculations that are based on energy balances and estimations of rates of heat transfer. Although some units operate in an unsteady state or cyclical mode, the calculational procedures are invariably based on steady-state conditions.

In heat transfer equipment, there is no shaft work, and potential and kinetic energy effects are small in comparison with the other terms in an energy balance equation. Heat flow to or from the surroundings is not usually desired in practice, and it is usually reduced to a small magnitude by suitable insulation. It is customary to consider this heat loss or gain negligible in comparison to the heat transfer through the walls of the tubes from the hot combustion gases to the water in the boilers. Thus, all the sensible heat lost by the hot gases may be assumed transferred to the steam. In this case, Eq. (8.2.8) may therefore be applied.

The heart of the heat transfer calculation is the heat flux, which is based on the area of the heating surface and is a function of the temperature difference driving force and the overall heat transfer coefficient, as shown in Eq. (8.2.11). This relatively simple equation can be used to estimate heat transfer rates, area requirements, and temperature changes in a number of heat transfer devices. However, problems develop if more exact calculations are required. The properties of the fluid (viscosity, thermal conductivity, specific heat, and density) are important parameters in these calculations. Each of these, especially viscosity, is temperature dependent. Since a temperature profile in which the temperature varies from point to point exists in a flowing stream undergoing heat transfer, a problem appears in the choice of temperature at which the properties should be evaluated. When temperature changes within the stream become large, the difficulty of calculating heat transfer quantities is increased. Because of these effects, the entire subject of heat transfer to fluids with phase change (as in a waste heat boiler) is complex and in practice is treated empirically rather than as a general theory.

The rigorous design and/or performance evaluation of a waste heat boiler is an involved procedure. Fortunately, several less rigorous methods are available in the literature. One approach that is fairly simple and yet reasonably accurate is that devised by Ganapathy.[4] This method provides a technique for sizing waste heat boilers of the fire-tube type and involves the use of a performance evaluation chart (see Fig. 8.4.1) that is based on fundamental heat transfer equations plus some simplifying assumptions.

Details regarding Ganapathy's method are now discussed. Equations (8.2.9), (8.2.13), and (8.2.14) may be combined and applied to the boiler where the temperature outside the tubes (T_c) is constant due to the water-to-steam phase change. The result is

$$\begin{aligned} q &= w_h c_{Ph}(T_{h1} - T_{h2}) \\ &= U_i A_i \frac{(T_{h1} - T_c) - (T_{h2} - T_c)}{\ln[(T_{h1} - T_c)/(T_{h2} - T_c)]} \end{aligned} \tag{8.4.1}$$

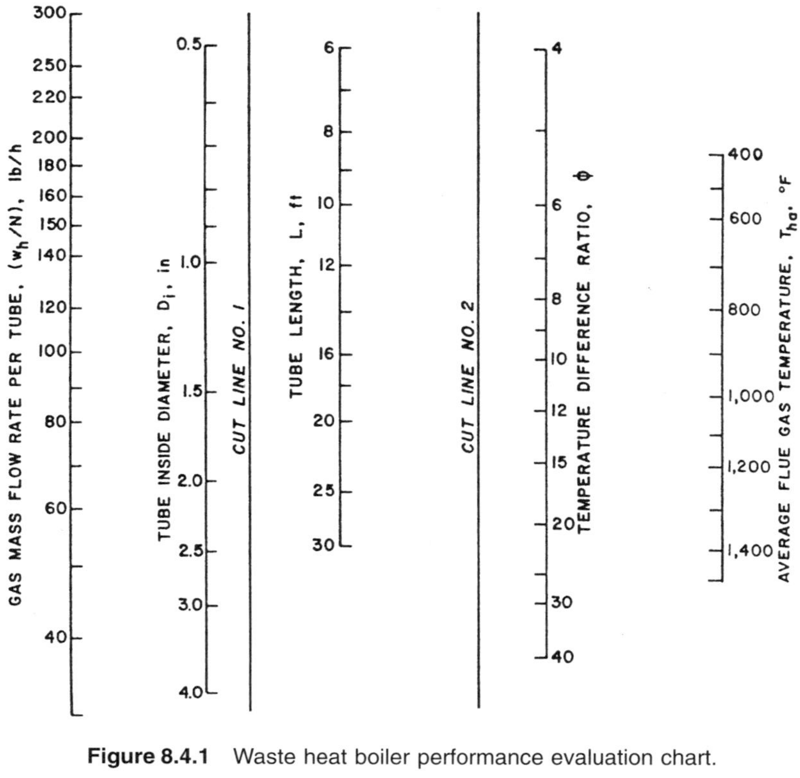

Figure 8.4.1 Waste heat boiler performance evaluation chart.

where U_i = overall heat transfer coefficient based on the inside area (Btu/h-ft^2-°F)
A_i = inside heat transfer area (ft^2)
T_c = boiling temperature of the water (°F)

Note that U_iA_i was used in Eq. (8.4.1) instead of U_oA_o. The two expressions are equal. In this procedure it is slightly more convenient to use the inside heat transfer area rather than the outside as the basis for U. Because the outside film coefficient (h_o) associated with a boiling liquid is much greater than the inside coefficient for the flue gas, the inside resistance predominates and is responsible for about 95% of the total resistance. Equation (8.2.17) may be simplified to the approximate relationship

$$U_i = 0.95h_i \tag{8.4.2}$$

The total inside heat transfer area is given by

$$A_i = \pi D_i NL \tag{8.4.3}$$

where D_i = tube inside diameter (ft)
N = total number of tubes
L = tube length (ft)

Equation (8.2.18) may be solved explicitly for h_i:

$$h_i = 0.023\,\frac{k^{0.6}G^{0.8}c_P^{0.4}}{D_i^{0.2}\,\mu^{0.4}} \tag{8.4.4}$$

and the gas mass flux (G) may be expressed in terms of w_h:

$$G = \frac{4w_h}{D_i^2 N} \tag{8.4.5}$$

Equations (8.4.1) through (8.4.5) may be combined to give

$$\ln\left(\frac{T_{h1} - T_c}{T_{h2} - T_c}\right) = \frac{U_iA_i}{w_h c_P} = \frac{CLF(T)}{D_i^{0.8}(w_h/N)^{0.2}} \tag{8.4.6}$$

where $F(T) = k^{0.6}/c_P^{0.6}\,\mu^{0.4}$
C = constant equal to 0.0833 when D_i is in feet, or 0.608 when D_i is in inches

Keeping in mind that this procedure is to be used for engineering design purposes and is not intended for rigorous analytical calculations, the value of

$F(T)$ does not, in practice, vary over a very large range for most incinerator flue gas streams. The performance evaluation chart shown in Fig. 8.4.1 is based on Eq. (8.4.6) for a typical value of $F(T)$.

The following describes a procedure for the use of the chart as a design tool. It is assumed in this procedure that the tube inside diameter and the number of tubes have been chosen and the tube length is to be determined. Illustrative problems in Section 8.5 further demonstrate the use of Fig. 8.4.1.

1. From the inlet and desired outlet gas temperatures, and the water saturation (boiling) temperature, calculate the arithmetic average gas temperature (T_{ha}) and the value of the temperature difference ratio (ϕ) where

$$\phi = \frac{T_{h1} - T_c}{T_{h2} - T_c} \tag{8.4.7}$$

   Mark those points on the appropriate axes.
2. Draw a straight line connecting these points, and extend the line to the left to cut line No. 2. Mark this point B.
3. Mark the value of the flue gas mass flow rate per tube (w_h/N) and the value of the inside diameter on the appropriate axes. [For design purposes, a good starting value of (w_h/N) is 80–150 lb/h.]
4. Draw a straight line connecting these points and extend the line to the right to cut line No. 1. Mark this point A.
5. Connect points A and B by a straight line. The intersection of this line with the L axis yields the appropriate tube length.

The chart could be used in a similar fashion as a performance evaluation tool. In this case, the outlet gas temperature would be unknown and hence would have to be estimated in order to determine the average gas temperature. In order to avoid a time-consuming trial-and-error procedure involving the gas outlet temperature and the average gas temperature, Eq. (8.4.8) may be used to estimate T_{ha} without too much loss of accuracy:

$$T_{ha} = 0.6(T_{h1} + T_c) \tag{8.4.8}$$

8.5 ILLUSTRATIVE EXAMPLES

Example 8.5.1 An incinerator is enclosed with flat walls each constructed of a 5-in. layer of a special firebrick (thermal conductivity = 0.028 Btu/h-ft-°F) backed by an 8-in. layer of brick (conductivity = 0.82 Btu/h-ft-°F). The inside and outside surface temperatures of the wall are 1900 and 190°F, respectively. (a) Calculate the heat loss per unit area (heat flux) through the wall, and (b) the temperature at the brick–firebrick interface.

SOLUTION

1. For 1 ft^2 of wall, the resistances are

$$R_f = \frac{L_f}{k_f A} = \frac{\frac{5}{12}}{0.028(1)} = 14.88$$

$$R_b = \frac{\frac{8}{12}}{0.82(1)} = 0.813$$

$$R(\text{total}) = R_f + R_b = 15.69 \text{ h-°F/Btu}$$

where the subscripts f and b refer to the firebrick and the brick, respectively.

The temperature drop across the wall is

$$\Delta T = 1900 - 190 = 1710\text{°F}$$

The heat loss for 1 ft^2 of area is given by

$$q = \frac{\Delta T}{\Sigma R} = \frac{1710}{15.69} = 109 \text{ Btu/h}$$

or

$$q/A = 109 \text{ Btu/h-ft}^2$$

2. For the brick layer,

$$\Delta T_b = qR_b = (109)(0.813) = 89\text{°F}$$

At the interface,

$$T = 190 + 89 = 279\text{°F}$$

Example 8.5.2 Estimate the heat loss in Btu/h from a hazardous waste incinerator that is rated at approximately 60×10^6 Btu/h. The incinerator is a 14-ft i.d., 40-ft long carbon steel rotary kiln. Six inches of firebrick line the inside of the incinerator followed by 4 in. of insulating brick. The outer shell is 1-in. thick carbon steel. If the flue gas velocity and incinerator operating temperature are 15 ft/s and 1600°F, respectively, calculate the heat loss from the unit. Neglect the rotational effect of the incinerator and assume the ambient temperature to be 45°F. Average thermal conductivity data (independent of temperature) are provided here by the vendor.

$$k \text{ (firebrick)} = 1.09 \text{ W/m-°C}$$

$$k \text{ (insulating brick)} = 0.26 \text{ Btu/h-ft-°F}$$

SOLUTION The describing equation is obtained from Eqs. (8.2.5) and (8.2.14):

$$q = \frac{\Delta T}{\Sigma R} = U_o A_o \, \Delta T$$

where ΔT = difference between incinerator operating temperature and ambient air temperature
ΣR = sum of five resistances to heat flow

The five resistances that need to be examined are

$$R \text{ (flue gas film)} = \frac{1}{h_i A_i} \tag{8.5.1}$$

$$R \text{ (firebrick)} = \frac{L_f}{k_f A_{lm,f}} = \frac{\ln(r_{fo}/r_{fi})}{2\pi L' k_f} \tag{8.5.2}$$

Note

$$A_{lm,f} = 2\pi L' \left(\frac{r_{fo} - r_{fi}}{\ln(r_{fo}/r_{fi})} \right)$$

$$R \text{ (insultating brick)} = \frac{\ln(r_{bo}/r_{bi})}{2\pi L' k_b} \tag{8.5.3}$$

$$R \text{ (carbon steel)} = \frac{\ln(r_{co}/r_{ci})}{2\pi L' k_c} \tag{8.5.4}$$

$$R \text{ (ambient air film)} = \frac{1}{h_o A_o} \tag{8.5.5}$$

where L_f = thickness of firebrick
k_f = thermal conductivity of firebrick
$A_{lm,f}$ = log-mean heat transfer area of the firebrick
L' = length of incinerator
r_{fi}, r_{fo} = distance from incinerator axis to inside and outside surfaces of firebrick, respectively
f, b, c = subscripts representing firebrick, insulating brick, and carbon steel, respectively

Based on the geometry of the incinerator and the thicknesses given in the problem statement:

$$r_{\text{fo}} = 6.667 \text{ ft} \qquad r_{\text{fi}} = 6.167 \text{ ft}$$

$$r_{\text{bo}} = 7.000 \text{ ft} \qquad r_{\text{bi}} = 6.667 \text{ ft}$$

$$r_{\text{co}} = 7.083 \text{ ft} \qquad r_{\text{ci}} = 7.000 \text{ ft}$$

The thermal conductivities are converted to engineering units as follows:

$$k \text{ (firebrick)} = 1.09 \text{ W/m °C} \qquad = 0.608 \text{ Btu/h-ft-°F}$$

$$k \text{ (insulating brick)} = 0.260 \text{ Btu/h-ft-°F}$$

The resistance associated with the steel wall [Eq. (8.5.4)] is obviously close to zero. The inside (flue gas) film resistance [Eq. (8.5.1)] is difficult to quantify. A convective coefficient can be calculated from the Dittus–Boelter equation. The companion radiation coefficient will be at least an order of magnitude greater than the convective coefficient, and nearly impossible to accurately calculate. The combined coefficients, however, will produce a resistance that will be negligible relative to that of the two bricks. The outside resistance [Eq. (8.5.5)] can be estimated from a free convection coefficient. A trial-and-error calculation is necessary since the outside temperature of the steel is required. This resistance may, however, also be neglected relative to the two main resistances. It is therefore concluded that only the firebrick [Eq. (8.5.2)] and insulating brick [Eq. (8.5.3)] resistances need be calculated. The two (primary) resistances are

$$R[\text{Eq. (8.5.2)}] = \frac{\ln(6.667/6.167)}{2\pi(0.608)(40)} = 5.10 \times 10^{-4} \text{ h-°F/Btu}$$

$$R[\text{Eq. (8.5.3)}] = \frac{\ln(7.000/6.667)}{2\pi(0.26)(40)} = 7.46 \times 10^{-4} \text{ h-°F/Btu}$$

The sum of the two resistances (ΣR) is given by

$$\Sigma R = R[\text{Eq. (8.5.2)}] + R[\text{Eq. (8.5.3)}] = 12.56 \times 10^{-4} \text{ h-°F/Btu}$$

The heat loss is therefore

$$q = \frac{\Delta T}{\Sigma R} = \frac{1600 - 45}{12.56 \times 10^{-4}} = 1{,}238{,}057 \text{ Btu/h}$$

This indicates that nearly 2% of the heat generated in the incinerator is lost across the walls. One can conclude that, based on these calculations, the unit is well insulated since losses from most industrial incinerators have been reported in the 2–4% range. Note that many designs today do not use a backup insulating brick behind the firebrick. This results in a thicker firebrick lining requirement. However, the resultant heat losses will be higher in the range of 3–5%.

Example 8.5.3 120,000 lb/h of flue gas from a hazardous waste incinerator is to be cooled from 1800 to 500°F in a waste heat boiler. If 2-in. i.d. tubes and a flow rate of 150 lb/h through each tube is to be used, estimate the required heat transfer area, tube length, heat duty, and water mass flow rate. Water at 200°F is available for the steam generator; saturated steam at 80 psia is needed. The average specific heat of the flue gas is 0.26 Btu/lb-°F. Use Ganapathy's method.

SOLUTION From steam tables or Table 4.5.1, the boiling point of water at 80 psia is 312°F. The temperature difference ratio (ϕ) is

$$\phi = \frac{T_{h1} - T_c}{T_{h2} - T_c} = \frac{1800 - 312}{500 - 312} = 7.9$$

The average gas temperature is

$$T_{ha} = 1/2(T_{h1} + T_{h2}) = 1/2(1800 + 500) = 1150°\text{F}$$

On the nomograph (Fig. 8.4.1), connect $(W_h/N) = 150$ with $D_i = 2$ and extend the line to intersect cut line No. 1 at point A. Connect ϕ (the temperature difference ratio) = 7.9 with $T_{ha} = 1150$ and extend the line to intersect cut line No. 2 at point B. The line connecting points A and B intersects the L line close to 21.5 ft.

The number of tubes (N) is

$$\frac{w_h}{w_h/N} = \frac{120{,}000}{150} = 800$$

The total heat transfer area (inside) is given by

$$A_i = NL\pi D_i = (800)(21.5)\left(\tfrac{2}{12}\right) = 9005 \text{ ft}^2$$

Note: This is only one design; several alternatives are possible by changing (w_h/N) or L. If L has to be limited because of space constraints, the nomograph may be used in reverse to calculate (w_h/N). The heat duty from Eq. (8.2.9) is

$$q = w_h c_P(T_{h1} - T_{h2}) = (120{,}000)(0.26)(1800 - 500)$$
$$= 4.06 \times 10^7 \text{ Btu/h}$$

The required water flow rate is obtained from an enthalpy balance. The enthalpy of water at 200°F is 168.1 Btu/lb. Note that, since pressure has very little effect on the enthalpy of liquids, the enthalpy of liquid water at 200°F and ambient air pressure is essentially the same as that of liquid water at 200°F and its saturation pressure. The value of h (168.1 Btu/lb) was obtained from the saturated steam tables. The enthalpy of steam at 80 psia is 1183.1 Btu/lb. The required water flow is, therefore,

$$w_c = \frac{q}{h_2 - h_1} = \frac{4.06 \times 10^7}{1183.1 - 168.1} = 40{,}000 \text{ lb/h}$$

Example 8.5.4 If 90,000 lb/h of flue gas at 1400°F is used in the waste heat boiler designed in Example 8.5.3, calculate the exit gas temperature, duty, and steam rate. Assume that the average specific heat of the flue gas is the same.

SOLUTION Calculate w_h/N.

$$\frac{w_h}{N} = \frac{90{,}000 \text{ lb/h}}{800 \text{ tubes}} = 112.5 \text{ lb/h per tube}$$

Connect $w_h/N = 112.5$ with $D_i = 2$ and extend the line to point A on cut line No. 1. Connect point A with $L = 21.5$ and extend the line to point B on cut line No. 2. Since the outlet gas temperature is not known, the average gas temperature can be approximated using Eq. (8.4.8):

$$T_{ha} = 0.6(T_{h1} + T_c) = 0.6(1400 + 312) = 1027°\text{F}$$

Connect point B with $T_{ha} = 1027$. This line intersects the ϕ line at about 8.7. Calculate T_{h2} from

$$8.7 = \frac{1400 - 312}{T_{h2} - 312}$$

which yields a value of 437°F. The duty (q) and steam mass flow rate (w_c) are calculated as in Example 8.5.3:

$$q = (900{,}000)(0.26)(1400 - 437) = 2.25 \times 10^7 \text{ Btu/h}$$

$$w_c = \frac{2.25 \times 10^7}{(1183.1 - 168.1) = 22{,}200 \text{ lb/h}}$$

PROBLEMS

1. Using the resistances calculated in Example 8.5.2, determine the maximum temperature of the 1-in. carbon steel incinerator wall.

2. A flat incinerator wall with a surface area of 480 ft^2 consists of 6-in. of firebrick with a thermal conductivity of 0.61 Btu/h-°F and an 8-in. outer layer of rock wool insulation with a density of 10 lb/ft^3 (see Table 8.2.1). If the temperature of the inside face of the firebrick and the outside surface of the rock wool insulation are 1900 and 140°F, respectively, calculate (a) the heat loss through the wall in Btu/h and (b) the temperature at the interface between the firebrick and the rock wool.

3. A rotary kiln incinerator is 30 ft long, has a 12-ft i.d., and is constructed of ¾-in. carbon steel. The inside of the steel shell is protected by 10 in. of firebrick (k = 0.608 Btu/h-ft-°F) and 5 in. of Sil-o-cel insulation (see Table 8.2.1) covers the outside. The ambient air temperature is 85°F and the average inside temperature is 1800°F. The present heat loss through the furnace wall is 6% of the heat generated by combustion of the waste. Calculate the thickness of Sil-o-cel insulation that must be added to cut the losses to 3%.

4. A 12-ft-long waste heat boiler contains 500 1-in. tubes through which 72,500 lb/h of hot incinerator flue gas at 1700°F (inlet temperature) flows in parallel. The boiler is to generate steam at 70 psia, and the available water may be assumed to be at its boiling point for that pressure. The average specific heat capacity of the flue gas is 0.30 Btu/h-ft-°F. Using Ganapathy's method, estimate (a) the outlet temperature of the flue gas and (b) the rate of steam generation in pounds per hour.

5. Using Ganapathy's method, determine the required length of a waste heat boiler to be used to cool hot gases (average specific heat = 0.279 Btu/h-ft-°F) from 2000 to 550°F and generate 30,000 lb/h of steam at 330°F from water at 140°F. The boiler contains 800 ½-in. i.d. tubes.

REFERENCES

1. F. W. Dittus and L. M. K. Boelter, *Univ. Calif. (Berkeley) Pub. Eng.*, **2**, 1930.
2. R. Perry and D. Green, *Perry's Chemical Engineering Handbook*, 7th ed., McGraw-Hill, New York, 1997.
3. J. P. Holman, *Heat Transfer*, 5th ed., McGraw-Hill, New York, 1963.
4. V. Ganapathy, "Size or Check Waste Heat Boilers Quickly," *Hydrocarbon Processing*, 169–170, September (1984).

9

QUENCHERS

9.1 INTRODUCTION

Hot gases from combustion operations must be cooled before entering air pollution control devices, which normally are not designed for very high temperature operation (>500°F). These gases are usually cooled either by recovering the energy in a waste heat boiler, as discussed in Chapter 8, or by quenching. Both methods may be used in tandem. For example, when scrubbing is used for air pollution control, a waste heat boiler can reduce the incinerator exit gas temperature down to about 500°F; a water quench can then be used to further reduce the gas temperature to around 200°F, as well as saturate the gas with water. This secondary cooling and saturation eliminates the problem of water evaporation in the scrubbing system and also alleviates potential problems associated with particulate generation in caustic scrubbers.

Although quenching and the use of a waste heat boiler are the most commonly used methods for combustion gas cooling in hazardous waste incinerator applications, there are several other techniques for cooling the hot gases. All methods may be divided into two categories: *direct-contact* and *indirect-contact* cooling. The direct-contact cooling methods include (a) dilution with ambient air, (b) quenching with water, and (c) contact with high heat capacity solids. Among the indirect contact methods are (d) natural convection and radiation from ductwork, (e) forced-draft heat exchangers, and (f) waste heat boilers. With the *dilution* method, the hot gaseous effluent from the incinerator is cooled by adding sufficient ambient air to result in a mixture of gases at the desired temperature. The *water quench* method uses the heat of vaporization of water to cool the gases. When water is sprayed into the hot gases under conditions conducive to evaporation, the energy contained in the gases evaporates the water, and this results in a cooling of the gases. The hot exhaust gases from an incinerator may also be quenched using submerged exhaust quenching. This is another technique employed by a few incinerator vendors. In the *solids contact* method, the hot gases are cooled by giving up heat to a bed of ceramic elements. The bed in turn is cooled by incoming air to be used in the incinerator. *Natural convection and radiation* occur whenever there is a temperature difference between the gases inside a duct and the atmosphere surrounding it. Cooling hot gases by this method requires only the provision of enough heat transfer area to obtain the desired amount of cooling. In *forced-draft heat exchangers,* the hot gases are cooled by forcing cooling fluid past the barrier separating the fluid from the hot gases.

Of these five methods (excluding waste heat boilers), the most commonly used is water quenching—either spray quench or submerged exhaust quench. The spray quench method is given extensive coverage in this chapter. Because the other four techniques are also occasionally used, some attention is given to them also.

9.2 DILUTION WITH AMBIENT AIR

The cooling of gases by dilution with ambient air is the simplest method that can be employed. Essentially, it involves the mixing of ambient air with a gas of known volume and temperature to produce a low-temperature mixture that can be admitted to the air pollution control device. In designing such a system, the volume and temperature of gas necessary to ensure particulate capture and the amount of ambient air required to provide a gaseous mixture of the desired temperature are first determined.

Although little instrumentation is required, a gas temperature indicator with a warning device, at the very least, should be used ahead of the air pollution control device to ensure that no damage occurs owing to sudden, unexpected surges of temperature. The instrumentation may be expanded to control either the fuel input to the incinerator or the volume of ambient air to the exhaust system.

The amount of air needed is that necessary to cool the gases to approximately 400°F, which then permits the use of high-temperature air pollution control devices. When the volume of the hot gases is small, this method may be used economically. When large volumes of hot gases require cooling, as is normally the case with hazardous waste incinerators, the size of the control device becomes excessively large with dilution cooling. As a result, this method is rarely used in hazardous waste incineration (HWI) applications.

The quantity of air required for cooling may be calculated directly from an enthalpy balance, that is, the heat "lost" by the combustion (flue) gas is equal to that "gained" by the dilution air. The design of the vessel or duct to accomplish this mixing process is based on a residence time of approximately 0.8–1.5 s (based on the combined flow rate) at the average temperature.

9.3 QUENCHING WITH LIQUIDS

When a large volume of hot gas is to be cooled, a method other than dilution with ambient air should be used. This is usually the case in HWI, and the cooling method most often used is quenching.

Cooling by liquid quenching is essentially accomplished by introducing the hot gases into a liquid (usually water) contacting device. When the water evaporates, the energy necessary to vaporize the water is obtained at the expense of the hot combustion gas, resulting in a reduction in the gas temperature. The temperature of the combustion gases discharged from the quencher is at the adiabatic saturation temperature of the combustion gas if the operation is adiabatic and the gas leaves the quencher saturated with water vapor. (A saturated gas contains the maximum water vapor possible at that temperature; any increase in water content will result in condensation.) Simple calculational and graphical procedures are available for estimating the adiabatic saturation temperature of a gas (see Chapter 4).

There are four types of quenchers that may be employed: spray towers, venturi scrubbers, submerged exhaust tanks, and packed towers (generally a poor choice). The venturi scrubber is actually an air pollution control device but can also be used simultaneously for both particulate removal from, and quenching of, the hot incinerator gases. When a venturi scrubber is the primary air pollution control device on an incinerator, precooling of the gases is desirable to prevent submicron particulate from forming as recycled liquors contact the high-temperature gases. One vendor has trademarked their design as Sub-X (submerged exhaust). The four types of quenchers are discussed next.

Spray Towers

In one spray tower design, the quench cooling is accomplished simply by spraying water at the top of the tower as the hot gases travel upwards through the tower. When the unvaporized water reaches the sump at the bottom of the tower, it is recirculated by pumping back up to the top. Since about 10% of the water stream is vaporized during contact with the hot gases, makeup water must constantly be added. In the wet bottom towers, a level control is used to maintain required flow. For efficient evaporation of the water, the gas velocity should be from 400 to 600 fpm (ft/min) and the entire cross section of the gas stream should be covered with a fine spray of water. This necessitates proper location, type, and number of spray nozzles to ensure adequate coverage of the gas-side flow area.

Water spray pressures generally range from 50 to 150 psig; however, to reduce the amount of moisture collected, some installations have employed pressures as high as 400 psig. These pressure levels may be used for a clean filtered water spray. Recycled streams containing solids will tend to plug orifices of spray nozzles that require high pressures. Recycled stream sprays should utilize larger orifices or a dual fluid spray similar to those described in

Chapter 7 for the aqueous waste atomizers. Since the moisture collected in spray towers readily corrodes steel, these devices are frequently lined with materials resistant to corrosion. Since the gases discharged from the incinerator are exceptionally hot, the first portion of the duct carrying the gases from the incinerator to the quench tower are usually refractory lined or made from high-temperature alloys resistant to the corrosion from the acid gases and the hot–cold interface that exists in these towers. If the gases contain HCl, stainless steel is not recommended. Initial quenching should be done with fresh filtered water rather than the recycle stream. By cooling the gases to levels below 1000°F initially, any ash or salts that may be contained in the recycle stream will not be impacted by the lower temperatures. Experience has shown that gases at 1800°F and higher will create submicron particulate from the salts contained in the recycle stream. This will only add to the requirements of any downstream air pollution control system.[1]

For controlling the gas temperature leaving the quench chamber, a flow controller is generally used to regulate the amount of water sprayed into the quench chamber. For emergency conditions, a high-high temperature switch on the exit temperature recording monitor should activate an emergency water valve (from a fire tower) or open an emergency vent valve to divert the hot gases away from the air pollution control device into the emergency bypass stack.

Cooling hot gases with a water quench is relatively simple and requires very little space. Quench towers are easy to operate and, with automatic temperature controls, only that amount of water is used that is needed to maintain the desired temperature of the gases at the discharge. Their installation and operating costs are generally considered to be less than that for other cooling methods. Quench towers require careful design review for applications where the gases to be cooled contain a large amount of gases or fumes containing acid gases. These will be highly corrosive when condensed in the tower or along the inside wall of the vessel behind any protective insulation. Towers handling these gases are usually brick lined to protect the shell from the high temperature with an elastomer lining on the steel shell to serve as corrosion protection from the condensing acids. This requires careful planning to minimize maintenance problems, not only in the tower itself but in the remainder of the ductwork, the control device, and the blower.

Design techniques of spray towers are available in the literature. The following procedure, due in part to Fair[2,3] and in part to Farrell,[4] is recommended.

Before the spray tower can be sized, the water flow rates necessary to bring the flue gas to the desired temperature must first be fixed. This is accomplished by means of energy and material balances around the tower, not only overall balances, but componential balances as well. Componential balances, where the mass and enthalpy of each component entering and exiting the tower is accounted for, are necessary because material will be exchanged between the water and gas streams. For example, some water from the quench water stream will end up as water vapor in the gas stream and any soluble gases in the gas stream will be dissolved by the water. In normal operations, about 10% of the water stream is vaporized. For design purposes, therefore, a 9:1 ratio between the water recirculation rate and the makeup water input rate may be assumed. It is also customary to fix the temperature differential between the incoming makeup water and the exiting cooled gases at 2–3°F.

Once the necessary flow rates and temperatures have been defined, the tower may be sized. The basic design parameters for the spray tower are the diameter and height. The diameter of a spray tower may be calculated from the superficial velocity (v_t), which is given by Eq. (9.3.1) in units of feet per second:

$$v_t = 0.2\left(\frac{\rho_L - \rho_G}{\rho_G}\right)^{0.5} \tag{9.3.1}$$

where ρ_L = density of liquid (lb/ft^3)
ρ_G = density of gas (lb/ft^3)

The area of the tower is obtained by dividing the volumetric flow rate of the combustion gases by v_t:

$$A = \frac{w_G}{3600\rho_G v_t} \tag{9.3.2}$$

where w_G = mass flow rate of the gas (lb/h).

The diameter is then given by

$$D = \left(\frac{4A}{\pi}\right)^{0.5} \tag{9.3.3}$$

In order to calculate the tower height (Z) (ft) a volumetric heat transfer coefficient (U) (Btu/h-ft^3-°F) is employed. This coefficient is then used to establish the required contact volume (V) via Eq. (9.3.4):

$$V = \frac{q}{U\,\Delta T_{lm}} \tag{9.3.4}$$

where V = tower volume (ft^3)
q = heat rate required to vaporize the water to the combustion gas discharge temperature (Btu/h)
ΔT_{lm} = log-mean temperature difference driving force across the quencher

If the liquid-side resistance to heat transfer is neglected, U is given by the hot gas volumetric film heat transfer coefficient (h) as provided in Eq. (9.3.5):

$$h = \frac{0.43G^{0.8}L^{0.4}}{Z^{0.5}} \tag{9.3.5}$$

where h = gas-side volumetric heat transfer coefficient (Btu/h-ft^3-°F)
G = gas superficial mass velocity at the bottom of tower (lb/h-ft^2)
L = liquid superficial mass velocity at the bottom of tower (lb/h-ft^2)
Z = height of tower (ft)

The *superficial* mass velocity of a stream is calculated by dividing the mass flow rate (at or near the tower bottom) by the cross-sectional area. If there were no other material in the tower besides that particular stream, the *superficial* and *true* mass velocities would be equal. The mass flow rates are specified at the bottom of the tower because these quantities change as the streams move through the tower (due to the exchange of material) and reach their highest values at the bottom. The tower volume is given by

$$V = AZ \tag{9.3.6}$$

Equations (9.3.4) through (9.3.6) may be solved simultaneously for Z.

Venturi Scrubber

Although primarily utilized as an air pollution control device (and hence described in greater detail in Chapter 10), the venturi scrubber is also sometimes used at incinerator facilities as a quencher. The device consists of three sections of ductwork: the first, a converging section, followed by the second, a short length of constant cross section called the *throat,* followed by the third, a diverging section. These scrubbers utilize the kinetic energy of a moving gas stream to atomize the scrubbing liquid into droplets. Liquid, almost always water, is injected into the high-velocity gas stream either at the inlet to the converging section or at the venturi throat. In the process, the water is atomized into extremely small liquid droplets. These droplets have extremely large surface areas, which is an ideal situation for heat and mass transfer. There is no satisfactory generalized design correlation for this type of quencher. Design should therefore be based on full-scale data or at least laboratory or pilot-scale data. Limited data[5] suggest that a water flow rate of 10 gpm/100 scfm of hot flue gas at approximately 2000°F is sufficient to cool the gases to the adiabatic saturation temperature (which is usually between 150 and 200°F). A minimum pressure drop of 5 in. H_2O is recommended. An average liquid droplet size of approximately 100 microns (μm) is produced, which has high enough surface area to accommodate the simultaneous heat and mass transfer quench process.

This method of water quenching has found wide application in hazardous waste incineration operations. Although pressure drop is a major (negative) consideration, there is particulate capture and some gaseous removal (despite the elevated temperatures). Most quench calculations do not include the "credit" associated with particulate and gaseous control.

Submerged Exhaust Tanks

The use of submerged quench systems is based on the submerged combustion devices designed and applied over 100 years ago. The first recorded patent on submerged combustion equipment was granted to an Englishman by the name of Collier, in the year 1886. Later, another Briton, Smethurst developed a similar device for the carbon dioxide treatment of chemical waste.[6] None of these early devices are of great significance except that they proved the possibility of direct flame and hot gas contact heating of liquids.

Industrial applications of this technique have been applied to specific process applications in the chemical, metallurgical, and refining industries. These major applications are heating, evaporation, inert gas generation, and gas absorption or stripping operations.[6]

With the emphasis placed on chemical waste disposal in the 1970s, installations of submerged combustion were developed and utilized for the chlorinated and fluorinated liquid and gaseous wastes from plant processes. Nittetu Chemical Engineering Ltd. of Japan developed a system to recover HCl as acid at 18 and 35% or as anhydrous (100% HCl).[7] The basic Sub-X process was licensed from a U.S. firm, Thermal Research and Engineering Corporation (now Selas T-Thermal) in 1964. This application was later expanded for use as an incinerator hot gas quench using direct-contact submerged heat exchange for quenching the exhaust gases. The gases are discharged from the bottom of a downcomer tube through orifices to create many small gas bubbles. The discharge varies from 10 to 30 in. below the water level.[8] Other vendors use a similar approach, but the gases flow into the tank and over the surface of the water. The primary cooling is by sprays at the entry to the quench tank above the water level. The deep submergence provides scrubbing of acid gases as well as large particulate in the gas stream. It serves as a first-stage air pollution control device (see Fig. 7.2.7). The sizing of downcomer tubes, gas orifices, and quench tank size is proprietary information that may be obtained from the vendors directly. Current suppliers of submerged exhaust quench systems are T-Thermal of Blue Bell, PA (Sub-X) Callidus Technologies, Inc. of Tulsa, OK and John Zink Co. of Tulsa, OK.

Packed Towers

Packed columns find their main HWI application in the cleaning of combustion gases and hence will be covered in detail in Chapter 10 (see Section 10.5). However, these devices may also be used for quenching the hot flue gases and, hence, some descriptive material plus design (as a quencher) information on this topic will be presented in this chapter.

Packed columns, used in general for continuous contacting of liquid and gas streams, are usually vertical columns that have been filled with packing or other artifacts of large surface area. The liquid is distributed over, and trickles down through the packed bed, thus exposing a large surface area to the gas stream. When a packed tower is used for quenching, the hot gas stream is introduced near the bottom and flows upward through the packed bed against the downward-flowing cooling water stream. The water stream is recirculated and a makeup water stream is continuously added at the top. The makeup water is needed to replace the evaporated water that is picked up by the gas stream.

Consider a packed tower operating at a given liquid rate, as the gas rate is gradually increased. After a certain point the gas rate becomes so high that the drag on the liquid is sufficient to keep the liquid from flowing freely down the column. Liquid begins to accumulate and tends to block the entire cross section for flow (so-called *loading*). This, of course, increases both the pressure drop and prevents the packing from mixing the gas and liquid effectively, and ultimately some liquid is even carried back up the column. This undesirable condition, known as *flooding,* occurs fairly abruptly, and the superficial gas velocity at which it occurs is called the *flooding velocity.* The calculation of column diameter is based on flooding conditions, the usual operating range being taken as 50–75% of the flooding rate.

Before the packed tower can be sized, the flow rates and temperatures must be determined. This is accomplished by using the energy and material balances described earlier for the spray tower. Equation (9.3.7) relates the liquid and gaseous flow rates in the tower with the column cross-sectional area at the flooding point.[9]

$$Y = -3.84 - 1.06X - 0.119X^2 \tag{9.3.7}$$

where $Y = \ln(w_G^2 F \psi \mu_L^{0.2} / \rho_G \rho_L g_c A_f^2)$
$X = \ln[(L/G)(\rho_G/\rho_L)^{0.5}]$
L = liquid superficial mass velocity at bottom of packing (lb/h-ft^2)
G = gas superficial mass velocity at bottom of packing (lb/h-ft^2)
ρ_L = liquid density (lb/ft^3)
ρ_G = gas density (lb/ft^3)
F = packing factor
A_f = cross-sectional area of column (ft^2)
μ_L = liquid viscosity (cP)
ψ = ratio, density of water/liquid density
g_c = constant = 4.173×108 lb-ft/lb$_f$-h^2

This equation is applicable in the range $0.01 < e^X < 10$ or $-4.6 < X < 2.3$.

In the following design procedure, Eq. (9.3.7) is used to determine the column diameter:

1. From the superficial mass velocities, calculate a value for the parameter Y.

2. From the value of Y, determine the cross-sectional area of the column that would cause flooding for the given flow rate. Equation (9.3.8), which is based on the definition of Y above, may be used for this purpose:

$$A_f = \left[\frac{w_G^2 F \psi \mu^{0.2}}{\rho_G \rho_L g_c e^Y}\right]^{0.5} \tag{9.3.8}$$

   The packing factor (F) is a parameter based on the *specific surface* of the packing used (ft^2 surface/ft^3 volume) and the *void fraction* of the packed region of the column. Its value is usually supplied by the vendor, although F's for most common types of packing can be found in the literature.

3. Calculate the column cross-sectional area (A) for the fraction of the flooding velocity (f) chosen for operation. Usually f is chosen as 0.60:

$$A = \frac{A_f}{f} \tag{9.3.9}$$

4. The tower diameter can then be calculated from Eq. (9.3.3).

Note: The constants in Eq. (9.3.7) are not dimensionless. The only units that may be used, therefore, are those designated in the list of symbol definitions below the equation.

A somewhat similar procedure to that employed for a spray tower is used to calculate the height of a packed tower quencher. Unfortunately, very little data are available for estimating the hot gas-side heat transfer coefficient. The recommended procedure is to calculate h (which is again assumed equal to U) using Eq. (9.3.10):

$$h = \left(\frac{\mathrm{Sc_G}}{\mathrm{Pr_G}}\right)^{2/3}\left(\frac{c_{PG}G}{H_G}\right) \tag{9.3.10}$$

where $\mathrm{Sc_G}$ = Schmidt number for the gas (dimensionless) = $\boldsymbol{u}_G/p_G D'$
Pr_G = Prandtl number for the gas (dimensionless) = $c_{PG}\boldsymbol{u}_G/k_G$
c_{PG} = combustion gas specific heat capacity (Btu/lb-°F)
G = gas superficial mass velocity (lb/h-ft^2)
H_G = height of a gas-phase transfer unit (ft)
μ_G = gas viscosity (lb/ft-h)
ρ_G = gas density (lb/ft^3)
D' = diffusion coefficient (ft^2/h)
k_G = gas conductivity (Btu/h-ft-°F)

The term H_G is calculated from

$$H_G = \phi \mathrm{Sc}^{0.5} \frac{D^n}{(Lf_1 f_3 f_3)^m} \tag{9.3.11}$$

where D = tower diameter (ft)
$f_1 = \mu_L^{0.16}$
$f_2 = (62.4/p_L)^{1.25}$
$f_3 = (72.8/\sigma)^{0.8}$
μ_L = liquid viscosity (lb/ft-s)
s = surface tension (lb_f/ft)
p_L = liquid density (lb/ft^3)
ϕ, n, m = empirical coefficients that depend on the type of packing and flow conditions

Once an estimate for U has been made, V is calculated directly from Eq. (9.3.4).

The packing height is given by

$$Z = V/A \tag{9.3.12}$$

A simplified approximate calculational procedure for designing either a spray or packed-tower water quencher is available.[10] The tower area (A) is given by

$$A = bw_{in}/\rho_G \tag{9.3.13}$$

where w_{in} = total mass flow rate of hot gas entering quencher (lb/s)
ρ_G = hot gas density (lb/ft^3)
b = dimensionless constant in the 0.1–0.3 range

The tower diameter is then

$$D = \left(\frac{4A}{\pi}\right)^{0.5} \tag{9.3.14}$$

The tower height may be estimated by

$$Z = 2.0D \tag{9.3.15}$$

with the tower volume (V) given by

$$V = 2D^3$$

The units of A, D, Z, and V are ft^2, ft, ft, and ft^3, respectively. This procedure provides an approximate 25–50% safety factor relative to the rigorous calculations.

9.4 CONTACT WITH HIGH HEAT CAPACITY SOLIDS

Inert ceramics with high heat capacities have, in a few cases, been used to cool hot incinerator effluent gases from fume incinerators to temperatures suitable for air pollution control devices. Besides cooling the gases by absorbing heat, this method also allows a certain amount of energy recovery by using the heat absorbed by the solids to subsequently preheat the Volatile Organic Compounds (VOC) laden airstream (fume) being introduced to the combustion zone. In one such system, the solid elements are packed into beds called *regenerative thermal oxidizers (RTOs).*[11] These chambers operate on a *regenerative* principle—alternately absorbing, storing, and recycling heat energy. The chambers are thus used in cyclic fashion. In the first half-cycle, the process air (VOC fume) is passed through the previously heated ceramic elements in a particular chamber, simultaneously cooling the bed and transferring heat into the fume that is then moved on to the incinerator (combustion) chamber. At any point during the cycle, the temperature at the bed varies over 1000°F from the cold side (i.e., the inlet side for the ambient air) to the hot side. In the second half-cycle, the hot clean oxidized gases are introduced to the hot side of the bed, flow through the bed in the opposite direction of the cold fume flow in the first half-cycle, and the cold side; the heat lost by the gases is absorbed by the bed that is brought to its maximum average temperature for the start of the next cycle. Typical cycle times range from 2 to 10 min. Manufacturers claim over 90% thermal energy recovery with this device. At the same time, the destruction efficiency of the system has attained levels of 98–99% of the incoming VOC stream.

9.5 NATURAL CONVECTION AND RADIATION

When a hot gas flows through a duct, the duct becomes hot and heats the surrounding air. As the air becomes heated, natural drafts are formed, carrying the heat away from the duct. This phenomenon is *natural convection.* Heat is also discharged from the hot duct to its surroundings by *radiant energy.* Both of these modes of heat transfer have already been discussed in Chapter 8.

Equation (8.2.14) shows that the rate of heat transfer is a function of the resistances to heat flow, the log-mean temperature difference between the hot gas and the air surrounding the duct, and the surface area of the duct. The rate of heat transfer (q) is determined by the amount of heat to be removed from the hot gaseous effluent entering the exhaust system. For any particular basic process, the mass flow rate of gaseous effluent and its maximum temperature are fixed. The cooling system must, therefore, be designed to dissipate sufficient heat to lower the effluent temperature to the operating temperature of the air pollution control device to be used. The rate of heat transfer can be calculated by the enthalpy difference of the gas at the inlet and outlet of the cooling system:

$$q = w_G \, \Delta h \tag{9.5.1}$$

where q = rate of heat transfer
w_G = mass flow rate of gas (lb/h)
Δh = enthalpy change between inlet and outlet conditions (Btu/lb)

The log-mean temperature difference in this case is the difference in temperature between the air surrounding the duct, and the inlet and outlet temperature of the gas. This term, too, is fixed for a particular process and is calculated by applying Eq. (8.2.13) as follows:

$$\Delta T_{\mathrm{lm}} = \frac{(T_{\mathrm{h1}} - T_{\mathrm{c}}) - (T_{\mathrm{h2}} - T_{\mathrm{c}})}{\ln[(T_{\mathrm{h1}} - T_{\mathrm{c}})/(T_{\mathrm{h2}} - T_{\mathrm{c}})]} \tag{9.5.2}$$

where T_{h1} = gas temperature of inlet (°F)
T_{h2} = gas temperature at outlet (°F)
T_{c} = air temperature (°F)

In many incinerator operations the temperature of the gaseous effluent is not constant but varies as different wastes are burned. The atmospheric temperatures also vary a great deal. In such cases, the cooling system must be designed for the worst conditions that prevail to ensure adequate cooling at all times. The inlet temperature (T_{h1}) chosen for the design calculations must be the maximum temperature of the gas entering the system; T_{h2} must be the maximum allowable temperature of the gas discharged from the cooling system; and T_{c} must be the maximum expected atmospheric temperature.

The overall heat transfer coefficient (U_{o}) is calculated from the individual film coefficients, h_{i} and h_{o} as shown in Eq. (8.2.17). Equation (8.2.18) may be employed to calculate the inside coefficient (h_{i}). As already described, this coefficient is a measure of the flow of heat through the inside film. An increase in h_{i} will, therefore, increase the rate of heat transferred from the gas to the atmosphere. It can be seen that an increase in the Reynolds number (Re) will increase the rate of heat transfer. Since the weight of gas flowing is fixed, Re can be increased only by increasing the velocity of the gas. An increase in velocity will increase the power required to move the gases through the exhaust system. Consequently, the optimum velocity for good heat transfer at reasonable blower-operating costs must be determined. It is known that a sacrifice in heat transfer rate to obtain lower blower horsepower can result in a more economical cooling system. Owing to the many variables involved, however, each system must be calculated on its own merits.

The outside film coefficient (h_o) is the sum of the coefficient due to natural convection (h_c) and the coefficient due to radiation (h_r):

$$h_o = h_c + h_r \tag{9.5.3}$$

An empirical equation for h_c for vertical pipes more than 1-ft high and for horizontal pipes is[12]

$$h_c = 0.27\left(\frac{\Delta T}{D_o}\right)^{0.25} \tag{9.5.4}$$

where ΔT = temperature difference between the outside duct wall and the air ($T_w - T_c$)
D_o = outside duct diameter (ft)

The radiation coefficient (h_r) may be calculated by Eq. (8.2.29). Since the emissivity of the surface is a function of the surface condition, and a black surface generally gives the highest emissivity, blackening of the ductwork is recommended.

To calculate h_o by Eqs. (8.2.29), (9.5.3), and (9.5.4), the temperature of the duct wall (T_w) must first be assumed and then checked. The following equation may be used for this purpose:

$$T_w = T_a - \left(\frac{h_o}{h_o + h_i}\right)(T_a - T_c) \tag{9.5.5}$$

where T_a = average flue gas temperature (°F)
T_c = ambient air temperature (°F)

If the calculated T_w is not the same as the assumed T_w, a new T_w must be estimated and h_o recalculated. When the assumed T_w and calculated T_w are the same, the corresponding h_o is then used in Eq. (8.2.17) to calculate U_o. Typical values for the overall heat transfer coefficient are in the 0.5–2.0 Btu/h-ft^2-°F range.

The heat transfer area (A_o) can now be found; the length of duct needed to give the necessary area is then calculated by using the outside diameter. If the length of duct needed is large, the ductwork will probably be arranged in vertical columns to conserve floor space. The columns require several 180° bends, which will offer a large resistance to the flow of gas. To minimize these losses, the gas velocity should be low, preferably less than normal dust-conveying velocities. By joining the bottoms of the columns with hoppers, any dust settling out as a result of low velocities can be collected without fouling the system. If the cooling area is such that a single loop around the plant or across a roof is sufficient, sharp bends should be avoided. When gases are cooled through a large temperature range, the volume will be considerably reduced, so that smaller diameter ductwork may be needed as the gases proceed through the cooling system. With cooling columns, the diameter of the duct joining the last column and the air pollution control device must be sized properly to provide suitable conveying velocities for the cooled effluent.

For most convection–radiation cooling systems, the only equipment used is sufficient ductwork to provide the required heat transfer area. Unless the temperature of the gases discharged from the incinerator is exceptionally high, or if there are corrosive gases or fumes present, black iron ductwork is generally satisfactory. The temperature of the duct wall can be determined for any portion of the ductwork by using the method previously described for determining T_w. If T_w proves to be greater than black iron can withstand, either a more heat-resistant material should be used for that portion of the system or a portion of the cooled gas should be recirculated to lower the gas temperature at the cooling system inlet.

With this type of cooling, flexibility in controlling the gas temperature is limited. When either the gas stream temperature or air temperature, or both, are lower-than-design values, the gases discharged from the cooling device will be less than that calculated, and condensation of moisture from the effluent within the control device might result. Conversely, when design temperatures are exceeded, the temperature of the gases discharged from the cooling system could become too high.

The radiation–convection cooling system is in operation whenever hot gases are flowing through the exhaust system. Since gases being cooled are not diluted with any cooling fluid, the air pollution control device need not be sized for an extra volume of gases. Since no water is used, there is no need for pumps, and corrosion problems are nonexistent. On the other hand, these cooling systems require considerable space, and system blower horsepower requirements are increased owing to the additional resistance to gas flow. This type of cooling system is seldom used in HWI, mainly because of the excessive space requirement and the high cost of the ductwork required for transport of the hot corrosive gases.

9.6 FORCED-DRAFT COOLING

As discussed in Chapter 8, heat transfer by convection is due to fluid motion. Cold fluid adjacent to a hot surface receives heat, which is imparted to the bulk of the fluid by mixing. With natural convection, the heated fluid adjacent to the hot surface rises and is replaced by colder fluid. By agitating the fluid, mixing occurs at a much higher rate than with natural currents, and heat is taken away from the hot surface at a much higher rate. In most process applications, this agitation is induced by circulating the fluid at a rapid rate past the hot surface. This method of heat transfer is called *forced convection.* Since forced convection transfers heat much faster than natural convection, most process applications use forced-convection heat exchangers. Whenever possible, heat is exchanged between hot and cold streams to reduce the heat input to the process. There are, however, many industrial applications where it is not feasible to exchange heat in this fashion, and so a cooling fluid such as water or air is used, and the heat removed from the stream by the coolant is dissipated to the atmosphere. When water is used, the heat is taken from the process stream in a shell-and-tube cooler, and the heat picked up by the water is dissipated to the atmosphere in a cooling tower. When air is used as the cooling medium in either shell-and-tube or fin-tube coolers, the heated air is discharged to the atmosphere and is not recirculated through the cooler.

With forced-convection cooling, the temperature of the cooled stream can be controlled within narrow limits even with widely varying atmospheric or water temperatures. Heat transfer area is greatly reduced from that needed with natural convection. Power requirements to force the process stream through the cooler are generally less. On the other hand, either a pump or a blower is needed to circulate the cooling fluid through the cooler. With water cooling, a cooling tower may be needed and additional maintenance is required to clean scale from the tubes.

Because there are so many different types of forced-convection heat exchangers, it is impossible to present a specific design procedure unless a specific type is chosen for discussion. The following describes a general approach for the calculation of the total heat transfer surface required ($A_{o,total}$), which is the most critical parameter in heat exchanger design. To use this approach, however, much information on the specific type of exchanger must be obtained either from the literature or from the vendor.

Generally, the hot gas flow rate, inlet and desired outlet gas temperatures, and the inlet coolant temperature are already known. The first three quantities can be used to calculate the heat duty. The coolant flow rate is chosen on the basis of pressure drop information and heat transfer coefficient data, both of which are highly dependent on the coolant velocity and heat exchanger configuration. These data are often available from the vendor. An enthalpy balance (Eq. 8.2.8) may then be applied to determine the coolant outlet temperature. In order to calculate the overall heat transfer coefficient (U_o) both the gas-side and coolant-side film coefficients must be determined. Procedures to evaluate each of these coefficients (except when the fluid is flowing inside tubes or pipes) must be obtained for the specific flow configuration either from the literature or from the vendor. If fouling is a potential problem, the appropriate fouling factors should also be obtained. Equation (8.2.20) is applied to calculate U_o. Note that in this equation, A_o, A_i, and A_{lm} are defined from the heat transfer surface for a single passage for one of the fluids, such as a pipe or tube. In an industrial heat exchanger, there are many such passages. The three areas in Eq. (8.2.20) are usually known at the start of the design process; the total heat transfer area ($A_{o,total}$) is not. This latter is obtained from

$$q = U_o A_{o,total} \phi \, \Delta T_{lm} \tag{9.5.6}$$

where q = heat duty (Btu/lb)
U_o = overall heat transfer coefficient (Btu/h-ft^2-°F)
$A_{o,total}$ = total heat transfer area (ft^2)
ϕ = correction (dimensionless)
ΔT_{lm} = log-mean temperature difference (°F)

The factor (ϕ) is a correction for the use of ΔT_{lm} as the average temperature difference driving force. If the flows are parallel (i.e., either cocurrent or countercurrent), this factor is unity. In industrial heat exchangers, this is usually not the case, and ϕ must be obtained either from the literature or the vendor for the particular flow patterns used in the exchanger.

9.7 ILLUSTRATIVE EXAMPLES

Example 9.7.1 For the system shown in Fig. 9.7.1, design an adiabatic quench spray tower to cool the combustion gases from the incinerator. Provide calculations based on (1) top, (2) bottom, and (3) average tower conditions. The composition of the combustion gas stream (stream 1) is given in the following list:

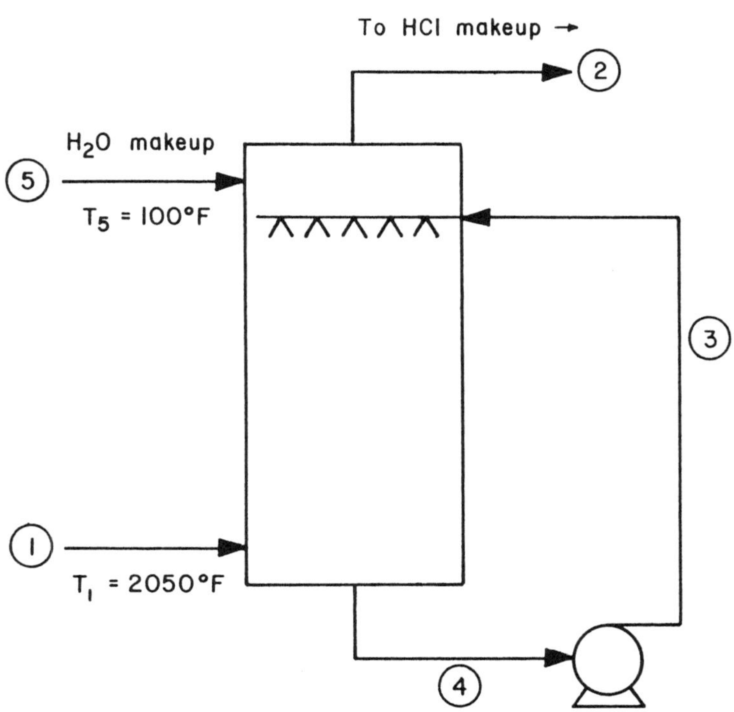

Figure 9.7.1 Quench spray tower.

	MW	(lb/h)	(lbmol/h)	Mole fraction
CO_2	44	12,023	273.3	0.0536
H_2O	18	9,092	505.1	0.0990
N_2	28	106,783	3,813.7	0.7476
HCl	36.5	193	5.3	0.0010
O_2	32	16,115	503.6	0.0988
Total		144,206	5,101.0	1.0

Average MW = 28.27

For a reference temperature (T_r) of 32°F:

$c_{P,\text{BDA}}$ = 0.271 Btu/lb-°F between 32 and 2050°F

$c_{P,\text{BDA}}$ = 0.248 Btu/lb-°F between 32 and 200°F

$c_{P,\text{H}_2\text{O}}$ = 1.0 Btu/lb-°F

$c_{P,\text{wv}}$ = 0.515 Btu/lb-°F between 32 and 2050°F; 0.447 between 32 and 200°F

$c_{P,\text{CG}}$ = 0.3 Btu/lb-°F between 32 and 2050°F

Δh_{vap} = 1075 Btu/lb at 32°F

(*Note* The subscript BDA stands for bone dry air, wv for water vapor, and CG for combustion gases. See Table 5.2.2 for calculation of c_P for gases in table above.) Assume (a) the water makeup temperature (T_5) equal to 100°F and (b) a 2°F approach temperature for T_2.

SOLUTION An enthalpy balance around the adiabatic quencher gives

$$H_1 - H_2 + H_3 - H_4 + H_5 = 0 \tag{9.7.1}$$

The enthalpy of an air–water vapor mixture is defined as

$$H_{\text{air},T} = w_{\text{BDA}} c_{P,\text{BDA}}(T - T_r) + w_{\text{H}_2\text{O}}[\Delta h_{\text{vap},T_r} + c_{P,\text{wv}}(T - T_r)] \tag{9.7.2}$$

A water balance on the recycle line yields

$$w_3 = w_4 \tag{9.7.3}$$

Equation (9.7.1) now becomes

$$H_1 - H_2 + w_3 c_{P,\text{H}_2\text{O}}(T_3 - T_r) - w_4 c_{P,\text{H}_2\text{O}}(T_4 - T_r) + w_5 c_{P,\text{H}_2\text{O}}(T_5 - T_r) = 0 \tag{9.7.4}$$

By definition, for an adiabatic quench, $T_3 = T_4$. The makeup temperature is now set at $T_5 = 100°F$, and the exit vapor temperature is allowed to approach the inlet liquid by 2°F, that is, $T_2 = T_3 + 2$. The enthalpy balance becomes

$$H_1 - H_2 + w_5 c_{P,H_2O}(T_5 - T_r) = 0$$

$$H_{1,2050} - H_{2,(T_3+2)} + w_5 c_{P,H_2O}(100 - 32) = 0 \qquad (9.7.5)$$

The enthalpy of the combustion gas is first calculated:

$$H_1 = (135{,}114)(0.271)(2050 - 32) + (9092)[1075 + 0.515(2050 - 32)]$$
$$= 93{,}113{,}817 \text{ Btu/h}$$

T_3 (or T_2) is needed in order to calculate H_2. A trial-and-error calculation must be performed in order to obtain T_3. This is accomplished by assuming values of T_3 until Eq. (9.7.5) is satisfied. After several trials, a T_3 of 180°F is assumed so that $T_2 = 178°F$. Since stream 2 is saturated with water, the mole fraction of water is given by the ratio of the vapor pressure of water at 178°F to the total pressure of the system:

$$y_{2,H_2O} = P'/P \qquad P' \text{ at } 178°\text{F} = 7.26 \text{ psia}$$
$$= 7.26/14.7 = 0.4940$$

If z is the number of moles of water in the vapor outlet stream (stream 2) and assuming no HCl removal in quench,

$$0.4940 = \frac{z}{z + (5101 - 505.1)}$$

Solving this equation for z gives

$$z = 4488 \text{ lbmol } H_2O \text{ is stream 2}$$

$$w_{2,H_2O} = (4448)\ (18)$$
$$= 80{,}785 \text{ lb/h}$$

Note: This result could have been obtained from use of the psychrometric chart provided in Fig. 4.5.2.

The number of moles of water in stream 5 is that in stream 2 minus that in stream 1. The mass of makeup water in stream 5 is

$$w_5 = (4488 - 505.1)\,(18)$$
$$= 71{,}692 \text{ lb/h}$$

The enthalpy of stream 2 is then

$$H_2 = (135{,}114)(0.248)(180 - 32) + 80{,}785[1075 + 0.447(178 - 32)]$$
$$= 97{,}075{,}290 \text{ Btu/h}$$

An enthalpy balance check on Eq. (9.7.5) is approximately accomplished for this assumed T_3 of 180°F. The composition of the effluent gas from the quencher is given in the following table:

	MW	(lb/h)	(lbmol/h)	Mole fraction
CO_2	44	12,023	273.3	0.0301
H_2O	18	80,785	4,488	0.4941
N_2	28	106,783	3,813.7	0.4198
HCl	36.5	193	5.3	0.0006
O_2	32	16,115	503.6	0.0521
Total		216,628	9,083.8	1.0

Average MW = 23.74

Note that the mole fraction of water vapor is approximately 50% of the total pound-moles. This serves as a good rule of thumb to provide a quick estimate of water vapor in saturated combustion gases knowing the dry gas volume in pound-moles. One can now proceed to the heat duty (q) calculation. For adiabatic operation, this is approximately given by

$$q = w_5(h_{2,H_2O} - h_{5,H_2O})$$
$$= 71{,}693\ [1075 + 0.447(180 - 32) - 1.0(100 - 32)] = (71{,}693)(1073.16)$$
$$= 76{,}937{,}773 \text{ Btu/h} \tag{9.7.6}$$

The log-mean temperature difference across the quench unit is

$$\Delta T_{\text{lm}} = \frac{(T_1 - T_4) - (T_2 - T_3)}{\ln[(T_1 - T_4)/(T_2 - T_3)]} = \frac{1871 - 2.0}{\ln(1871/2.0)}$$
$$= 273.2°\text{F}$$

With the material and energy balances complete, one can proceed to the physical design of the quencher. The recirculating liquid flow rate must be set. This information is also required to size the recirculating pump. Approximately

8–10% of the liquid entering at the top is vaporized. Select 9% liquid vaporization across the unit:

$$w_5 = 71{,}693 \text{ lb/h}$$

$$w_3 = (71{,}693)\,(9) = 645{,}237 \text{ lb/h}$$

The total liquid flow at the top of the tower is then 716,930 lb/h. At 178°F, the liquid density and viscosity are 60.6 lb/ft^3 and 0.347 cP, respectively. The inlet vapor density at 2050°F (MW = 28.27) is 0.0154 lb/ft^3. The outlet vapor density at 180°F (MW = 23.74) is 0.0508 lb/ft^3.

Top Conditions (1)

The diameter of the spray tower is calculated from the superficial velocity (v_t) provided in Eq. (9.3.1):

$$v_t = 0.2\left(\frac{\rho_L - \rho_G}{\rho_G}\right)^{0.5} \tag{9.3.1}$$

If this is evaluated at the gas exit (top),

$$v_t = 0.2\left(\frac{60.6 - 0.0508}{0.0508}\right)^{0.5} = 6.90 \text{ ft/s}$$

The required area (A) and diameter (D) are then

$$A = \frac{216{,}628}{(3600)(0.0508)(6.9)}$$

$$= 171.7 \text{ ft}^2$$

$$D \cong 15 \text{ ft}$$

At this diameter, the real area becomes 176.6 ft^2.

In order to calculate the tower height (Z), the volumetric heat transfer coefficient (U) is employed. If the liquid side resistance to heat transfer is small, as is the case with adiabatic humidification, U is equal to the hot gas film heat transfer coefficient (h) as given in Eq. (9.3.5):

$$h = \frac{0.043G^{0.8}L^{0.4}}{Z^{0.5}} \tag{9.3.5}$$

The terms G and L in Eq. (9.3.5) are superficial mass velocities for the gas and liquid, respectively, with units of pounds per hour per square foot (lb/h-ft^2). Thus,

$$G = 216{,}628/176.6 = 1227 \text{ lb/h-ft}^2$$

$$L = 716{,}930/176.6 = 4059 \text{ lb/h-ft}^2$$

Equation (9.3.5) is rearranged to solve for $hZ^{0.5}$:

$$hZ^{0.5} = (0.043)(1227)^{0.8}(4059)^{0.4} = 353$$

The design equation for the spray tower volume (V) is given by Eq. (9.7.7):

$$\begin{aligned} V = AZ &= \frac{qZ^{0.5}}{hZ^{0.5}\ dT_{\text{lm}}} \\ &= 176.6Z = \frac{76{,}937{,}773(Z)^{0.5}}{(353)(273.3)} \\ &= 176.6Z = 797.5(Z)^{0.5} \\ Z^{0.5} &= 4.51 \\ Z &= 20.39 \text{ ft} \\ V &= 3601 \text{ ft}^3 \end{aligned} \tag{9.7.7}$$

Note: The volumetric heat transfer coefficient (U or h) is 79.6 Btu/h-ft^3-°F.

Bottom and Average Conditions (2,3)

The key calculated results for a design based on either bottom conditions or average (arithmetic) tower conditions—as well as the results for top conditions—are presented here in tabular form:

	Top	Bottom	Average
w_L (lb/h)	716,930	645,237	681,083
w_G (lb/h)	216,628	144,206	180,417
ρ_L (lb/ft^3)	60.5	60.5	60.5
ρ_G (lb/h)	0.0508	0.0154	0.0331
μ_L (cP)	0.345	0.345	0.345
v_t (ft/s)	6.90	12.53	9.71
A (ft^2)	178	208	181
D (ft^2)	15(15)	16.3(17)	15.2(15)
A_{real} (ft^2)	176.6	227	176.6
L (lb/h-ft^2)	4059	2842	3450
G (lb/h-ft^2)	1227	635	931
V (ft^3)	3601	12,222	7911
Z (ft)	20.39	53.84	37.115
U (Btu/h-ft^3-°F)	79.6	25.7	56.2

A tower design based on average conditions is recommended. Thus, a 15-ft-diameter column with approximately 30 ft between gas inlet and spray nozzles should be employed. Because of the assumptions made in the calculations, the design should allow for the installation of additional nozzles lower in the column and/or for the capability of increasing the water recirculation rate.

Example 9.7.2 With reference to Example 9.7.1, obtain an approximate design of this unit using the procedure outlined in Eqs. (9.3.13)–(9.3.15).

SOLUTION The hot gas density (2050°F) is first calculated:

$$\begin{aligned}\rho &= PM/RT = (1.0)(28.27)/(0.73)(2510)\\ &= 0.0154 \text{ lb/ft}^3\end{aligned}$$

The tower area (A) and the diameter (D) are then

$$\begin{aligned}A &= (0.2)(144{,}206)/(3600)(0.0154)\\ &= 520 \text{ ft}^2\\ D &= 25.7 \text{ ft}\end{aligned}$$

The tower height is estimated from Eq. (9.3.15):

$$\begin{aligned}Z &= 2.0D = 2.0(25.7)\\ &= 51.4 \text{ ft}\end{aligned}$$

The results are in reasonable agreement with those in Example 9.7.1.

Example 9.7.3 It is proposed to cool the waste flue combustion gas in Example 9.7.1 by using ambient air at 70°F. Calculate the quantity (mass, mole, and volume basis) of air required to cool the gases to an acceptable temperature of 560°F.

SOLUTION Under adiabatic conditions,

$$\begin{aligned}q_{\text{flue}} &= q_{\text{air}}\\ q_{\text{flue}} &= w_{\text{flue}}\bar{c}_P\,\Delta T = (144{,}206)(0.3)(2050 - 560)\\ &= 64.5 \times 10^6 \text{ Btu/h}\end{aligned}$$

Since $q_{\text{flue}} = q_{\text{air}}$,

$$64.5 \times 10^6 = w_{\text{air}}(0.243)(560 - 70)$$

or

$$\begin{aligned} w_{\text{air}} &= 541{,}698 \text{ lb/h} \\ &= 18{,}679 \text{ lb-mol/h} \\ &= 7{,}215{,}456 \text{ ft}^3\text{/h (70°F)} \end{aligned}$$

Note: The combined mass (and volume) flow of the gas is now significantly higher. This will adversely impact on the economics for the air pollution control equipment. Note that with water cooling (see Example 9.7.1), the total tower exit mass flow is 216,628 lb/h at an outlet temperature of 178°F compared to more than triple [(541,698 + 144,206)/(216,628) = 3.16] the flow at a temperature of 560°F with air cooling.

Example 9.7.4 Design the air quench tank in Example 9.7.3 if a 1.5-s residence time is required.

SOLUTION The total mass flow rate of the gas is now

$$\begin{aligned} w_t &= 144{,}206 + 541{,}698 \\ &= 685{,}904 \text{ lb/h} \end{aligned}$$

The volumetric flow at 560°F (assuming air) is

$$\begin{aligned} Q_a &= w_t RT/PM = (685{,}904)(0.73)(1020)/(1.0)(29) \\ &= 1.76 \times 10^7 \text{ ft}^3\text{/h} \\ &= 4892 \text{ ft}^3\text{/s} \end{aligned}$$

The volume of the tank is, therefore,

$$\begin{aligned} V_t &= (4892)(1.5) \\ &= 7338 \text{ ft}^3 \end{aligned}$$

The physical dimensions of the tank are usually set by minimizing surface (materials) cost. The total surface is given by

$$S = 2(\pi D^2/4) + \pi DH$$

To minimize the total surface, the derivative of S with respect to D is set equal to zero:

$$\frac{dS}{dD} = \pi D + \pi H = 0 \quad \text{or} \quad D = H$$

The dimensions of the tank may now be calculated:

$$V_t = (\pi D^2/4)(H) = (\pi D^3/4$$

$$D = 21.05 \text{ ft}$$

$$H = 21.05 \text{ ft}$$

Note: The size of the tank is excessive. This is another reason why air quenching is rarely used.

Example 9.7.5 It is proposed to reduce the temperature of 10% of the 144,206 lb/h of flue gas described in Example 9.7.1 to 550°F, using a *solids contact* method that operates on a 1-h cooling cycle. Assume the average heat capacities of the flue gas and solid to be 0.3 and 0.88 Btu/lb-°F, respectively, over the temperature range in question. The initial temperature of the solids is 70°F. Assuming an approach temperature of 40°F, what mass of solid must be provided in order to cool the flue gas to the required temperature during each hour of operation.

SOLUTION Set up an enthalpy balance on both the flue gas and solid.

$$\Delta H_{\text{flue}} = \Delta H_{\text{solid}}$$

For the flue gas, the enthalpy change for one hour of operation is

$$\Delta H_f = mc_p(T_2 - T_1) = (0.1)(144{,}206)(0.3)(2050 - 550)$$
$$= 6{,}489{,}000 \text{ Btu}$$

For the solids,

$$\Delta H_s = m_s(0.88)(510 - 70)$$

Since $\Delta H_s = \Delta H_f$,

$$m_s = 6{,}489{,}000/(0.88)(440) = 16{,}759 \text{ lb solids}$$

Example 9.7.6 With reference to Example 9.7.1, design a radiative heat exchanger to cool the combustion gases from 2050 to 180°F. The ambient air temperature is 60°F and an overall heat transfer coefficient for the cooler of 1.5 Btu/h-ft^2-°F may be assumed to apply.

SOLUTION Based on the solution in Example 9.7.1,

$$q = (144{,}206)(0.3)(2050 - 180)$$
$$= 80.9 \times 10^6 \text{ Btu/h}$$

The log-mean temperature difference is given by

$$\Delta T_{\text{lm}} = [(2050 - 60) - (180 - 60)]/\ln[(2050 - 60)/(180 - 60)]$$
$$= 666°\text{F}$$

The radiative surface area required is

$$A = q/U\,\Delta T_{\text{lm}} = 80.9 \times 10^6/(1.5)(666)$$
$$= 80{,}980 \text{ ft}^2$$

The volumetric flow (Q_a) at inlet conditions is given by

$$Q_a = 5101(379)(2050 + 460)/(60 + 460)$$
$$= 9{,}331{,}789 \text{ ft}^3/\text{h}$$

Duct or pipe velocities of 60 ft/s for this type of application are typical. Assume this value to apply at inlet (2050°F) conditions. The duct area and diameter may now be calculated:

$$A_d = (9{,}331{,}789)/(3600)(60)$$
$$= 43.2 \text{ ft}^2$$
$$D = 7.4 \text{ ft}$$

The length of required heat exchange ducting is then

$$L = A/\pi D$$
$$= 3475 \text{ ft}$$

The practicality of this method of cooling is questionable.

PROBLEMS

1. For the system described in Example 9.7.1, propose a design if the quencher is a packed tower. Perform the calculation based on bottom conditions.

2. Using the simplified approximate procedure presented in Section 9.3, estimate the packing height and tower diameter if the waste flue gas flow rate in Example 9.7.1 is 166,500 lb/h.

3. Using ambient air at 60°F, calculate the quantity of dilution air required to cool the gases in Problem **2** to a temperature of 520°F.

4. Design the air quencher in Problem **2** if a 1.2-s residence time is recommended.

5. With reference to Problem **2** calculate the mass of 0.91 Btu/lb-°F heat capacity solids required to cool the gases to 180°F for one hour of operation. The initial temperature of the solids is 60°F, and an approach temperature of 20°F may be assumed.

6. With reference to Problem **2**, design a radiative cooler to reduce the flue gas temperature from 2050 to 130°F. Ambient air is at 70°F.

REFERENCES

1. J. J. Santoleri, "Retrofit Experience with Interim RCRA Incinerators," *Proceedings of APCA International Specialty Conference—Thermal Treatment of Industrial and Hospital Wastes*, Pittsburgh, PA, Nov., 1987, pg. 8–9.
2. J. Fair, "Designing Direct Contact Coolers/Condensers," *Chem. Eng.,* 91–100, June 12 (1972).
3. J. Fair, "Process Heat Transfer by Direct Fluid-Phase Contact," *AIChE Heat Transfer Symposium,* Series No. 118, **68**, 1–11 (1971).
4. R. Farrell, private communication, 1999.
5. W. H. McAdams, *Heat Transmission*, 2nd ed., McGraw-Hill, New York, 1942.
6. Thermal Research & Engineering Corp. Staff and S. L. Soo Associates, *Manual of Submerged Combustion*, Conshohocken, PA, 1961.
7. J. J. Santoleri, "Chlorinated Hydrocarbon Waste Recovery and Pollution Abatement," ASME National Incineration Conference, New York, June, 1972.
8. J. J. Santoleri, "Incineration of Waste Liquids Containing Organic Compounds and Inorganic Salts," AIChE 2nd National Conference on Complete Water Reuse, Chicago, May, 1975.
9. J. Reynolds, personal notes, 1998.
10. T. Shen, U. Choi, and L. Theodore, *Hazardous Waste Incineration Manual*, U.S. EPA Air Pollution Training Institute, Research Triangle Park, NC, 1985.
11. W. S. Montaz, T. J. Truppi, and J. J. Seiwert, Jr., "Sizing up RTO and RCO Heat Transfer Media," *Pollution Engineering*, 34–38, Dec. (1997).
12. D. Q. Kern, *Process Heat Transfer*, McGraw-Hill, New York, 1950.

10

AIR POLLUTION CONTROL EQUIPMENT

10.1 INTRODUCTION

One major concern associated with all types of waste incineration is the emission of air pollutants. The greatest mass of air contaminants consists primarily of the following criteria pollutants: oxides of nitrogen, oxides of sulfur, and particulate matter (PM). Trace levels of noncriteria pollutants such as chlorinated by-products, benzene, heavy metals, and acid gases are also of great concern. Available data indicate that criteria pollutant emissions from hazardous waste, medical waste, and municipal solid waste incinerators are about the same as those from various industrial combustion processes. The emission rates are dependent on the waste incineration rate, chemical composition of the waste, incinerator type, air pollution control equipment, and incinerator system operating parameters.

The primary end products in the combustion of hydrocarbons in air are mostly carbon dioxide (CO_2) and water vapor (H_2O). However, air pollutants may result from two sources when wastes are incinerated:

1. *Products of Incomplete Combustion (PICs)* These result from the incomplete combustion of organic waste constituents and conversion of certain constituents present in the waste, auxiliary fuel, and/or combustion air. The PICs include carbon monoxide, carbon, hydrocarbons, aldehydes, amines, organic acids, polycyclic organic matter (POM), and any other waste constituents or their partially degraded products that escape complete thermal destruction in the incinerator. (In well-designed and well-operated incinerators, these incomplete combustion products are emitted only in trace amounts.)

2. *Combustion Products from Hazardous, Medical, and Municipal Solid Wastes* Waste compounds commonly contain other elements in addition to carbon and hydrogen. When such wastes or some auxiliary fuels are burned, those other elements form combustion products other than carbon dioxide and water vapor. These other products that are generally considered as air pollutants include:
   a. Hydrogen chloride (HCl) and small amounts of chlorine (Cl_2) from the incineration of chlorinated hydrocarbons
   b. Acid halogens from the incineration of organic halogens
   c. Sulfur oxides, mostly as sulfur dioxide (SO_2) but also including small amounts of sulfur trioxide (SO_3) formed from sulfur or sulfur compounds present in the waste and/or fuel mixture
   d. Phosphorus pentoxide and/or phosphoric acid formed from the incineration of organophosphorus compounds
   e. Nitrogen oxides (NO_x) from the nitrogen in the combustion air and/or from organic nitrogen compounds present in the waste
   f. Particulate matter consisting of metal salts from the waste, metal oxides formed by combustion, and fragments of incompletely burned material (primarily carbonaceous)
   g. PICs, which include unburned hydrocarbons [byproducts of incomplete combustion of the POHCs (principal organic hazardous constituents)], dioxins, and furans resulting from incomplete combustion or de novo reformation in the downstream cooler zones of the air pollution control systems[1]

If the incineration system is operated properly, 99.99% DRE (destruction and removal efficiency) of the hazardous wastes can be achieved. In addition, the PICs are usually not a problem and, of the combustion products in the second source just listed, only HCl, Cl_2, and particulates are regularly encountered.

Regarding gaseous pollutants, Cl_2 is present in conjunction with HCl, but equilibrium favors HCl formation at the high temperatures employed in chlorinated waste incinerators (see Chapters 5 and 6). The amount of Cl_2 generated with relation to HCl varies based on the operating temperature of the final combustion zone and the excess air or oxygen available for the reaction.[2] At lower or higher temperatures, Cl_2 emission levels will increase and combined with HCl will provide a source for reaction with either copper or iron compounds (if present) to form copper chloride and iron chloride. These have been found to be catalyst surfaces for the formation of polychlorinated dibenzo-*p*-dioxins (PCDD) and dibenzofurans (PCDF). The critical reformation temperature of 300°C (572°F) must be avoided in the quench and air pollution control (APC) systems to prevent PCDD/F emissions.

The NO_x emissions are not economically amenable to control by scrubbing or other postgeneration removal techniques. If the waste does not contain nitrogen compounds, NO_x emissions can be minimized by staged combustion when nitrogen is contained in the organics or low-temperature combustion to minimize NO_x generated from the nitrogen in the combustion air. The SO_x emissions are naturally a strong function of the sulfur content of the waste–fuel mixture. As with NO_x, this has not surfaced as an emission problem in hazardous waste incineration (HWI) applications. Thus, the only gaseous emission of concern is HCl and the accompanying trace quantities of Cl_2. Recent federal regulations (MACT standards of Sept. 30, 1999) limits stack emission of HCl/Cl_2 from a hazardous waste combustor [HWI, cement kiln (CK), lightweight aggregate kilns (LWAK)] to levels from 21 to 230 ppm [corrected to 7% O_2 dry volume (d.v.)] dependent upon the combustor type and whether the unit is an existing or new installation. Reference Chapter 3 for the individual limits. These gaseous pollutants are usually neutralized and removed from the combustion gases by absorbers.

Regarding particulates, these emissions are strongly influenced by the chemical composition of the waste being incinerated and the auxiliary fuel, the type of incinerator and its operating parameters, and the air pollution control system. PM primarily consists of inert ash, various salts, and condensed gaseous contaminants. Most of the pollutants of concern, other than criteria pollutants such as heavy metals and organic toxic by-products, are collected as PM. The metals and toxic by-products condense as or on fine particles as the exhaust gas stream cools. Concentrations of uncontrolled emissions of PM from HWIs can normally be expected to be between 0.2 and 10 gr/dscf (grains per dry standard cubic feet) of exhaust gases. Controlled concentrations typically range from 0.002 to 0.08 gr/dscf. Incinerators burning waste with high solids or ash content emit higher grain loadings and larger particle sizes. Most engineers in the HWI field define ash as the solid residue that remains after the waste–fuel mixture is incinerated. Ash produced during incineration is primarily inorganic and falls into two basic categories: (a) *Bottom ash* refers to the ash discharged from the exhaust end of the primary combustion chamber after incineration and is normally associated with inerts. The composition of the ash depends on the composition of the waste being incinerated and can, therefore, vary greatly. (b) Fly *ash* consists of the ash that is entrained in exhaust gases leaving the incinerator and that is usually captured in the APC equipment.

Because HWIs are designed for complete destruction of these organic compounds, the ash normally contains very little carbonaceous material. Solid materials not susceptible to oxidation (e.g., glass or ceramic) constitute the major ash species. The amount of ash produced is very small in relation to total mass of waste incinerated. The relative proportion of fly ash to bottom ash is influenced by the waste composition and the incinerator design and operation. As expected, relatively little bottom ash and fly ash result when liquid or gaseous hazardous wastes are incinerated. It is common practice for purposes of emission calculations for liquid injection incinerators to assume that all the ash present in the waste–fuel mixture will *fly*, that is, will be carried out and appear in the flue gas.

There are four classes of air pollution equipment that are employed for acid gas and PM control: wet scrubbers, dry scrubbers, wet–dry scrubbers, and combinations of wet–dry–wet scrubbers. The selection procedure for choosing one of these control devices depends on the inlet acid gas loading, potential acid removal capability, particulate loading, particle size distribution, and regulatory requirements. In any event, the technologies for these devices are well established.[3]

Most waste combustion facilities today employ one of four possible process schemes for air pollution control:

1. *Wet Scrubber System*
   a. Venturi scrubber (VS for particulates) followed by a packed-bed absorber scrubber (PBS) for gases. The VS could be a low ΔP (LVS) or high ΔP (HVS) scrubber based on the PM loading and size. If the VS is used only as a primary quench upstream of the PBS, an LVS is used.
   b. Wet electrostatic precipitator (WESP) for PM combined with a PBS (for gases).
   c. Ionizing wet scrubber (IWS) for PM followed by a PBS (for gases).
   d. Packed-bed absorber towers (PBS) used for absorbing and neutralizing acid gases directly after an HVS, which removes the PM. It is also used in a wet–dry–wet system downstream of a dry scrubber for more efficient removal of acid gases.
   e. Future trends in the wet scrubber area including the Dyna-Wave Unit, the Multi-Micro-Venturi (MMV), the condensing scrubber, and the collision scrubber will be briefly described in this section.
2. *Dry Scrubber System* Installed after boiler or quench system: Dry lime or caustic injection system followed by either
   a. Electrostatic precipitator (ESP) for PM removal only.
   b. Fabric filter (FF) baghouse for PM removal with a percentage of acid gas neutralization on bag surface.
3. *Wet–Dry Scrubber System*
   a. Spray dryer absorber (SDA) for quench and neutralization of acid gases followed by an ESP or FF for PM removal.
4. *Wet–Dry–Wet Scrubber System* These systems are a combination of the wet–dry scrubber followed by a wet scrubber system. They are briefly described here, but no detailed description covering this design is included in this chapter.
   a. SDA for quench and neutralization followed by PM control device as noted above followed by PBS tower for quenching and final acid gas removal to meet emission standards for HCl and SO_2 not attained by systems in 2 or 3 above.
   b. SDA for quench and neutralization followed by ESP or FF for PM removal followed by low-pressure drop quencher (ST or VS) followed by a PBS for final acid gas removal.

Municipal solid waste (MSW) incinerators, on the other hand, have traditionally employed either an ESP or FF after the waste heat boiler. Venturi scrubbers have been used on medical waste incinerators. When burning hazardous wastes the APC is located between the incinerator and the induced-draft fan or stack. If a quencher or waste heat boiler is employed, the APC is located downstream of these units.

10.2 WET SCRUBBERS

Wet scrubbers have found widespread use in cleaning contaminated gas streams because of their ability to effectively remove both particulate and gaseous pollutants. Specifically, *wet scrubbing* describes the technique of bringing a contaminated gas stream into intimate contact with a liquid. In this section, the term *scrubber* will be restricted to those systems that utilize a liquid, usually water, to achieve or assist in the removal of particulate matter from a carrier gas stream.

The particulate collection mechanisms involved in the wet scrubbing operation may include some or all of the following:

- Inertial impaction
- Direct interception
- Diffusion (Brownian motion)
- Electrostatic forces
- Condensation
- Thermal gradients

The types of scrubbers for particulate control reviewed in this section are spray towers (ST), PBS units, wet electrostatic precipitators (WESP), ionizing wet scrubbers (IWS), and venturi scrubbers (LVS or HVS). More extensive details are presented for the venturi since this control device is often used in HWI facilities for high-efficiency particulate control.

Spray Towers

The most common low-energy scrubbers are *gravity spray towers* in which liquid droplets are made to fall through rising exhaust gases and are drained at the bottom of the chamber (see Fig. 10.2.1). The droplets are usually formed by liquid atomized in spray nozzles. The spray is directed into a chamber shaped to conduct the gas through the finely divided liquid. In a vertical tower, the relative velocity between the droplets and the gas is eventually the terminal settling velocity of the droplets. To avoid spray droplet reentrainment, however, the terminal settling velocity of the droplets must be greater than the velocity of the rising gas stream. In practice, the vertical gas velocity typically ranges

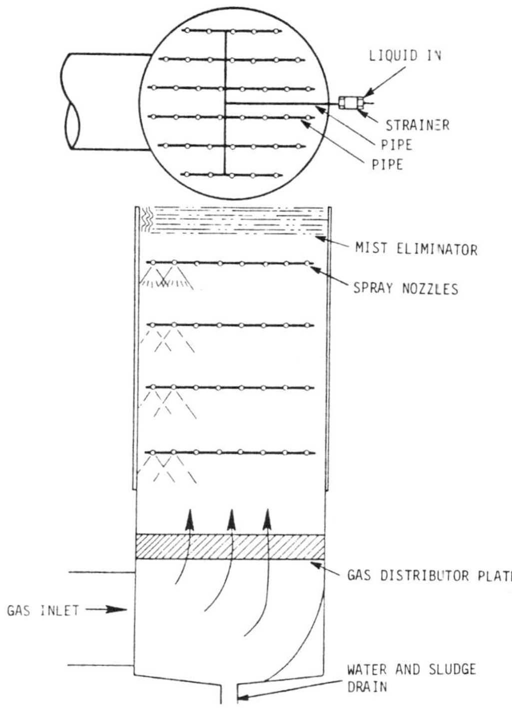

Figure 10.2.1 Spray scrubber.

from 2 to 5 ft/s. For higher velocities, a mist eliminator must be used in the top of the tower.

Spray towers (ST) are suited for both particle collection and mass transfer (gas absorption), as are all wet scrubbers. Operating characteristics include low-pressure drop (typically <1–2 in. H_2O exclusive of mist eliminator and gas distribution plate), ability to handle liquids having a high solids content, and liquid requirements ranging from 3 to 30 gal/1000 ft^3 of gas treated. Spray towers are capable of handling large gas volumes and are often used as quenchers (precoolers) as described in Chapter 9 to reduce gas stream temperatures. Gas rates from 800 to 2500 lb/h-ft^2 are typical. Gas retention times within the tower typically range from 2 to 5 s. The chief disadvantage of spray towers is their relatively low scrubbing efficiency for particles in the 0- to 5-μm range.

Venturi Scrubbers

To achieve high collection efficiency of particulates by *impaction,* a small droplet diameter and high relative velocity between the particle and droplet are required. In a VS this is often accomplished by introducing the scrubbing liquid at right angles to a high-velocity gas flow in the venturi throat (vena contracta). Very small water droplets are formed, and high relative velocities are maintained until the droplets are accelerated to their normal velocity. Gas velocities through the venturi throat typically range from 12,000 to 24,000 ft/min. The velocity of the gases alone causes the atomization of the liquid. The liquid may be introduced through atomizers or by overflow of a weir at the upper end of the VS and flow by gravity along the inside surface of the entrance cone. This is preferable to atomizers that tend to plug with recycled scrubber liquids. The energy expended in the scrubber is accounted for by gas stream pressure drop through the scrubber. Another factor important to the effectiveness of the VS is the conditioning of the particulates by condensation. If the gas in the reduced pressure region in the throat is fully saturated, or superheated (preferably), some condensation will occur on the particulates in the throat due to the cooling effect of the scrubbing liquid.

The venturi itself is only a gas conditioner and must be followed by a separating section for the elimination of entrained droplets (see Fig. 10.2.2). Water is injected into the venturi in quantities ranging from 6 to 20 gal/1000 actual ft^3 of gas. Very high collection efficiencies are achievable with operating pressure drops of 25–100 in. H_2O and are not at all uncommon. For example, a 10 in. H_2O pressure drop VS can typically remove particles larger than 2 μm with virtually 100% efficiency, while a 100 in. H_2O pressure drop venturi is often required to remove particles as small as 0.3–0.4 μm. Since collection efficiency is directly related to pressure drop, variable-throat venturis have been introduced to maintain pressure drop with varying gas flows.

One of the more popular and widely used collection efficiency equations is that originally suggested by Johnstone et al.[4]

$$E = 1 - e\ -KR(\Psi)^{0.5} \tag{10.2.1}$$

where E = efficiency (fractional)
Ψ = inertial impaction parameter (dimensionless)
R = liquid-to-gas ratio (gal/1000 acf or gpm/1000 acfm)
K = correlation coefficient, the value of which depends on the system geometry and operating conditions (typically 0.1–0.2 acf/gal)

The term ψ is given by

$$\Psi = \frac{C d_p^2 \rho_p \nu_t}{9 \mu_G d_1} \tag{10.2.2}$$

where d_p = particle diameter (feet)
ρ_p = particle density (lb/ft^3)
ν_t = throat velocity (ft/s)
μ_G = gas viscosity (lb/ft-s)
d_1 = mean droplet diameter (ft)
C = Cunningham correction factor

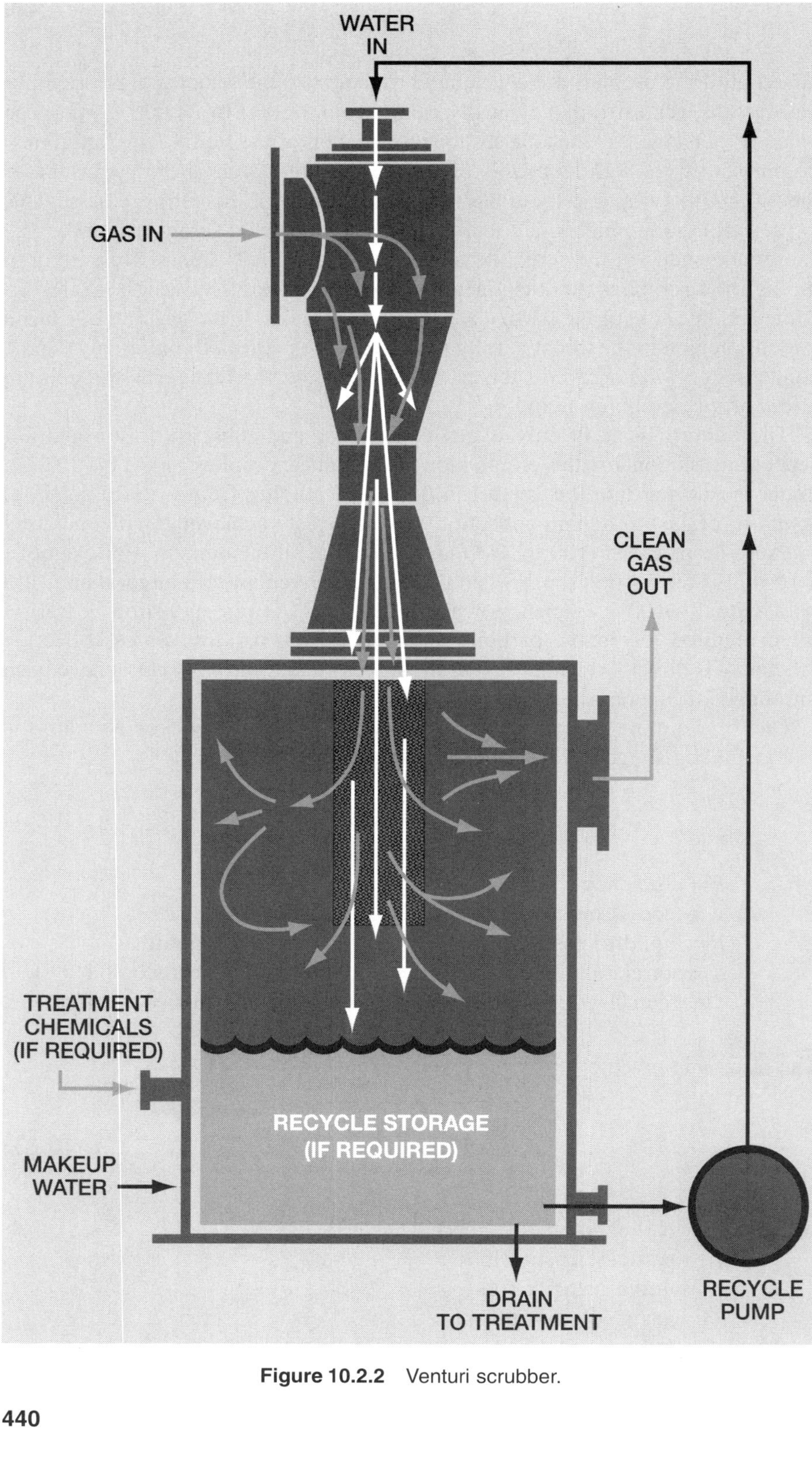

Figure 10.2.2 Venturi scrubber.

Values for the Cunningham correction factor are provided in Table 10.2.1. This correction is usually neglected in scrubber calculations, but the effect becomes more pronounced as the particle size decreases, particularly below 1 μm. The mean droplet diameter, d_l, for standard air and water in a venturi scrubber is given by the Nukiyama and Tanasawa relationship:

$$d_l = \frac{16{,}400}{v_t} + 1.45R^{1.5} \qquad (10.2.3)$$

The pressure drop for gas flowing through a venturi scrubber can be estimated from knowledge of liquid acceleration and frictional effects along the wall of the equipment. Frictional losses depend largely on the scrubber geometry and are usually determined experimentally. The effect of liquid acceleration is, however, predictable. Equation (10.2.4) for estimating pressure drop through venturi scrubbers (given as a function of throat gas velocity and liquid-to-gas ratio) assumes that all the energy is used to accelerate the liquid droplets to throat velocity:

$$\Delta P' = 5 \times 10^{-5} v_t^2 R \qquad (10.2.4)$$

where $\Delta P'$ = pressure drop (in. H_2O).

Another somewhat simpler equation that applies over a fairly wide range of R's is[5]

$$\Delta P^* = 0.8 + 0.12R \qquad (10.2.5)$$

where ΔP^* is a dimensionless pressure drop equal to the pressure drop divided by a velocity head.

Studies by Hesketh[6] show that the pressure drop predictions obtained from throat velocity measurements may be subject to error at low velocities if Eq. (10.2.4) is applied for all ranges of velocities. Hesketh's equation for venturi scrubbers that have liquid injected before the throat[7] is given by

Table 10.2.1 Cunningham correction factors for air at atmospheric pressure

Particle diameter (μm)	70°F	212°F	500°F
0.1	2.88	3.61	5.14
0.25	1.682	1.952	2.525
0.5	1.325	1.446	1.711
1.0	1.160	1.217	1.338
2.5	1.064	1.087	1.133
5.0	1.032	1.043	1.067
10.0	1.016	1.022	1.033

$$\Delta P' = \frac{v_t^2 \rho_G A_t^{0.133} R^{0.78}}{1270} \tag{10.2.6}$$

where ρ_G = gas density (lb/ft^3)
A_t = throat cross-sectional area (ft^2)

Contact power theory is an empirical approach relating particulate collection efficiency and pressure drop in wet scrubber systems. The concept is an outgrowth of the observation that particulate collection efficiency in spray-type scrubbers was mainly determined by pressure drop for the gas plus any power expended in atomizing the liquid. Contact power theory assumes that the particulate collection efficiency in a scrubber is solely a function of the total pressure loss in the unit. The total power loss (p_t) is assumed to be composed of two parts: the power loss of the gas passing through the scrubber (p_G) and the power loss of the spray liquid during atomization (p_L). These two terms are given here in equation form:

$$p_G = 0.157\Delta P' \tag{10.2.7}$$

where p_G = contacting power based on gas stream energy input (hp/1000 acfm)
$\Delta P'$ = pressure drop across the scrubber (in. H_2O)

and

$$p_L = 5.83 \times 10^{-4} P_L R \tag{10.2.8}$$

where p_L = contacting power based on liquid energy input (hp/1000 acfm)
P_L = liquid inlet pressure (psia)
R = liquid-to-gas ratio (gal/1000 acf)

Then,

$$p_t = p_G + p_L$$

To correlate contacting power with scrubber collection efficiency, the latter is best expressed as the number of transfer units. The number of transfer units is defined by analogy to mass transfer and given by

$$N_t = \ln\left(\frac{1}{1 - E}\right) \tag{10.2.9}$$

where N_t = number of transfer units (dimensionless)
E = fractional collection efficiency (dimensionless)

The relationship between the number of transfer units and collection efficiency is by no means unique. The number of transfer units for a given value of contacting power (hp/1000 acfm) or vice versa, varies over nearly an order of magnitude. For example, at 2.5 transfer units ($E = 0.918$), the contacting power ranges from approximately 0.8 to 10.0 hp/1000 acfm, depending on the scrubber and the particulate.

For a given scrubber and particulate properties, there will usually be a very distinct relationship between the number of transfer units and the contacting power. Semrau plotted the number of transfer units (N_t) for a series of scrubbers and particulates against total power consumption; a linear relation on a log–log plot was obtained.[8] This relationship is independent of the type of scrubber and can be expressed by[9]

$$N_t = \alpha p_t^{\beta} \tag{10.2.10}$$

where α, β = parameters for the type particulates being collected (see Table 10.2.2)

When a combustion gas contains a high particulate loading, as well as one or more of the gaseous pollutants discussed previously, a venturi scrubber is often used in conjunction with a packed-bed or plate-tower scrubber. The venturi scrubber removes the particulates from the stream to prevent fouling of the packed-bed or plate-tower absorber and may also remove a significant fraction of gases highly soluble in water. However, venturi scrubbers alone are not

Table 10.2.2 Constants for use with Eq. (10.2.10)

Aerosol	Scrubber type	α	β
Raw gas (lime dust and soda fume)	Venturi and cyclonic spray	1.47	1.05
Prewashed gas (soda fume)	Venturi, pipe line, and cyclonic spray	0.915	1.05
Talc dust	Venturi	2.97	0.362
Black liquor recovery furnace fume	Orifice and pipe line	2.70	0.362
Cold scrubbing water humid gases	Venturi and cyclonic spray	1.75	0.620
Hot fume solution for scrubbing (humid gases)	Venturi, pipeline, and cyclonic spray	0.740	0.861
Hot black liquor for scrubbing (dry gases)	Venturi evaporator	0.522	0.861
Phosphoric acid mist	Venturi	1.33	0.647
Foundry cupola dust	Venturi	1.35	0.621
Open-hearth steel furnace fume	Venturi	1.26	0.569
Talc dust	Cyclone	1.16	0.655
Copper sulfate	Solivore		
	(A) with mechanical spray generator	0.390	1.14
	(B) with hydraulic nozzles	0.562	1.06
Ferrosilicon furnace fume	Venturi and cyclonic spray	0.870	0.459
Odorous mist	Venturi	0.363	1.41

considered suitable for the removal of low solubility gases; when water is used as the scrubbing medium, estimated efficiencies are <50–70%.[10] Venturi scrubbers using water are not suitable for highly efficient (more than 99%) removal of either HCl or HF. Thus, although venturi scrubbers are usually designed for particulate collection, they can be used for simultaneous gas absorption as well. There is no satisfactory generalized design correlation for these types of scrubbers, especially when absorption with a chemical reaction is involved. Reliable design should be based on full-scale data or at least laboratory- or pilot-scale data.

Available data indicate that venturis at HWI facilities operate with pressure drops in the 30–100 in. H_2O range. Liquid-to-gas ratios for venturi scrubbers are usually in the range of 5–20 gal/1000 ft^3 of gas. At existing HWI facilities, liquid-to-gas ratios ranging from 7 to 45 gal/1000 ft^3 of gas have been reported. In many cases, a minimum ratio of 7.5 gal/1000 ft^3 is needed to ensure that adequate liquid is supplied to provide good gas sweeping. Gas velocities for venturi scrubbers operating at HWI facilities are in the 100–400 ft/s range, similar to other industrial applications. The low end of this range, 100–150 ft/s, is typical of power plant applications, while the upper end of the range has been applied to lime kilns and blast furnaces.

If water is the scrubbing fluid for an incinerator burning chlorinated organic compounds, the wastewater effluent will contain suspended particulates, some dissolved HCl (hydrochloric acid), and other soluble constituents that may be present. The venturi scrubbing process also involves either a single pass or recirculation of the scrubbing fluid. If recirculation is used, scrubber fluid is recycled through the venturi scrubber until the total dissolved solids (TDS) content reaches approximately 5%. When this occurs, a portion of the scrubbing fluid is removed (*blowdown*) and new scrubbing fluid is added to make up for the fluid lost as blowdown. The blowdown from the single pass or recirculation scrubbing system is neutralized (as needed) before delivery to on-site wastewater treatment processes, on-site storage facilities (e.g., vaporization ponds), municipal sewer, or a receiving water body.

Wet Electrostatic Precipitators

The wet electrostatic precipitator (WESP) is a variation of the dry ESP design described later in this chapter. The two major added features in a WESP system are

1. A preconditioning step, where inlet sprays in the entry section are provided for cooling, gas absorption, and removal of coarse particles. A more recent application utilizes a vertical tube, water cooled on the outside causing condensation of water vapor carried in the flue gases.
2. A wetted collection surface, where the liquid (either sprayed into the gases or condensed from the gases) is used to continuously flush away collected materials.

Particle collection is achieved by the introduction of liquid to the gas stream through either sprays located above the electrostatic field sections or condensation of the water contained in the saturated gases, and migration of the charged particles and sprayed liquid droplets to the collection plates. The collected liquid droplets or the condensed liquid in the gas stream form a continuous downward-flowing film over the collection plates, and keep them clean by removing the collected particles. In the condensing design, the liquid is that formed by condensation on the water-cooled surface of the collector tube.

The WESP overcomes some of the limitations of the dry ESP. Its operation is not influenced by the resistivity of the particles. The resistivity of the water film (which is very low) is the governing factor in the dust discharging process. The particles carry a stronger charge in a WESP, enhancing the performance of the device.[11] Furthermore, since the internal components are continuously being washed with liquid, buildup of tacky particles is controlled, and there is some capacity for removal of gaseous pollutants. In general, applications of the WESP fall into two areas: removal of fine particles and removal of condensed organic fumes. Outlet particulate concentrations are typically in the 10^{-3} to 10^{-2} gr/ft^3 range.[12] There is limited data on the acid gas collection efficiency of a WESP. This device has been installed to control HF emissions. Using a liquid/gas (L/G) ratio of 5 gal/1000 acf and a liquid pH between 8 and 9, fluoride removal efficiencies >98% have been measured; outlet concentrations of HF were found to be <1 ppm.[13]

At the present time, there are WESP installations at HWI facilities worldwide. A typical application is the use of the WESP in conjunction with a low-pressure-drop VS upstream, where a major portion of the gaseous contaminants and heavy particles will be removed. The WESP will then serve as a second-stage control device for removal of the submicron particles, in particular the metal fumes, and remaining gaseous pollutants. Installation of a Lurgi WESP has been operating in Germany since 1991 with excellent results. A U.S. vendor, Croll Reynolds, Co., Inc., Westfield, NJ, indicates that pilot units are available for testing at owner's facilities on a slip stream from the existing stack gases. With the MACT standards issued in September 1999, many facilities will require retrofit to meet the lower PM emission standards. Pilot testing is recommended prior to to full-scale design and installation.[14]

Some of the advantages of a WESP include

- Simultaneous gas absorption and PM removal
- Low energy consumption
- No PM resistivity problems
- Efficient removal of submicron particles

Disadvantages of the WESP are

- Low gas absorption efficiency

- Sensitive to changes in flow rate
- PM collection is wet

Ionizing Wet Scrubbers

The ionizing wet scrubber (IWS) was a relatively new development in the technology of removal of particulate matter from a gas stream in the early 1980s. These devices have been incorporated in on-site as well as off-site commercial incineration facilities.[10,15]

In the IWS, high-voltage ionization in a charge section places a static electrical charge on the particles in the gas stream, which then passes through a cross-flow packed-bed scrubber. The packing is polypropylene in the form of circular-wound spirals and gearlike wheel configurations, providing a large surface area. Particles with sizes of 3 μm or larger are trapped by inertial impaction within the packed bed. Smaller charged particles pass close to the surface of either the packing material or a scrubbing water droplet. An opposite charge in that surface is induced by the charged particle, which is then attracted to, and attached to, the surface. All collected particles are eventually washed out of the scrubber. The scrubbing water also functions to absorb gaseous pollutants.

According to Ceilcote Air Pollution Control, Strongsville, OH, (an IWS vendor), collection efficiency of a two-stage IWS is greater than that of a baghouse or a conventional ESP for particles in the 0.2- to 0.6-μm range. For 0.8 μm and above, the efficiency of bag collectors is greater; at 1.8 μm and above, the ESP is as effective as the IWS.[12] Scrubbing water can include caustic soda or soda ash when needed for efficient absorption of acid gases. Corrosion resistance of the IWS is achieved by fabricating its shell and most internal parts from glass-fiber-reinforced plastic (FRP) and thermoplastic materials. Pressure drop through a single-stage IWS is around 5 in. H_2O (primarily through the wet scrubber section). All internal areas of the ionizer section are periodically deluge flushed with recycled liquid from the scrubber recycle system.

The advantages of an IWS are those of the combination of a WESP and a packed-bed absorber. The disadvantages include

- Need for separation of PM from the scrubbing medium, potential scaling and fouling problems (aggravated by recycle neutralizing solution)
- Possible damage to the scrubber if scrubber solution pumps fail
- Additional cooling air over the electrical equipment, which increases the volume flow into the induced-draft fan
- Maintaining a maximum velocity through the electrode/plate section, which sets the requirement for a number of parallel stages
- Need for multiple stages to ensure collection efficiency and the need for a downstream mist eliminator

Despite some of these limitations, it is a particulate control device that has worked efficiently at a number of HWI facilities.

Absorbers

Packed-bed scrubbers (PBS), a major type of wet scrubber applied as the APC in waste combustion systems, find normal application in the removal of acid gases and vapors. It is not considered as a PM control device, especially for submicron particulates. To date, the major problems in using packed-bed scrubbers for particulate removal are plugging of the packing and the maintenance problems subsequently incurred. A general rule of thumb is to allow a maximum of 0.2 gr/ft^3 of dust to enter a packed tower.

The process of absorption conventionally refers to the intimate contacting of a mixture of gases with a liquid so that part of one or more of the constituents of the gas will dissolve in the liquid. In an HWI facility, the contact usually takes place in some type of packed column. (The use of the packed column as a quenching device was discussed in Chapter 9. In this section, its use as a gas-cleaning device will be covered.)

Gas absorption as applied to the control of air pollution is concerned with the removal of one or more pollutants from a contaminated gas stream by treatment with a liquid. Consider, for example, the process taking place when a mixture of air and gaseous HCl is brought into contact with H_2O. The HCl is soluble in H_2O, and those molecules that come into contact with the water surface dissolve immediately. However, the HCl molecules are initially dispersed throughout the gas phase, and they can only reach the water by gas-phase diffusion. When the HCl at the water surface has dissolved, it is distributed throughout the water phase by a second diffusional process. Consequently, the rate of absorption is determined by the rates of diffusion in both the gas and the liquid phases. Equilibrium is another important factor to be considered in controlling the operation of absorption systems. The rate at which the pollutant will diffuse into an absorbent liquid will depend on the departure from equilibrium that is maintained. The rate at which equilibrium is established is then essentially dependent on the rate of diffusion of the pollutant through the nonabsorbed gas and through the absorbing liquid.

Packed columns, as described in Chapter 9, are used for the continuous contact between liquid and gas. The countercurrent packed column (see Fig. 10.2.3) is the most common type of unit encountered in gaseous pollutant control for the removal of the undesirable gas, vapor, or odor. This type of column has found widespread application in both the chemical and pollution control industries. The gas stream containing the pollutant moves upward through the packed bed against an absorbing or reacting liquid that is injected at the top of the packing. This results in the highest possible efficiency. Since the pollutant concentration in the gas stream decreases as it rises through the column, there is constantly fresher liquid available for contact. This provides maximum average driving force for the diffusion process throughout the bed.

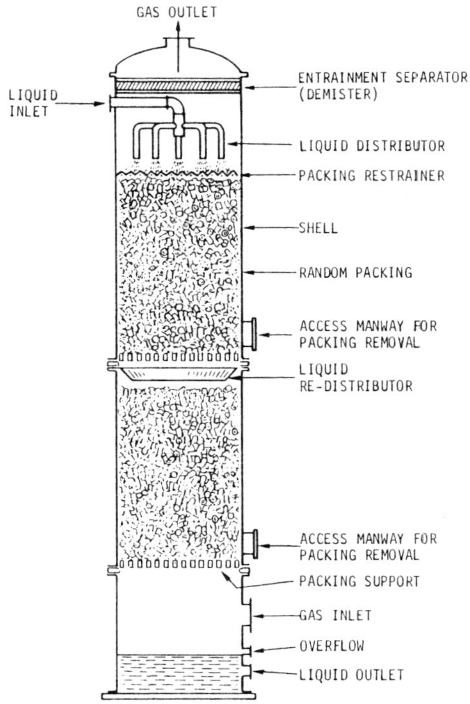

Figure 10.2.3 Countercurrent packed column.

Liquid distribution plays an important role in the efficient operation of a packed column. A good packing from the process viewpoint can be reduced in effectiveness by poor liquid distribution across the top of its upper surface. Poor distribution reduces the effective wetted packing area and promotes liquid channeling. The final selection of the mechanism of distributing the liquid across the packing depends on the size of the column, type of packing, tendency of packing to divert liquid to column walls, and materials of construction for distribution. For stacked packing, the liquid usually has little tendency to cross distribution and thus moves down the column fairly uniformly in the cross-sectional area that it enters. In the dumped condition, most packings follow a conical distribution down the column, with the apex of the cone at the liquid impingement point. For well-distributed liquid flow and reduced channeling of gas and liquid to produce efficient use of the packed bed, the impingement of the liquid onto the bed must be as uniform as possible. The liquid coming down through the packing and on the inside wall of the column should be redistributed after a bed depth of approximately 3 column diameters for Raschig rings and 5–10 column diameters for saddle packing. As a guide, Raschig rings usually have a maximum of 10–15 ft of packing per section, while saddle packing can use a maximum of 12–20 ft. As a general rule of thumb, however, the liquid should be redistributed every 10 ft of packed height. The redistribution brings the liquid off the wall and outer portions of the column and directs it toward the center area of the column. Redistribution is usually not necessary for stacked bed packings, as the liquid flows essentially in vertical streams.

Packed columns are characterized by a number of features to which their widespread popularity may be attributed:

- *Minimum Structure* The packed column needs only a packing support and liquid distributor about every 10 ft along its height.
- *Versatility* The packing material can be changed by simply dumping it and replacing it with a type giving better efficiency, lower pressure drop, or higher capacity. The depth of packing can also be easily changed if efficiency turns out to be less than anticipated, or if feed or product specifications change.
- *Corrosive-Fluids Handling* Ceramic packing is common and often preferable to metal or plastic because of its corrosion resistance. It may also be preferred at the base or inlet of a tower when handling hot combustion gases, as in a hazardous waste incineration facility.
- *Low Pressure Drop* Unless operated at very high liquid rates, where the liquid becomes the continuous phase as its films thicken and merge, the pressure drop per linear foot of packed height is relatively low.
- *Low Investment* When plastic packings are satisfactory or when the columns are less than 3–4 ft in diameter, cost is relatively low.

The packing is the heart of the performance of this type of equipment. Its proper selection entails an understanding of packing operational characteristics. The main points to be considered in choosing the column packing include

- *Durability and Corrosion Resistance* The packing should be chemically inert to the fluids being processed.
- *Free Space per Cubic Foot of Packed Space* This controls the liquid hold-up in the column, as well as the pressure drop across it.
- *Wetted Surface Area per Unit Volume of Packed Space* This affects the pressure drop across the column.
- *Packing Resistance to the Flow of Gas* This affects the pressure drop across the column.
- *Packing Stability and Structural Strength to Permit Easy Handling and Installation*
- *Weight per Cubic Foot of Packed Space*
- *Cost per Square Foot of Effective Surface*

The pressure drop through a packed column for any combination of liquid and gas flows in the operable range is an important economic consideration in the design of such columns. For most random packings, the pressure drop suffered by the gas is influenced by the gas and liquid flow rates. At constant gas rate, an increase in liquid throughput—which takes up more room in the packing (increased holdup) and so leaves less room for the gas (greater restriction)—is accompanied by an increase in pressure drop until the liquid flooding rate is reached. At this point, any slight excess that cannot pass through remains at the top of the packing, building up a deeper and deeper head (or pressure drop), hypothetically reaching an infinite value. Similarly, at constant liquid downflow, increasing gas flow is again accompanied by rising pressure drop until the flooding rate is reached, whereupon the slightest increase will cause a decline in permissible liquid throughput. This causes the remainder to again accumulate above the packing, so that pressure again increases. For a particular packing, the most accurate pressure drop data will be those available directly from the manufacturer. However, for the purposes of estimation, Figure 10.2.4 is simple to use and usually gives reasonable results.

Some general rules of thumb in the design of packed columns do exist, but these must be applied discriminately because there are other considerations that might have to be taken into account, for example, allowable pressure drop, possible column height restrictions, and so on. For approximation purposes, if

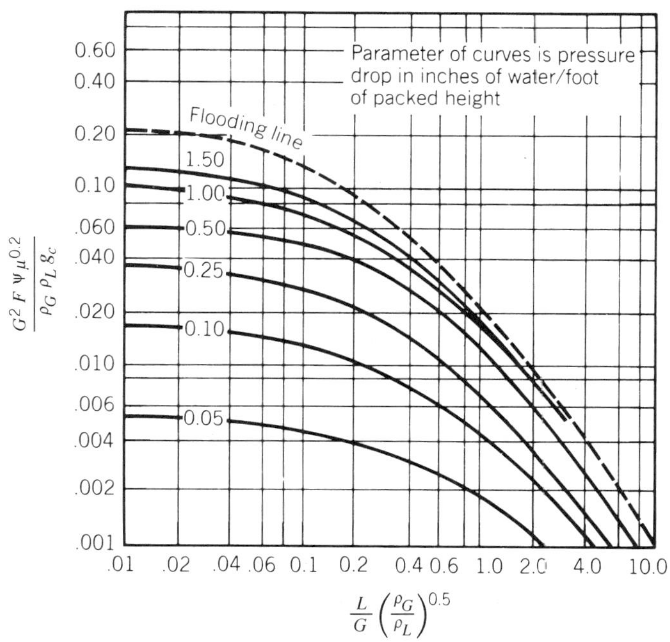

Figure 10.2.4 Generalized pressure drop correlation for packed towers: L = liquid rate (lb/s-ft^2); G = gas rate (lb/s-ft^2); ρ_L = liquid density (lb/ft^3); ρ_G = gas density (lb/ft^3); F = packing factor; μ = viscosity of liquid (cP); ψ = ratio of densities of water to liquid; and g_c = constant = 32.2 lb-ft/lb$_f$-s^2.

the gas rate is greater than about 500 acfm, a nominal packing size <1 in. would probably not be practical; similarly, at about 2000 acfm, sizes <2 in. would likely also be impractical. The nominal size of the packing should never exceed about $\frac{1}{20}$th of the column diameter. The usual practice is to design so that the operating gas rate is approximately 60% of the rate that would cause flooding. If possible, column dimensions should be in readily available sizes (i.e., diameters to the nearest foot and heights to the nearest 0.5 ft). If the column can be purchased "off the shelf" as opposed to being specially made, a substantial savings can be realized.

Plate columns may also be employed as absorbers, although they are rarely used. These devices are essentially vertical cylinders in which the liquid and gas are contacted in stepwise fashion on plates or trays, in a manner shown schematically in Figure 10.2.5. The liquid enters at the top and flows downward via gravity. On the way, it flows across each plate and through a downspout to the plate below. The gas passes upward through openings in the plate, then bubbles through the liquid to form a froth, disengages from the froth, and passes on to the next plate above. The overall effect is a multiple countercurrent contact of gas and liquid. Each plate of the column is a stage since on the plate the fluids are brought into intimate contact, interphase diffusion occurs, and the fluids are separated. The number of theoretical plates is dependent on the difficulty of the separation to be carried out and is determined solely from material balances and equilibrium considerations. The diameter of the column,

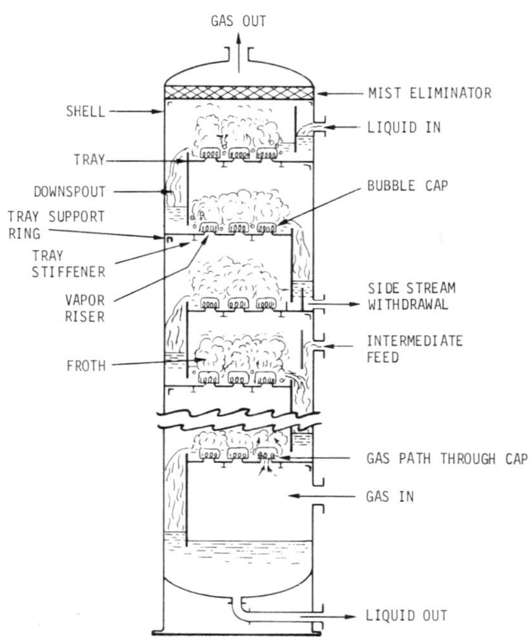

Figure 10.2.5 Plate column.

on the other hand, depends on the quantities of liquid and gas flowing through the column per unit time. The *actual* number of plates required for a given separation is greater than the *theoretical* number due to plate inefficiency.

In bubble-cap plates, the vapor moves upward through *risers* into the bubble cap, out through the slots as bubbles, and into the surrounding liquid on the plates. Figure 10.2.6 demonstrates the vapor–liquid action for a bubble-cap plate. The bubble-cap plate design is the most flexible of plate designs for high and low vapor and liquid rates. On the average, plates are usually spaced approximately 24 in. apart.

The design of the absorption tower is similar in some aspects to that for the water-quenching packed tower discussed in Section 9.3. Assuming that the liquid and gas superficial mass velocities (again as measured near the bottom of the packing where these quantities are at their maximum values) are known, Eq. (9.3.9) can be used to determine the column area and diameter (9.3.3). The

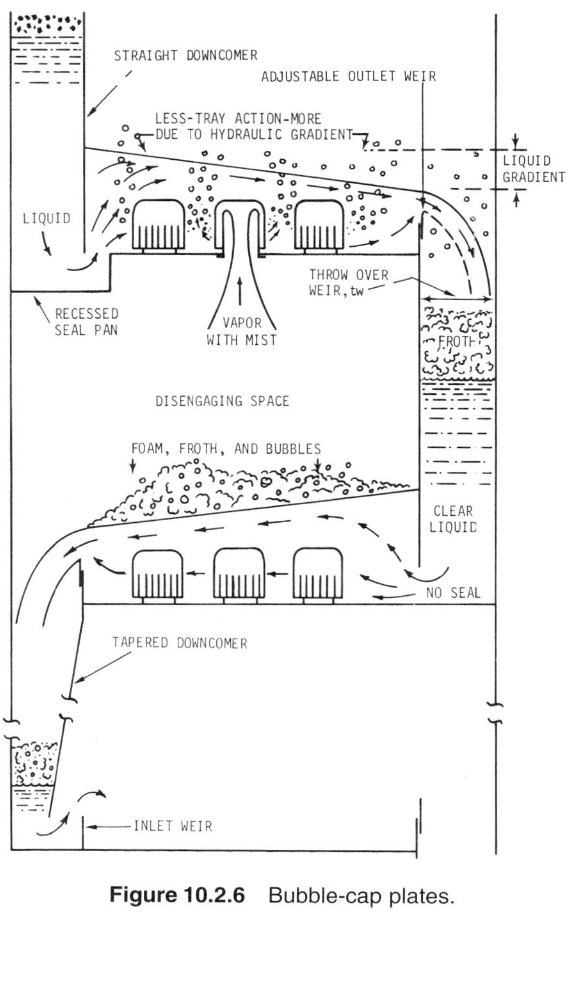

Figure 10.2.6 Bubble-cap plates.

pressure drop may be evaluated directly from Figure 10.2.4. The column height is given by

$$Z = N_{OG} H_{OG} \tag{10.2.11}$$

where N_{OG} = number of overall transfer units (dimensionless)
H_{OG} = height of overall transfer unit (ft)
Z = height of packing (ft)

In actual gas absorption design practice, the number of transfer units (N_{OG}) is obtained experimentally or calculated using any of the methods to be explained later in this section. The height of a transfer unit (H_{OG}) is also usually determined experimentally for the system under consideration. Information on many different systems using various types of packings has been compiled by the manufacturers of gas absorption equipment and should be consulted prior to design. The data are usually in the form of graphs depicting, for a specific system and packing, the H_{OG} versus the gas rate (in lb/h-ft^2) with the liquid rate (in lb/h-ft^2) as a parameter. The packing height Z is then simply the product of the H_{OG} and the N_{OG}.

In incineration operations, the pollutant to be absorbed (usually HCl) is in the very dilute range. If Henry's law applies (see Section 4.5), the number of transfer units is given by

$$N_{OG} = \ln\left[\left(\frac{y_1 - mx_2}{y_2 - mx_2}\right)\left(1 - \frac{1}{A}\right) + \frac{1}{A}\right] \Big/ \left(1 - \frac{1}{A}\right) \tag{10.2.12}$$

where y_1 = inlet mole fraction of pollutant in vapor stream (dimensionless)
y_2 = outlet mole fraction of pollutant in vapor stream (dimensionless)
m = Henry's law constant (dimensionless)
x_2 = outlet mole fraction of pollutant in liquid stream (dimensionless)
A = absorption factor = L_m/mV_m (dimensionless)
L_m = liquid superficial mole velocity (lbmol/h-ft^2)
V_m = vapor superficial mole velocity (lbmol/h-ft^2)

The solution to this equation can conveniently be found graphically from Figure 10.2.7.

The term m represents the slope of the *equilibrium* line, which according to Henry's law should appear as a straight line on an x-y plot. A line describing the actual (x, y) operating points within the absorber may also be graphed on the same coordinates; this line is determined by means of a material (mol) balance around the column. (The slope of the *operating* line is given by the ratio of the liquid-to-gas molar flow rates.)[3] For most air pollution applications where small quantities of pollutants are absorbed, both the equilibrium and operating lines may be assumed to be straight. When a highly soluble gas like

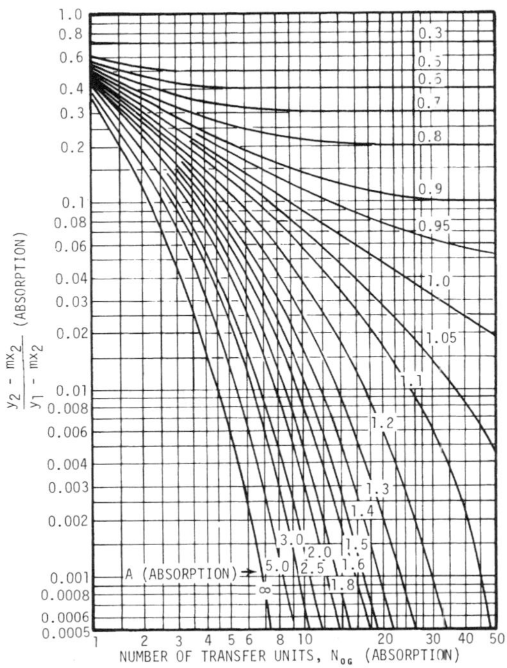

Figure 10.2.7 N_{OG} for absorption columns with constant absorption factor (A).

HCl is absorbed or when a chemical reaction consumes the absorbate (pollutant), the value of m approaches zero, and Eq. (10.2.12) reduces to

$$N_{OG} = \ln\left(\frac{y_1}{y_2}\right) = \ln\left(\frac{w_1}{w_2}\right) \tag{10.2.12a}$$

where w_1 = inlet mass flow rate of solute in vapor stream (lb/h)
w_2 = outlet mass flow rate of solute in vapor stream (lb/h)

The pressure drop through a packed column, which is a function of tower size, flow rates, type of packing, and so on, is also an important design consideration for packed columns. The reader is again referred back to Section 9.3 for a discussion on this topic.

The most important design considerations for plate columns include calculation of the column diameter, type and number of plates to be used, actual plant layout and physical design, and plate spacing (which, in turn, determines the column height).

The column diameter and consequently its cross section must be sufficiently large to handle the gas and liquid at velocities that will not cause flooding or excessive entrainment. The superficial gas velocity for a given type of plate at flooding is given by the relation

$$\nu_f = C_f[(\rho_L - \rho_G)/\rho_G]^{0.5} \tag{10.2.13}$$

where ν_f = gas velocity through the net column cross-sectional area for gas flow (ft^3/s-ft)

ρ_L, ρ_G = liquid and gas densities, respectively (lb/ft ^{3})

C_f = empirical coefficient that depends on the type of plate and operating conditions

The net cross section is the difference between the column cross section and the area taken up by downcomers and unused plate space. In actual design, some percentage of v_f is usually used for nonfoaming liquids, 80–85% of ν_f, and 75% or less for foaming liquids.

For cases where both the operating line and the equilibrium line curve may be considered straight (dilute solutions), the number of theoretical plates may be determined directly without recourse to graphical techniques. This will frequently be the case for a relatively dilute gas (as usually encountered in air pollution control) and liquid solutions where, more often than not, Henry's law is usually applicable. Since the quantities of gas absorbed are small, the total flows of liquid and gas entering and leaving the column remain essentially constant. Hence the operating line will be substantially straight. For such cases, the Kremser–Brown–Souders equation for determining the number of theoretical plates (N_p) applies:

$$N_P = \left[\left(\frac{y_{N_P+1} - mx_0}{y_1 - mx_0}\right)\left(1 - \frac{1}{A}\right) + \frac{1}{A}\right] \Big/ \ln A \tag{10.2.14}$$

where y_{N_P+1} = inlet mole fraction of solute in vapor stream

y_1 = outlet mole fraction of solute in vapor stream

x_0 = inlet mole fraction of solute in liquid stream

A = absorption factor = L_m/mV_m

Here mx_0 is the gas composition in equilibrium with the entering liquid (m is Henry's law constant slope of equilibrium curve). If the entering liquid contains no solute gas, then $x_0 = 0$ and this equation can be simplified further. The solute concentration in the gas stream, y_{N_P+1} and y_1 represent inlet and outlet conditions, and L_m and V_m the total molar flow rates of liquid and gas per unit column cross-sectional area. Equation (10.2.14) has been plotted in Figure 10.2.8 for convenience and may be used for solution to this equation.

Water is often employed in absorbers to control emissions. When water is used as the scrubbing medium, an acidic stream is produced that must be neutralized prior to discharge, unless the HCl concentration in the incinerator off-gases is high enough to make HCL recovery economically attractive.[10,15] Hydrochloric acid concentration in the scrubbing liquor is normally limited to 1–2% by adjusting the makeup water and gas flow rates. However, higher acid concentrations can be achieved through some modest innovative design pro-

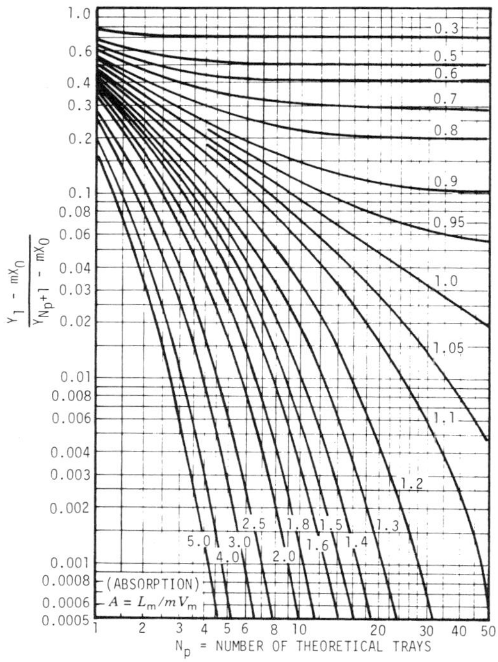

Figure 10.2.8 Number of theoretical stages for absorption columns with constant absorption factor (*A*).

cedures that can include stage operations and liquid recycling.[16] The scrubber must also be lined with or made of acid-resistant material.

Caustic solution (typically 10–20 wt % caustic soda in water) is also commonly used in packed-bed absorbers to control HCl and HF emissions. Because these compounds react with caustic, the driving force for mass transfer is increased and more efficient removal is achieved at the same liquid-to-gas ratio and packing depth. Neutralization is also achieved in situ if sufficient caustic is supplied for complete conversion of HCl to NaCl. Unlike water scrubbing, caustic scrubbing can also achieve high removal efficiencies for SO_2 and HBr, which are less soluble in water than HCL or HF. When gases such as SO_2 are being scrubbed, the caustic addition rate is adjusted to maintain an alkaline scrubbing medium. Alternatively, the caustic addition rate can be adjusted to substoichiometric levels. This reduces the scrubber water makeup rate needed to maintain a specified acid concentration in the scrubber liquor.

Lime slurry, typically 10–32 wt % $Ca(OH)_2$ in water, can also be used to control emissions of HCl, HF, and SO_2. However, lime slurries are not often used as the scrubbing liquid in packed-bed designs because of plugging problems. The use of lime slurries can lead to plugging of the spray nozzles and cause scale formation on the surfaces of the scrubber equipment, particularly scrubber internal and mist eliminator surfaces. The magnitude of the scaling problem will depend on the levels of HCl, HF, and SO_x in the incinerator exhaust gases. Lime solutions are used in plate tower scrubbers, however, because lime is less expensive than caustic. At several HWI facilities, spray dryers with lime slurry injection are used to control emissions of HCl and SO_2.

Materials of construction are selected, in part, on the basis of the degree of acidity or alkalinity provided by the scrubbing liquid during operation. Deviation from the design pH condition or range may result in deterioration of the scrubber structure in contact with the liquid. Furthermore, maintenance of the pH design condition is important to scrubber liquid absorption efficiency when removing gaseous contaminants.

Quench water and scrubber (venturi and/or absorber) effluents are normally combined for treatment and ultimate disposal. Depending on the scrubbing liquids used and gaseous contaminants removed, the wastewater may contain chlorides, fluorides, sulfites, sulfates, phosphates, bromides, and bromates, as well as particulate matter. Liquid waste streams containing sodium fluoride can be treated with lime or limestone slurry to yield insoluble calcium fluoride. Sulfates and phosphates can also be readily removed from the wastewater stream because of the low solubility of their calcium salts. Therefore, treatment normally includes clarification (to remove particulates), neutralization (to take care of any residual acid or base that may still be present), and dilution (to help control TDS levels). Particulates that are insoluble in the scrubber fluid become suspended solids in the scrubber wastewater. If the particulates dissolve in the scrubber fluid, they contribute to the wastewater's TDS level. Suspended solids in scrubber wastewater generally present few, if any, problems[10] because their concentrations are usually <2.2 gr/ft^3. Suspended solids are usually removed by on-site settling ponds. Overflow from settling ponds can be recycled to the scrubber. Wastewater with either high or low pH levels is usually neutralized by adding acid or base, as necessary, prior to the final discharge to a municipal sewer or receiving stream. A high concentration of total dissolved solids (NaCl, $CaCl_2$, and in some cases, the excess NaOH not used to neutralize HCl) is commonly reduced by piping scrubber effluents to in-plant treatment systems or by diluting with other plant process streams and storing in a holding pond or lagoon.

Future Trends

The basic design of APC systems has remained relatively unchanged since first used in the early part of the twentieth century. Some modest equipment changes and new types of devices have appeared in the last 20 years, but all have essentially employed the same capture mechanisms used in the past. One area that has recently received some attention is hybrid systems—equipment that in some cases operate at higher efficiency more economically than conventional devices. Tighter regulations and a greater concern for environmental control by society has placed increased emphasis on the development and application of these systems.[17]

Several specialty air pollution control devices, particularly in the scrubber area, are receiving more attention. Although on the market for many years, their unique collection mechanism has recently attracted more interest. Four such units include the Dyna-Wave Unit, the Multi-Micro-Venturi (MMV), the Electro-Dynamic Venturi (EDV), the condensing scrubber, and the collision scrubber. These are briefly described below.

The Dyna-Wave Unit operates in a manner that creates a "froth" zone through which contaminated gas must pass. There is extreme turbulence in the zone and efficient collection of submicron particles occurs.[18] The MMV scrubber consists of a "stacked" bank of staggered tubular elements arranged in a pattern similar to a bank of heat exchanger tubes. The thin spacing causes successive micro-venturi flows at the "pinch" sections leading to high efficiencies. The EDV concept consists of a sequence of fundamental functions, namely saturation, condensation, ionization, and filtration.[19] In condensing scrubbers, water vapor in the flue gas stream is condensed into particulates causing the particle to grow to a size that enables easier collection by conventional means. At the same time, the temperature and volume of the contaminated gases is reduced providing a lower capital and operating cost for the downstream collection system and induced-draft fan. Finally, in the collision scrubber two gas streams are made to collide head-on at the discharge of opposed venturi throats containing water droplets. The collision action shreds the water droplets into finer ones that can more effectively collect submicron particles. The process also produces a larger liquid surface area for gas absorption.[20]

10.3 DRY SCRUBBER SYSTEMS

The success of electrostatic precipitators (ESP) and fabric filters (FF) in removing fine particles from flue gas streams has encouraged the use of combined dry-scrubbing/ESP or FF systems for the dual purpose of removing both particulates and acid gases simultaneously. Dry scrubbers offer potential advantages over their wet counterparts, especially in the areas of energy savings and capital costs. Furthermore, the dry-scrubbing process design is relatively simple, and the product is a dry waste rather than a wet sludge.

There are two major types of so-called dry scrubber systems: spray drying and dry injection. The first process is often referred to as a *wet–dry* system. When compared to a conventional wet scrubber, it uses significantly less scrubbing liquid. This will be covered in more detail in Section 10.4. The second process may be referred to as a *dry–dry* system since no liquid scrubbing is involved. The spray drying system is predominantly used in utility and industrial applications. However, both processes have had success in the HWI field.

Dry injection processes generally involve pneumatic introduction of a dry, powdery alkaline material, usually a sodium-base sorbent, from a storage silo and pneumatically conveyed and injected into the dirty combustion gas stream by means of an inductor or diffuser. A reaction chamber allows time for the partial neutralization and particle drying to take place prior to the capture by the FF. The injection point in such processes can vary from the boiler-furnace area all the way to the flue gas entrance to the baghouse, depending on operating conditions and design criteria. The temperature of the gases into the FF is usually in the 400–450°F range. Because of the higher operating temperature used for the dry injection, the acid gas neutralization is not as efficient as that for the wet/dry technology with spray dryer that can be easily controlled at a nominal 285°F. Acid gas neutralization and particle filtration continues on the filter cake.

Dry acid gas scrubbers are characterized by a zero liquid discharge and low water usage, with a resultant lack of steam plume. The pressure drop across the entire APC is considerably less than that required for wet scrubbers using a venturi. This results in an appreciable reduction in the induced-draft fan motor size.

The first U.S. studies of the dry injection method were made with a dry-powdered nahcolite (a mineral form of sodium bicarbonate) injected both into the flue gas stream and onto the filter bags for SO_2 removal. The system chemistry is based on the following reaction:

$$Na_2CO_3 + SO_2 \rightarrow Na_2SO_3 + CO_2$$

Efficiencies of 90% with 60% sorbent utilization were achieved in pilot-plant tests on flue gas from a lignite-fired boiler. This was encouraging, but the outlook for ready availability of nahcolite was not. Also, as was the case with spray-dryer systems, the waste solids from this sodium-based process are water soluble, posing potential disposal problems. There are some data available on HCl emission control with dry scrubbers. Interestingly, ETS, Inc., an air pollution consulting firm with headquarters in Roanoke, Virginia, has successfully installed a dry reactor system as a pilot plant on a slip stream of flue gas from a polychlorinated biphenyl (PCB) incinerator, removing more than 98% of the HCl emissions. The reactor system consists of a dry reactor and baghouse. Flue gas is diverted cyclonically into the reactor where the rotating slinger unit delivers the dry reactant (either hydrated lime or a sodium-based compound) perpendicularly to the flue gas flow, thus creating maximum mixing and intimate contact with the gaseous pollutants. The gaseous pollutants combine with the alkaline reactant forming dry particles. This dry material is then collected in a baghouse. This system was originally designed to remove SO_2 emissions from boiler flue gas. (ETS, Inc., has also installed a spray cooler and baghouse pilot plant on a glass melting furnace to remove both gaseous and particulate arsenic emissions.)

More recently, tests have been conducted on two 120 TPD Enercon Incinerators with dry sorbent injection of lime with activated carbon. Results show marked reductions (75–85%) of mercury, dioxins with a combination of lower temperatures <340°F, and activated carbon injection.[21]

Electrostatic Precipitators

Electrostatic precipitator technology was developed in the United States by Dr. Fredrick Gardner at the turn of the century. Frederick Cottrell's precipitator was successfully applied in 1907 to the collection of sulfuric acid mist and shortly thereafter proven in a number of ore processing, chemical, and cement plants.

Electrostatic precipitators are satisfactory devices for removing small particles from moving gas streams at high collection efficiencies. They have been used almost universally in power plants for removing fly ash from the gases prior to discharge. Although electrostatic precipitators have found limited application in treating gases from incinerators in the United States because of their high cost and the varying nature of the electrical resistivity of the particulate matter, they do have the capability of fine particulate control. Resistivity plays an important role in determining whether a particle can be readily collected in this device.

There are three classes of electrostatic precipitators:

- Dry electrostatic precipitators (ESP)
- Wet electrostatic precipitators (WESP) (described in Section 10.2)
- Ionizing wet scrubbers (IWS) (described in Section 10.2)

The ESP is by far the most popular; it has been used successfully to collect both solid and liquid particulate matter from many operations including smelters, steel furnaces, petroleum refineries, and utility boilers.

Dry High-Voltage Electrostatic Precipitators

Dry ESPs come in two varieties: (a) the high-voltage single-state unit and (b) the low-voltage two-stage unit. The low-voltage precipitator, where the particles are charged and collected in separate sections, is used only for small applications and is not generally employed at hazardous waste incineration facilities; this type will not be discussed here. The two major high-voltage ESP configurations currently used are tubular and plate. A tubular precipitator consists of a cylindrical collection electrode with the discharge electrode located on the axis of the cylinder, as schematically shown in Figure 10.3.1. However, the vast majority of ESPs installed are of the plate type. Particles are collected on flat, parallel collecting surfaces spaced 8–12 in. apart, with a series of discharge electrodes spaced along the centerline of adjacent plates, as shown schematically in Figure 10.3.1. A typical arrangement of a commercial plate-type ESP is shown in Figure 10.3.2. The gas to be cleaned passes horizontally between the plates or vertically up through the plates. Collected particles are usually removed by rapping and deposited in bins or hoppers at the base of the precipitator. Typical design parameters for plate-type precipitators are presented in Table 10.3.1.

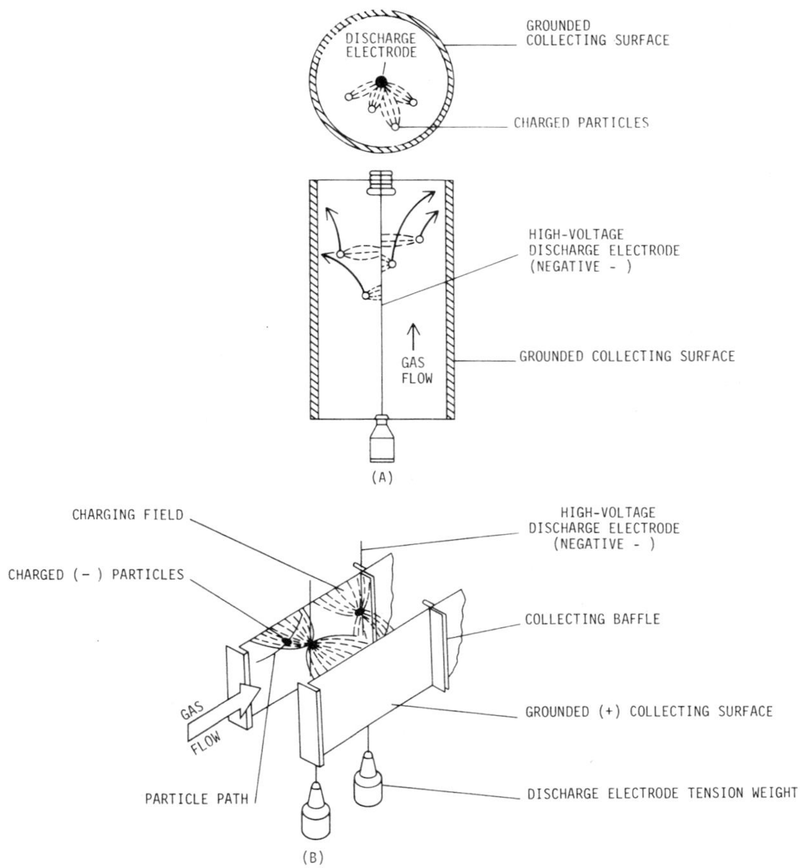

Figure 10.3.1 Electrostatic precipitator: (*a*) tubular-type precipitator; (*b*) plate-type precipitator.

Recently, for increased performance and reliability, precipitators have been divided into a number of independently energized bus sections. Each bus section has its own transformer rectifier, voltage stabilization controls, and high-voltage conductors that energize the discharge electrodes within that section. The main reasons for sectionalization are to offset the dampening effects on corona power input of heavy flue dust loadings and to reduce the effect of bus section failures. These heavy flue gas dust loadings occur mainly in the inlet sections of a precipitator. By sectionalization, corona power input and particle charging can be increased in the inlet sections, thereby raising the overall precipitator efficiency.

The approach taken by industry to size ESPs for various applications makes use of the Deutsch–Anderson equation [Eq. (10.3.1)] or a modified form of it:

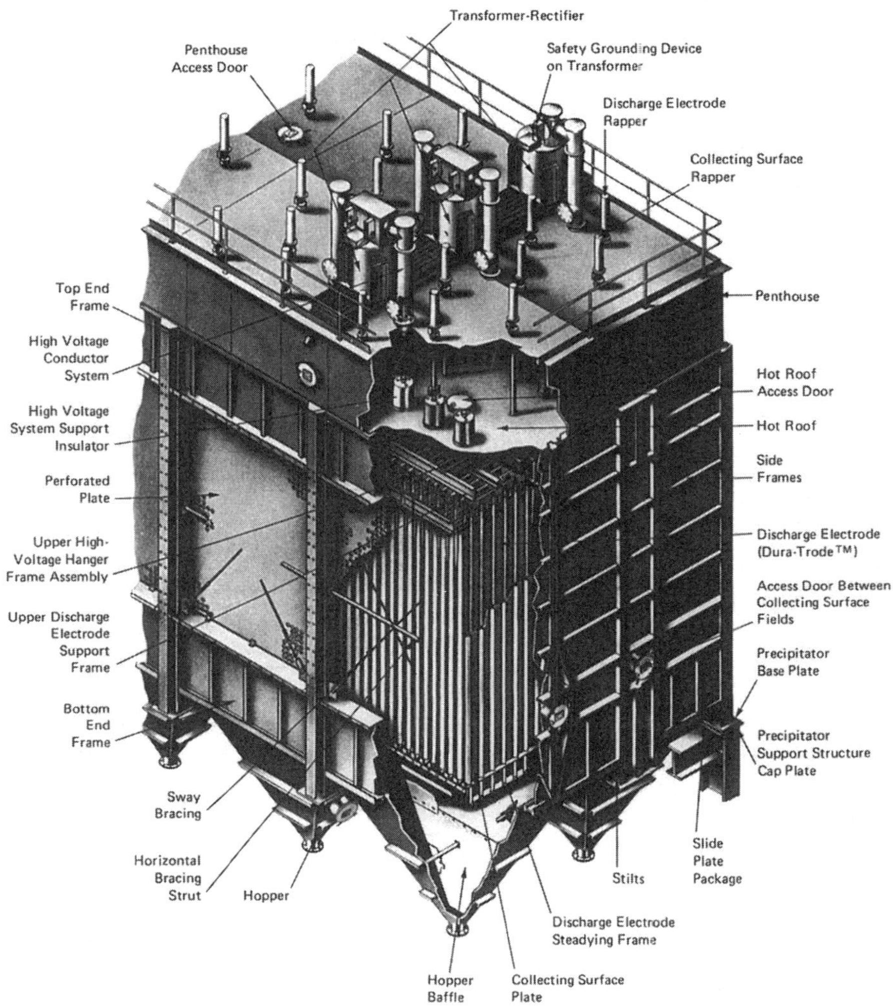

Figure 10.3.2 Plate-type electrostatic precipitator.

$$E = 1 - e^{-uA/Q_a} \tag{10.3.1}$$

where E = collection efficiency (dimensionless)
u = particle migration velocity (ft/s)
A = collecting area (ft^2)
Q_a = gas flow rate (ft^3/s)

In practice, the term u in Eq. (10.3.1) represents the *effective* migration velocity or precipitation rate parameter that is selected on the basis of experience with a particular dust. Since the desired collection efficiency and gas flow

Table 10.3.1 Typical design parameters for electrostatic precipitators

Parameter	Range of values
Precipitation rate (effective migration velocity)	0.1–0.7 ft/s
Plate spacing	8–11 in.
Gas velocity	2–8 ft/s
Plate height	12–45 ft
Plate length	0.5–2.0 × height
Applied voltage	30–75 kV
Corona strength	0.01–1.0 mA/ft of wire
Field strength	7–15 kV/in.
Residence (treatment) time	2–10 s
Draft loss (pressure drop)	0.1–0.5 in. HO
Efficiency	To 99.9+ %
Gas temperature	To 700°F (standard) 1000°F (high temperature) 1300°F (special)

rate are usually specified, the required collecting area can be determined once an appropriate precipitation rate parameter has been chosen. Recently, better correlations with field data on high-efficiency ESPs have been obtained by raising the exponential term in Eq. (10.3.1) to a power m using existing values of u. Typical values of m range between 0.4 and 0.7, with 0.5 as the norm.

The power required for a particular application is usually determined on an empirical basis, with past experience playing the key role. Performance curves for a given application typically relate power requirements to the efficiency and gas volume handled. The degree of sectionalization required is also determined on an empirical basis. The type and number of rappers for the collection and discharge electrodes depend on the properties of the dust and gas, the current densities, and the configuration of the electrodes and electrode support structures. High-resistivity dust is usually more difficult to remove because of the increased electrical force holding it to the plate.

Dust thickness buildup of 0.5–1.0 in. prior to dislodging is common. Typical rapper accelerations range from 10 to 100 times that due to gravity (g). With fly ash, for example, accelerations of 10–30 g's may be satisfactory, while highly resistive dust (difficult to remove) may require accelerations as high as 200 g's. Table 10.3.2 shows the number of rappers per unit area of collecting electrode and the number per unit length of discharge wire for various applications.

Depending on gas and dust conditions, and the required collecting efficiency, the gas velocities in an industrial ESP are between 2.5 and 8.0 ft/s. A uniform gas distribution is of prime importance for precipitator operation; it should be accomplished with a minimum expenditure of pressure drop. This is not always easy, since gas velocities in the duct ahead of the precipitator may be 30–100 ft/s in order to prevent dust buildup.

Table 10.3.2 Typical rapping practices for various applications

Application	Collection electrodes (rappers/1000 ft^2)	Discharge electrodes (rappers/1000 ft)
Utilities	0.25–0.90	0.09–0.66
Pulp and paper	0.25–0.99	0.21–0.32
Metals	0.11–0.82	0.28–0.50
Cement	0.33–0.52	0.19–0.33

Fabric Filters (FF)–Baghouses

One of the oldest, simplest, and most efficient methods for removing solid particulate contaminants from gas streams is by filtration through fabric media. The fabric filter is capable of providing high collection efficiencies for particles as small as 0.1 μm and will remove a substantial quantity of those particles as small as 0.01 μm. In its simplest form the industrial fabric filter consists of a woven or felted fabric through which dust-laden gases are forced. A combination of factors results in the collection of particles on the fabric filters. When woven fabrics are used, a dust cake eventually forms; this, in turn, acts predominantly as a sieving mechanism. When felted fabrics are used, this dust cake is minimal or nonexistent and the primary filtering mechanisms are a combination of inertial forces, impingement, and so on. These are essentially the same mechanisms that are applied to particle collection on wet scrubbers, wherein the collection media is in the form of liquid droplets rather than solid fibers.

As particles are collected, the pressure drop across the fabric filtering media increases. Due in part to fan limitations, the filter must be cleaned at predetermined intervals. Dust is removed from the fabric by gravity and/or mechanical means. The fabric filters or bags are usually tubular or flat. The structure in which the bags hang is referred to as a *baghouse*; the number of bags in a baghouse may vary from a couple to several thousand. Quite often, when great numbers of bags are involved, the baghouse is compartmentalized so that one compartment may be cleaned while others are still in service.

The basic filtration process may be conducted in many different types of fabric filters in which the physical arrangement of hardware and the method of removing collected material from the filter media will vary. The essential differences may be related, in general, to the following:

- Type of fabric
- Cleaning mechanism(s)
- Equipment geometry
- Mode of operation

Depending on these factors the equipment will follow one of three systems as shown in Figure 10.3.3. Bottom-feed units are characterized by the intro-

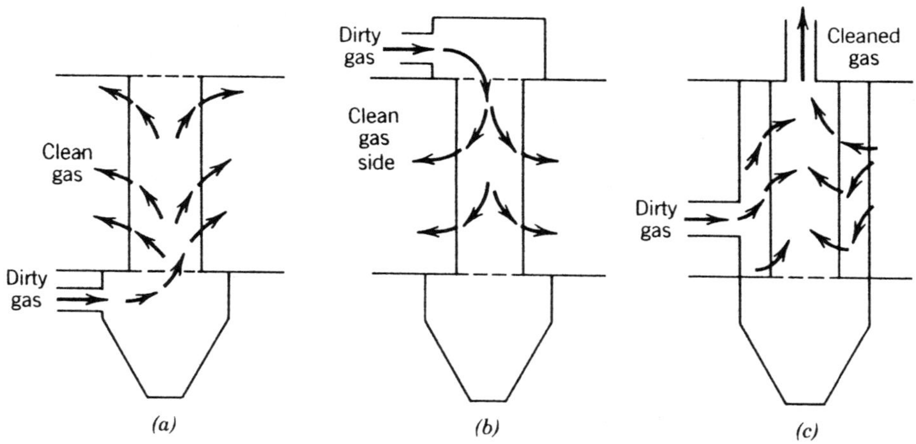

Figure 10.3.3 Baghouse filtration methods: (*a*) bottom feed; (*b*) top feed; (*c*) exterior filtration.

duction of dust-laden gas through the baghouse hopper and then to the interior of the filter tube. In top-feed units, dust-laden gas enters the top of the filters to the interior or clean-air side. When the gas flow is from inside the bag to the outside, by virtue of the pressure differential, the internal area of the filter element will be open and self-supporting; unsupported filter elements are tubular. When the filtration process is reversed, with the gas flow from outside the bag to inside, it is necessary to support the media against the developed pressures so that the degree of collapse is controlled. Supported filter elements are either of the tubular or envelope shape.

Baghouse collectors are available for either intermittent or continuous operation. Intermittent operation is employed where the operational schedule of the dust generating source permits halting the gas cleaning function at periodic intervals (regularly set by time or by pressure differential) for removal of collected material from the filter media *(cleaning)*. Collectors of this type are primarily utilized for the control of small-volume operations such as grinding and polishing and for aerosols of a very coarse nature. For most air pollution control installations and major dust control problems, however, it is desirable to use collectors that allow for continuous operation. This is accomplished by arranging several filter areas in a parallel flow system and cleaning one area at a time according to some preset mode of operation (see Fig. 10.3.4).

Baghouses may also be characterized and identified according to the method used to remove collected material from the bags. Particle removal can be accomplished in a variety of ways, including shaking the bags, blowing a jet of air on the bags from a reciprocating manifold, or rapidly expanding the bags by a pulse of compressed air. In general, the various types of bag-cleaning methods can be divided into those involving flexing and those involving a

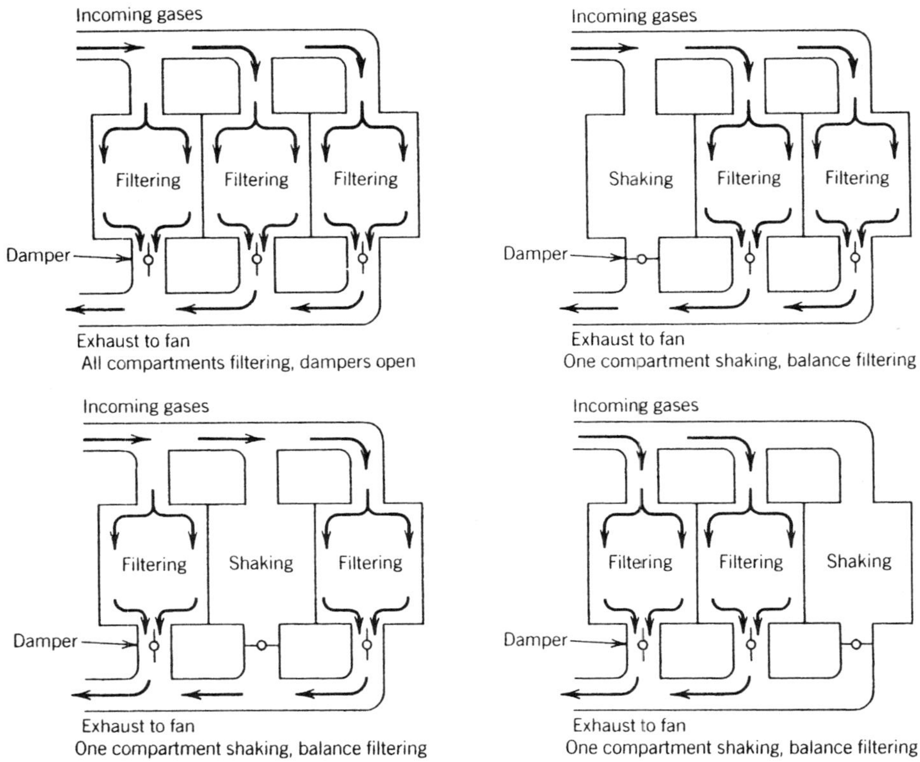

Figure 10.3.4 Baghouse parallel flow system.

reverse flow of clean air. Figure 10.3.5 illustrates some of these fabric flexing cleaning methods.

There are two basic types of filtration that occur in commercial fabric filters; the first is referred to as *media* or *fiber filtration* and the second is *layer* or *cake filtration.* In fiber filtration the dust is retained on the fibers themselves by settling, impaction, interception, and diffusion. In cake filtration the fiber acts as a support on which a layer of dust is deposited to form a microporous layer capable of removing additional particles by sieving as well as other basic filtration mechanisms (impaction, interception, diffusion, settling, and electrostatic attraction). In practical industrial cloth filters, both methods occur, but cake filtration is the more important process after the new filter cloth becomes thoroughly impregnated with dust.

Perhaps the outstanding characteristic of fabric filters is their inherent very high collection efficiency with even the finest particles. These units usually have the capability of achieving efficiencies of 99% almost automatically—provided they are properly constructed and maintained in satisfactory operating condition.

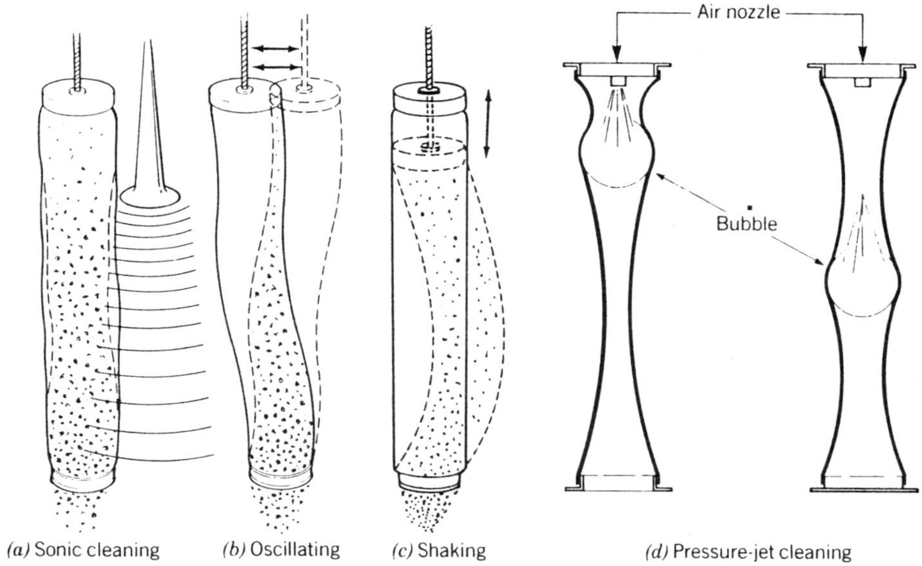

Figure 10.3.5 Fabric flexing cleaning methods.

The number of variables necessary to design a fabric filter is very large. Since fundamentals cannot treat all of these factors in the design and/or prediction of performance of a filter, this determination is basically left up to the experience and judgment of the design engineer. In addition, there is no one formula that can determine whether or not a fabric filter application is feasible. A qualitative description of the filtration process is possible, although quantitatively the theories are far less successful. Theory, coupled with some experimental data, can help predict the performance and design of the unit. The state of the art of engineering process design is the selection of filter medium, superficial velocity, and cleaning method that will yield the best economic compromise. Industry relies on certain simple guides and calculations, which are usually considered proprietary information, to achieve this. Despite the progress in developing pure filtration theory, and in view of the complexity of the phenomena, the most common methods of correlation are based on predicting a form of a final equation that can be verified by experiment.

Pressure drop will be considered first. The pressure drop equation for a fluid flowing through a fabric in a baghouse is given by

$$\Delta P = \Delta P(\text{fabric}) + \Delta P(\text{cake}) = K_1 v + K_2 W_i v^2 \theta \qquad (10.3.2)$$

where ΔP = total pressure drop across both fabric and cake (in. H_2O)
V = superficial velocity through the bag/cake (ft/min)
W_i = inlet particulate loading in gas (lb/ft^3)

θ = elapsed time in filtering cycle (min)
K_1 = resistance coefficient for the bag (fabric) [(in. H_2O)-ft/min]
K_2 = resistance coefficient for the deposited dust (cake) [(in. H_2O)/(lb dust/ft^2 cloth area)-(ft/min)]

Thus, a plot of ΔP versus θ yields a straight line. At $\theta = 0$, the only resistance to flow is that of the filter medium; the pressure drop, however, is a linear function of time, and as time increases, the resistance due to the dust cake may predominate.

An equation that can be used for determining the collection efficiency of a baghouse is[3]

$$E = 1 - e^{-(\psi L + \phi\theta)} \tag{10.3.3}$$

where ψ = constant based on fabric (ft^{-1})
ϕ = constant based on cake (s^{-1})
θ = time of operation to develop the cake thickness (s)
L = fabric thickness (ft)
E = collection efficiency (dimensionless)

The exit concentration (W_e) for the combined resistance system (the fiber and the cake) is

$$W_e = W_i e^{-(\psi L + \phi\theta)} \tag{10.3.3a}$$

where W_e = exit concentration (lb/ft^3)
W_i = inlet concentration (lb/ft^3)

Design Practice

The design of industrial baghouse equipment requires consideration of many factors. The most important design considerations include the operational pressure drop, cloth area, cleaning mechanism, fabric material and fabric life, baghouse configuration, and costs. Exhaust volume through the usual single-compartment fabric collector will not be constant because of the increasing resistance to airflow as the dust cake accumulates. The reduction in flow rate will be a function of the system pressure relationships, the exhaust fan characteristics, and the point of rating. Drop-off of exhaust volume is usually not severe in practice because pressure losses for the system of ducts and hoods usually equal or exceed that of the fabric collector; therefore a reduction in exhaust volume will cause a corresponding reduction in the pressure needs for that portion of the system. Most fabric collectors have the centrifugal fan on the clean-air side of the collector, where the more efficient backward-curved blade fan can be applied. This fan construction has a steeply rising pressure–volume characteristic, providing a significant increase in available static pressure as exhaust volume is reduced. This fan characteristic prevents overloading when collectors with new, clean media are first placed in service and where, for a short time, the pressure loss through the clean fabric is low. When fabric collectors are designed for continuous operation, pressure relationships become more complex, but approach a constant value. The smaller the fraction of the elements taken out of service at any one time for dust cake removal, the more uniform the pressure drop of the fabric collector. For reverse-jet types of continuous collectors, minor pressure variations occur, depending on the position and direction of travel of the blow-back device. Pressure variations are minimal in pulse-jet designs because the cleaning cycle is extremely short and relatively few elements are taken out of service at any one time. Typical pressure drops for woven cloths range between 3 and 6 in. H_2O. However, the use of high-velocity filtration, felted fabrics, or the presence of a sticky or low-porosity dust cake often requires pressure drops on the order of 6–10 in. H_2O.

The size of a baghouse is primarily determined by the area of filter cloth required to filter the gases. The choice of a filtration velocity [or its equivalent, the *air-to-cloth ratio* (ACR) in cubic feet per minute of gas filtered per square foot of filter area] must take certain factors into consideration. Although the higher velocities are usually associated with the greater pressure drops, they also reduce the filter area required. Practical experience has led to the use of a series of ACR ratios for various materials collected and types of equipment. Ratios in current use range from <1:1 to >20:1. The choice depends on cleaning method, fabric, and characteristics of the particles. The rule of thumb for ACR ratios for conventional fabric filter baghouses with woven cloth is 1.5–3.0 ft^3/min-ft^2. Remember that during the cleaning cycle, the ACR is increased if compartmentalization is used with one compartment always off-stream. This increased value is then given by (ACR) $(N - 1)/N$, where N is the total number of compartments. When dust is fine or loadings are high, it is recommended that the filtering velocity be <3 acfm/ft^2. Experience has also demonstrated that more cloth is required with increased temperature.

It is best to select the cleaning mechanism and the filter fabric together, since both items are closely related. For example, felted fabrics are almost exclusively cleaned by pulse or reverse-jet air, whereas most woven fabrics are cleaned by other means. By the process of elimination, a review of past successful filter operations will usually show that only a few cleaning mechanism–fabric combinations are compatible and sufficiently attractive to warrant economic evaluation.

Fabric deterioration often results from the combined assault of several factors (rather than from any single effect) such as thermal erosion, mechanical stress through repeated flexing, chemical attack, and abrasion. All possible modes of failure should be considered and again, previous experience is usually the best guide. For example, extrapolating from experience might show that the reduction in fiber life through thermal erosion might double for a 20°F rise in temperature or that the mechanical attrition rate doubles when the frequency of cleaning is doubled.[22] Selecting the number of compartments for a fabric filter installation requires information such as the allowable variation in gas flow, the availability of sizes of commercial units, and the expected frequency of maintenance. A single compartment unit is the most foolproof, predictable, and least expensive fabric collector design. It may be necessary, however, to provide additional compartments for emergencies, extended maintenance, or unexpected increases in process effluent.

10.4 WET–DRY SCRUBBER SYSTEMS

Spray Dryer—ESP or Fabric Filter (Baghouse)

The use of fabric filters for particulate control has increased in the late 1980s and into the 1990s. Although an ESP may be used downstream from the dry scrubber or spray dryer, the choice in most installations thus far has been the baghouse. Pulse-jet units with some type of acid-resistant glass bags is the preferred choice.

The method of operation of the spray dryer absorber (SDA) is relatively simple, requiring only two major equipment items—a spray dryer similar to those used in the chemical food-processing and mineral preparation industries and a fabric filter (baghouse) or ESP to collect the fly ash and entrained solids. In the SDA, the sorbent solution, or slurry, is atomized into the incoming flue gas stream to increase the liquid–gas interface and to promote the mass transfer of the HCl or SO_2 from the gas to the slurry droplets, where it is absorbed. The lime slurry solution may be mechanically atomized by either a rotary atomizer or by spray nozzles and injected into the combustion gas stream to be cleaned. The atomized lime droplets react with the HCl and SO_2 in the combustion gas stream, forming salts. The water droplet evaporates because of the 400–500°F temperature of the combustion gases leaving a dry particulate, provided a sufficient drying time is allowed. The rotary atomizer results in smaller droplet size and size distribution and is less subject to plugging and wear than the spray nozzle, but is higher in cost. This atomizer projects the droplets radially outward and perpendicular to the gas flow. The droplets decelerate rapidly owing to the drag forces of the downward moving flue gas and eventually attain the velocity of the flue gas. The radial distance between the atomizer and the dryer wall must be sufficient to allow for adequate drying of the largest droplets. The length to diameter (L/D) ratio of the cylindrical section of the SDA is typically 0.8:1. In a SDA with dual-fluid pneumatic nozzle, atomization is in the direction of the gas flow. The L/D is typically 2:1.

Optimum SDA performance is achieved through proper choice of L/D, droplet size, and residence time. The designed residence time for most SDAs is 10–12 sec at maximum design. Process monitoring includes concentration and pH of lime slurry, accumulation of solids within the scrubber, induced-draft fan horsepower consumption, reagent preparation system operation, and temperatures of inlet and outlet gas. Inspection and maintenance is critical to continued good operation of the scrubber. Operating staff must routinely monitor and record readings from all instruments used to measure flows, pressures, and temperatures associated with the scrubber. Any abnormal readings require immediate investigation. Gas outlet temperature is an indication of gas flow and slurry flow. Low readings indicate a high slurry flow or failed atomizer. High readings indicate a drop in gas flow or a drop in slurry flow. Gas flows are determined by induced-draft fan amps, draft pressure, and baghouse pressure drop. A low slurry flow could be due to pump failure, line, valve, or filter blockage or atomizer blockage.

Simultaneously, the thermal energy of the gas evaporates the water in the droplets to produce a dry powdered mixture of chloride, sulfite-sulfate, and/or some unreacted alkali. Since the flue gas is not saturated and contains no liquid carryover, potentially troublesome mist eliminators are not required. In the case of SO_2, the reaction with the lime slurry is as follows:

$$SO_2 + Ca(OH)_2 \rightarrow CaSO_3 + H_2O$$

$$2CaSO_3 + O_2 \rightarrow +2CaSO_4$$

After leaving the SDA, the solid-bearing gas passes through the fabric filter (or ESP) where the dry product is collected and where a percentage of the unreacted alkali reacts with SO_2 for further removal. The cleaned gas is then discharged through the fabric filter plenum to an induced-draft (ID) fan and the stack (see Fig. 10.4.1).

Among the inherent advantages that the SDA enjoys over wet scrubbers are

- Lower capital costs
- Lower draft losses
- Reduced auxiliary power
- Reduced water consumption, with liquid-to-gas (L/G) ratios significantly lower than those of wet scrubbers
- Continuous, two-stage operation, from liquid feed to dry product

Because the SDA operates near the flue-gas saturation point, no reheat is required, and there is little effect of water–vapor plume problems.

The sorbent of choice for most SDA systems is a lime slurry. Some systems use a sodium carbonate solution. Although the latter will generally achieve a higher level of SO_2 removal than a lime slurry at similar operating conditions, the significant cost advantage that lime has over sodium carbonate makes it the overwhelming favorite. Also, when sodium alkalis are used, the products are highly water soluble and may create disposal problems.

When compared to limestone, lime has the higher SO_2 removal efficiency (85 vs. 60% for limestone, at a stoichiometric ratio of 1.0). Hand in hand with removal efficiency is the concept of alkali utilization; increased utilization allows for lower stoichiometric ratios and generates less waste product. While

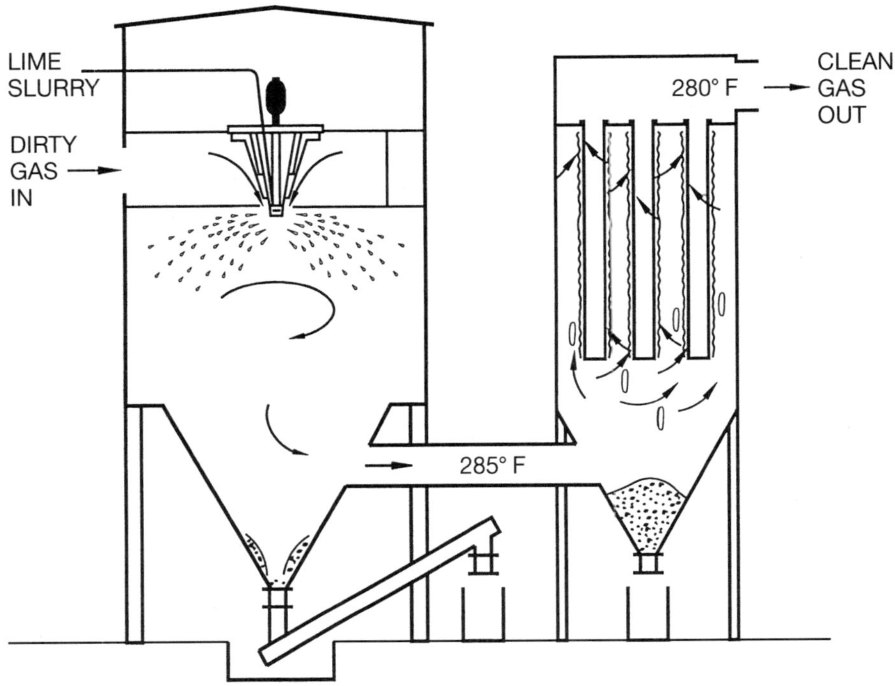

Figure 10.4.1 Spray dryer-baghouse system.

100% utilization is attainable with lime by using performance-enhancing modifications, such as sorbent recycle and hot-gas bypass, limestone typically gives 60–70% utilization.

SDA design can be affected by the choice of the particulate collection device. Here, fabric filters have an inherent advantage over ESPs in that a percentage of unreacted alkali in the collected solids on the bag surface can react with the remaining SO_2 in the flue gas. This percentage depends mainly on the temperature of the gas entering the fabric filter; in some cases it can be as much as 10%. One disadvantage of using a fabric filter is that, since the fabric is somewhat sensitive to wetting, a margin above saturation temperature (about 25–35°F) must be kept for bag protection. Increased utilization allows for lower stoichiometric ratios and generates less waste products.

Limitations to the SDA system do exist, of course. Efficiencies of 85% can be obtained with good sorbent utilization, but anything >85% removal efficiency requires too high a consumption of expensive alkali to be economical. To reach these higher removal efficiencies, a PBS is usually installed downstream of the FF. This is the fourth type of system described earlier in the chapter. Also, the higher L/G ratios required at higher sulfur content can result in cooling of the gas below the acid dew point.

Mercury is frequently present in quantities sufficient to generate high concentrations, often of short duration, in the flue gas from medical waste incinerators and municipal solid waste incinerators. Experience with HWIs indicate much lower levels of mercury than found in MWC and MSW systems. Control of mercury compounds and other volatile metals, along with the trace organic PCDD/PCDF (dioxin/dibenzofuran) compounds, has been effectively accomplished in SD/FF systems by the addition of a relatively small quantity of activated carbon of approximately 0.003 lb/lb of medical waste. Tests conducted in the late 1980s and early 1990s were reported at an Air and Waste Management Association (AWMA) meeting in Kansas City, Missouri, in June 1992. Results showed reduction of PCDD/Fs from 7.3 ng/m^3, dry volume at 10% CO_2, to levels of 0.014–0.027 ng/m^3 (a 99% reduction) with addition of 4.5–7.5 kg/h of activated carbon. A sorbent consisting of calcium hydroxide with surface-activated substances like activated carbon or lignite coke has been produced and patented as Sorbalit in the United States and Europe. It has successfully been applied to SDA/FF systems reducing emissions of mercury and PCDD/PCDFs.[23,24] More recently, results of tests conducted at Waste Technologies Industries, in East Liverpool, Ohio, have shown that activated carbon injection is extremely effective in reducing dioxin emissions in APC systems operating at higher temperatures than is typical in a rapid quench system.[25] Comparisons have been made between HWIs that have used rapid quench systems (wet scrubbers) versus those using waste heat boilers or SDA/FFs downstream of the high-temperature combustion zone. Systems that are able to quench the high-temperature combustion gases from 1000 to 1200°C (1832 to 2192°F) to adiabatic (75–90°C) in fractions of a second have emissions well below the levels established by the recent MACT standards [0.2 ng/dscm Total Toxic Equivalent (TEQ)]. This rapid quenching action freezes the reaction from the combustion zone preventing changes to the reaction products. Gradual cooling through the critical temperature zone of 300°C (572°F) with contact with PM, precursors, copper or iron, and Cl_2 results in the reformation of PCDD/Fs. Table 10.4.1 provides data from tests conducted on HWIs from the late 1980s to 1997. The results of these tests are shown in Figure 10.4.2 with the differences related to the type of downstream quench and/or APC system. Note that systems utilizing rapid quench have the lowest levels of PCDD/PCDF emissions. Most are well below the 0.2 ng TEQ/dscm limit established by the MACT standards for new incinerators. This provides a major advantage to systems using rapid quench wet scrubbers immediately following the high-temperature combustion zones. This includes spray towers, venturi scrubbers, and submerged exhaust quenchers.[2] Note that in the MACT standards of September 30, 1999, discussed in Chapter 3, the allowable emission level for existing incinerators is 0.4 ng TEQ/dscm if the flue gas temperature does not exceed 400°F at the inlet to the PM control device. This applies as well to both existing and new cement kilns and lightweight aggregate kilns. However, the standards for new incinerators must meet the 0.20 ngTEQ/dscm. (See Table 3.10.1)

Table 10.4.1 PCDD/PCDF data in TEQ (ng/dscm @ 7 % O_2) for HW incinerators in HWC MACT Standards Emissions Database, Nov., 1995

Company	Code	Location	Report date	Facility type	APCS	Stack temp (°F)	PM (gr/dscf) min/max	Cl_2 (ppm min/ max)	HCl (ppm min/ max)	# Points	TEQ (ng/dscm) 7% O_2 Min.	Max.
3M	334	Cottage Grove, IL	Sep 90	RK	WS/ESP/PT	75	0.0395 0.107	N/A	7.87 26.8	8	0.337	4.53
Aptus	325	Coffeyville, KS	Jan 91	RK	SD/FF/WS/IWS	100	0.001 0.006		0.225 9.1	11	1.75	2.48
Aptus	327	Aragonite, UT	Aug 92	RK	SD/FF/WS/ESP	150	0.000288 0.0027	0.122 0.314	0.205 2.96	12	0.57	27.9
Bros	807	Bridgeport, NJ	Jul 91	RK	C/WHB/VQ/PT/HS/DM	190	0.00195 0.00492	0.1 0.473	0.894 3	9	0.162	0.99
Chevron	500	Richmond, CA	Jul 93	LI	QC/VS/KOV/DM	180	0.0008 0.0028	0.267 13.7	0.501 2.9	6	0.03	0.0467
Ciba Geigy	706	McIntosh, AL	Jul 88	LI	QT/HS/C	180	0.0565 0.066		0.21 1.21	3	0.0133	0.0333
Dept/Army	346	Johnson Atoll	Jun 92	RK	C/QC/VS/PT/DM	190	0.000485 0.00199		0.703 0.859	4	0.114	0.125
Dept/Army	347	Tooele, UT	Oct 93	RK	C/QC/VS/S/DM	230	0.000586 0.0152		0.0746 3.51	10	0.00473	0.0404
Dow	808	Plaquemine, LA	Feb 88	RK	QT/PBS/ESP	140	0.0089 0.0134		0.278 0.687	2	0.131	0.178
Dow	353	Midland, MI	Jun 92	RK703	QC/VS/DM/ESP	90	0.0096 0.0115		0 0	3	0.122	0.287
Dow	354	Midland, MI	Mar 92	RK830	QC/AS/VS/DM/IWS	90	0.00038 0.0018		0.325 0.357	4	0.00564	0.0164
General Electric	330	Pittsfield, MA	Apr 91	LI	QT/WS/DM	170			30 58.7	5	3.85	76.5
New Bedford Superfund	903	New Bedford, MA	Jan 92	RK	VS/PT/CA/HEPA	140	0.0345 0.0486		0.231 2.52	3	0.0599	0.119

Pfizer	502	Groton, CT	Jul 90	RH	WHB/QC/PBC/VS/ES	73	0.033 0.04		1.16 33.1	3	0.0117	0.0231
Rocky Mountain Arsenal	902	Adams City, CO	Dec 93	LI	QT/VS/PT	183	0.0194 0.0238		2.28 5.59	3	0.00572	0.0612
Rollins Env. Services	214	Baton Rouge, LA	May 88	RK	QS/ABS/CT/IWS/DM	100	0.00859 0.0241		1.58 1.77	3	0.0443	0.19
Rollins Env. Services	216	Bridgeport, NJ	Dec 89	RK	QS/ABS/CT/VS/DM	100	0.0162 0.0286		7.84 10.8	6	0.01	0.15
Rollins Env. Services	221	Deer Park, TX	Aug 88	RK	QS/ABS/CT/VS/DM	125	0.0131 0.0205	3.25 10.7	1.23 9.9	5	0.102	0.776
Ross	331	Grafton, OH	Mar 93	RK	PT/WS	125	0.007 0.01			3	0.0167	0.106
Vertac	914	Jacksonville, AR	Dec 91	RK	C/SD/FF/VS/WS	202	0.0038 0.0042	16.1 24.2		3		4.39
Vulcan	229	Wichita, KS	Apr 91	LI	WHB/ACS/HS/CS	150	0.00917 0.0126		69.7 257	6	1.05	11.2
WTI	222	East Liverpool, OH	Apr 94	RK	WHB/SD/ESP/Q/PBS	200	0.00105 0.0035	0.046 6 1.11	0.181 2.21	28	0.0249	4.1
Zeneca	725	Bayonne, NJ	Jun 90	LI	WS/QT	160	0.0199 0.0288	31 85.2	1.7 7.8	3	0.0589	0.247
Confidential Client*			Apr 97	LI	WHB/QT/ABS/DM	149	0.0137 0.022		65.7 316	3	5.35	10.9
Confidential Client*			Apr 97	LI	WHB/QT/ABS/DM	160			69 84.8	3	3.5	4.39

No. of facilities 21; no. of units 24; no. of data sets 149.

Facility type: RK, rotary kiln; LI, liquid injection; RH, rotary hearth.

Abbreviations: ACS, acid scrubber; AS, absorber; C, cyclone; CA, carbon absorber; CS, caustic scrubber; DI, dry injection; DM, demister; ESP, electrostatic precipitator; FF, fabric filter; GC, gas cooler; HEPA, high-efficiency particulate air filter; HES, high-energy scrubber; HP, HEPA filter; HS, hydrogen chloride scrubber; IWS, ionizing wet scrubber; KOV, knock out vessel; MC, multiple cyclone; PBC, packed-bed condition; PBS, packed-bed scrubber; PT, packed tower; Q, quencher; QS, quench separator; QC, quench column; QT, quench tower; S, scrubber; SD, spray dryer; SS, spray saturator; VS, venturi scrubber; VQ, venturi quench; WHB, waste heat boiler; WS, wet scrubber.

*Tests conducted by RMT/Four Nines in 1997.

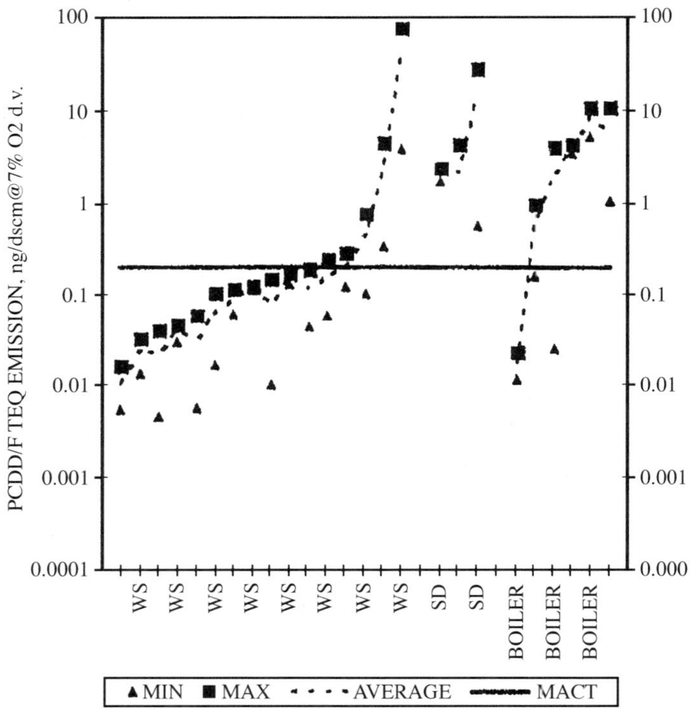

Figure 10.4.2 APC system vs. TEQ.

10.5 ILLUSTRATIVE EXAMPLES

Example 10.5.1 A horizontal parallel-plate ESP consists of a single duct 24 ft high and 20 ft deep with an 11-in. plate-to-plate spacing. A collection efficiency of 88.2% is obtained with a flow rate of 4200 acfm. The inlet loading is 2.82 gr/ft^3. Calculate the following:

1. The bulk velocity of the gas (assume a uniform distribution)
2. The outlet loading
3. The drift velocity for this system
4. A revised collection efficiency if the flow rate is increased to 5400 acfm
5. A revised collection efficiency if the plate spacing is decreased to 9 in.

SOLUTION

1. The bulk velocity is given by $v = Q_a/A_d = 4200/24\left(\frac{11}{12}\right) = 191$ ft/min, where A_d = duct cross-sectional area (ft^2).
2. The outlet loading is $2.82(1 - 0.882) = 0.333$ gr/ft^3.

3. ϕ is first calculated from Eq. (10.3.1).

$$E = 1 - e^{-\phi}$$
$$0.882 = 1 - e^{-\phi}$$

Solving for ϕ

$$\phi = 2.14$$

Since

$$\phi = uA_d/Q_a$$
$$2.14 = (u)(24)(20)(2)/(4200/60)$$
$$u = 0.156 \text{ ft/s}$$

4. If Q_a = 5400 acfm, then a new ϕ can be calculated assuming the same drift velocity.

$$\phi = (156)(24)(20)(2)/(5400/60) = 1.67$$

Calculate the revised collection efficiency:

$$E = 1 - e^{-1.67} = 0.812$$

5. Since Q_a, u, and A are all constant, the Deutsch–Anderson equation predicts that the efficiency does not change if the plate spacing is 9 in.

Example 10.5.2 A duct-type electrostatic precipitator is to be used to clean 100,000 acfm of a gas stream containing particulates from a hazardous waste incinerator. The proposed precipitator consists of three bus sections (fields) arranged in series, each with the same collection surface. The inlet loading has been measured as 40 gr/ft^3 and a maximum outlet loading of 0.08 gr/ft^3 is allowed by local EPA regulations. The drift velocity for the particulates has been experimentally determined in a similar incinerator installation with the following results:

- First section (inlet) 0.37 ft/s
- Second section (middle) 0.35 ft/s
- Third section (outlet) 0.33 ft/s

1. Calculate the total collecting surface required based on the *average* drift velocity and the required total efficiency.
2. Find the total mass flow rate (lb/min) of particulates captured by each section using these drift velocities.

SOLUTION

1. Calculate the required total collection efficiency based on the given inlet and outlet loading:

$$E = 1 - [(\text{outlet loading})/(\text{inlet loading})] = 1 - [(0.08)/(40)] = 0.998$$
$$= 99.8\%$$

Calculate the average drift velocity (u):

$$u = (0.37 + 0.35 + 0.33)/3$$
$$= 0.35 \text{ ft/s}$$

Calculate the total surface area required using the following equation for the electrostatic precipitator [from Eq. (10.3.1)]:

$$A = -\ln(1 - E)/(u/Q_a)$$

where A = total surface area
E = collection efficiency
u = drift velocity (average)
Q_a = gas volumetric flow rate

$$A = -\ln(1 - 0.998)/(0.35 \times 60/100{,}000)$$
$$= 29{,}593 \text{ ft}^2$$

2. Calculate the collection efficiencies of each section. Assume that each section has the same surface area:

$$E_1 = 1 - \exp(-Au_1/3Q_a) = 1$$
$$- \exp[-(29{,}593)(0.37)/3(100{,}000/60)] = 0.888$$
$$E_2 = 1 - \exp(-Au_2/3Q_a) = 0.874$$
$$E_3 = 1 - \exp(-Au_3/3Q_a) = 0.858$$

Calculate the mass flow rate of particulates captured by each section using the collection efficiencies calculated previously:

$$w_1 = (E_1)(\text{inlet loading})(Q_a)$$
$$= 3.552 \times 106 \text{ gr/min}$$
$$= 507.4 \text{ lb/min}$$

$$w_2 = (1 - E_1)(E_2 \text{ inlet loading})(Q_a)$$
$$= 3.915 \times 105 \text{ gr/min}$$
$$= 55.9 \text{ lb/min}$$
$$w_3 = (1 - E_1)(1 - E_2)(E_3)(\text{inlet loading})(Q_a)$$
$$= 4.843 \times 104 \text{ gr/min}$$
$$= 6.9 \text{ lb/min}$$

Example 10.5.3 A venturi scrubber is employed to reduce the discharge of particulates from the stack of a hazardous waste incineration facility to the atmosphere. The unit is presently treating 80,000 acfm of gas with a concentration of 8.2 gr/ft^3 of particulates and operating at a pressure drop of 5 in. H_2O. Experimental studies have yielded the following particle-size collection efficiency data:

Particle Diameter (μm)	Collection Efficiency (%)
20	58
30	76
50	91
75	94
100	98.9
150	99.4
200	99.9+

Determine the overall collection efficiency of the venturi scrubber assuming the following particle size distribution: 2% by weight of 20-μm particles; 3%, 30 μm; 7%, 50 μm; 10%, 75 μm; 13%, 100 μm; 19%, 150 μm; 46%, 200 μm.

SOLUTION The calculations are presented in tabular form:

Particle Diameter (μm)	Weight Fraction (x_i)	(E_i)	Collection Efficiency (x_iE_i)
20	0.02	58	1.16
30	0.03	76	2.28
50	0.07	91	6.37
75	0.10	94	9.40
100	0.13	98.9	12.85
150	0.19	99.4	18.89
200	0.46	99.9	46.00
			E = 96.95%

The overall collection efficiency is 96.95%.

Example 10.5.4 A fly-ash-laden gas stream is to be cleaned by a venturi scrubber using a L/G ratio of 8.5 gal/1000 ft^3. The efficiency can be calculated from Eq. (10.2.1).

$$E_i = 1 - e^{-kR\psi^{0.5}}$$

where E = fractional collection efficiency of particle size d_{pi}
R = liquid-to-gas ratio (gal/1000 acf)
k = a constant (1000 ft^3/gal)
ψ = inertial impaction number (dimensionless)

The fly ash has a particle density of 43.7 lb/ft^3 and k = 200 ft^3/gal. The throat velocity is 272 ft/s and the gas viscosity is 1.5×10^{-5} lb/ft-s. The particle size distribution is

d_{pi} (μm)	% by Weight
<0.10	0.01
0.1–0.5	0.21
0.6–1.0	0.78
1.1–5.0	13.0
6.0–10.0	16.0
11.0–15.0	12.0
16.0–20.0	8.0
>20.0	50.0

Make use of the Nukiyama and Tanasawa relationship and neglect the Cunningham correction factor.

SOLUTION Use the Nukiyama and Tanasawa relationship [Eq. (10.2.3)] to calculate to the mean droplet diameter (d_1):

$$d_1 = (16{,}400/v_t) + 1.45R^{1.5}$$

where v_t = throat velocity (ft/s)
R = liquid-to-gas ratio (gal/1000 ft^3)
d_1 = mean droplet diameter (μm) = $(16{,}400/272) + 1.45(8.5)^{1.5} = 96.2$ μm

Express the inertial impaction number (ψ) in terms of d_p using Eq. (10.3.2) (assuming $C = 1$):

$$\psi = \frac{d_{pi}^2 \rho_{pi} v_t}{9\mu_G d_1}$$

$$= d_{pi}^2(43.7)(272)/[(9)(1.5 \times 10^{-5})(96.2)(3.048 \times 10^5)]$$

$$= 3.003 d_{pi}^2 (d_{pi}^2 \text{ in } \mu\text{m})$$

where d_{pi} = particle diameter (average for the ith size range)
ρ_{pi} = particle density
v_t = throat velocity
μ_G = gas viscosity
d_1 = mean droplet diameter

Consistent units must be employed; note that there are 3.048×10^5 μm/ft.

Obtain the individual collection efficiency in terms of d_{pi} using Eq. (10.3.1):

$$E_i = 1 - e^{(-kR\psi_1^{0.5})}$$

$$= 1 - e^{-(2.94)d_{pi}}$$

Note: The numerical value of k in this equation is 0.2.

Comment The previously given collection efficiency is referred to in industry as the size efficiency or grade efficiency equation for the system. One can use this type of an equation to calculate efficiencies for different particle sizes. (Similar equations also are available for other particulate control devices.) These individual collection efficiencies are used to calculate an overall efficiency. This is demonstrated in the next step.

Calculate the overall collection efficiency by completing the following table:

d_p (μm)	E_i	x_i (%)	x_iE_i (%)
0.05	0.1367	0.01	0.001367
0.30	0.586	0.21	0.123
0.80	0.905	0.78	0.706
3.0	0.9998	13.0	12.998
8.0	0.9999	16.0	15.999
13.0	0.9999	12.0	12.000
18.0	0.9999	8.0	8.000
20.0	0.9999	50.0	50.000

$$E = \Sigma x_iE_i = 99.928\%$$

Comment The Johnstone equation assumes particle capture to occur by inertial impaction, that is, the particles, because of their inertia, strike the liquid droplets and are captured. This does not apply to submicron particulates. However, another collection mechanism defined as *molecular diffusion* comes into play. The Cunningham correction factor should be included in the analysis if a significant quantity of these very small particles are in the gas stream. Generally, venturi scrubbers are not overly effective with submicron particles.

Example 10.5.5 A vendor proposes to use a spray tower in a hazardous waste incineration operation to reduce the discharge of solids to the atmosphere. The inlet loading of the gas stream from the kiln is 5.0 gr/ft^3 and is to be reduced to 0.05 in order to meet state regulations. The vendor's design calls for a water pressure drop of 80 psi and a pressure drop across the tower of 5.0 in. H_2O. The gas flow rate is 10,000 acfm and a water rate of 50 gal/min is proposed. Assume the contact power theory to apply.

1. Will the spray tower meet regulations?
2. What total power loss is required to meet regulations?
3. Propose a set of operating conditions that will meet the standard. The maximum gas and water pressure drop across the unit are 15 in. H_2O and 100 psi, respectively.
4. What conclusions can be drawn concerning the use of a spray tower for this application?

SOLUTION

1. Using Eqs. (10.2.7)–(10.2.9),

$$N_t = \ln(1/1 - E)$$

The term p_t is calculated as follows:

$$p_t = p_G + p_L$$

$$p_G = 0.157\Delta P' = 0.157(5) = 0.785$$

$$p_L = 5.83 \times 10^{-4} P_L R$$

$$= 5.83 \times 10^{-4}(80)(50/10) = 0.233$$

Therefore,

$$p_t = 1.018 \text{ hp/1000 acfm}$$

For an HWI dust and/or ash, Table 10.2.2 gives $\alpha = 1.47$ and $\beta = 131.05$. Thus,

$$N_t = 1.47(0.18)^{1.05} = 1.50$$

$$1.5 = \ln(\ 1/1\ -\ E)$$

$$E = 77.7\%$$

Calculate the required efficiency:

$$E = (5.0\ -\ 0.05)/5.0 = 0.99 = 99\%$$

Since the regulations require 99% efficiency, the spray tower will not meet the regulations.

2. Calculate p_t at the required efficiency:

$$E = 0.99$$

$$N_t = \ln(1\ -\ 0.99) = 4.605$$

$$4.605 = 1.47(p_t)^{1.05}$$

$$p_t = 2.96 \text{ hp/1000 acfm}$$

3. Assume the maximum gas and water pressure drops across the unit to be 15 in. H_2O and 100 psi, respectively. Calculate p_G and p_L:

$$p_G = 2.36$$

$$p_L = 0.60$$

Calculate R in gal/1000 acf:

$$R = p_L/(5.83 \times 10^{-4}P_L) = 0.6/[5.83 \times 10^{-4}(100)] = 10.3 \text{ gal /1000 acf}$$

Determine the new water flow rate:

$$(10.3 \text{ gal/1000 acf})(10{,}000 \text{ acfm}) = 103 \text{ gal/min}$$

4. The unit has limited applicability, at best, for high collection efficiency operations.

Example 10.5.6 It is proposed to install a pulse-jet fabric filter system to clean a 10,000-scfm (32°F) air stream at 250°F, containing 4.0 gr/ft^3 of pollutant from a hazardous waste incinerator facility. To reduce the outlet loading to below 0.01 gr/dscf, an average ACR of 1.5 cfm/ft^2 cloth is required. The following information, given by filter bag manufacturers, is available at the beginning of the selection process:

	Filter bag			
	A	B	C	D
Tensile strength	Excellent	Above average	Fair	Excellent
Recommended maximum operation temperature (°F)	260	275	260	220
Resistance factor	0.9	1.0	0.5	0.9
Relative cost/bag	2.6	3.8	1.0	2.0
Standard size	8 in. × 16 ft	10 in. × 16 ft	1 ft × 16 ft	1 ft × 20 ft

1. Determine the filtering area required for this operation.
2. Based on the required area and this information, select the most suitable filter bag and calculate the number of bags that should be used.

SOLUTION

1. $$Q_a = 10{,}000(250 + 460)/(460 + 32) = 14{,}430 \text{ acfm}$$

   $$\text{Filtering area} = 14{,}430/1.5 = 9620 \text{ ft}^2$$
2. Bag D is not good since the maximum temperature is 220°F. Bag C is not good because one needs either excellent or above average tensile strength for pulse units. Therefore only A or B is acceptable.

Calculate the area/bag for bag A and bag B.

$$\text{Area/bag for A} = (\pi)\left(\tfrac{8}{12}\right)(16) + \left(\tfrac{18}{2}\right)^2(\pi/4) = 34 \text{ ft}^2$$

$$\text{Area/bag for B} = (\pi)\left(\tfrac{10}{12}\right)(16) + \left(\tfrac{10}{12}\right)^2(\pi/4) = 42 \text{ ft}^2$$

Compare the relative cost of bags A and B:

Bag type	Area/bag	No. of bags	Relative cost
A	34.0	283	736
B	42.0	229	870

Type A bags should be used.

Example 10.5.7 A hazardous waste incinerator plant has an exhaust flow of 175,000 acfm following a dry scrubber with a loading of 6.0 gr/ft^3. The ACR is 8 and the efficiency is 99.3%. The allowable pressure drop is 10 in. H_2O.

How many 18 in. × 20 ft bags are needed? What is the frequency of cleaning if the vendor's equation for the pressure drop is

$$\Delta P = 0.3v + 4.0\ W_i v^2 \theta$$

$$v = \text{ft/min}$$

$$W_i = \text{lb/ft}^3$$

$$\theta = \text{min}$$

where ΔP = pressure drop (in. H_2O)
v = filtration velocity (ft/min)
W_i = dust concentration (lb/ft^3 of gas)
θ = time after bags were cleaned (min)

SOLUTION

$$\text{Bag area} = \left(\tfrac{18}{12}\right)(\pi)(20) + \left(\tfrac{18}{12}\right)^2(\pi/4) = 96\ \text{ft}^2$$

$$\text{Filtering area required} = 175{,}000/8 = 21{,}875\ \text{ft}^2$$

$$\text{Number of bags} = 21{,}875/96 = 230\ \text{bags}$$

$$\text{Frequency of cleaning, } \theta = (\Delta P - 0.3v)/4W_i v^2 \quad \text{where}$$

$$W_i = (6.0\ \text{gr/ft}^3)(\text{lb/7000 gr}) = 8.57 \times 10^{-4}\ \text{lb/ft}^3$$

$$= [10 - (0.3)(8)1/4(8.57 \times 10^{-4})(8^2) = 34.6\ \text{min}$$

Example 10.5.8 Consider the situation where 50,000 acfm of gas with a dust loading of 5.0 gr/ft^3 flows through a baghouse with an average filtration velocity of 10 ft/min. The pressure drop is given by

$$\Delta P = 0.20v + 5.0\ W_i v^2 \theta$$

The fan can maintain the volume flow rate up to a pressure drop of 8 in. H_2O. Show that the baghouse can be operated for 16.8 min between cleanings.

In an attempt to determine the efficiency of this unit at "terminal" conditions, both the fabric and deposited cake (individually) were subjected to laboratory experimentation with the same gas and same flow velocity. The following data were recorded:

Fabric alone

Thickness = 0. 1 in.

Concentration at the inside surface of the cloth = 1.0 g/cm^3

Concentration at the outside surface of the cloth = 0.1245 g/cm^3

Cake alone

Time = 16.8 min (end of cleaning cycle)

Concentration at the inside surface of the cake = 1.0 g/cm^3

Concentration at the outside surface of the cake = 0.0778 g/cm^3

Using the "Theodore" collection efficiency model, determine the overall efficiency at the start and end of a cleaning cycle.

SOLUTION First show that the baghouse can be operated for 16.8 min:

$$(\Delta P - 0.2v)/5W_i v^2 = [8 - (0.2)(10)]/[(5)(5/7000)(10^2)]$$

$$= 16.8 \text{ min}$$

Consider the collection efficiency of the fabric alone:

$$\ln(W_o/W_i) = -\psi L$$

$$\ln(0.1245/1.0) = -\psi(0.1/12)$$

$$\psi = 250 \text{ ft}^{-1}$$

For cake alone,

$$\ln(W_o/W_i) = -\phi\theta$$

$$\ln(0.0778/1.0) = -\phi(16.8)$$

$$= 0.152 \text{ min}^{-1}$$

Calculate the collection efficiency with the cake and fabric combined [Eq. (10.3.2)]:

$$E = 1 - e^{-[\psi L + \phi\theta]}$$

$$E = 1 - e^{-[(250)(0.1/12)+(0.152)(16.8)]}$$

$$= 0.9903 = 99.03\% \text{ (at the end of cleaning cycle)}$$

$$E = 87.55\% \text{ (at the start of the cycle)}$$

Example 10.5.9 A waste incinerator emits 300 ppm HCl with peak values of 500 ppm 15% of the time. The airflow is a constant 25,000 acfm at 75°F and 1 atm. Only sketchy information was submitted with the scrubber permit application for a spray tower. Determine if the spray unit is satisfactory.

Data

Emission limit = 25 ppm HCl

Maximum gas velocity allowed through the tower = 3 ft/s

Number of sprays = 6
Diameter of the tower = 14 ft

The plans show a countercurrent water spray tower. For a very soluble gas (Henry's law constant approximately zero), the number of transfer units (N_{OG}) can be determined by Eq. (10.2.12a):

$$N_{OG} = \ln(y_1/y_2)$$

where y_1 = concentration of inlet gas
y_2 = concentration of outlet gas

In a spray tower, the number of transfer units (N_{OG}) for the first (or top) spray will be ~0.7. Each tower spray will have approximately 60% of the N_{OG} of the spray above it. The final spray, if placed in the inlet duct, has a N_{OG} of 0.5. The spray sections of a tower are normally spaced at 3-ft intervals. The inlet duct spray adds no height to the column.

SOLUTION

$$\begin{aligned} V &= Q_a/A = Q_a/(\pi D^2/4) = (25{,}000)(4)/(60)(3.14)(14)^2 \\ &= 2.7 \text{ ft/s} \end{aligned}$$

The gas velocity meets the requirement, since the maximum allowable gas velocity is 3 ft/s.

The number of transfer units (N_{OG}) to meet the regulations is

$$N_{OG} = \ln(y_1/y_2) = \ln(500/25) = 2.99$$

The total number of transfer units provided by a tower with six spray sections is

Spray section	N_{OG}
Top	0.7
Second	0.7 × 0.6 = 0.42
Third	0.42 × 0.6 = 0.25
Fourth	0.25 × 0.6 = 0.15
Fifth	0.15 × 0.6 = 0.09
Inlet	0.5
	Total = 2.11

$$2.11 = \ln(500/y_2)$$

$$y_2 = 60.6 \text{ ppm}$$

The spray tower does not meet the HCl regulation that limits the emission of HCl to 25 ppm.

Example 10.5.10 A packed column is to be installed in a hazardous waste incinerator facility as an air pollution control device. It is designed to absorb HCl from a gas stream. Given the operating conditions and type of packing, calculate the height of packing and column diameter.

Operating Data and Assumptions

Gas mass flow rate = 5,000 lb/h
HCl concentration in inlet gas stream = 2.0 mol %
Scrubbing liquid = pure H_2O
Packing type = 1 in. Raschig rings
Packing factor (F) for 1 in. Raschig rings = 160
H_{OG} of the column = 2.5 ft Henry's law constant, m = 0.20
Density of gas (air) = 0. 075 lb/ft^3
Density of liquid = 62.4 lb/ft^3
Viscosity of liquid = 1.8 cP

The unit operates at 60% of the flooding gas mass velocity, the actual liquid flow rate is 500% more than the minimum, and 99% of HCl is to be collected based on state regulations.

SOLUTION According to Henry's law

$$x_l^* = y_1/m$$

where x_l^* = equilibrium outlet liquid composition
y_1 = inlet gas composition
m = Henry's constant

$$x_1^* = 0.02/0.20 = 0.10$$

Since it is required to remove 99% of the HCl, there will be 1% of HCl remaining in the outlet gas stream:

$$y_2 = (0 - 01y_1)/[1 - y_1) + (0 - 01y_1)]$$

where y_2 = outlet gas composition
$y_2 = 0.01(0.02)/[(1 - 0.02) + 0.01(0.02)] = 0.0002$

The minimum ratio of molar liquid flow rate to molar gas flow rate, $(L_m/G_m)_{min}$, by a material balance, is

$$(L_m/G_m)_{min} = (y_1 - y_2)/(x_1^* - x_2)$$

where x_2 = inlet liquid composition

$$(L_m/G_m)_{min} = (0.02 - 0.0002)/(0.1 - 0) = 0.198$$

For a liquid flow rate 500% more than the minimum,

$$(L_m/G_m) = 0.198 \times 6.00 = 1.188$$

N_{OG} is determined from Eq. (10.2.12) or Figure 10.2.7:

$$(y_1 - mx_2)/(y_2 - mx_2) = 0.02/0.0002 = 100$$

$$A = L_m/mG_m = 1.188/0.2 = 5.94$$

$$N_{OG} = 5.3$$

The height of packing is

$$Z = H_{OG} \times N_{OG} = (2.5)(5.3) = 13.2 \text{ ft}$$

The generalized flooding correlation [Eq. (9.3.7)] is applied to determine the cross-sectional area:

$$(L/G)(\rho_G/\rho_L)^{0.5} = (L_m/G_m)(18/29)(\rho_G/\rho_L)^{0.5}$$

where 18/29 = ratio of MW of H_2O to air
L = liquid mass velocity (lb/s-ft^2)
G = gas mass velocity (lb/s-ft^2)

From Eq. (9.3.7),

$$X = \ln\left[\frac{L}{G}\left(\frac{p_G}{p_L}\right)^{0.5}\right] = \ln\,[(1.188)(18/29)(0.075/62.4)^{0.5}]$$

$$= -3.67$$

$$Y = -3.84 - 1.06X - 0.119X^2 = -3.85 - 1.06(-3.67) - 0.119(-3.67)^2$$

$$= -1.55$$

Equation (9.3.8) may be used to calculate the cross-sectional area for flooding (A_f):

$$e^{Y} = e^{-1.55} = 0.212$$

$$A_f = \left[\frac{w_G^2 F \psi \mu^{0.2}}{\rho_G \rho_L g_c e^{Y}} \right]^{0.5} \qquad (9.3.8)$$

$$= \left[\frac{(5000)^2(160)(1)(1.8)^{0.2}}{(0.075)(62.4)(4.173 \times 10^8)(0.212)} \right]^{0.5} = 3.3 \text{ ft}^2$$

The cross section for 60% flooding is given by Eq. (9.3.9):

$$A = \frac{A_f}{f} = \frac{3.3}{0.60} = 5.5 \text{ ft}^2$$

The column diameter is then

$$D = \left(\frac{4A}{\pi} \right)^{0.5} = 2.6 \text{ ft}$$

Approximately 40 ft of packing in a 2.6-ft i.d. column is required. Four 10-ft packed columns with redistributors is recommended. The liquid flow rate should be checked with the vendor to determine if it is above the minimum wetting rate.

PROBLEMS

1. A horizontal parallel-plate ESP consists of a single duct 24 ft high and 20 ft deep with an 11-in. plate-to-plate spacing. A collection efficiency of 88.2% is obtained with a flow rate of 4200 acfm. The inlet loading is 2.82 gr/ft^3. Calculate the following using a modified form of the Deutsch–Anderson equation, with the exponent $m = 0.5$ (i.e., exponent on ϕ).

a. The outlet loading
b. The drift velocity for this system
c. A revised collection efficiency if the flow rate is increased to 5400 acfm
d. A revised collection efficiency if the plate spacing is decreased to 9 in.

2. A horizontal parallel-plate ESP consisting of four passages 24 ft high and 20 ft deep with an 11-in. plate spacing is used in an incineration process to treat 41,800 acfm of air containing 4.3 gr/ft^3 of fly ash. Using an average drift velocity of 36.3 ft/min for the system, calculate the collection efficiency of the unit.

3. As a recently hired engineer in a hazardous waste incineration facility, you have been assigned the job of submitting a preliminary design of an ESP to treat 180,000 acfm of gas laden with solids from the incinerator burning dry sewage sludge. The inlet loading of 3.47 gr/ft^3 is to be reduced to 0.07.

A previous (similar) design employed a drift velocity, throughput velocity, and plate height of 0.2 ft/s, 4.0 ft/s, and 25 ft, respectively. Submit your design.

4. A fly-ash-laden stream is to be cleaned by a venturi scrubber using an L/G ratio of 8.5 gal/1000 ft^3. The fly ash has a particle density of 0.7 g/cm^3. Use a throat velocity of 272 ft/s and a gas viscosity of 1.5×10^{-5} lb/ft-s. The particle size distribution is

Particle size range (μm)	% by Weight
0.10	0.01
0.10–0.50	0.21
0.60–1.00	0.78
1.10–5.00	18.00
6.00–10.00	16.00
11.00–15.00	12.00
16.00–20.00	8.00
20.00	45.00

Assume a correlation coefficient (k) of 250 ft^3/gal. Determine the overall collection efficiency.

5. A vendor proposes to use a spray tower in a hazardous waste incineration operation to reduce the discharge of solids to the atmosphere. The inlet loading of the gas stream from the kiln is 5.0 gr/ft^3 and is to be reduced to 0.08 gr/ft^3 in order to meet state regulations. The vendor's design calls for a water pressure drop of 80 psi and a pressure drop across the tower of 5.0 in. H_2O. The gas flow rate is 10,000 scfm and a water rate of 50 gal/min is proposed. Assume the contact power theory to apply.

a. Will the spray tower meet regulations?

b. What total pressure loss is required to meet regulations?

c. What conditions can be drawn concerning the use of a spray tower for this application?

6. A hazardous waste incineration facility plans to install a fabric filter as an air-cleaning device.

a. How many bags, each 8 in. in diameter and 12 ft long, must be used to treat the exhaust gas that has a particulate loading of 2 gr/ft^3; the exhaust fan is rated at 7000 acfm? Assume the filtering velocity is 2 ft/min.

b. Estimate the pressure drop after 4 h of operation if the resistance coefficients of the filter and dust cake are, respectively, K_1 = 0.8 in. H_2O/(ft/min) and K_2 = 3.0 in. H_2O/(lb dust/ft^2 cloth area)-(ft/min). [See Eq. (10.3.2).]

c. Calculate the fan horsepower (fan/motor efficiency of 63%) at the maximum pressure drop.

7. A hazardous waste incineration plant has an exhaust flow of 175,000 acfm following a dry scrubber with a loading of 8.0 gr/ft^3. The ACR ratio is 5.0 and the efficiency is 99.0%. The allowable pressure drop is 8 in. H_2O. How many 18-in. × 20-ft bags are needed? What is the frequency of cleaning if the vendor's equation for the pressure drop is

$$\Delta P = 0.2v + 5.0W_i v^2 \theta$$

$$v = \text{ft/min}$$

$$W_i = \text{lb/ft}^3$$

$$\theta = \text{min}$$

where ΔP = pressure drop (in. H_2O)
v = filtration velocity (ft/min)
W_i = dust concentration (lb/ft^3 of gas)
θ = time after bags were cleaned (min)

8. Consider the situation where 50,000 acfm of gas with a dust loading of 5.0 gr/ft^3 flows through a baghouse with an average filtration velocity of 10 ft/min. The pressure drop is given by

$$\Delta P = 0.20v + 5.0W_i v^2 \theta$$

The fan can maintain the volumetric flow rate up to a pressure drop of 8 in. H_2O. Show that the baghouse can be operated for 3.4043 min between cleanings.

In an attempt to determine the efficiency of this unit at "terminal" conditions, both the fabric and deposited cake (individually) were subjected to laboratory experimentation. The following data were recorded:

Fabric alone
Thickness = 0.05 in.
Concentration at inlet side of cloth = 1.0 g/cm^3
Concentration at outlet side of cloth = 0. 185 g/cm^3
Cake alone
Time = 4.0 min (end of cleaning cycle)
Concentration at inlet side of cake = 1.0 g/cm^3
Concentration at outlet side of cake = 0.08 g/cm^3

Using the "Theodore" collection efficiency model, determine the overall efficiency at the start and end of a cleaning cycle.

9. A waste incinerator emits 300 ppm HCl with peak values of 600 ppm. The airflow is a constant 5000 acfm at 75°F and 1 atm. Only sketchy information was submitted with the scrubber permit application for a spray tower. You are requested to determine if the spray unit is satisfactory.

DATA

Emission limit = 30 ppm HCl
Maximum gas velocity allowed through the tower = 3 ft/s
Number of sprays = 6
Diameter of the tower = 10 feet

The plans show a countercurrent water spray tower.

10. A packed column is to be installed in a hazardous waste incineration facility as the gaseous pollution control equipment. It is designed to absorb HCl from a gas stream. Given the operating conditions and type of packing, calculate the height of packing and column diameter.

Operating Data and Assumptions

Gas mass flow rate = 4000 lb/h
HCl concentration in inlet gas stream 2.5 mol %
Scrubbing liquid = pure water
Packing type = 1-in Raschig rings
H_{OG} of the column = 2.5 ft
Henry's law constant, m = 0.20
Density of gas (air) = 0.075 lb/ft^3
Density of water = 62.4 lb/ft^3
Viscosity of water = 1.8 cP

The unit operates at 60% of the flooding gas mass velocity, the actual liquid flow rate is 200% more than the minimum, and 99% of the HCl is to be collected based on state regulations.

REFERENCES

1. J. J. Santoleri, "Dioxin Emissions—Effect of Chlorine, Time, and Temperature at 300 Degrees C," 14th International Incineration Conference, Bellevue, WA, May 8–12, 1995.
2. J. J. Santoleri, "The Impact of Thermal Oxidation Quench Systems on Dioxin Formation," 17th International Conference on Incineration and Thermal Technologies, Salt Lake City, UT, May 11–15, 1998.
3. J. Mycock, J. McKenna, and L. Theodore, *Handbook of Air Pollution Control Engineering and Technical Equipment,* CRC Press, Boca Raton, FL, 1995.
4. H. F. Johnstone, R. B. Field, and M. C. Tassler, *Ind. Eng. Chem.*, **46**, 1601 (1954).

5. L. Theodore, C. LoPinto, L. Llamas, and C. Murray, *Proc. Environ. Eng. Sci. Conf.*, **4**, 365 (1974).
6. H. E. Hesketh, "Fine Particle Collection Efficiency Related to Pressure Drop, Scrubbant and Particle Properties and Contact Mechanism," *J. Air Pollut. Control Assoc.*, **24**, 938–942 (1974).
7. H. E. Hesketh, "Atomization and Cloud Behavior in Wet Scrubbers," US–USSR Symposium on Control of Fine Particulate Emissions, January 15–18, 1974.
8. K. T. Semrau, "Correlation of Dust Scrubber Efficiency," *J. Air Pollut. Control Assoc.*, **10**, 200–207 (1960).
9. W. Strauss, *Industrial Gas Cleaning*, Pergamon, New York, 1966.
10. Revised U.S. EPA, Engineering Handbook for Hazardous Waste Incineration, (1991).
11. P. Knisley, J. Evans, R. J. McCormick, and W. Buckley, "Waste Incinerator Upgrades at the Eastman Chemical Company, Kingsport, TN," IT3 Conference, Orlando, FL, May, 1999.
12. The Ceilcote Corporation, *IWS Bulletin*, Strongsville, OH, Bulletin 12–19, 2000.
13. J. W. MacDonald, private communication, 1998.
14. Telecon with Croll-Reynolds, October, 1999.
15. U.S. EPA,"Engineering Handbook for Hazardous Waste Incineration," prepared by Monsanto Research Corporation, Dayton, OH, EPA Contract No. 68-03-3025, September, 1981.
16. J. J. Santoleri, "Chlorinated Hydrocarbon Waste Recovery and Pollution Abatement," ASME National Incineration Conference, New York, June, 1972.
17. M. K. Theodore and L. Theodore, *Major Environmental Issues Facing the 21st Century*, Prentice-Hall, Upper Saddle River, NJ, 1998.
18. P. B. Goossens and W. Kint, "Incineration Flue Gas Scrubbing Using Reverse Jet Scrubbers," 14th International Incineration Conference, Bellevue, WA, May, 1995.
19. S. T. Eagleson and K. Gilman, "Compliance Strategies for CETRED," 14th International Incineration Conference, Bellevue, WA, May, 1995.
20. G. H. Bacon, R. Li, and K. Y. Liang, "Control Particulate and Metal HAPS," *Chemical Engineering Progress*, Dec. (1997).
21. G. Rigo and J. Chandler, "Retrofitting ESP Equipped MWCs to Meet 1995 Emission Guidelines," ASME Research Committee on Industrial and Municipal Waste Meeting, EPA, Research Triangle Park, NC, April 21, 1997.
22. J. J. Santoleri, "Hazardous Waste Incineration Course," AWMA—Course No. WST-258, Kansas City, MO, March 27, 1995.
23. O. B. Christiansen and B. Brown, "Control of Heavy Metals and Dioxins from Hazardous Waste Incinerators by Spray Dryer Absorption Systems and Activated Carbon Injection," 85th Annual Meeting, AWMA, Paper No. 92-41.06, Kansas City, MO, June, 1992.
24. J. Blumbach and L.-P. Nethe, "SORBALIT—A New Economic Approach Reducing Mercury and Dioxin Emissions," 85th Annual Meeting, AWMA, Paper No. 92-41.09, Kansas City, MO, June, 1992.
25. D. R. Roeck and A. Sigg, "Evaluation of Carbon Injection for Controlling PCDD/PCDF Emissions at WTI's Commercial Hazardous Waste Incineration Facility," 14th International Incineration Conference, Bellevue, WA, May, 1995.

11

MATERIALS HANDLING AND ANCILLARY EQUIPMENT

Contributing author: Richard D. Ross

11.1 INTRODUCTION

The major equipment items used in waste combustion systems—incinerators, combustion chambers, waste heat boilers, quenchers, and air pollution control devices—are covered in previous chapters of Part III. This chapter deals with the storage of various waste materials and equipment used for transporting and controlling flow of wastes, combustion gases, air, water, scrubber reagents, recycled scrubber liquids, and the like, from the point at which they enter the system until they leave through a stack or outfall to a wastewater treatment plant, sewer, or stream.

The equipment discussed in this chapter is involved in the storage and movement of these materials. Storage systems involve tanks, drums, and impoundments for liquids, bins and/or boxes for solids, tanks, sumps, and drums for sludges. Waste combustor systems utilize a variety of conveyor types for feed and ash removal systems. Pipes, ducts, fittings, and stacks are employed for transporting liquids, sludges, and gases within the facility. Valves are used for controlling the flow of these materials. A pump is used as the motive force for the flow of liquids and sludges. A fan or compressor is used for gases.

There are many other devices that could appropriately be considered under the title "Ancillary Equipment" for waste combustion systems. For example, blenders and evaporators are used in the pretreatment of waste, and activated carbon canisters are used for capturing fugitive gases in process vents. The devices are too numerous to include and describe here. The interested reader is directed to the literature for additional and more detailed information on this equipment.

11.2 WASTE STORAGE

It is important to have sufficient storage for the variety of waste streams to be treated in a hazardous waste facility. Storage is required not only for the expected waste flow through the facility but more importantly for wastes that accumulate during periods when the treatment equipment is not operating. Each site will vary in its storage capacity and the types of storage vessels.

Waste combustion systems consist of many components and auxiliaries that are considered high maintenance. This results in scheduled as well as unscheduled downtime for the waste treatment processing equipment. Storage systems should be designed with sufficient capacity to provide for these downtime periods.

Liquid and Sludge Storage

Liquid and sludge wastes enter the waste treatment facility in tank trucks, railroad tank cars, totes, trailers containing drums, boxes, or similar containers. Bulk liquid or sludges are usually delivered to the facility in tank trucks or railroad tank cars that must be unloaded into on-site storage tanks or fed directly to the combustor. Direct feed has proven advantageous when handling high-viscosity materials. These materials are transported in heated tank trucks. Waste streams that may polymerize if cooled in transfer or in storage are typical of those directly fed to the combustor. Others that may react adversely with water or oxygen in the atmosphere resulting in air emissions are also candidates for the direct feed system.

Storage tanks used in waste combustion facilities are of all physical design types and materials of construction. The tank design depends on the characteristics of the waste material—corrosivity, reactivity, volatility, density, and viscosity. Waste treatment facilities must include storage for all types of materials. The stored volume, vessel configuration, the need for agitation, heating and air emission control as well as inert blanketing, flame arrestors, pressure/vacuum relief valves, and other fire protection devices are factors that must be considered in the storage facility design. Storage tanks must be designed for ease of leak detection and with proper access for internal inspection and cleaning. Proper number, size, and location of nozzles and connections must be provided for internal level, pressure, and temperature measurement. Provision should also be made for external insulation when necessary and appropriate.

At commercial disposal sites such as hazardous waste incinerators (HWIs), cement kilns, and lightweight aggregate kilns, special alloy tanks or tanks requiring special materials of construction are often installed when there is a continuous supply of a high disposal revenue corrosive waste stream. If this is not the case, the facility should avoid the costs of these special alloy tanks. Interchangeability of storage tanks at a facility, that is, tanks that can be used for different waste streams, are desirable and economical. Tanks of identical capacity and configuration are also more economical to purchase and install. Hazardous waste storage can usually be accommodated by carbon steel tanks with a few stainless steel, glass-lined, and fiber-reinforced plastic (FRP) tanks for special wastes. Plastic-lined carbon steel tanks are generally not practical because they will sooner or later be misused. Since most waste streams delivered to a waste combustion system are not corrosive, carbon steel is usually satisfactory.

In-ground tanks are not permissible. Tanks with flat bottoms sitting on a concrete pad are not advisable because visible leak detection is impossible. Vertical tanks supported on legs or horizontal tanks supported on saddles are the most reliable. The capacity of these tanks must be commensurate with the liquid waste throughput of the waste combustion system but normally will be under 10,000 gal. Liquids containing solids often require storage in an agitated tank. Smaller capacity tanks should be used for this service. Smaller tanks are also desirable when higher viscosity materials are stored and require heat to maintain pumping viscosity.

Large tanks are difficult to agitate. Tanks that store liquids containing solids must be cleaned periodically. Small tanks are easier to clean. It has been common practice to store the "difficult" wastes in smaller tanks. Horizontal tanks on saddles are also preferred for large storage capacity. Saddles used for horizontal tanks are better supports than steel columns needed for vertical tanks. Saddles are also more economical. A large horizontal tank can also be shop fabricated.

Mixing or agitation in tanks containing multiphase liquids or liquids and solids increases in difficulty with the tank volume. Therefore, if continuous mixing is required for a specific waste stream, it is best accomplished in a small vessel, either by using a conventional agitator or by a high-volume recirculation of the contents through a pump. The required nozzles and connections should be provided for the mixer and recirculation lines.

Many liquid wastes will increase in viscosity with a decrease in temperature and may even become solid. One should note that water freezes at 32°F. Storage of aqueous streams will require a heated tank design in areas where the ambient temperature drops below 32°F. Steam or hot oil jackets and/or coils will be required as an external jacket or internal coil within the tanks.

Storage tanks are usually confined to a single area designated as a *tank farm*. Regulations for hazardous waste storage require that tank farms provide secure containment of all spills or leaks from the tanks within the tank farm. This means that an impervious enclosure usually made of concrete must be built under and around the storage tanks. This enclosure must have the capacity to contain *all* the liquid from the largest tank in the containment area.

Many bulk sludges received at waste combustion facilities are not suitable for tank storage because they are not fluid enough to be pumped. Those that are pumpable can be fed directly from the transport source to the combustion zone by a diaphragm pump. Those that are not pumpable need to be handled as bulk solids and fed through the solids feeder mechanisms.

Storage considerations are different for materials that are not received in bulk, and many tons of hazardous waste are received in totes and drums. Totes usually contain liquid wastes because they cannot be emptied if they contain sludges or solids. While totes vary in size and configuration as well as materials of construction, they hold about 475 gal of liquid waste. Some means of pumping the waste into storage tanks or directly to the combustor is a critical step in the process. Steel drums of 55- and 80-gal capacity containing liquid must be similarly handled unless the incinerator is a slagging-type kiln. With such a kiln, it is possible if the contents have appropriate thermal properties, to feed a steel container to the incinerator and have it destroyed in the incinerator. The wastes will be combusted and the steel will be converted into a molten slag. A more acceptable drum disposal method is to place wastes in burnable plastic or fiber drums, usually of 20- or 30-gal capacity. These drums can be fed directly into the primary combustion zone that will burn both the drum and the contents.

Drum storage as well as tote storage should also provide for easy leak detection and recovery of leaked materials. Totes used for storage of hazardous wastes should never be placed with totes more than one tote high. If metal, plastic, or fiber drums are used for storage, drums should not be placed in rows with more than two drums high. Adequate drainage and collection area must be provided in the storage area so that any leakage from totes or drums can be contained within the facility, repackaged and properly treated.

Liquid storage design must take into account the volatility and flammability of the materials stored. Maximum emissions from storage tanks occur during loading, not during storage. The vapor pressure, flash point, autoignition temperature, and calorific value of the material is vital to the loading and storage method prescribed. High vapor pressure materials need to be stored under pressure. If materials are flammable, storage must be done under a blanket of nitrogen gas at a nominal pressure. If the waste stream vapor pressure is moderate and the emissions from the waste material are below allowable emission standards, then a pressure/vacuum breather is all that is required. If not, refrigeration to condense the vapors or carbon adsorption of the vapors may be utilized.

The choice of a vapor control system for volatile hazardous waste storage is primarily an economic consideration. It is always preferable and usually more economical to use slight pressurization with a nitrogen blanket than to provide refrigeration, activated carbon, or other methods to prevent volatile organic compounds (VOC) emissions. This may not always be possible and will often depend upon the vapor pressure of the material in the storage tank. This becomes an operational choice. High vapor pressure wastes should be destroyed in the waste combustor on a priority basis due to storage problems and fire potential.

Refrigeration of vapors to condense the VOCs is only practical to about 40°F unless dry nitrogen is used as a blanketing gas since atmospheric moisture can form ice on the refrigeration coils. Lower refrigeration temperatures are possible for VOC mixtures with low boiling points if dry nitrogen is used as a blanket. This approach is usually expensive to install and more expensive to operate.

High vapor pressure VOC storage systems have successfully used activated carbon adsorbers to capture emissions with or without a nitrogen blanket. As the vapor pressure of the stored liquid increases, more liquid will vaporize at ambient temperatures, so the amount of vapor to be captured by activated carbon or other means needs to be accurately estimated a priori. Fifty-five-gallon carbon canisters are readily available from a number of suppliers, but these may last only a short time if the VOC emission and/or concentration rate is high. In this case, a larger system will be necessary. As a rule of thumb, the vapor phase activated carbon weight required is five times the weight of VOC adsorbed, but this varies with each VOC (depending upon its adsorption isotherm). One should note that some VOCs, especially ketones, contain oxygen that may cause spontaneous combustion of the activated carbon over time. While activated carbon can be regenerated on-site with steam, most waste combustion facilities replace spent carbon with fresh material.

It is not without precedent that tank farms handling large quantities of high vapor pressure VOCs might choose to utilize a thermal oxidizer for VOC destruction, but the quantity of emissions, even during loading, is low and not at such a rate to warrant the expense of this type of control.

Solid Waste Storage

Solid wastes that are hazardous are usually not volatile and often consist of contaminated soils that contain hydrocarbons. These usually arrive in trucks and roll-off boxes. Trucked waste must be dumped into *bunkers* or other enclosures. Roll-off boxes may be utilized as temporary storage prior to incineration if properly covered to prevent particulate or vapor emissions.

11.3 FEED SYSTEMS

Liquid Wastes

Feeding *liquid waste* to the incinerator can include a wide variety of techniques and equipment. Liquid wastes that have a high calorific value (>10,000 Btu/lb) are usually employed as fuel in the auxiliary burners of the system. This is the best solution if these liquid materials have reasonably low viscosity and can be pumped and atomized. The liquids are fed under pressure by pumps.

Aqueous wastes listed as hazardous air pollutants (HAPs) in the Clean Air Act Amendments (CAAA) of 1990 [Maximum Achievable Control Technology

(MACT) Rules, Sept. 30, 1999] are usually injected into the combustion zones (primary and/or secondary) via atomizing nozzles. The heat generated in the combustion zones will evaporate the water and oxidize the organics and solids present.

Containerized Wastes

Containerized sludges and solids present a major challenge to the feed system design. These materials, containerized in fiber-paks, plastic drums, cardboard boxes, or other containers need first to be categorized as to their total heat content. A 55-gal drum containing a waste of 5000 Btu/lb material has a potential heat release of approximately 2 million Btu. If this is a very volatile material, it could burn completely in several minutes or less. If burned in 5 min, the heat release rate would be 24 million Btu/h, and there probably is insufficient combustion air supplied to the incinerator to burn it. Consequently, container feeds must match the ability of the incinerator to provide air for combustion, and this requires close control of the heat content of every container and the rate at which containers are fed. To comply with this thermal restriction, most HWIs limit the heat content of containers accordingly and through "bar" coding identify each and every container that enters the incinerator.

The feed methods for containers vary slightly. Powered roller conveyors are used to bring the containers from storage to the incinerator. At this point several options are available. One is the "ram" feeder. The container is introduced into a ram feeder, which closes after accepting the container. The hydraulic feed ram then pushes the container into the feed chute as the airlock door on the front of the incinerator opens. As the ram retracts, the door closes to minimize inlet air to the incinerator. Similarly, a "skip" hoist or similar mechanism can be employed to lift the container to a point where it may be fed by gravity through an airlock on a chute. In order for these container feed systems to operate properly without causing the incinerator to "puff" or register high CO in the stack, the heat content of each container must be known and the feed rate adjusted accordingly.

Sludges and Solid Wastes

These wastes are the most difficult to feed to an incinerator and depending upon their characteristics might utilize a variety of conveyor and feed options. Sludges and contaminated solids (most likely to be soils) are usually not pumpable. Some sludges are pumpable and can be pumped into the incinerator at rates compatible with their heat release. For soils and sludges that are not pumpable, there are screw conveyors, belt conveyors, and bulk methods such as a feed hopper or chute on the incinerator that is loaded by a *front end loader* or a *clam shell.* In most circumstances, a *weigh belt* is used to measure the feed rate to the incinerator.

The physical characteristics of the sludge/soil are more important than the heat content in choosing a feed method because the heat content is generally very low. Screw conveyors and belt conveyors work well with dry, nonsticky feed materials. Wet materials need to be mixed with inert materials such as sand or clean soils to enhance processing. Batch ram feeding can be employed where the material is compatible, that is, not too wet and not too dry, but this is seldom achieved, so batch feeding is not normally practical. The screw feeder or auger is usually best for solids and sludges because it can feed continuously at variable feed rates. Dry solids can be belt fed to a feed chute on a continuous basis, but few systems need to handle dry solids. Designing a universal feed system for solids and sludges is essentially impossible unless the owner/operator is willing to prepare and *condition* the feed to make it compatible to the feed system. This requires significant flexibility.

11.4 ASH HANDLING SYSTEMS

Removing ash from the incinerator is often as difficult as injecting the feed into the chamber, except that the ash characteristics do not vary as much as the feed characteristics. Since the HWI may be handling a variety of liquids, containerized and solid wastes at the same moment, the ash will reflect this diversity; therefore, the ash removal system must also cope with this same diversity.

At the end of the incineration process there are three methods to handle the ash. The ash may be collected "hot" in a disposal container, sprayed with water and allowed to cool by natural convection and radiation. The hot ash could pass into a wet quench where it is immersed in water, evaporating the water while it cools. The ash may also be cooled indirectly by water sprayed onto a screw conveyor or rotary drum cooler containing the ash. Indirect cooling of ash from HWI is usually found in small kilns. Most large kilns in operation today use the direct ash quench with a drag chain to remove the ash from the water.

The ash generated from HWI systems is often benign but may occasionally contain metal components that render the ash a hazardous waste. In these cases, ash must be mixed with other materials that will solidify the ash in a manner that it will meet toxicity characteristic leaching procedure (TCLP) regulations. The ash from an incinerator will usually meet disposal standards without further treatment but, if not, can be further treated by mixing in a solids mixing machine with lime, portland cement, or a variety of other proprietary materials available for solidification. The resulting mixture solidifies and may be landfilled without leaching if it passes the TCLP test. Processed ash that is not hazardous may be stored in bunkers, bins, or roll-off boxes and subsequently landfilled.[1,2]

11.5 PIPES, DUCTS, FITTINGS, AND VALVES

Pipes

The most common conduits for fluids are pipes and tubing. Both have circular cross sections, but pipes tend to have larger diameters and thicker walls. Because of the heavier walls, pipes can be threaded, while tubing is not normally threaded. Because HWI systems usually handle large flow rates that require the larger diameters associated with pipes, the use of tubing in these systems is primarily limited to instrument air, chemical metering systems, lubrication systems, or the internals of heat exchangers.

Pipes and tubing are manufactured from a wide variety of materials. The selection of the material depends on the temperature and pressure of the fluid and its corrosive characteristics. Materials available are both metallic and non-metallic. Metallic piping can be steel, a variety of stainless steels, chromium and nickel alloys, aluminum, brass and a variety of copper alloys, monel, inconel and many others. Non-metallic materials are often used where corrosion resistance is required and temperatures are compatible. These include PVC, polyethylene, polypropylene, and fiberglass reinforced polyester products. Where pressure and corrosion are factors, rubber and plastic lined steel pipe can be used effectively. Steel pipe can also be lined with tin, lead, or other metallic coatings for special purposes. Glass pipe and glass lined steel pipe are used for corrosive applications where contamination may be a problem but are not often seen in the HWI industry.

Pipe connections may be made in several ways. Small pipes (diameters <2 in.) usually have threaded or socket weld joints depending upon codes and requirements. Larger pipes normally have flanged or welded joints.

Since pipes are manufactured with many different diameters and wall thickness, the following method of standardization has been established by the American National Standards Institute (ANSI). By convention, pipe and fitting sizes are characterized by nominal diameter and wall thickness. For pipes over 12 in. in diameter, the nominal diameter is the actual outside diameter. For pipes from 3 to 12 in., the nominal value is approximately equal to the inside diameter. For pipes greater than 3 in. the nominal size can be considerably smaller than the actual inside diameter.

Typical pipe diameters for various liquid flow capacities in HWI applications are given in Table 11.5.1. The wall thickness of a pipe is specified by a schedule number, which is a function of the internal pressure and allowable stress, as follows:

$$\text{Schedule number} = 1000\ P/S \tag{11.5.1}$$

where P = internal working pressure (lb/in^2 gauge)
S = allowable stress (lb/in^2)

There are 10 schedule numbers in use and these range from 10 to 160; the

Table 11.5.1 Nominal pipe diameters for liquid waste flows

Capacity (gpm)	Nominal Pipe Diameter (in.)
0–15	1
15–70	2
70–150	3
150–250	4

thickness of the pipe wall increases with the schedule number. Schedule 40 is the most commonly used pipe thickness for normal temperature and pressure applications.

Piping in HWI facilities include liquid waste feeding and transfer within the storage complex, fuel oil and caustic solutions, and nitrogen blanketing and purging lines. The total length of pipe employed depends on a number of factors, including overall facility size, number of storage tanks, and storage facility layout.

Ducts

Ducts are used only to convey air or other gases. Pipes may also be used for this purpose if higher pressures are required. Ducts normally have thin walls and are generally used for conveying gases at low pressures and temperatures. Pipes are usually circular in cross section, but ducts may be fabricated in many different cross sections. Ducts are usually round, square, or rectangular. Ducts are generally larger in cross section than pipes because they carry fluids with low densities at high volumetric flow rates. In HWI applications, ducts are used primarily to move exhaust gases from the incinerator to the boiler or quencher and then to the air pollution control equipment and subsequently to the stack.

Ducts, like piping, can be fabricated from a wide variety of metallic and nonmetallic materials as discussed previously. Large ducting is often field fabricated whereas small ducts may be fabricated in a shop. Material costs of ducting generally increase along with corrosion resistance, so the selection of material must be determined by the desired life in the anticipated environment. This environment is a function of the characteristics of the waste being incinerated and the operating conditions of the incinerator. For maximum resistance to moisture or corrosive gases, stainless steel is used when the cost can be justified.

Gases traveling through a duct impose two loads on its structure: one related to pressure, the other to velocity. The dominant load is normally the difference in static pressure across the duct wall; the second load is that caused by turbulent airflow, which generates relatively small but pulsating forces on the duct wall.

Duct construction is classified in terms of operating pressure. The low-pressure range is from 0 to 2 in. H_2O (0.072 psig) and the high-pressure range up to 10 in. H_2O (0.36 psig). The wall thickness depends on the material of construction, operating pressure, and the cross-section shape. Typical wall thickness for a 3-ft-diameter round duct made of galvanized steel is about 0.003 in. for low-pressure and 0.004 in. for high-pressure operation.

Fittings

A fitting is a piece of equipment that has for its function one or more of the following:

- To join two pieces of straight pipe—couplings and unions
- To change direction—elbows (90° and 45°) and tees, wyes, and crosses
- To change pipe diameter—reducers and bushings
- To terminate a pipe—plugs and caps
- To join two or more pipes—tees, wyes, and crosses

A *coupling* is a short piece of pipe threaded on the inside and used to connect straight sections of pipe. A *union* is also used to connect two straight sections but differs from a coupling in that it can be opened conveniently without disturbing the rest of the pipeline. When a coupling is opened, a considerable amount of piping must usually be dismantled. An *elbow* is an angle fitting used to change flow direction, usually by 90°. A *tee* (shaped like the letter T) can also be used to change flow direction but is more often used to combine two streams into one. A *reducer* is a coupling for two pipe sections of different diameter. A *bushing* is also a connector for pipes of different diameter, but, unlike the reducer coupling, it is threaded on both inside and outside. The larger pipe screws onto the outside of the bushing, and the smaller pipe screws into the inside of the bushing. To terminate a pipeline, *plugs,* which are threaded on the outside, and *caps,* which are threaded on the inside, are used. A *wye* (shaped like the letter Y) is similar to the *tee* and is used to combine two streams.[3]

Valves

Valves are used to control the flow of a fluid or to cut off the flow completely. Valves may be used in both pipes and ducts. In ducts they are usually called "dampers."

There are many different types of valves, the most commonly used are the gate valve and the globe valve. The *gate* valve consists of a simple disk on the end of a threaded stem that moves perpendicular to the pipe in which it is installed. Gate valves are used primarily for *on–off* control of a liquid. As the fluid passes through the gate valve, only a small amount of turbulence is generated; the direction of flow is not altered and the flow cross-sectional area inside the valve is only slightly smaller than that of the pipe causing only a minor pressure drop. Problems with abrasion and erosion of the disk arise when the valve is used in positions other than fully open or fully closed.

The *globe* valve is designed for flow control. In the globe valve, the liquid passes through a horizontal opening, or valve seat, into which a plug with a slightly beveled edge is inserted when the stem is closed. Good control of flow is achieved with this type of valve, but at the expense of a higher pressure loss than a gate valve.

Today, standard gate and globe valves are being replaced with *ball valves.* These provide the tight shutoff of the gate valve and to some extent the control of a globe valve, accompanied by low pressure loss. The ball valve consists of a ball or sphere that has a hole through its center that can be smaller than the pipe or the same size. The ball rotates 90° in the valve body from fully open to fully closed.

The *check* valve is used to prevent "backflow" in a pipeline and is often employed on the discharge side of centrifugal pumps. The *butterfly* valve operates in a horizontal plane like a ball valve, but instead of a ball it consists of a flat disk for control. The *plug* valve utilizes a rotating tapered plug in the same manner, usually for on–off service. The *needle* valve, a variation of the globe valve, utilizes a narrow tapered plug and gives close flow control for small flow volumes. The *diaphragm* valve was designed to handle very viscous liquids, slurries, or corrosive liquids that may tend to clog the moving parts of other valves and use plastic or rubber diaphragms to close the flow area.

In modern HWI facilities flow is often controlled automatically. The valves are then driven by pneumatic or electric operators using valve positioners for repeatability

11.6 FANS, PUMPS, AND COMPRESSORS

To move fluids through the various pieces of equipment at the incinerator facility, including piping and ductwork, requires mechanical energy, not only to impart an initial velocity to the material, but more importantly to overcome pressure losses that occur throughout the flow path.

In this section, three devices that convert electrical energy into the mechanical energy that is to be applied to the various streams are discussed. These devices are *fans or blowers,* which move low-pressure air and other gases; *compressors,* which move high-pressure air and other gases; and *pumps,* which move liquids and liquid–solid mixtures such as suspensions, slurries, and sludges. In HWI operations, fans or blowers are used to move the gases, and pumps to move the liquids and semiliquids. These two devices will be discussed in some detail here. Compressors are mainly used in these operations for air and sometimes for nitrogen. The compressed air is generally used for liquid atomization and as an instrument air source and the nitrogen for blanketing and purging.

Fans

The terms *fan* and *blower* are often used interchangeably. They are essentially the same rotating device so no distinction will be made between the two in the discussion of their operation. In generic terms fans are low-pressure devices and blowers are generally employed when generating pressure heads of 1 psig or more. Pressures above 3 psig generally require compressors or positive-displacement-type blowers, which are essentially low-pressure compressors.

Fans are usually classified into centrifugal or axial-flow types. In a centrifugal fan, the gas is introduced into the center of the revolving wheel (the eye) and discharged tangentially from the rotating blades. In an axial-flow fan, the gas moves through the axis of rotation of the fan blades or propeller. Both types are used in industry, but the centrifugal fan is more widely used at incineration facilities.

The gas in a centrifugal fan is subjected to centrifugal force. This force compresses the gas giving it additional static pressure. Centrifugal fans are enclosed in a scroll-shaped housing that helps convert kinetic energy to static pressure. Gas rotating between the fan blades is compressed in the fan scroll, which increases the static pressure. Centrifugal fans are classified by blade configuration as radial, forward curved, backward curved, air foil, and radial tip–forward curved heel.

Radial or *straight blade* fans physically resemble a paddle wheel with radial blades attached to the rotor. This is the simplest design of all centrifugal fans. This enables most radial blade fans to be built with great mechanical strength and to be easily repaired. These fans can be used in a variety of situations, especially heavy-duty applications. This fan can handle errosive and corrosive gases as well as very sticky gases. It is particularly well suited for high static pressure operations and can generate pressures in excess of 50 in. H_2O. When operated properly, the horsepower efficiency range is 55–70%, with 65% as a typical value.

Forward curved fans are the most popular for general ventilation purposes (high flow rates and low static pressures). These fans have both the heel and the tip of the blade curved forward in the direction of rotation. Blades are smaller and spaced much closer together than in other blade designs. Forward curved fans are generally not used with dirty gases when dust or sticky materials are present because contaminants easily accumulate on the blades and cause imbalance. Efficiencies range from 50 to 70%, with 65% being typical.

Backward curved or *backward inclined* fans have blades inclined in a direction opposite to that of the direction of rotation. This feature causes the gas to leave the tip of the blade at a lower velocity than the wheel-tip speed, a factor that improves the mechanical efficiency. These types of fans are not suitable for a heavily particulate-laden gas, sticky material, or abrasive dust.

Centrifugal forces tend to build up particulate matter on the backside of the fan blades. Backward curved fans are rarely used in air pollution control despite typical efficiencies of 70%.

The airfoil fan is similar to the backward curved fan, except that the blade has been contoured to increase stability and operating efficiency. These fans are more expensive to construct than backward curved fans but have lower power requirements. They are used only occasionally in pollution control. The gas must be clean and noncorrosive. Since the blades are hollow, abrasion and wear could allow dust, water vapor, and so on, to enter the blade and cause imbalance. This is the most efficient of the various fan types, with typical efficiencies at the 85% level.

A modification of the radial fan is the *radial tip–forward curved wheel.* The blades are curved forward with this unit. This fan is reportedly more dependable than the radial or high tip speed applications (i.e., high static pressures) due to better vibrational characteristics and its ability to resist fatigue. It finds application in the processing of large flow rates (>200,000 acfm) with light to medium particulate loadings. However, sticky material and particulates, in general, can accumulate in the slight curvature of the blades and cause imbalance. Efficiencies here typically range from 50 to 75%, with 70% being common.

Generally, centrifugal fans are easier to control, more robust in construction, and less noisy than axial units. They have a broader operating range at their highest efficiencies. Centrifugal fans are better suited for operations in which there are flow variations, and they can handle dust and fumes better than axial fans.

Fan laws are equations that enable the results of a fan test (or operation) at one set of conditions to be used to calculate the performance at another set of conditions, including differently sized but geometrically similar models of the same fan design. The fan laws can be written in many different ways. The three key fan laws are provided in the following equations:

$$Q_a = k_1(\text{RPM})D^3 \tag{11.6.1}$$

$$P_s = k_2(\text{RPM})^2 D^2 \rho \tag{11.6.2}$$

$$\text{HP} = k_3(\text{RPM})^3 D^5 \rho \tag{11.6.3}$$

where Q_a = volumetric flow rate
P_s = static pressure
HP = horsepower
RPM = revolutions per minute
D = wheel diameter
ρ = gas density
k_1, k_2, k_3 = proportionality constants

Thus, these three laws may be used to determine the effect of fan speed, fan size, and gas density on the flow rate, static pressure head, and horsepower. For two conditions, where the constants k remain unchanged, Eqs. (11.6.1) through (11.6.3) become

$$(Q'_a/Q_a) = (\text{RPM}'/\text{RPM})(D'/D)^3 \tag{11.6.4}$$

$$(P'_s/P_s) = (\text{RPM}'/\text{RPM})^2(D'/D)^2(\rho'/\rho) \tag{11.6.5}$$

$$(\text{HP}'/\text{HP}) = (\text{RPM}'/\text{RPM})^3(D'/D)^5(\rho'/\rho) \tag{11.6.6}$$

Note: The *prime* refers to the new condition. It is also important to note that the fan laws are approximations and should not be used over wide ranges or changes of flow rate, size, and so on.

A rigorous, extensive treatment on fan selection is beyond the scope of this text. It is common practice among fan vendors to publish voluminous data in tabular form providing flow rate, static pressure, speed, and horsepower at a standard temperature and gas density. These are often referred to as multirating tables.

Note: These tables should not be used for fan selection except by those who have experience in this area. For those who do not, the proper course of action to follow is to provide the fan manufacturer with a complete description of the system and allow the manufacturer to select and guarantee the optimum fan choice.

To help in the actual selection of fan size, a typical fan rating table is given in Table 11.6.1. The fan size and dimensions are usually listed at the top of the table. Values of static pressure are arranged as columns that contain the fan speed and brake horsepower required to produce various volume flows. The point of maximum efficiency at each static pressure is usually underlined or printed in special type. In order to select a fan for the exact condition desired, it is sometimes necessary to interpolate between values presented in the multirating tables. Straight-line interpolation can be used with negligible error for multirating tables based on a single fan size. Some multirating tables attempt to show ratings for a whole series of geometrically similar (homologous) fans in one table; in this case, interpolation is not advised.[4]

The selection procedure is, in part, an examination of the fan curve and the system curve. A fan curve for air, relating static pressure with flow rate, is provided in Figure 11.6.1. Note that each type of fan has its own characteristic curve. Also note that fans are usually tested in the factory or laboratory with open inlets and long smooth straight discharge ducts. Since these conditions are seldom duplicated in the field, actual operation often results in lower efficiency and reduced performance. A system curve is also shown in Fig. 11.6.1. This curve is calculated prior to the purchase of the fan and provides a best estimate of the pressure drop across the system through which the fan must

Table 11.6.1 Typical fan rating table[a–e]

Q_a	Static Pressure (in. H_2O)											
	$\frac{1}{2}$		1		$1\frac{1}{2}$		2		$2\frac{1}{2}$		3	
(acfm)	rpm	bhp[e]	rpm	bhp	rpm	bhp	rpm	bhp	rpm	bhp	rpm	bhp
5,727	216	0.74	278	1.34	330	2.01	378	2.75	421	3.50	460	4.28
6,873	236	0.97	291[d]	1.67	339	2.42	384	3.24	424	4.06	462	4.91
8,018	250	1.27	305	2.05	352	2.87	393	3.76	430	4.69	467	5.66
9,164	271	1.63	320	2.49	366	3.42	405	4.35	441	5.36	475	6.40
10,309	293	2.12	338	3.05	381	4.06	419	5.06	453	6.14	486	7.26
11,455	315	2.72	356	3.65	396	4.76	432	5.88	468	7.03	499	8.19
12,600	337	3.46	377	4.39	413	5.58	448	6.81	482	8.04	514	9.31
13,746	360	4.39	399	5.25	430	6.48	465	7.82	496	9.16	527	10.53
14,891	382	5.43	421	6.25	451	7.52	481	8.93	512	10.35	542	11.80
16,037	405	6.66	442	7.48	473	8.67	501	10.16	529	11.69	557	13.25
17,182	429	8.08	463	8.97	496	10.05	521	11.54	547	13.18	574	14.81
18,328	451	9.64	486	10.61	517	11.61	543	12.99	566	14.78	591	16.49
19,473	474	11.46	510	12.47	539	13.47	565	14.85	587	16.56	610	18.35
20,619	497	13.51	532	14.55	560	15.60	587	16.82	610	18.54	630	20.40
21,764	520	15.82	556	16.82	584	17.98	608	19.17	632	20.77	652	22.59

Source: Bayler Blower Co.

[a]Wheel style: backward inclined.

[b]Wheel diameter: 50½ in.

[c]Maximum fan speed: 1134 rpm.

[d]Performances underlined are those at maximum efficiency.

[e]Brake horsepower = bhp.

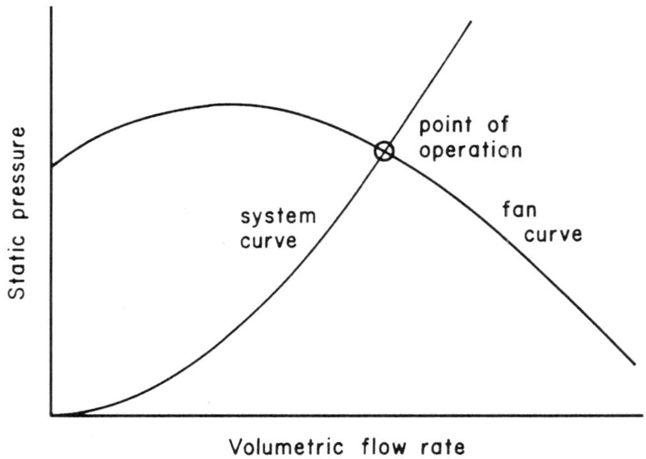

Figure 11.6.1 System and fan characteristic curves.

deliver the gas. (This curve should approach a straight line with an approximate slope of 1.8 on log-log coordinates.) The system pressure (drop) is defined as the resistance through ducts, fittings, equipment, contractions and expansions, and so on. A number of methods are available to estimate the total system pressure change. These vary from very crude approximations to detailed, rigorous calculations. The simplest procedure is to obtain estimates of the pressure change associated with the movement of the gas through all of the resistances described previously. The sum of these pressure changes represents the total pressure drop across the system, and represents the total pressure (change) that must be developed by the fan on both the inlet and exit sides of the fan. This calculation becomes more complex if branches (i.e., combining flows) are involved. However, the same stepwise procedure should be employed. When the two curves are superimposed, the intersection is defined as the *point of operation*. The fan should be selected so that it operates just to the right of the peak on the fan curve. The fan operates most efficiently and with maximum stability at this condition. If a fan is selected for operation too close to its peak, it will surge and oscillate. Thus, the point of intersection of the two curves determines the actual volumetric flow rate. If the system resistance has been accurately specified and the fan properly selected, the two performance curves will intersect at the design flow rate. If system pressure losses have not been accurately specified or if undesirable inlet and outlet conditions exist, design conditions will not be obtained. Dampers and fan speed changes can provide some variability on operating conditions.

There are a number of process and equipment variables that are classified as part of a fan specification. These include

- Flow rate (acfm)
- Temperature
- Density
- Gas stream characteristics
- Static pressure that needs to be developed
- Motor type
- Drive type
- Materials of construction
- Fan location
- Noise controls

Fans may be driven by a direct motor drive at motor speed either by mounting the fan wheel on the motor shaft or through a coupling between the motor shaft and fan shaft. While mounting the wheel directly on the motor shaft is economical, it usually is advisable only for small fans in light service applications. Large or severe service fans use coupled motor drives because direct drive units have lower maintenance costs and lower power transmission losses; however, the speed cannot be varied and if a speed change is required, the cost can be expensive.

Belt drives can provide this flexibility if the power is <200 hp. For larger or smaller fans a variable-speed motor drive or a hydraulic fan drive can be used to achieve the same flexibility as a belt drive but at significantly increased cost. The main induced-draft (ID) fan on many HWI systems is often designed with a variable-speed drive for flexibility and power savings.

Materials of construction can vary as they do for piping and ductwork. Carbon steel is commonly employed when treating dry air up to temperatures approaching 1000°F. Fiberglass may be used for corrosive conditions, but the temperature should not exceed 250°F. Fan wheels are often made of aluminum so all materials of construction of a fan should be examined before selection.

The fan may be located downstream or upstream of a particular piece of equipment in the process. Fans located downstream are referred to as *induced-draft* or *negative pressure* fans. Air or gas leakage is into, rather than out of, the unit. Equipment costs could be higher because of the need of heavier construction material when operating under negative (vacuum) pressure.

Fans located upstream are referred to as *forced-draft* or *positive pressure fans.* Forced-draft fans are seldom employed in "dirty" particulate-laden systems for obvious reasons. The type of incinerator employed will probably dictate the choice between induced-draft and forced-draft fans.

Fans are often noisy. Noise controls may be dictated by their location in the plant or adjacent property. To minimize noise effects, the fan should be properly designed and properly operated. A fan operated close to the point of maximum efficiency will reduce noise levels. A fan in proper balance is less noisy than one not in balance. If noise problems are expected or are already present, acoustical insulation should be applied to the fan housing or the fan entirely enclosed in an acoustical housing. Insulation can also be added to the ductwork if necessary.

Gas (or air) horsepower (GHP) and brake horsepower (BHP) are the two terms of interest. These may be calculated from Eqs. (11.6.7) and (11.6.8):

$$\text{GHP} = 0.0001575Q_a\ \Delta P = Q_a\ \Delta P/6356 \tag{11.6.7}$$

$$\text{BHP} = 0.0001575Q_a\ \Delta P/E \tag{11.6.8}$$

where Q_a = volumetric flow rate (acfm)
ΔP = SP = static pressure head developed (in H_2O)
E_f = fractional fan efficiency

For cold startups, the static pressure can be significantly higher than the normal operating pressure drop, and the fan must be able to handle the increased resistance. The correction for the cold static pressure may be calculated from

$$\Delta P_c = \Delta P(\rho_c/\rho) \tag{11.6.9}$$

where ρ = gas density (lb/ft^3)
c = subscript denoting cold gas

For HWI systems, the ID fan (i.e., located downstream) is almost always used so that any leakage will be into rather than out of the unit. Either a single ID fan, or two ID fans in series, are used to create *draft* through the waste heat boiler (if there is one), quencher, and the air pollution control equipment, and to maintain a slight negative pressure in the incinerator. Typical pressure drops through the incinerator, waste heat boiler, quencher, and adsorber are 4, 4, 1, and 7 in. H_2O, respectively. The pressure drop across a venturi scrubber may range from 10 to 100 in. H_2O. Pressure drops across dampers and ductwork may be neglected. Typical fan-motor efficiencies are in the 50–60% range.[5]

All cooling and combustion air, excluding atomizing air (unless a low-pressure air atomizing burner is used) is supplied to the incinerator by one or more forced-draft fans or blowers. This assumption is somewhat conservative for certain designs, particularly rotary kiln incinerators, in which a variable fraction of the combustion airflow is induced through negative pressure operation. The inlet air is generally assumed to be at ambient temperature and pressure.

Pumps

In an HWI facility, pumps transfer liquid wastes, liquid–solid mixtures such as slurries and sludges, auxiliary fuel, water for cooling, quenching and scrubbing, and liquid reagents for neutralization.

Pumps are broadly classified as either *centrifugal* or *positive displacement.* Centrifugal pumps are widely used in the process industry because of simplicity of design, low initial cost, low maintenance, and flexibility of application. Centrifugal pumps have been built to move as little as a few gallons per minute against a small back pressure (head) and as much as several thousand gallons per minute against a pressure of several hundred pounds per square inch. In its simplest form, the centrifugal pump consists of a circular impeller rotating within a casing. In *single end-suction* centrifugal pumps the fluid enters the pump near the center of the rotating impeller and is thrown outwards by centrifugal force. The kinetic energy of the fluid increases from the center of the impeller to the tips of the impeller vanes. This high velocity is converted to a high pressure as the fast-moving fluid leaves the impeller and is driven into slower moving fluid in the volute or diffuser.[3] Impellers are either *closed type (shrouded) or open.* The closed impeller has plates on both sides of the impeller vanes where the open type has a plate on only one side of the vanes. The closed impeller is generally more efficient than the open type but more susceptible to plugging or fouling. Centrifugal pumps may be horizontal or vertical. The vertical centrifugal pump is usually immersed in the liquid it is pumping.

Not all of the centrifugal pumps described previously produce the radial flow (directed away from the axis of rotation); many produce an axial flow (directed along the axis of rotation) and others a combination of the two. The turbine type of centrifugal pump has smaller, straighter vanes, is driven at high speed, and generates a highly radial flow to produce higher pressures at lower flow rates. The axial-flow type employs multibladed propellers that generate a highly axial flow, resulting in large flow rates at lower pressures. For a specific balance between flow rate and pressure, impellers can be shaped to provide results between that of the turbine and axial-flow type; these are referred to as *mixed-flow* impellers.

The impeller is the heart of the centrifugal pump. It consists of a number of curved vanes or blades that are shaped to give smooth fluid flow between the blades. In the *straight-vane, single-suction closed* impeller, the surfaces of the vanes are defined by straight lines parallel to the axis of rotation. The *double-suction* impeller is, in fact, two single-suction impellers arranged back to back in a single casing. Centrifugal pump casings may be of several designs, but their main function is to convert kinetic energy imparted to the fluid by the impeller into a higher pressure. In addition, the casing provides an inlet and an outlet for the pump and contains the fluid. Casings may be either the *volute* or the *diffuser* type. The *volute* type has a continuous flow area that allows the velocity to decrease gradually, thereby reducing eddy formation; this minimizes the loss of energy due to turbulence. The *diffuser-type casing* has stationary guides that offer the liquid a widening path from impeller to casing; this also keeps turbulence at a minimum.[6]

Centrifugal pumps may be made with more than one impeller or stage for conditions that require high discharge pressures without excessive tip speeds. Each stage in a multistage pump incrementally increases the pressure.

Positive displacement pumps move the fluid by displacing small portions of the liquid as it passes through the pump. These can be roughly divided into *reciprocating pumps* and *rotary pumps.*

Reciprocating pumps operate by the direct action of a piston on liquid contained in a cylinder. As the liquid is compressed by the piston, the pressure forces it through a discharge valve into a pipeline at the pump outlet. As the piston retracts in the cylinder, liquid is drawn into the cylinder from the source or reservoir. The cycle is then repeated.

The piston is driven or moved by a rotating crankshaft through a crosshead. The rate of liquid delivery is a function of the volume swept out by the piston and the number of strokes per unit time. A fixed volume is delivered for each stroke, but the actual delivery may be less due to leakage past the piston and/or failure to fill the cylinder when the piston retracts. The volumetric efficiency of the pump is defined as the ratio of the actual volumetric discharge to the pump displacement. The volumetric efficiency for well-maintained pumps is at least 95%.

Reciprocating pumps can deliver the higher pressure of any type of pump (20,000 psig); but, their capacities are relatively small compared to centrifugal pumps. Also, because of the nature of the operation of the reciprocating pump, the discharge flow rate tends to be somewhat pulsating. Liquids containing abrasive solids can damage the machined surfaces of the piston and cylinder. Because of its positive displacement operation, reciprocating pumps can be used to meter liquid volumetric flow rates. Flow from reciprocating pumps can be controlled only by changing the speed or interval of the piston. There is the possibility of pump or pipe failure if one closes or throttles the discharge.

Another form of the reciprocating pump, widely used in the waste industry for sludges, is the *diaphragm* pump. This pump operates by filling a flexible diaphragm with air that, in turn, expands into a chamber containing the liquid to be pumped, forcing the liquid out of the pump and into the discharge pipe. The reverse cycle pulls more liquid into the pump chamber for the next pressure cycle. Diaphragm pumps may be mechanically driven employing a piston, or plunger immersed in oil to transfer its pressure to the liquid through the diaphragm, or it may be operated by a compressed air source where each cycle is controlled by opening and closing solenoid valves in the air line.

Rotary pumps combine a rotational device with positive displacement of the liquid. The rotating elements mesh with elements of the stationary casing in much the same way that two gears mesh. As the rotating elements come together, a pocket is created that first enlarges, drawing in liquid from the inlet or suction line. As rotation continues, the pocket of liquid is trapped, reduced in volume, and then forced into the discharge line at a higher pressure. The flow rate of liquid from a rotary pump is a function of size and speed of rotation and is slightly dependent on the discharge pressure. Unlike reciprocating pumps, rotary pumps deliver nearly constant flow rates. Rotary pumps are used on liquids of almost any viscosity, but the liquids must not contain abrasive solids. For this reason, they are very effective with high-viscosity waste mixtures. They operate in moderate pressure ranges (5000 psig), have small to medium capacities, and, like the reciprocating pump, can be used for metering liquids.

Rotary pumps consist of *gear* and *screw*-type pumps and *progressive cavity* pumps. The latter uses a single-shaft fluted impeller (rotor) that rotates within a flexible liner (stator) located within the pump body. As the material is fed into the driven end of the rotor, it is progressively squeezed in the direction of flow by the contact between the rotor and stator building pressure at the outlet. Progressive cavity pumps are used to move or feed highly viscous wastes in HWI installations.

The operation of a pump may result in the release of air contaminants. Both the reciprocating and centrifugal pumps can be sources of hazardous air emissions. The opening in the cylinder through which the connecting rod drives the piston is the major source of contaminant release from a reciprocating pump. In centrifugal pumps, the only potential source of leakage occurs where the drive shaft passes through the casing. Several methods have been devised for sealing the clearance between the pump shaft and fluid casing. For most applications, packed seals and mechanical seals are widely used. Packed seals can be used on both positive displacement and centrifugal pumps. A typical packed seal consists of a stuffing box filled with a sealing material or packing that encases the moving shaft. The stuffing box is fitted with an adjustable ring that compresses the packing and causes it to tighten around the shaft. Materials used for packing vary with the fluid temperature, physical and chemical properties, pressure, and pump type. Some commonly used materials are metal, rubber, leather, and plastics. For cases where a mechanical seal is not feasible, the diaphragm pump mentioned above and specialized pumps such as canned-motor or electromagnetic pumps are used. These specialty pumps are used where no leakage can be tolerated and are available in a limited range of sizes. They have been used to handle both high-temperature and very low temperature liquids. These pumps follow the same hydraulic principles as the traditional centrifugal pump. Because of their small size, they operate with rather low efficiencies, but in dangerous applications, efficiency must be sacrificed for safety.[5]

Pumps are employed at HWI sites for one and/or a combination of the following:

1. Pumping wastes from trucks or delivery vehicles to storage
2. Pumping wastes from storage to the incinerator
3. Pumping waste from trucks or delivery vehicles directly into the incinerator
4. Pumping fuel oil (where applicable) to the incinerator
5. Pumping water to the quench unit and waste heat boiler (where applicable)
6. Circulating water in absorbers and scrubbers (where applicable)
7. Pumping and circulating caustic solutions in absorbers (where applicable)

A general relationship that may be used to determine pump HP requirements is[2]

$$\mathrm{HP} = (7.27 \times 10^{-5} w\ \Delta P)/\rho E \qquad (11.6.10)$$

where w = liquid flow rate (lb/h)
ΔP = developed pressure drop (psi)
ρ = liquid density (lb/ft)
E = pump-motor efficiency (fraction basis)

Compressors

Compressors, unlike fans and pumps, find only limited and specialized applications in HWI plants. In most incinerators, liquid wastes are atomized into fine droplets in the burner or in a separate atomizer in the combustion chamber. Dual fluid atomizers that employ compressed air or steam provide the best atomization. Air compressors are required to provide the atomizing air. Compressed air is also used for atomization of water in quench chambers and spray towers, packed towers, and wet electrostatic precipitators. Here, the fine water droplets cool the gases and remove particles from the flue gas stream. Compressors also supply instrument air for HWI plants. The compressed air must be dried and filtered before use.

Compressors operate similarly to pumps except that they handle a compressible fluid instead of a liquid. Compressors have essentially the same classifications as pumps: rotary, reciprocating, and centrifugal.

Centrifugal compressors are employed when large volumes of gases are to be handled at low to moderate pressure increases (0.5–50 psi). *Rotary* compressors have smaller capacities and can achieve discharge pressures up to 100 psig. *Reciprocating* compressors are the most common type used in industry and are capable of compressing small gas flows to as much as 3500 psig. With specially designed compressors, discharge pressures as high as 25,000 psig can be reached, but these devices are capable of handling only very small volumes and do not work well for all gases. The reciprocating compressor is the most common supply of compressed air for atomization and instrument air normally used.

The following equation may be used to calculate compressor power requirements when compressor operation is adiabatic and the gas (usually air) follows ideal gas behavior:

$$W_s = \frac{\gamma RT}{\gamma - 1}\left[\left(\frac{P_2}{P_1}\right)^{(\gamma-1)/\gamma} - 1\right] \qquad (11.6.11)$$

where W_s = compressor work required per lb mol of air
R = 1.987 Btu/lbmol-°R
T = air temperature at compressor inlet conditions (°R)

P_1, P_2 = air inlet and discharge pressures
γ = ratio of the heat capacity at constant pressure to that at constant volume—typically 1.3 for air

11.7 STACKS

A stack is a hollow conduit used to conduct gases from a process to the atmosphere. The stack is often referred to as a *chimney* if constructed of masonry. Chimneys may contain and support one or more separate stacks or *flues*. Incinerator stacks may be fabricated of steel or pipe, carbon steel and stainless steel, fiberglass reinforced plastic (FRP), or lined steel pipe and duct depending upon the temperature and corrosive aspects of the gas conveyed. Chimneys constructed of structural concrete or brick and lined with refractory are also utilized. Most stacks in the HWI industry are designed as forced-draft stacks; that is, the flue gases are forced through the stack by a fan upstream from the stack. Some chemical and utility applications use metal stacks that are made of a double wall with an air space between the metal sheets. This double wall provides an insulating air pocket to prevent condensation on the inside of the stack and thus avoid corrosion of the metal.

Tall stacks often provided the natural draft for incinerators in the past, but natural draft is no longer employed to any extent in the HWI (except for emergency stacks) because of the limited capability of generating sufficient draft for operation. The natural draft from a tall stack is nevertheless useful in reducing forced-draft pressure requirements. Tall stacks also provide more effective dispersion of the flue gas into the atmosphere.

Whenever stack gas temperatures are higher than the ambient air temperature, draft will be produced by the buoyancy of the hot flue gases. The theoretical natural draft may be estimated by

$$\Delta P_{\mathrm{nd}} = 0.019PL[(1/T_{\mathrm{o}}) - (1/T_{\mathrm{s}})] \qquad (11.7.1)$$

where ΔP_{nd} = theoretical natural draft (psi)
P = barometric (ambient) pressure (psia)
L = height of stack above the breeching (ft)
T_{o} = absolute ambient temperature (°R)
T_{s} = absolute temperature of stack gases (°R)

The value for T_{s} should be an average value that takes into account heat losses and, for tall stacks, the temperature drop on expansion as the atmospheric pressure declines with increasing altitude. As the flue gases pass up the stack, frictional losses reduce the fraction of the theoretical draft that is available. Frictional losses may be estimated from the following equation:

$$\Delta P_{\mathrm{f}} = 4.01 \times 10^{-5}\,(Lu^2 p/S) \qquad (11.7.2)$$

where ΔP_f = frictional loss (psi)
u = mean stack gas velocity (ft/s)
p = stack perimeter (ft)
S = stack cross-sectional area (ft^2)

In addition, some of the draft is lost to the velocity head in the rising gases. This loss is referred to as the *expansion loss* and may be estimated by

$$\Delta P_e = 2.59 \times 10^{-3} (u^2/S) \qquad (11.7.3)$$

where ΔP_e = the expansion loss (atm).

Typical natural draft stacks[7] operate with gas velocities of about 30 ft/s. The two process design variables for the stack are the height and diameter. The stack height is primarily dictated by atmospheric dispersion considerations. The diameter (D) may be determined by the equation[7]

$$D = (0.03Q_a)^{0.5} \qquad (11.7.4)$$

where D = stack diameter (ft)
Q_a = flue gas flow rate (acfs)

An important factor in handling acid gases in a stack involves maintaining a high internal temperature. This often retards the detrimental effect on the masonry without the necessity for other precautions. If the flue gases are such that high temperatures alone are not sufficient, it may be necessary to protect the walls by using an independent lining for the full height of the stack. In large chimneys with support masonry around the flue, a 3- to 4-in. air space between the flue and the support walls is often used. The independent lining may be an acid-proof brick with acid-proof mortar or an acid-resistant castable or gunnite material. The mortar should be carefully chosen for its resistance not only to the particular acid involved, but to moisture and expansion. In addition, the top of the chimney should be protected by a cap covering both the lining and main walls and made of a material not affected by the flue gas. While room should be allowed for expansion, fumes and moisture must not be allowed to penetrate under the cap.

11.8 ILLUSTRATIVE EXAMPLES

Example 11.8.1 Fan A has a blade diameter of 46 in. It is operating at about 1575 rpm while transporting 16,240 acfm of flue gas and requires 47.5 bhp. Fan B is to replace fan A. It is to operate at 1625 rpm with a blade diameter of 42 in. and is of the same homologous series as fan A. What is the power requirement of fan B?

SOLUTION Using the fan law [Eq. (11.6.6)], the power requirement for fan B is calculated:

$$\mathrm{HP}'/\mathrm{HP} = (\mathrm{RPM}'/\mathrm{RPM})^3(D'/D)^5(\rho'/\rho) = (1625/1575)^3(42/46)^5(1)$$
$$= 0.697$$
$$\mathrm{HP}' = (0.697)(47.5) = 33.1 \text{ bhp}$$

Example 11.8.2 A fan operating at a speed of 1694 rpm delivers 12,200 acfm of flue gas at 5.0 in. H_2O static pressure and requires 9.25 bhp. What will be the new operating conditions if the fan speed is increased to 2100 rpm?

SOLUTION The new fan flow rate is calculated using Eq. (11.6.4):

$$Q_a' = Q_a(\mathrm{RPM}'/\mathrm{RPM})(D'/D)^3 = 12{,}200(2100/1694)(1)$$
$$= 15{,}124 \text{ acfm}$$

Using Eq. (11.6.5), the new static pressure is calculated:

$$P_s' = P_s(\mathrm{RPM}'/\mathrm{RPM})^2(D'/D)^2(\rho'/\rho) = (5.0)(2100/1694)^2(1)(1)$$
$$= 7.68 \text{ in. } H_2O$$

The required HP is calculated using Eq. (11.6.6):

$$\mathrm{HP}' = \mathrm{HP}(\mathrm{RPM}'/\mathrm{RPM})^3(D'/D)^5(\rho'/\rho) = (9.25)(2100/1694)^3(1)(1)$$
$$= 17.62 \text{ bhp}$$

Example 11.8.3 Calculate the HP required to process a 6500-acfm gas stream from an incinerator. The pressure drop across various pieces of equipment has been estimated to be 6.4 in. H_2O. The pressure loss for ductwork, elbows, valves, and so on and expansion–contraction losses are estimated at 4.4 in. H_2O. Assume an overall fan-motor efficiency of 63%.

SOLUTION The total pressure drop, ΔP (in. H_2O), is

$$\Delta P = (6.4 + 4.4) = 10.8 \text{ in. } H_2O$$

The brake horsepower required is calculated by Eq. (11.6.8):

$$\mathrm{BHP} = Q_a(\text{acfm})\Delta P \text{ (in. } H_2O)(1.575 \times 10^{-4})/(\text{fan efficiency})$$
$$= (6500)(10.8)(1.575 \times 10^{-4})/(0.63)$$
$$= 17.55 \text{ bhp}$$

Note: The term 1.575×10^{-4} is a conversion factor to obtain units of horsepower (hp).

Example 11.8.4 It is necessary to deliver a liquid waste with density and viscosity similar to that of water to an incinerator at a rate of 80 gal/min. The pump must operate against a pressure of 180 psi. A pump with characteristics shown in Fig. 11.8.1 is available with variable-speed drive. At what speed should the pump be operated? What horsepower is needed to maintain flow?

SOLUTION Plotting this information in Fig. 11.8.1 indicates that the speed is between the 400- and 600-rpm lines (point A). Interpolation gives a speed of about 425 rpm. Interpolating again on the HP curve gives a value of about 17 hp (point B). Using this data with Eq. (11.6.10) with an efficiency of 50% results in a pump horsepower of approximately 15.

Example 11.8.5 The exhaust flow rate from a small hazardous waste incineration facility is 1200 scfm (70°F). All of the gas is discharged through a small stack, which has an inlet area of 1.2 ft^2. The exhaust gas has a temperature of 300°F. What is the velocity of the gas through the stack inlet in ft/s? Neglect the pressure drop across the stack.

SOLUTION Using Charles' law to calculate the flow rate,

$$Q_a = Q_s(T_a/T_s) = 1200(760/530)$$
$$= 1721 \text{ acfm}$$

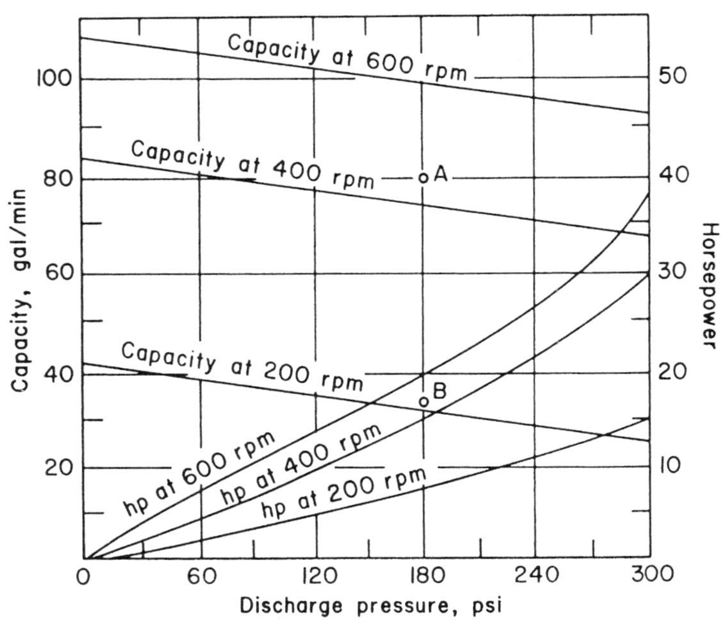

Figure 11.8.1 Performance characteristics of an external gear pump.

The velocity is calculated from

$$v = Q_a/A = 1721/1.2 = 1434 \text{ ft/min}$$
$$= 23.9 \text{ ft/s}$$

Example 11.8.6 Compressed air is to be employed in the nozzle of a liquid injection incinerator to assist the atomization of a liquid hazardous waste. The air requirement for the nozzle is 7.5 lb/min at 40 psia. If atmospheric air is available at 60°F and 1.0 atm, calculate the power requirement.

SOLUTION The ideal gas law will apply at these conditions. Set the coefficient for air γ equal to 1.3. The compressed energy requirement (delivered to the air) is given by Eq. (11.6.11):

$$W_s = \frac{\gamma R T_1}{(\gamma - 1)}\left[\left(\frac{P_2}{P_1}\right)^{(\gamma-1)/\gamma} - 1\right]$$
$$= \frac{(1.3)(1.987)(520)}{1.3 - 1}\left[\left(\frac{40}{14.7}\right)^{(1.3-1)/1.3} - 1\right]$$
$$= 1163 \text{ Btu/lbmol of air}$$

The power is

$$\text{HP} = -(1163)(7.5/29)(778)$$
$$= -234{,}000 \text{ ft-lb}_f\text{/min}$$

This result may be divided by 33,000 to yield a HP requirement of 7.1 hp. The power required has the opposite sign. The reader should also note that this represents the *minimum* power required to accomplish this job.

PROBLEMS

1. Fan A has a blade diameter of 48 in., operates at 1650 rpm, delivers 18,450 acfm, and requires 50.0 bhp. Fan B operates at 1590 rpm and has a blade diameter of 45 in. If fan B is moving a gas of the same density and is of the same homologous series as fan A, what is its power requirement?

2. Calculate the HP needed to process a 8500 acfm gas stream from an incinerator. The pressure drop across the various pieces of equipment has been estimated to be 8.4 in. H_2O. The pressure loss for ductwork, elbows, valves, and so on and expansion–contraction losses are estimated at 5.8 in. H_2O. Assume an overall fan-motor efficiency of 58%.

3. A hazardous waste incinerator facility consists of a quench tower, scrubber, demister, and fans. The pressure drop due to bends, valves, elbows, and expansion–contraction losses is estimated to be 12.5 in. H_2O. The combustion gas to be treated has a flow rate of 23,000 acfm. The pressure drop across each device is given as follows:

Location	ΔP (in. H_2O)
Quench tower	5.1
Scrubber	6.8
Demister	3.9

Calculate the overall pressure drop across the facility.

4. A fan operating at a speed of 1750 rpm delivers 13,250 acfm at 5.5 in. H_2O static pressure and requires a BHP of 10.5. What will be the new operating conditions if the fan speed is increased to 2000 rpm?

5. It is necessary to pump a constant-flow stream of liquid with a density and viscosity similar to that of water in a reactor at a rate of 70 gal/min. The pump must operate against a pressure of 200 psi. A pump with characteristics shown in Fig. 11.8.1 is available with variable-speed drive. At what speed should the pump be operated? What HP is needed to maintain this flow?

6. The exhaust gas flow rate from a facility is 1500 scfm. All of the gas is vented through a small stack, which has an inlet area of 1.3 ft^2. The exhaust gas temperature is 350°F. What is the velocity of the gas through the stack inlet in ft/s? Assume standard conditions to be 70°F and 1 atm. Neglect the pressure drop across the stack.

7. Compressed air is to be employed in the nozzle of a liquid injection incinerator to assist the atomization of a liquid hazardous waste. The air requirement for the nozzle is 0.5 lbmol/min at 50 psia. If atmospheric air is available at 50°F and 1.0 atm, calculate the power requirement.

REFERENCES

1. M. LaGrega, P. Buckingham, J. Evans, and Environmental Resources Management, *Hazardous Waste Management*, McGraw-Hill, New York, 1994.
2. W. Blackman, *Basic Hazardous Waste Management*, 2nd ed., CRC Press, Lewis Publishers, Boca Raton, FL, 1996.
3. J. McKetta, *Encyclopedia of Chemical Processing and Design*, Vol. 63, Marcel Dekker, New York, 1998.
4. T. Shen, U. Choi, and L. Theodore, *Hazardous Waste Incineration Manual*, U.S. EPA Air Pollution Training Institute, Research Triangle Park, NC, 1985.

5. U.S. EPA, *Engineering Handbook for Hazardous Waste Incineration*, prepared by Monsanto Research Corporation, Dayton, OH, EPA Contract No. 68-03-3025, Sept., 1981.

6. R. H. Perry and D. W. Green, *Perry's Chemical Engineers' Handbook*, 7th ed., McGraw-Hill, New York, 1997.

7. R. Niessen, *Combustion and Incineration Processes*, 2nd ed., Dekker, New York, 1995.

IV

FACILITY DESIGN

The purpose of Part IV is to expose the reader to some aspects of the design of an overall waste incineration facility. The complete engineering design of such an installation is a complex task involving not only the application of engineering principles of the type covered earlier in this text, but also the consideration of economic, legal, O&M (operation and maintenance), environmental, and often political and social factors. Most of these factors are too involved and too installation specific to be treated in this text. The authors feel, however, that an exposure to the basics of engineering design, even if only a rather sketchy one, can provide a better understanding of many of the technical (engineering) aspects associated with a waste incineration facility.

The mechanics of the design process—the collecting of required information, preparation of schematics, application of material and energy balances plus other calculations, writing of a design report, and so on—are presented in Chapter 12. Economic analysis, an obviously important facet of any design procedure, is discussed in Chapter 13. Since illustrative examples often paint a clearer picture for the reader than textual presentation, the final chapter of this text, Chapter 14, consists mainly of two rather extensive plant design problems and their solutions. Since both problems involve overall facility design, these examples serve the function of merging many of the equipment design procedures presented earlier in the text into a more complete plant design package.

12

DESIGN PRINCIPLES

Partial contribution: Donald L. Corwin

12.1 INTRODUCTION

Current design practices for waste incineration systems usually fall into the category of state of the art and pure empiricism. Past experience with similar applications is commonly used as the sole basis for the design procedure. The vendor maintains proprietary files on past installations; these files are periodically revised and expanded as new installations are evaluated. In designing a new unit, the files are consulted for similar applications and old designs are heavily relied on.

By contrast, the engineering profession in general, and the chemical engineering profession in particular, has developed well-defined procedures for the design, construction, and operation of chemical plants. These techniques, tested and refined for better than a half-century, are routinely used by today's engineers. These same procedures may also be used in the design of hazardous waste incineration (HWI) facilities.

The purpose of this chapter is to introduce the reader to some of these process design fundamentals. Such an introduction to design principles, however sketchy, can provide the reader with a better understanding of the major engineering aspects of an HWI facility, including some of the operational, economic, controls and instrumentation for safety and regulatory requirements, and environmental factors associated with the process. No attempt is made in the sections that follow to provide extensive coverage of this topic; only general procedures and concepts are presented and discussed.

12.2 PRELIMINARY STUDIES

A process engineer is usually involved in one of two activities: building a plant or deciding whether to do so. The skills required in both cases are quite similar, but the money, time, and detail involved are not as great in the latter situation. It has been estimated that only one out of 15 proposed new processes ever achieves the construction stage. Thus, knowledge at the preliminary stage is vital to prevent financial loss on one hand and provide opportunity for success on the other. In well-managed process organizations, the engineer's evaluation is a critical activity that usually involves considerable preliminary research on the proposed process. Successful process development consists of a series of actions and decisions, the most significant of which takes place well before plant construction.

It is important to determine whether a project has promise as early in the development stage as possible. This section discusses some of the preparatory work required before the design of an HWI facility can be formally initiated. In the chemical process industry, there may be an extended period of preparatory work required if the proposed plant/facility is a unique or first-time application. This can involve bench-scale work by chemists to develop and better understand the process chemistry. This is often followed by pilot experimentation by process and/or development engineers to obtain scale-up and equipment performance information. However, these two steps are usually not required in the design of an established system. This is presently the situation with most HWI designs although some bench-scale or pilot work may be necessary and deemed appropriate by management.

In general design practice, there are usually five levels of sophistication for evaluating and estimating. Each is discussed in the following list.

1. The first level requires little more than identification of products, raw materials, and utilities. This is what is known as an *order of magnitude estimate* and is often made by extrapolating or interpolating from data on similar existing processes. The evaluation can be done quickly and at minimum cost, but with a probable error exceeding $\pm 50\%$.
2. The next level of sophistication is called a *study estimate* and requires a preliminary process flow sheet (to be discussed in the next section) and a first attempt at identification of equipment, utilities, materials of construction, and other processing units. Estimation accuracy improves to within $\pm 30\%$ probable error, but more time is required and the cost of the evaluation can escalate to over \$30,000 for a \$5 million plant. Evaluation at this level usually precedes expenditures for site selection, market evaluation, pilot plant work, and detailed equipment design. If a positive evaluation results, pilot plant and other activities may also begin.
3. A *scope* or *budget authorization*, the next level of economic evaluation, requires a more defined process definition, detailed process flow sheets,

and prefinal equipment design (discussed in a later section). The information required is usually obtained from pilot plant, marketing, and other studies. The scope authorization estimate could cost upwards of $80,000 for a $5 million facility with a probable error exceeding ±20%.

4. If the evaluation is positive at this stage, a *project control estimate* is then prepared. Final flow sheets, site analyses, equipment specifications, and architectural and engineering sketches are employed to prepare this estimate. The accuracy of this estimate is about ±10% probable error. A project control estimate can serve as the basis for a corporate appropriation, for judging contractor bids, and for determining construction expenses. Due to increased intricacy and precision, the cost for preparing such as estimate for the process can approach $150,000.
5. The final economic analysis is called a *firm* or *contractor's estimate*. It is based on detailed specifications and actual equipment bids. It is employed by the contractor to establish a project cost and has the highest level of accuracy, ±5% probable error. A cost of preparation results from engineering, drafting, support, and management/labor expenses. Because of unforeseen contingencies, inflation, and changing political and economic trends, it is impossible to assure actual costs for even the most precise estimates.

In the case of an HWI design, data on similar existing processes are normally available and economic estimates or process feasibility are determined from these data. It should be pointed out again that most processes in real practice are designed by duplicating or *mimicking* similar existing systems. Simple algebraic correlations that are based on past experience are the rule rather than the exception. This stark reality is often disappointing and depressing to students and novice engineers involved in design. For HWI facilities, the only preparatory work normally required is the gathering of all existing physical and chemical property data of the waste(s) and auxiliary materials and chemicals. Process chemistry information may also be needed, but this too may be obtained directly from the literature, company files, or a similar type of application.

12.3 PROCESS SCHEMATICS

To the practicing engineer, particularly the chemical engineer, the process flow sheet is the key instrument for defining, refining, and documenting a chemical process. The process flow diagram is the authorized process blueprint, the framework for specifications used in equipment designation and design; it is the single, authoritative document employed to define, construct, and operate the chemical process.[1]

Beyond equipment symbols and process stream flow lines, there are several essential constituents contributing to a detailed process flow sheet. These include equipment identification numbers and names, temperature and pressure designations, utility designations, volumetric or molar flow rates for each process stream, and a material balance table pertaining to process flow lines. The process flow diagram may also contain additional information such as energy requirements, major instrumentation, and physical properties of the process streams. When properly assembled and employed, a process schematic provides a coherent picture of the overall process. It can point up some deficiencies in the process that may have been overlooked earlier in the study, for example, by-products (undesirable or otherwise) and recycle needs. But basically, the flow sheet symbolically and pictorially represents the interrelation between the various flow streams and equipment, and permits easy calculations of material and energy balances. These two topics are considered in the next sections. Controls and instrumentation must also be considered in the overall requirements of the waste combustion system. These concerns are covered later in this chapter.

As one might expect, a process flow diagram for a chemical or petroleum plant is usually significantly more complex than that for an HWI facility. As described earlier, a typical facility consists of the incinerator, followed by a waste heat boiler and/or a quench. This in turn is followed by the air pollution control equipment that may consist of separate devices for particulate and gaseous pollutant control (e.g., venturi scrubber–absorber) or a dry scrubber system (e.g., spray dryer–baghouse). Discharge gases from the control equipment are directed to an induced draft fan and then released to the atmosphere through a stack; occasionally a bypassing system will be included. There may be recycling and blowdown of the scrubbing liquid in the absorber or the combined quench–absorber system. The flow sequence and determinations thus approach a "railroad" or sequential type of calculation that does not require iterative calculations.

There are a number of symbols that are universally employed to represent equipment, equipment parts, valves, piping, and so on. These are depicted in the schematic in Fig. 12.3.1. Although a significant number of these symbols are used to describe some of the chemical and petrochemical waste processes, only a few are needed for even the most complex hazardous waste facilities. These symbols obviously reduce, and in some instances replace, detailed written descriptions of the process. Note that many of the symbols are pictorial, which helps in better describing process components, units, and equipment.

The degree of sophistication and details of a flow sheet usually vary with time. The flow sheet may initially consist of a simple free-hand block diagram with limited information that includes only the equipment; later versions may include line drawings with pertinent process data such as overall and componential flow rates, utility and energy requirements, and instrumentation. During the later stages of the design project, the flow sheet will consist of a highly detailed P&I (piping and instrumentation) diagram; this aspect of the design procedure is beyond the scope of this text; for information on P&I diagrams, the reader is referred to the literature.[2]

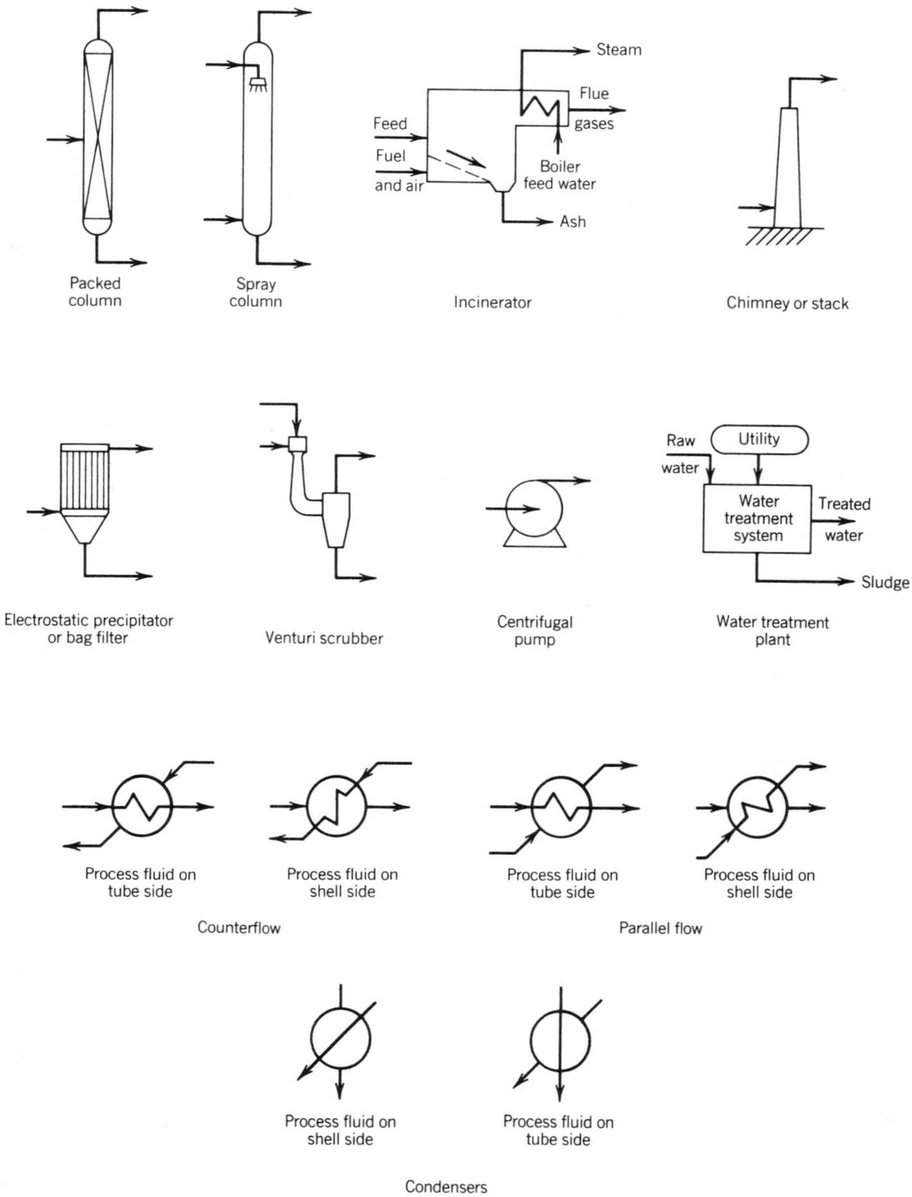

Figure 12.3.1 Flow sheet symbols.

In a sense, flow sheets are the international language of the engineer, particularly the chemical engineer. Chemical engineers conceptually view a (chemical) plant as consisting of a series of interrelated building blocks that are defined as *units* or *unit operations*. The plant essentially ties together the various pieces of equipment that make up the process. Flow schematics follow the successive steps of a process by indicating where the pieces of equipment are located and the material streams entering and leaving each unit.

12.4 MATERIAL AND ENERGY BALANCES

Overall and componential material balances have already been described in rather extensive detail in Part II. Material balances may be based on mass, moles, or volume, usually on a rate (time rate of change) basis. Care should be exercised here since moles and volumes are *not* conserved, that is, the quantities change during the course of the combustion reaction. Thus, the initial material balance calculation should be based on *mass*. Mole balances and molar information are important in chemical reaction and phase equilibria calculations. Volume rates play an important role in equipment sizing calculations.

The units (pound, kilogram, etc.) employed may also create a problem. Despite earlier efforts by the Environmental Protection Agency (EPA) and other government agencies, industry still primarily employs the British and/or American engineering units in practice; there has been a reluctance to accept Standard International Units, commonly referred to as SI, that employ a modified set of metric units. As indicated earlier, this text uses primarily engineering units. However, for those individuals who are more comfortable with the SI system, a short writeup on the conversion of units has been prepared by the authors and included in the appendix.

Most design calculations in the chemical process industry today include transient effects that can account for process upsets, startups, shutdowns, and so on. The describing equations for these time-varying (unsteady-state) systems are differential. The equations usually take the form of a first-order derivative with respect to time, where time is the independent variable. However, design calculations for most HWI facilities assume steady-state conditions, with the ultimate design based on worst-case or maximum flow conditions. This greatly simplifies the calculations since the describing equations are no longer differential, but rather algebraic. Thus, these equations provide an accounting or inventory of all mass entering and leaving one or more pieces of equipment, or the entire process.

The heart of any material balance analysis is the *basis* selected for the calculation. The usual basis is a unit of time, for example, minutes, hours, days, or year. For more complex calculations, and this may include multicomponent systems and recycle streams, one may choose as a basis a convenient amount of a "key" component or element. Note also that the calculation may be based on either a feed stream, an intermediate stream, or a product stream. Selecting a feed stream as a basis is preferred since it often allows one to follow the calculations through the process in a "railroad" manner, that is, in sequential order.[2]

The number of material balance equations can be significant depending on the number of components in the system, process chemistry, and pieces of equipment. These are critical calculations since the size of the equipment is often linearly related to the quantity of material being processed. This can then significantly impact—often linearly or even exponentially—on capital and operating costs. In addition, componential rates can impact on (other) equipment needs, energy considerations, materials of construction, and so on. With respect to HWI facilities, material balance calculations are required to: (1) size all equipment, (2) predict flue gas components and rates from the incinerator, and (3) determine particulate and gaseous air pollution control efficiencies. Topic 2 has already been treated extensively in Chapters 5 and 6; and topic 3 was discussed in Chapter 10.

Topics 2 and 3 are also interrelated. Componential balances in conjunction with chemical reaction equilibria calculations can provide information on Cl_2 and SO_3 generation (see Chapters 5 and 6). These individual flow rates must be treated analytically in control equipment calculations.

Once the material balance is completed, one may then proceed directly to the energy calculations, some of which play a significant role in the design of an incinerator facility. As indicated earlier, energy calculations for incinerator facilities are also based on steady-state conditions. An extensive treatment of this subject has already been presented in an earlier part of the text, and need not be repeated here. However, a thorough understanding of thermodynamic principles—particularly the enthalpy calculations—is required for most of the energy (balance) calculations. Entropy calculations are employed in all meaningful energy conservation analysis. This subject is beyond the scope of this presentation but is a topic that may be given more serious attention by the engineering community in the not-to-distant future. Industry has always recognized that wasting energy reduces profits. Before the Arab oil embargo, the cost of energy was often a negligible part of the overall process cost and immense operational inefficiencies were tolerated. The sudden decrease in the availability of natural gas and oil resulting from the embargo raised the cost of energy and encouraged the elimination of unnecessary energy consumption. These energy-saving measures have continued to be employed even when energy costs are low.

One of the principal jobs of an engineer involved in the design of an incinerator facility is to account for the energy that flows into and out of each process unit and to determine the overall energy requirement for the process. This is done by performing energy balances on each process unit and on the overall process. These balances play an important role in HWI facility design. They find particular application in determining incinerator fuel requirements, in heat exchanger design, in heat recovery systems, in specifying materials of construction, and in calculating fan and pump power requirements.

The heat energy calculation is that associated with the entire incinerator system, primary combustion chamber (PCC) and secondary combustion chamber (SCC). Both the conservation laws for mass and energy and the principles of thermodynamics must be applied in order to determine fuel requirements, operating temperatures, and air (excess) requirements for proper operation of the incinerator. This calculational procedure can be extremely tedious and detailed. An overall and componential material balance is applied around the incinerator that is then followed by an enthalpy balance. The enthalpy balance takes into account waste–fuel mixture, stoichiometric and excess air, product (flue) gas, latent (phase), and combustion (reaction) enthalpy effects. Fortunately, algorithms have been developed and are readily available to perform these calculations (see Chapter 6) and provide information on the interrelationship among incinerator operating temperature, waste–fuel heating value, and excess air requirements. Since there may be several operating scenarios to be reviewed, the energy requirements for the cases with wastes having the lowest heat content will be greatest. This affects the sizing of the fuel systems, availability from the suppliers and impacts that may arise based on the seasons of the year. For example, plants are often curtailed on natural gas during severe weather in the winter. The availability of the maximum rate of fuel oil required during these peak periods must be examined as well as the storage needed to take care of these situations.

12.5 EQUIPMENT AND PROCESS DESIGN

Equipment design calculations are usually based on maximum throughput capacities for a new process. The reason for this is to enable the system to perform satisfactorily under the most extreme conditions. Material and energy balances based on these conditions are required before the individual equipment design. Each piece of equipment is designed individually and independently of all others. Equipment design is the first major step toward assessing the capital costs or purchase price of the entire plant.

As noted previously, chemical engineers describe the application of any piece of equipment that operates on the basis of mass, energy, and/or momentum transfer as a *unit operation*. A combination of two or more of these unit operations is defined as a unit process. A whole chemical process can be described as a coordinated set of unit operations and unit processes. This subject matter has received much attention in recent years and, as a result, is well covered in the literature.[3–5] From details on these unit operations and processes, it is therefore possible to design new plants more efficiently by coordinating a series of *unit actions*, each one of which operates according to certain laws of physics regardless of the other operations being performed along with it. The unit operation of waste combustion, for example, is used in many different types of industries; many of the critical design parameters for the combustion zones, however, are common to all combustion systems and independent of the particular industry.

There are usually five conceptual steps to be considered with design of equipment. These are:

1. Identification of the parameters that must be specified
2. Application of the fundamentals underlying theoretical equations or concepts
3. Enumeration, explanation, and application of simplifying assumptions
4. Possible use of correction factors for nonideal behavior
5. Identification of other factors that must be considered for adequate equipment specification

For an HWI plant, the major pieces of equipment that must be considered are all of the following:

- Storage and handling facilities (feed and residuals)
- Incinerator (rotary kiln or liquid injection)
- Waste heat boiler (primary quench system or energy recovery if economically practical)
- Quench system
- Wet scrubber–venturi scrubber (particulate removal scrubber)
- Absorber (packed tower for acid gas absorption)
- Spray dryer (quench and acid gas absorption)
- Baghouse or electrostatic preciptators (ESP)
- Fan(s) and blower(s)
- Stack
- Pumps (feed, recycle, and scrubber)

Calculation procedures for most of these have been presented in Part III. The design of a complete HWI facility is the topic of Chapter 14. Since design calculations are generally based on the maximum throughput capacity for the proposed process or for each piece of equipment, these calculations are never completely accurate. It is usually necessary to apply reasonable safety factors when setting the final design. Safety factors vary widely and are a strong function of the accuracy of the data involved, calculational procedures, and past experience. Attempting to justify these is a difficult task.

Unlike many of the problems encountered and solved by the engineer, there is no absolutely correct solution to a design problem; however, there is usually a *better* solution. Many alternative designs when properly implemented will function satisfactorily, but one alternative will usually prove to be economically more efficient and/or attractive than the others.

Although all engineers approach design problems somewhat differently, six major steps are generally required. These six steps are discussed here and may also be applied to the design of an HWI facility:

1. The first step is to conceptualize and define the process. A designer must know the bases and assumptions that apply, the plant capacity, and the process time allowed. Some of the answers to a host of questions pertaining to the process operation will be known from past experience. Many times, a series of cases with variations in flow of solids, slurries, and liquids must be reviewed. Waste feed flow variations affect the storage facility design as well as equipment sizing.
2. After the problem has been defined, a method of solution must be sought. Although a method is seldom obvious, a good starting point is the construction of a process flow sheet. This effort usually produces several advantages. For example, it may suggest to the designer ways of reducing the complexity of the problem; it can allow for easier execution of material and energy balances, which in turn can point up the most important process variables; it is an efficient way to become familiar with the process; and information that is lacking often becomes evident.
3. The third step is the actual design of the process equipment that involves the numerous calculations needed to arrive at specifications of operating conditions, equipment geometry, size, materials of construction, controls, monitors, instrumentation, safety equipment (automatic waste feed cutoff), and so on. As part of this step, equipment costs must be established. Cost estimating precision is dependent on the desired accuracy of the estimate. If the decision based on an estimate is positive, a detailed project control or contractor's estimate will follow.
4. An economic analysis must also be performed in order to determine the process feasibility. The main purpose of this step is to answer the question of whether a process will ultimately be profitable or not. To answer this, raw material, labor, equipment, and other processing costs are estimated to give an accurate economic forecast for the proposed operation.
5. In a case where alternate design possibilities exist, economics and engineering optimization is necessary. (This topic is discussed in the next section.) Since this is often the case, optimization calculations are usually applied several times during most design projects.
6. The final step of this design scheme is the compiling of a design report. A design report may represent the only relevant product of months or even years of effort. This is also discussed in the next section. At a minimum the report should utilize the latest computer-generated layouts and graphics to provide an impressive final product.

These six activities are prominent steps in the development of all modern chemical processes, including hazardous waste incineration.

12.6 INSTRUMENTATION AND CONTROLS

The control of a hazardous waste handling unit requires careful consideration of all operational and regulatory requirements. The hazardous waste system is designed to process hazardous waste and reduce toxicity or other hazardous characteristic of the material. These characteristics were covered in earlier chapters. Safety should be a primary concern of all individuals involved with the handling, treatment, or movement of the hazardous material.

The safe operation of the hazardous waste unit requires that the controls keep the system operating within a safe operating envelope. The envelope is based on many of the design, process, and regulatory constraints. These are placed on the unit to ensure the proper destruction of the wastes. Additional controls may be installed to operate additional equipment needed for energy recovery, neutralization, or combustion products concentration. Acid furnaces occasionally are used to recover the acid generated by the combustion of chlorinated waste streams.

As the regulatory community expands its standards on waste combustion systems, there will be a need for additional controls. These controls must be installed on the systems to assure that operations are within the bounds of test burns discussed in Chapter 3.

Since many systems destroy the waste by oxidizing the material, the residence time at a predetermined temperature is of great importance. The method of ensuring that the proper conditions exist are a function of the control system. The control system must be designed to ensure that the proper amount of waste material is placed into the unit at an appropriate operational condition and minimize any upsets in waste feed rates or the established operating conditions of the system.

The control system should also be designed to vary one or more of the process variables to maintain the appropriate conditions with the unit. These variations are programmed into the system based on the past experience of the unit manufactured. The operational parameters that may vary include the waste flow rate, system airflow rate, temperatures in combustion zones and air pollution control (APC) system, system pressure, and flow rates in the APC system.

The proper control system has also been subjected to extensive analysis on operational problems and items that could go wrong. A hazardous operational (*HAZOP*) analysis is conducted on the control system to examine and identify all possible failure mechanisms. It is important that all of the failure mechanisms have appropriate response reactions by the control system. Several of the failure mechanisms that must be addressed within the appropriate control system response are excess or minimal temperature, excessive or subnormal flow rate, equipment failure, component failure, sticking or inoperable components, and broken circuits. All are examined including the response time of the control system to the problem and the appropriate response of the system to the problem. The system must identify the problem and integrate necessary actions and alarms into the system.

A complete review of the maximum and minimum process rate of change and the equipment rate of change must be identified before the control system can be defined. Liquid incinerators, solid waste processing units, and other waste combustion systems will require the same general consideration in their design. The system must be reviewed to define the primary control parameters for operation and safety. The regulatory limitations that are imposed on the unit must be identified and monitored to ensure that they are not exceeded.

The purpose of the control system is to ensure that the system is operating in a safe reliable manner within the guidelines of the design. The control system is provided for all of the variables that occur during operation of the total incineration—APC system.

Safety and Regulatory Concerns

The destruction of hazardous waste by its very nature is hazardous. The purpose of the treatment is to render the hazardous waste harmless or to contain the hazardous portion of the material. The processing of the material requires that some of the operational parameters of the equipment be closely monitored.

Safety The safe operation of equipment requires that some of the operational parameters be constrained within specific bounds. Each system has parameters that must remain within the appropriate bounds to assure that the system is stable. All systems will have safety equipment to prevent the system from being operated at a condition outside of the safe limits. The equipment may be electronic or mechanical.

The maintenance cycle for all safety equipment should be determined after consulting the local, state, and federal regulatory requirements. Insurance companies Industrial Risk Insurers (IRI) and Factory Mutual (FM) and national codes groups National Fire Protection Agency (NFPA) will have recommended regulations.

The most important item is to assure that all personnel are properly trained on the operational limitations of the equipment.

Regulatory The regulations require that each of the operational limits be monitored to assure that the system has not been operated when the parameter has been exceeded. Many of the permit parameters cannot be monitored on a continuous basis. The design of a control system must include both standard and nonstandard operational conditions. Most design conditions are based on the best guess at a maximum flow or operational condition. Variations in the steady-state conditions must be incorporated into the design. As part of the regulatory review, the flow rate, organic composition, batch size, and other physical and chemical characteristics of the waste stream must be reviewed to assess the maximum instantaneous rate. The process generating the waste material should also be reviewed to determine the variations that occur on a regular and also an infrequent basis. The magnitude and duration of these variations in character should be accommodated in the control system. The operation of a waste treatment unit must be monitored continuously to ensure that it is operating properly. The placement of monitor must be carefully selected to ensure that the variable is accurately measured or sampled. The sensing system must also accommodate normal fluctuations of the system. The failure mechanism of each component must be factored into the design. A properly instrumented system will have sensors to monitor all of the important parameters. It is important to define the critical process variables and then to supply the proper monitoring of those variables. These control system design features must be presented in the body of the permit application for review and approval by the regulatory personnel.

The Design Report

As pointed out in Step 6, a comprehensive plant design project report is often required. This material should be written up in a clear and concise fashion. In addition, the project leader might be requested to make informal and/or formal presentations to management on the study. The report and presentation should explain what has been accomplished and how it has been carried out.There are many different formats for design reports; the format will vary with the organization and the project. One possible outline for a project report is given in the following list:

- Title page
- Table of contents
- Executive summary
- Prefatory comments (optional)
- Introduction
- Discussion
- Comprehensive process design, process flow diagrams, including mass and energy balances and annualized costs
- Calculation and design limitations
- Optimization studies, pilot plant studies
- Illustrative calculations
- Appendix

Great care should be exercised with the preparation of the executive summary. It is recommended that the executive summary be no longer than one single-spaced typewritten page. It should contain a short introduction, important results, and pertinent recommendations and conclusions. It should *not* refer to the body of the report. In many instances, the executive summary is the only portion of the report that upper-level management will initially review.

The use of this somewhat crude approach to a plant design presented in this chapter can produce valuable information. Perhaps its greatest value is that it can provide early signals when the proposed process is technically unfeasible or economically prohibitive.

REFERENCES

1. G. D. Ulrich, *A Guide to Chemical Engineering Process Design and Economics*, Wiley, New York, 1984.
2. T. Shen, U. Choi, and L. Theodore, *Hazardous Waste Incineration Manual*, U.S. EPA Air Pollution Training Institute, Research Triangle Park, NC, 1985.
3. R. M. Felder and R. W. Rousseau, *Elementary Principles of Chemical Processes*, 2nd ed., Wiley, New York, 1986.
4. R. Perry and D. Green, *Perry's Chemical Engineering Handbook*, 7th ed., McGraw-Hill, New York, 1997.
5. W. L. McCabe, J. C. Smith, and P. Harriott, *Unit Operations of Chemical Engineering*, 5th ed., McGraw-Hill, New York, 1993.

13

ECONOMIC CONSIDERATIONS

Partial contribution: Richard L. Kratz

13.1 INTRODUCTION

Incineration has been recognized as one of the highest cost options for hazardous waste management. Hence, an understanding of the economics involved in incineration is quite important in making decisions at both the engineering and management levels. Every engineer should be able to execute an economic evaluation of a proposed project. If the project is not profitable, it should obviously not be pursued, and the earlier such a project can be identified, the fewer are the resources that will be wasted.

Before the cost of a hazardous waste incineration (HWI) facility can be evaluated, the factors contributing to the cost must be recognized. There are two major contributing factors: capital costs and operating costs; these are discussed in Sections 13.2 and 13.3, respectively. Once the total cost of the facility has been estimated, the engineer must determine whether or not the project will be profitable. This involves converting all cost contributions to an annualized basis, a method that is discussed in Section 13.4. If more than one project proposal is under study, this method provides a basis for comparing alternate proposals and for choosing the best proposal. Project optimization is the subject of Section 13.5, where a brief description of a perturbation analysis is presented. The chapter concludes with examples and problems that illustrate the material presented.

Detailed cost estimates are beyond the scope of this text. Such procedures are capable of producing accuracies in the neighborhood of $\pm 5\%$; however, such estimates generally require many months of engineering work. This chapter is designed to give the reader a basis for a *preliminary cost analysis* only, with an expected accuracy of approximately $\pm 20\%$.

13.2 CAPITAL COSTS

Capital Equipment Costs

Equipment cost is a function of many variables, one of the most significant of which *is capacity.* Other important variables include incinerator type, residence time, operating temperature, and degree of equipment sophistication. Preliminary estimates are often made from simple cost–capacity relationships that are valid when the other variables are confined to narrow ranges of values; these relationships can be represented by approximate linear (on log–log coordinates) cost equations of the form[1]

$$C = \alpha Q^{\beta} \tag{13.2.1}$$

where C = cost
Q = some measure of equipment capacity
α, β = empirical "constants" that depend mainly on equipment type

It should be emphasized that this procedure is suitable for rough estimation only; actual estimates from vendors are more preferable. Only major pieces of equipment such as the incinerator itself, waste heat boiler, and air pollution control (APC) equipment are included in this analysis; smaller peripheral equipment such as pumps and compressors are not discussed.

The most common types of HWI systems in use are the liquid injection and rotary kiln. Cost–capacity relationships for two types of liquid injection systems are presented here. Both include a waste heat boiler and a flue gas scrubbing system, in addition to the incinerator.

The first type, designated as case A in Table 13.2.1, is designed to burn relatively clean liquids that contain an insignificant amount of chlorine or salts. Normal operating temperatures seldom exceed 1800°F and residence times are usually short (0.3–0.5 s). The incinerator (described in Chapter 7) consists of a horizontally aligned cylindrical steel shell, combustion air blower, flame safeguard, and combustion control instrumentation. Case B is a liquid injection system designed to handle problematic liquid wastes such as highly chlorinated materials, salty aqueous wastes, and liquefied tars. Units in this category are obviously more costly and are designed to accommodate operating temperatures up to 2200–2400°F; residence times range from 1.5 to 2.5 s. These units can be either vertical or horizontal cylindrical steel shells and include combustion air blowers, flame safeguards, and combustion control instrumentation. The applicable capacity range for both case A and case B systems is from 1 to 100 million Btu/h.

Table 13.2.1 Values of α and β for use with Eq. (13.2.1)[a]

	α [dollar/Q]	β
Incinerator[b]		
Liquid injector (case A)	30.1	0.503
Liquid injector (case B)	41.6	0.512
Rotary kiln	220	0.478
Waste Heat Boiler[c]		
Case A	462	0.515
Case B	1944	0.401
Case C	9231	0.278
Scrubber[d]		
Case A	145	0.717
Case B	46	0.816

[a]Costs calculated using these constants are based on the value of the dollar in 1982.
[b]Units of Q are Btu/h.
[c]Units of Q are lb/h.
[d]Units of Q are acfm.

The rotary-kiln system in Table 13.2.1 includes the kiln—and secondary combustion chamber (SCC) and auxiliaries plus waste heat boiler (optional) and an air pollution control (APC) system. The cost of this system is considerably higher than that of the liquid injection systems. The applicable capacity range for the constants in Table 13.2.1 is 5–100 million Btu/h. The kiln temperatures range from 1400 to 1600°F while the afterburner temperatures range from 1800 to 2200°F with a residence time of approximately 2 s.

Note: A venturi scrubber is usually required for particulate control.

Waste heat boilers, discussed in Chapter 8, are primarily used in facilities that can accommodate 10–50 million Btu/h, where the steam can be used elsewhere. Their use is less likely in smaller facilities and in commercial facilities where there is usually no market for the steam.[1] In Table 13.2.1, waste heat boilers are divided into three categories. Case A is for smaller facilities using a fire-tube boiler with a gas flow rate ranging from 2000 to 100,000 lb/h. Case B is for water-tube boilers of standard design with a gas flow rate range from 15,000 to 150,000 lb/h. Water-tube boilers operating with extremely high particulate loadings and high acid gas environments make up case C, where the applicable range is 15,000–150,000 lb/h. The constants α and β for all three boiler types are based on inlet temperatures of 1800–2200°F and outlet temperatures of approximately 500°F. For the fire-tube boiler, the steam pressure is in the neighborhood of 150 psig. Water-tube steam pressures range from 150 to 200 psig.

Scrubbing system designs and costs also vary considerably from one system to the next. Typical scrubbing and flue gas handling systems include quenchers, venturi scrubbers, acid gas absorbers, caustic recycle systems, induced-draft (ID) fans, stacks, and auxiliaries. System costs calculated from Table 13.2.1 are based on a venturi pressure drop of 30 in. H_2O.

In Table 13.2.1, scrubbers are divided into two categories based on whether or not a waste heat boiler is involved. Case A is for systems receiving hot combustion gas directly from the incinerator (no waste heat boiler) at a temperature of 1800–2200°F. The applicable range of capacities for the constants listed in Table 13.2.1 for this case is 1000–200,000 acfm. Case B involves scrubbing systems receiving gas from waste heat boilers at a temperature of approximately 550°F. Because of the much lower operating temperatures, less costly materials of construction and smaller-sized equipment can be employed. The applicable capacity range is from 1000 to 50,000 acfm. Because scrubbing systems vary so much in type, *cost adjustment factors* (CAF) are used to account for this variation[1]; these are listed in Table 13.2.2. To use the CAFs for any of the four cases listed, the cost resulting from Eq. (13.2.1) is multiplied by the appropriate CAF.

Similar methods for estimating the costs of storage tanks, waste conveyors, pumps, compressors, site development, and so on are available in the literature.[1] If greater accuracy is needed, however, actual quotes from vendors should be used.

Again, the equipment cost estimation model just described is useful for a very preliminary estimation. If more accurate values are needed and if old price data are available, the use of an indexing method is better, although a bit more time consuming. The method consists of adjusting the earlier cost data to present values using factors that correct for inflation. A number of such indices are available; some of the most commonly used are:

1. *Chemical Engineering Fabricated Equipment Cost Index* (FECI),[2] past values of which are listed in Table 13.2.3
2. *Chemical Engineering Plant Cost Index,*[3] listed in Table 13.2.4
3. *Marshall and Swift (M&S) Equipment Cost Index,*[4] listed in Table 13.2.5

Table 13.2.2 Cost adjustment factors for scrubbing systems[1]

Adjustment from base line scenario	CAF
Extremely high venturi ΔP (100 in. H_2O)	2.0
Same but no acid gas absorption	1.2
No venturi required	0.85
No absorption system required	0.6

Table 13.2.3 Fabricated equipment cost index[2]

Year	Index
1999	434.1
1998	435.6
1997	430.4
1996	425.5
1995	425.4
1994	401.6
1993	391.2
1957–1959	100

Table 13.2.4 Plant cost index[3]

Year	Index
1999	390.6
1998	389.5
1997	385.5
1996	381.7
1995	381.1
1994	368.1
1993	359.2
1992	358.2
1991	361.3
1990	357.6
1989	355.4
1957–1959	100

Table 13.2.5 Marshall & Swift equipment cost index[4]

Year	Index
1999	1068.3
1998	1061.9
1997	1056.8
1996	1039.1
1995	1027.5
1994	993.4
1993	964.2
1992	943.1
1991	930.6
1990	915.1
1926	100

Other indices for construction, labor, buildings, engineering, and so on are also available in the literature.[2–4] Generally, it is not wise to use past cost data older than 5–10 years, even with the use of the cost indices. Within that time span, the technologies used in the processes may have changed drastically. The use of the indices could cause the estimates to be much greater than the actual costs. Such an error might lead to the choice of an alternative proposal other than the least costly. Note that the three indices listed in Tables 13.2.3, 13.2.4 and 13.2.5 range over the years 1989–1999. The base years for Tables 13.2.3 and 13.2.4 are 1957–1959 with an index of 100. One may compare the fabricated equipment index to costs available from the time frame between 1993 and 1999. Another comparison would be the plant cost index from 1989 to 1999. The Marshall and Swift index uses 1926 as a base year with a cost index of 100. Comparing the ratios for any 2 years, calculated from Tables 13.2.4 and 13.2.5, one notes that the change in index varies from 1 to 3%. Since these methods are for preliminary cost estimates with an expected accuracy of approximately ±20%, use of either index will not significantly impact the final estimate.

In September 1995, the EPA report, *Escalation Indexes for Air Pollution Control Costs* (EPA-452/R-95-006; NTIS #PB 96-100862) was installed on the OAQPS TTN. This report discussed why and how a set of air pollution control indices Vatavuk air pollution control cost indexes (the "VAPCCI") were developed and presented index values for several periods, from first quarter 1994 through first quarter 1995 (preliminary). Since then, data have become available to allow the updating of these indices through fourth quarter 1999 (preliminary). In addition, new indices have been developed for three more control device categories: electrostatic precipitators, fabric filters, and mechanical collectors.[5]

Capital Costs

The usual technique for determining the *capital costs* (i.e., *total* capital costs, which include equipment design, purchase, and installation) for the incineration facility is based on the *factored method* of establishing direct and indirect installation costs as a function of the known equipment costs. This is basically a *modified Lang method,* whereby cost factors are applied to known equipment costs.[6,7]

The first step is to obtain from vendors (or, if less accuracy is acceptable, from one of the estimation techniques previously discussed) the purchase prices of the primary and auxiliary equipment. The total base price, designated by X, which should include instrumentation, control, taxes, freight costs, and so on, serves as the basis for estimating the direct and indirect installation costs. The installation costs are obtained by multiplying X by the cost factors, which are available in the literature.[6–11] For more refined estimates, the cost factors can be adjusted to more closely model the proposed system by using adjustment factors that take into account the complexity and sensitivity of the system.[6,7]

The second step is to estimate the direct installation costs by summing all the cost factors involved in the direct installation costs, which include piping, insulation, foundation and supports, and so on. The sum of these factors is designated as the DCF *(direct installation cost factor).* The direct installation costs are then the product of the DCF and *X*.

The third step consists of estimating the indirect installation costs. The procedure here is the same as that for the direct installation costs; that is, all the cost factors for the indirect installation costs (engineering and supervision, start-up, construction fees, etc.) are added; the sum is designated by ICF (indirect installation cost factor). The indirect installation costs are then the product of ICF and *X*.

Once the direct and indirect installation costs have been calculated, the *total capital cost* (TCC) may be evaluated as

$$\text{TCC} = X + (\text{DCF})(X) + (\text{ICF})(X) \tag{13.2.2}$$

This cost is then converted to *annualized* capital costs with the use of the *capital recovery factor* (CRF), which is described in Section 13.5. The *annualized capital cost* (ACC) is the product of the CRF and TCC and represents the total installed equipment cost distributed over the lifetime of the facility.

Some guidelines in purchasing equipment are listed below:

- Do not buy or sign any documents unless provided with certified independent test data.
- Previous clients of the vendor company should be contacted and their facilities visited.
- Payment schedules should be established based on targeted goals; for example, 10% on receipt of approved engineering drawings, 15% upon receipt of major fabrication materials, and so on.
- Finally, 10–15% of the cost should be withheld until the installation is completed.
- Permit approval from state and local regulatory officials should be obtained. This may sometimes require federal approvals as well.
- A process and mechanical guarantee from the vendors involved should be required. Startup assistance should be included and assurance of prompt technical assistance should be obtained in writing. A complete and coordinated operating manual should be provided.
- Vendors should provide key spare parts necessary at startup and replacement parts for a minimum of the first 6 months to 1 year of operation. This should also consider special refractory brick or shapes that may be long delivery items.

13.3 OPERATING COSTS

Operating costs can vary from site to site since these costs, in part, reflect local conditions, for example, staffing practices, labor, and utility costs. Operating costs, like capital costs, may be separated into two categories: direct and indirect costs. *Direct* costs are those that cover material and labor and are directly involved in operating the facility. These include labor, materials, maintenance labor and maintenance supplies, refractories, replacement parts, waste (i.e., residues after incineration) disposal fees, utilities, and laboratory costs. *Indirect* costs are those operating costs associated with but not directly involved in operating the facility; costs such as overhead (e.g., building–land leasing and office supplies), administrative fees, local property taxes, and insurance fees fall into this category.

The major direct operating costs are usually those associated with labor and materials. *Materials* costs for incineration systems involve the cost of chemicals needed for the operation of the gas absorbers and/or spray dryers in the APC systems.[12] *Labor* costs differ greatly, depending on whether or not the facility is located on-site or off-site and the degree of controls and/or instrumentation; on-site facilities are usually owned and operated by the company generating the waste; off-site facilities are usually run by commercial incinerator companies. Off-site facilities require a greater amount of operating labor. Typically, there are three working shifts per day plus the standby shift used for rotation on a weekly basis with one supervisor per shift. On the other hand, on-site incinerators at manufacturing plants, particularly the liquid injection type, may be manned by a single operator for only one third or one half of each shift; usually only an operator, supervisor, and site manager are necessary to run the facility.[13] Salary costs vary from state to state and depend significantly on the location of the facility. The cost of *utilities* generally consists of that for electricity, water, fuel, compressed air, and steam. The annual costs are estimated with the use of material and energy balances. Costs for *waste disposal,* that is, for the incinerator (ash), baghouse or electrostatic precipitator (ESP) (fly ash) and scrubber (water) residues, can be estimated on a per ton capita basis. Costs of landfilling the ash can run significantly upwards of \$150/ton if the material is hazardous, to \$50/ton if it is nonhazardous. The cost of handling the scrubber effluent can vary depending on the method of disposal. For example, if conventional sewer disposal is used, the effluent probably has to be cooled and neutralized before disposal; the cost for this depends on the solids concentration. Annual maintenance costs can be estimated as a percentage of the capital equipment cost; the following values have been recommended: 5% for liquid injection incinerators, 7% for hearth types, and 10% for rotary kiln facilities.[12] In some applications, special high-temperature refractories are required that will impact the percentage used for maintenance. The annual costs of refractories and replacement parts can be computed by dividing the cost of the individual item by its expected lifetime. The life expectancies can be found in the literature.[7] Laboratory costs depend on the number of samples tested and the extent of these tests; these costs can be estimated as 10–20% of the operating labor costs.

The *indirect* operating costs consist of overhead, local property tax, insurance, administration, less any credits. The overhead comprises payroll, fringe benefits, social security, unemployment insurance, and other compensation that is indirectly paid to the plant personnel. This cost can be estimated as 70–90% of the operating labor, supervision, and maintenance costs.[11,12] Local *property taxes* and *insurance* can be estimated as 2–3% of the TCC while administration costs can be estimated as 2% of the TCC.

The total operating cost is the sum of the direct operating costs and the indirect operating costs less any credits that may be recovered [e.g., the value of recovered by-products (HCl acid) and/or steam]. Unlike capital costs, operating costs are always calculated on an annual basis.

13.4 PROJECT EVALUATION

In comparing alternate processes or different options of a particular process from an economic point of view, it is recommended that the total capital cost be converted to an annual basis by distributing it over the projected lifetime of the facility. The sum of both the annualized capital costs (ACC) and the *annual operating costs* (AOC) is known as the *total annualized cost* (TAC) for the facility. The economic merit of the proposed facility, process, or scheme can be examined once the total annual cost is available. Alternate facilities or options (e.g., a baghouse versus an ESP for particulate control, or two different processes for accomplishing the same degree of waste destruction) may also be compared. Note, a small flaw in this procedure is the assumption that the operating costs remain constant throughout the lifetime of the facility. However, since the analysis is geared to comparing different alternatives, the changes with time should be somewhat uniform among the various alternatives, resulting in little loss of accuracy.

The conversion of the total capital cost to an annualized basis involves an economic parameter known as the capital recovery factor (CRF). These factors can be found in any standard economics text[13,14] or can be calculated directly from

$$\mathrm{CRF} = \frac{(i)(1 + i)^n}{(1 + i)^n - 1} \tag{13.4.1}$$

where n = projected lifetime of the system (years)
i = annual interest rate (expressed as a fraction)

The CRF is a positive, fractional number. The ACC is computed by multiplying the TCC by the CRF. The annualized capital cost reflects the cost associated in recovering the initial capital outlay over the depreciable life of the system. Investment and operating costs can be accounted for in other ways, such as a *present worth* analysis. However, the capital recovery method is preferred because of its simplicity and versatility. This is especially true when comparing control systems having different depreciable lives. There are usually other considerations in such decisions besides the economics, but if all the other factors are equal, the proposal with the lowest total annualized cost should be the most viable.

If a plant (on-site) incineration system is under consideration for construction, the total annualized cost should be sufficient to determine whether or not the proposal is economically attractive as compared to other proposals. If, however, a commercial incineration process is being considered, the profitability of the proposed operation becomes an additional factor. The method presented here assumes a facility lifetime of 10 years and that the land is already available, the public has been informed, and state and local regulatory agencies have agreed that there is a need for the facility.

One difficulty in this analysis is estimating the revenue generated from the facility because both technology and costs can change from year to year. Also affecting the revenue generated is the availability and amount of waste to be handled by the facility. Naturally, the more waste handled, the greater are the revenues. One must be cautious in today's market with other options that are now available for specific waste streams [cement kilns, lightweight aggregate kilns, boilers, and other industrial furnaces permitted under the 1992 boiler and industrial furnace (BIF) regulations]. There has also been a reduction of plant process wastes due to pollution prevention efforts in process optimization at many plants reducing the amount in the waste stream, recycling between plants where one's waste stream becomes another's raw materials, and the combination of these which has resulted in waste minimization (see Chapter 2). If a reasonable estimate as to the revenue that will be generated from the facility can be made, a rate of return can be calculated.

This method of analysis is known as *the discounted cash flow method using an end-of-year convention,* that is, the cash flows are assumed to be generated at the end of the year, rather than throughout the year (the latter obviously being the real case). An expanded explanation of this method can be found in any engineering economics text.[13] The data required for the analysis are the TCC, the annual after-tax cash flow (A), and the working capital (WC). Generally, for HWI facilities, WC includes the on-site fuel inventory, caustic soda solution, maintenance materials (refractories, spare parts, etc.), and wages for approximately 30 days. The WC is expended at the startup of the plant (time = 0 years) and is assumed to be recoverable after the life of the facility (time = 10 years). For simplicity, it is assumed that the WC is 10% of the TCC and that the TCC is spread evenly over the number of years used to construct the facility. (For this example, this construction period is assumed to be 2 years.)

Usually, an after-tax rate of return on the initial investment of at least 30% is desirable. The method used to arrive at a rate of return will be discussed briefly. An annual after-tax cash flow can be computed as the annual revenues

(R) less the AOC and less income taxes (IT). Income taxes can be estimated at 50% (this number may vary depending upon the current tax laws) of taxable income (TI):

$$\text{IT} = 0.5(\text{TI}) \tag{13.4.2}$$

The taxable income is obtained by subtracting the AOC and the depreciation of the plant (D) from the revenues generated (R) or

$$\text{TI} = R - \text{AOC} - D \tag{13.4.3}$$

For simplicity, straight-line depreciation is assumed, that is, the plant will depreciate uniformly over the life of the plant. For a 10-year lifetime, the facility will depreciate 10% each year:

$$D = 0.1(\text{TCC}) \tag{13.4.4}$$

The annual after-tax cash flow (A) is then

$$A = R - \text{AOC} - \text{IT} \tag{13.4.5}$$

This procedure involves a trial-and-error solution. There are both positive and negative cash flows. The positive cash flows consist of A and the recoverable working capital in year 10. Both should be discounted *backward* to time = 0, the year the facility begins operation. The negative cash flows consist of the TCC and the initial WC. In actuality, the TCC is assumed to be spent evenly over the 2-year construction period. Therefore, one half of this flow is adjusted *forward* from after the first construction year (time = −1 year) to the year the facility begins operating (time = 0). The other half, plus the WC, is assumed to be expended at time = 0. Forward adjustment of the 50% TCC is accomplished by multiplying by an economic parameter known as the *single-payment compound amount factor F/P*, given by

$$F/P = (1 + i)^m \tag{13.4.6}$$

where i = rate of return (fraction)
m = number of years (in this case, 1 year)

For the *positive* cash flows, the annual after-tax cash flow (A) is discounted backward by using a parameter known as the *uniform series present worth factor* (P/A). This factor is dependent on both interest rate (rate of return) and the lifetime of the facility and is defined by

$$\frac{P}{A} = \frac{(1 + i)^n - 1}{i(1 + i)^n}$$

where n = lifetime of facility (in this case, 10 years).

Note: The P/A is the inverse of the CRF (capital recovery factor). The recoverable working capital at year 10 is discounted backward by multiplying WC by the *single present worth factor* (P/F) which is given by

$$\frac{P}{F} = \frac{1}{(1 + i)^n} \tag{13.4.7}$$

where n = lifetime of the facility (in this case, 10 years).

The positive and negative cash flows are now equated and the value of i, the rate of return, may be determined by trial and error from

$$\text{Term 1} + \text{term 2} = \text{term 3} + \text{term 4} \tag{13.4.8}$$

where term 1 = $[((1 + i)^{10} - 1)/i(1 + i)^{10}]A$; worth at year = 0 of annual after-tax cash flows

term 2 = $(1/(1 + i)^{10})$WC; worth at year = 0 of recoverable WC after 10 years

term 3 = (WC + $\frac{1}{2}$ TCC); assumed expenditures at year = 0

term 4 = $\frac{1}{2}$(TCC)$(1 + i)^1$; worth at year = 0 of assumed expenditures at year = −1

This method (i.e., the discounted cash flow method using an end-of-year conversion) is demonstrated in Example 13.6.5.

13.5 PERTURBATION STUDIES IN OPTIMIZATION

Once a particular process scheme has been selected, it is common practice to optimize the process from a capital cost and O&M (operation and maintenance) standpoint. There are many optimization procedures available, most of them too detailed for meaningful application to a hazardous waste facility. These sophisticated optimization techniques, some of which are routinely used in the design of conventional chemical and petrochemical plants, invariably involve computer calculations. Use of these techniques in HWI analysis is not warranted, however.

One simple optimization procedure that is recommended is the *perturbation study*. This involves a systematic change (or *perturbation*) of variables, one by one, in an attempt to locate the optimum design from a cost and operation viewpoint. To be practical, this often means that the engineer must limit the number of variables by assigning constant values to those process variables that are known beforehand to play an insignificant role. Reasonable guesses and simple or short-cut mathematical methods can further simplify the procedure. Much information can be gathered from this type of study since it usually identifies those variables that significantly impact on the overall performance of the process and also helps identify the major contributors to the total annualized cost.

13.6 ILLUSTRATIVE EXAMPLES

Example 13.6.1 Plans for the construction of a hazardous waste incinerator were initiated in 1993. The HCl emission control device was to be an absorber. The cost for this unit at that time was determined to be \$245,000. Estimate the cost of the absorber in terms of both 1995 and 1999 dollars.

SOLUTION This problem involves the use of the CE (*Chemical Engineering*) Fabricated Equipment Cost Index to obtain the FECI factors for the years 1993, 1995, and 1999. The cost for a particular year is the product of the cost in 1993 and the ratio of the FECI factor for that year to that for 1993. From Table 13.2.3,

$$1993 \text{ FECI} = 391.2$$

$$1995 \text{ FECI} = 425.4$$

$$1999 \text{ FECI} = 434.1$$

Cost in 1995 dollars is therefore

$$\begin{aligned}\text{Cost}(1995) &= \text{cost}(1993)(1995 \text{ FECI}/1993 \text{ FECI}) = (\$245{,}000)(425.4/391.2)\\ &= \$266{,}419\end{aligned}$$

Similarly, the cost in 1999 is

$$\begin{aligned}\text{Cost}(1999) &= \text{cost}(1993)(1999 \text{ FECI}/1993 \text{ FECI}) = (\$245{,}000)(434.1/391.2)\\ &= \$271{,}867\end{aligned}$$

Example 13.6.2 A vendor has provided a total installed capital cost estimate for a small proposed incinerator facility of \$30/acfm (350°F) of flue gas at the fan inlet. A baghouse is to be installed at the site for particulate control employing Teflon felt bags, each with an area of 12 ft^2 at an air-to-cloth ratio (acfm/ft^2) of 3.0 to 1.0. The cost per bag is \$90.00. The total pressure drop across the system is 16.5 in. H_2O, including 3.0 in. H_2O for the baghouse. Determine the installed capital, operating, and maintenance costs for the proposed facility on an annualized basis. The following data are available and are applicable at the time of the purchase:

- Estimated flue gas flow rate (at 350°F), Q_a = 70,000 acfm
- Overall fan efficiency, E_f = 60% (at 350°F)
- Operating time = 7000 h/yr
- Electrical power cost = \$0.06/kW-h
- Fuel requirements for incineration of waste = none
- Yearly maintenance cost (entire facility) = \$100,000/yr
- Replacement parts = 25% of bags each year
- Lifetime of facility, n = 15 yr
- Interest rate, i = 8%
- Salvage value = 0

Note: The salvage value is the amount that can be obtained, in dollars, for the equipment at the end of the facility or equipment lifetime.

SOLUTION The problem involves the determination of the total capital cost, the operating costs, and the maintenance costs. The CRF is used to convert the total capital cost to an annualized basis. The TCC is

$$\begin{aligned}\text{TCC} &= (\$30/\text{acfm})(Q_a) = (\$30/\text{acfm})(70{,}000\ \text{acfm})\\ &= \$2{,}100{,}000\end{aligned}$$

To convert this cost to an annualized basis, the CRF is calculated from Eq. (13.4.1):

$$\begin{aligned}\text{CRF} &= (i)(1+i)^n/[(1+i)^n - 1] = (0.08)(1+0.08)^{15}/[(1+0.08)^{15} - 1]\\ &= 0.11683\end{aligned}$$

The ACC is the product of TCC and CRF:

$$\begin{aligned}\text{ACC} &= (\text{TCC})(\text{CRF}) = (\$2{,}100{,}000)(0.11683)\\ &= \$245{,}343/\text{yr}\end{aligned}$$

The AOC is

$$\begin{aligned}\text{AOC} &= \frac{(Q_a)(\Delta P)(\text{operating time})(\text{cost of electricity})}{E_f}\\ &= \frac{(70{,}000)(16.5)(7000)(0.06)(1.175 \times 10^{-4})}{0.6}\\ &= \$94{,}998/\text{yr}\end{aligned}$$

Note: 1.175×10^{-4} is a conversion factor used to obtain the correct units. It is obtained as

$$\frac{[5.20(\text{lb}_f/\text{ft}^2)/(\text{in. H}_2\text{O})](60 \text{ min/h})}{2.6552 \times 10^6 \text{ (ft-lb}_f/\text{kW} - \text{h)}} = 1.175 \times 10^{-4}$$

The annual maintenance cost can be computed as the sum of the yearly maintenance cost and the cost of replacement parts. The total number of bags (N) necessary for operating the baghouse at the waste facility is calculated as follows:

$$\begin{aligned} N &= (Q_a)/(\text{ACR})(\text{area of bag}) \\ &= (70{,}000 \text{ acfm})/(3.0 \text{ acfm/ft}^2)(12 \text{ ft}^2) \\ &= 1944 \text{ bags} \\ \text{MC} &= \$100{,}000 + (0.25)(\text{cost of a bag})\ (N) \\ &= \$100{,}000 + (0.25)(\$90.00/\text{bag})(1944 \text{ bags}) \\ &= \$143{,}740 \text{ yr} \end{aligned}$$

The total annualized cost of the facility is calculated as the sum of the ACC, AOC, and annual maintenance cost (MC).

$$\begin{aligned} \text{TAC} &= \text{ACC} + \text{AOC} + \text{MC} = \$245{,}343/\text{yr} + \$94{,}998/\text{yr} + \$143{,}740/\text{yr} \\ &= \$484{,}082/\text{yr} \end{aligned}$$

Example 13.6.3 A stream of 100,000 acfm of particulate-contaminated gas from a waste facility is to be cleaned according to environmental regulations. You have been requested to find the best way to install a pollution control device at the site. A reputable vendor has provided information on the cost of the equipment, as well as installation, operating, and maintenance costs according to the condition of the gas treated. Given are all the data you have collected. Determine what equipment you would select for this work in order to minimize costs on an annualized basis.

	Electrostatic Precipitator	Venturi Scrubber	Baghouse
Equipment cost	\$3.1/acfm	\$1.9/acfm	\$2.5/acfm
Installation cost	\$0.80/acfm	\$1.4/acfm	\$1.0/acfm
Operating cost	\$0.06/acfm-yr	\$0.06/acfm-yr	\$0.095/acfm-yr
Maintenance cost	\$14,000/yr	\$28,000/yr	\$9500/yr
Lifetime of equipment	20 yr	15 yr	20 yr

Costs are based on comparable overall collection efficiencies. Interest rate is 10% and there is zero salvage value.

SOLUTION The first step is to convert the equipment, installation, and operating costs to total costs by multiplying each by the total gas flow, 100,000 acfm. Hence, for the electrostatic precipitator, the total costs are

$$\text{Equipment cost} = 100{,}000 \text{ acfm } (\$3.1/\text{acfm}) = \$310{,}000$$

$$\text{Installation cost} = 100{,}000 \text{ acfm } (\$0.80/\text{acfm}) = \$80{,}000$$

$$\text{Operating cost} = 100{,}000 \text{ acfm } (\$0.06/\text{acfm-yr}) = \$6000/\text{yr}$$

Note that the operating costs are on an annualized basis. The equipment cost and the installation cost must then be converted to an annual basis using the CRF. From Eq. (13.4.1),

$$\text{CRF} = (0.1)(1 + 0.1)^{20}/[(1 + 0.1)^{20} - 1] = 0.11746$$

The annual costs for the equipment and the installation is the product of the CRF and the total costs of each.

$$\begin{aligned}\text{Equipment annual cost} &= \$310{,}000(0.11746)\\ &= \$36{,}412/\text{yr}\\ \text{Installation annual cost} &= \$80{,}000(0.11746)\\ &= \$9{,}397/\text{yr}\end{aligned}$$

The calculations for the scrubber and the baghouse are performed in the same manner. The three devices can be compared after all the annual costs are added. The tabulated results follow:

	Electrostatic Precipitator	Venturi Scrubber	Baghouse
Equipment cost	\$310,000	\$190,000	\$250,000
Installation cost	\$80,000	\$140,000	\$100,000
CRF	0.11746	0.13147	0.11746
Annual equipment	\$36,412	\$24,980	\$29,365
Annual installation	\$9,397	\$18,405	\$11,746
Annual operating	\$6,000	\$6,000	\$9,500
Annual maintenance	\$14,000	\$28,000	\$9,500
Total annual	\$65,809	\$77,385	\$60,111

According to the analysis, the baghouse is the most economically attractive device since the annual cost is the lowest.

Example 13.6.4 Plans are underway to construct and operate a commercial hazardous waste facility in Trashtown in the state of Ecoarkana. The company is still undecided as to whether to install a liquid injection or rotary kiln incinerator at the waste site. The liquid injection unit is less expensive to purchase and operate than a comparable rotary kiln system. However, projected waste treatment income from the rotary kiln unit is higher since it will handle a larger variety and quantity of wastes. In addition, the rotary kiln will be treating solids as well as liquid wastes.

Based on the economic and financial data provided here, select the incinerator that will yield the higher annual profit.

Costs/Credits	Liquid Injection	Rotary Kiln
Capital ($)	2,625,000	2,975,000
Installation ($)	1,575,000	1,700,000
Operation ($/yr)	400,000	550,000
Maintenance ($/yr)	650,000	775,000
Income ($/yr)	2,000,000	2,500,000

Calculations should be based on an interest rate of 12% and a process lifetime of 12 years for both incinerators.[15]

SOLUTION For both units:

$$\text{CRF} = (0.12)(1 + 0.12)^{12}/[(1 + 0.12)^{12} - 1]$$
$$= 0.\ 1614$$

Annual capital and installation costs for the liquid injection (LI) unit are

$$\text{LI costs} = (2{,}625{,}000 + 1{,}575{,}000)(0.1614)$$
$$= \$677{,}880/\text{yr}$$

Annual capital and installation costs for the rotary kiln (RK) unit are

$$\text{RK costs} = (2{,}975{,}000 + 1{,}700{,}000)(0.1614)$$
$$= \$754{,}545/\text{yr}$$

A comparison of costs and credits for both incinerators is given in the following table:

	Liquid Injection	Rotary Kiln
Total installed ($/yr)	678,000	755,000
Operation ($/yr)	400,000	550,000
Maintenance ($/yr)	650,000	775,000
Total annual cost ($/yr)	1,728,000	2,080,000
Income credit ($/yr)	2,000,000	2,500,000
Profit ($/yr)	272,000	420,000

A rotary kiln incinerator is recommended.

Example 13.6.5 Two small commercial incineration facility designs are under consideration. The first design involves a liquid injection incinerator and the second a rotary kiln incinerator. For the liquid injection system, the total capital cost (TCC) is \$2.5 million, the annual operating costs (AOC) are \$1.2 million, and the annual revenue generated from the facility (R) is \$3.6 million. For the rotary kiln system, TCC, AQC, and R are \$3.5, 1.4, and 5.3 million, respectively. Using the discounted cash flow method, which design is more attractive? Assume a 10-year facility lifetime and a 2-year construction period.

SOLUTION The solution involves the calculation of the rate of return for each of the two proposals.

1. *The Liquid Injection System.* The working capital (WC), the depreciation of the plant (D), the taxable income (TI), the income tax to be paid (IT), and the annual after-tax cash flow (A) must first be calculated.

The depreciation is

$$D = 0.1(\text{TCC}) = 0.1(\$2{,}500{,}000) = \$250{,}000$$

The WC is set at 10% of the TCC:

$$\text{WC} = 0.1(\text{TCC}) = 0.1(\$2{,}500{,}000) = \$250{,}000$$

The TI is calculated using Eq. (13.4.3):

$$\text{TI} = R - \text{AOC} - D = \$3{,}600{,}000 - \$1{,}200{,}000 - \$250{,}000 = \$2{,}150{,}000$$

The IT is calculated from Eq. (13.4.2):

$$\text{IT} = 0.5(\text{TI}) = 0.5(\$2{,}150{,}000) = \$1{,}075{,}000$$

The after-tax cash flow is calculated using Eq. (13.4.5):

$$A = R - \text{AOC} - \text{IT} = 3{,}600{,}000 - \$1{,}200{,}000 - \$1{,}075{,}000 = \$1{,}325{,}000$$

The rate of return can be computed by solving Eq. (13.4.8) by trial and error:

$$\left[\frac{(1+i)^{10}-1}{i(1+i)^{10}}\right] A + \left(\frac{1}{(1+i)^{10}}\right) \text{WC} = \text{WC} + \frac{1}{2}\,\text{TCC} + \frac{1}{2}\,\text{TCC}\,(1+i)^{1}$$

or

$$\left[\frac{(1+i)^{10}-1}{i(1+i)^{10}}\right] 1.325 \times 10^{6} + \left[\frac{1}{(1+i)^{10}}\right] 0.250 \times 10^{6}$$
$$= 0.250 \times 10^{6} + 1.250 \times 10^{6} + 1.250 \times 10^{6}\,(1+i)^{1}$$

A rate of return (i) is assumed until both sides of the equation are equal. Initially assuming a rate of 30%, substitution of $i = 0.30$ in this equation yields

$$\$4{,}115{,}034 = \$3{,}125{,}000$$

The difference between the two sides of the equation is \$990,036. For a second try, assume $i = 0.40$. This equation becomes

$$\$3{,}207{,}192 = \$3{,}250{,}000$$

with a difference between the two sides of $-$\$42,807. The rate of return is very close to 40%. Linearly interpolating,

$$\frac{40 - i}{40 - 30} = \frac{(\$42{,}807) - 0}{(-\$42{,}807) - \$990{,}036}$$

$$i = 39.6\%$$

2. *The Rotary Kiln System.* As with the liquid injection system,

$$\text{WC} = D = 0.1\ (\$3{,}500{,}000) = \$350{,}000$$

$$\text{TI} = \$5{,}300{,}000 - \$1{,}400{,}000 - \$350{,}000 = \$3{,}550{,}000$$

$$\text{IT} = 0.5(\$3{,}550{,}000) = \$1{,}775{,}000$$

The annual after-tax cash flow is

$$A = \$5{,}300{,}000 - \$1{,}400{,}000 - \$1{,}775{,}000 = \$2{,}125{,}000$$

Equation (3.4.8) becomes

$$\left[\frac{(1 + i)^{10} - 1}{i(1 + i)^{10}}\right] 2.125 \times 10^6 + \left[\frac{1}{(1 + i)^{10}}\right] 0.350 \times 10^6$$
$$= 0.350 \times 10^6 + 1.750 \times 10^6 + 1.750 \times 10^6\ (1 + i)^1$$

Assuming $i = 0.40$ yields

$$\$5{,}141{,}850 = \$4{,}550{,}000$$

The difference is \$591,850. Assuming $i = 0.45$ yields

$$\$4{,}615{,}519 = \$4{,}637{,}500$$

with a difference of $-$\$21,981. Once again using linear interpolation, the interest rate is found to be 44.8%

Hence, by the *discounted cash flow method,* the rate of return on the initial capital investment is ~5% greater for the rotary kiln system than the liquid injection system. From a purely financial standpoint, the rotary kiln system is the more attractive.

PROBLEMS

1. A hazardous waste incinerator cost $985,000 in 1982. You intend to install a similar type incinerator in your facility in 1986. What is the cost of the new incinerator? If the total (direct plus indirect) installation cost is 60% of the cost of the incinerator, what is the annualized capital cost of this unit? The expected life of the incinerator is 10 years; assume an interest rate of 10%.

2. A 200,000-acfm stream of contaminated air is to be treated using one of three devices, an electrostatic precipitator, a venturi scrubber, or a baghouse. The following data were obtained from a reliable vendor. Which air pollution control device should be selected? Assume an interest rate of 10%.

	Electrostatic Precipitator	Venturi Scrubber	Baghouse
Total capital cost	$3.5/acfm	$2.8/acfm	$3.2/acfm
Operating cost	$0.03/acfm	$0.035/acfm	$0.08/acfm
Maintenance costs	$12,000/yr	$15,000/yr	$13,000/yr
Lifetime	10 yr	10 yr	10 yr

3. There are two options available to an engineer designing a hazardous waste facility. The engineer can use either a "conventional" incinerator or an incinerator with heat recovery. The engineer compiled the data given here. From an economic point of view, which incinerator should the engineer select? The lifetime of each incinerator is 10 years; the interest rate is 10%.

	Incinerator with Heat Recovery	Incinerator w/o Heat Recovery
Incinerator cost	$1,294,000	$1,081,000
Installation cost	$786,000	$659,000
Operating labor	$39,900/yr	$8,500/yr
Maintenance	$43,000/yr	$17,000/yr
Utilities	$958,000/yr	$821,000/yr
Overhead	$51,300/yr	$13,900/yr
Taxes, insurance, and administration	$86,200/yr	$72,600/yr
Credits (from steam recovery)	$380,000/yr	0

4. Resolve Example 13.5.4 if the equipment lifetime and interest rate are 20 years and 10%, respectively.

REFERENCES

1. R. J. McCormick and R. J. DeRosier, "Capital and O&M Cost Relationships for Hazardous Waste Incineration," Acurex Corp., Cincinnati, OH, EPA Report 600/2-84-175, Oct., 1984.
2. Economic Indicators, "Chemical Engineering Plant Cost Index," Equipment, *Chemical Engineering*, **99**, 178, July (1992); **101**, 206, May (1994); **102**, 180, April (1995); **102**, 162, May (1995); **103**, 168, May (1996); **104**, 229, May (1997); **105**, 189, June (1998); **107**, 164, May (2000).
3. Economic Indicators, "Chemical Engineering Plant Cost Index," *Chemical Engineering*, **103**, 172, Nov. (1996); **107**, 164, May (2000).
4. Economic Indicators, "Marshall and Swift Equipment Cost Index," *Chemical Engineering*, **103**, 172, Nov. (1996); Chemical Engineering, 164, May 2000.
5. W. M. Vatavuk, "Escalation Indexes for Air Pollution Control Costs," (EPA-452/R-95-006; NTIS # PB 96-100862), Sept., 1995.
6. R. B. Neveril, "Capital and Operating Costs of Selected Air Pollution Control Systems," Gard, Inc., Niles, IL, EPA Report 450/5-80-002, Dec., 1978.
7. W. M. Vatavuk and R. B. Neveril, "Factors for Estimating Capital and Operating Costs," *Chem. Eng.*, 157–162, Nov. 3, 1980.
8. G. A. Vogel and E. J. Martin, "Hazardous Waste Incineration Part I—Equipment Sizes and Integrated-Facility Costs," *Chem. Eng.*, 143–146, Sept. 5, 1983.
9. G. A. Vogel and E. J. Martin, "Hazardous Waste Incineration Part 2—Estimating Costs of Equipment and Accessories," *Chem. Eng.*, 75–78, Oct. 17, 1983.
10. G. A. Vogel and E. J. Martin, "Hazardous Waste Incineration Part 3—Estimating Capital Costs of Facility Components," *Chem. Eng.,* 87–90, Nov. 28, 1983.
11. G. D. Ulrich, *A Guide to Chemical Engineering Process Design and Economics*, Wiley, New York, 1984.
12. G. A. Vogel and E. J. Martin, "Hazardous Waste Incineration Part 4—Estimating Operating Costs," *Chem. Eng.*, 97–100, Jan. 9, 1984.
13. E. P. DeGarmo, J. R. Canada, and W. G. Sullivan, *Engineering Economy*, 6th ed., Macmillan, New York, 1979.
14. C. Hodgman, S. Selby, and R. Weast, ed., *CRC Standard Mathematical Tables*, 12th ed., Chemical Rubber Company, Cleveland, OH, 1961 (presently CRC Press, Boca Raton, FL).
15. L. Theodore and E. Moy, *Hazardous Waste Incinerators*, A Theodore Tutorial, AWMA Bookstore, Pittsburgh, 1999.

14

DESIGN OF A HAZARDOUS WASTE INCINERATION FACILITY

14.1 INTRODUCTION

This chapter presents two problems that illustrate design techniques associated with hazardous waste incineration (HWI). The example problem in Section 14.2 requires the application of some of the design and calculational procedures developed by the authors. The second example problem, Section 14.3, requires the use of several of the Environmental Protection Agency (EPA) worksheets presented in Chapter 6. Although most of the procedures used to design the equipment are approximate, they will provide fairly accurate results. In both examples, incineration calculations are first performed to determine the combustion flue products and the operating temperatures. These are followed by quench and/or waste heat boiler calculations, which in turn are followed by design calculations for the air pollution control equipment. Both examples conclude with horsepower requirements and economic analyses. This chapter will expose the reader to the type of engineering calculations that are involved in the design of modern HWI facility.

14.2 ILLUSTRATIVE DESIGN PROBLEM I

A chemical company is generating appreciable amounts of chlorobenzene (C_6H_5Cl) at one of its plants. Management has decided that it would be economically advantageous to design and construct an HWI facility on-site. The company's corporate engineering staff is in the process of preparing the preliminary process design for this facility. Since the waste is a liquid, a liquid injection incinerator is to be employed. Pilot studies indicate that burning with

50% excess air will satisfy the *four nines* [99.99% destruction and removal efficiency (DRE)] regulatory requirement.

Data and information pertinent to the incinerator are given in the following list:

1. The optimum length-to-diameter ratio (L/D) has been determined to be 4.
2. The net heating value (NHV) for C_6H_5Cl is assumed to be high enough so that no auxiliary fuel is required. The burner selected should provide for auxiliary fuel firing, either natural gas or fuel oil that will be used during initial lightoff and warmup of the system. A low fire amount of fuel may often be recommended by the incinerator vendor to allow continuous operation of the system in the event of a malfunction or cutoff of the waste supply. This eliminates the need for sophisticated controls to light off on auxiliary fuel during a waste feed cutoff.
3. The C_6H_5Cl contains 1.0 wt % of particulates in the form of fly ash.
4. The maximum capacity of the unit is to be 5000 lb C_6H_5Cl/h.

Economic studies have shown that a waste heat boiler should be used to recover energy from the flue gas. Steam at 80 psia is needed to drive a turbine elsewhere in the plant; water at 100°F is available for this purpose. A boiler of the fire-tube variety is to be designed to cool the flue gases to 550°F and generate the required steam. The tubes are to be 1.5 in. o.d., Birmingham Wire Gauge (BWG) 18; the optimum mass flow rate of flue gas is 100 lb/h through each.

A venturi scrubber is to be installed to collect the particulates using water as the scrubbing medium. Previous studies on a similar process indicate that the optimum L/G ratio and throat velocity should be 9.5 gal/1000 ft^3 and 250 ft/s, respectively. The correlation coefficient k (which is a function of the system geometry and operating conditions) is 0.200 acf/gal. Assume that the average particle size of the fly ash discharged is 1.2 μm and that the particle density is 46 lb/ft^3. Once the particulates are collected, the wastewater is pumped to an on-site wastewater treatment system.

The HCl in the flue gas is to be removed with water at 100°F using a packed absorption tower. The absorber operating temperature and pressure are 150°F and 1 atm, respectively, and the packing is 3.5 in. ceramic Raschig rings.

System pressure drop data have also been provided from preliminary studies. Estimated pressure drops across the incinerator, waste heat boiler, and absorber are 3, 5, and 6 in. H_2O, respectively. The pressure drops for valves, fittings, elbows, and ductwork (including expansion and contraction losses) totals 5 in. H_2O. Several manufacturers of induced-draft fans have suggested an overall fan motor efficiency of 65%.

Cost estimates indicate that the annual operating cost (AOC) will be $800,000 per year. The total capital (TCC) and maintenance costs (MC) have been estimated to be \$4,000,000 and \$100,000 per year, respectively. For an economic analysis, a 25-year facility lifetime and an 8% interest rate are assumed.

You and several other engineers have been called on to prepare a preliminary process design for the proposed HWI facility. The analysis should include:

- Combustion calculations
- Incinerator design
- Waste heat boiler design
- Venturi scrubber calculations
- Packed tower absorber design
- Horsepower requirements
- Economic evaluation, including annualized costs

SOLUTION

Preliminary Calculations

Basis: 5000 lb of chlorobenzene, C_6H_5Cl. (The 1% ash in the waste will be neglected in the stoichiometric calculations.) The NHV of C_6H_5Cl can be estimated by a modified form of Du Long's equation that includes the chlorine content:

$$\text{NHV} = 14{,}000m_{\text{C}} + 45{,}000[m_{\text{H}} - (m_{\text{O}}/8)] - 760m_{\text{Cl}} + 4500m_{\text{S}} \quad (5.4.9)$$

where NHV = net heating value (Btu/lb)

m_{C} = weight fraction of C (carbon) in the waste–fuel mixture

C, H, O, Cl, S = subscripts indicating carbon, hydrogen, oxygen, and sulfur, respectively

Using the basis of 5000 lb of C_6H_5Cl, the mass fractions are determined:

$$5000 \text{ lb of } C_6H_5Cl \, \frac{6 \times 12 \text{ lb of C}}{112 \text{ lb of } C_6H_5Cl} = 3214 \text{ lb of C}$$

$$5000 \text{ lb of } C_6H_5Cl \, \frac{5 \times 1 \text{ lb of H}}{112 \text{ lb of } C_6H_5Cl} = 223 \text{ lb of H}$$

$$5000 \text{ lb of } C_6H_5Cl \, \frac{1 \times 35 \text{ lb of Cl}}{112 \text{ lb of } C_6H_5Cl} = 1563 \text{ lb of Cl}$$

The weight percentage of each compound is 64.28% C, 31.26% Cl, and 4.46% H. The NHV is calculated directly by Eq. (5.4.9):

$$\begin{aligned}\text{NHV} &= 14{,}000(0.6428) + 45{,}000(0.0446) - 760(0.3126)\\ &= 10{,}769 \text{ Btu/lb}\end{aligned}$$

This number is slightly lower than that predicted in Example 5.7.9. To be conservative, this (lower) value will be used. Because a value of 10,769 Btu/lb is reasonably high for the NHV, no auxiliary fuel is required. However, in an actual installation, auxiliary fuel capacity should be provided for during lightoff and warmup of the system. No hazardous wastes may be injected or burned in a Resource Conservation and Recovery Act (RCRA) permitted incinerator unless the combustion zone temperature reaches the permit level. Most systems will reduce the auxiliary firing rate to low fire as a minimum during the waste burning cycle. The calculations that follow do not include the flow of auxiliary fuel and combustion air necessary for the auxiliary fuel. The incinerator should be sized for this added flow as will be noted in the calculations that follow for incinerator design.

Combustion Calculations

The chemical equation for the combustion of 1.0 lbmol of chlorobenzene with stoichiometric air is

$$C_6H_5Cl + 7O_2 + 26.3N_2 \rightarrow CO_2 + 2H_2O + HCl + 26.3N_2 \quad (14.2.1)$$

From Eq. (14.2.1), the stoichiometric amount of air is 33.3 lbmol (O_2 + N_2). Since 50% excess air is to be employed, the total amount of air is 1.5 (33.3) = 50 lbmol, and the amount of excess air is 16.7 lbmol. Equation (14.2.1) is now rewritten for 50% excess air. Also shown are the molecular weight and mass flow rate (w_i) of each component (based on the maximum capacity of 5000 lb of C_6H_5Cl/h):

	C_6H_5Cl +	$10.5O_2$ +	$39.5N_2$ →	$6CO_2$ +	H_2O +	HCL +	$39.5N_2$ +	$3.5O_2$
MW	112	32	28	44	18	36	28	32
w_i	5000	14,986	49,328	11,774	1606	1606	49,328	4995

(14.2.2)

Table 14.2.1 shows flue gas quantities and gas partial pressures (assuming a total pressure of 1 atm) of each component in the flue gas. The temperature of the flue gas may be estimated by Eq. (6.4.20).

$$\begin{aligned}T &= 60 + \frac{\text{NHV}}{(0.3)[1 + (1 + \text{EA})(7.5 \times 10^{-4})(\text{NHV})]}\\ &= 60 + \frac{10769}{0.3[1 + (1 + 0.50)(7.54 \times 10^{-4})(10{,}769)]}\\ &= 2784°\text{F}\end{aligned}$$

Table 14.2.1 Flue gas composition (Illustrative Problem I)

Component	Mass Flow (lb/h)	Molar Flow (lbmol/h)	Mole Fraction		Partial Pressure (atm)
			Wet	Dry	
CO_2	11,774	268	0.1155	0.12	0.1155
H_2O	1,606	89.2	0.0384	0	0.0384
HCl	1,606	44.6	0.0192	0.0199	0.0192
N_2	49,328	1,762	0.7595	0.7898	0.7595
O_2	4,995	156	0.0672	0.0698	0.0672

where T = temperature (°F)
NHV = net heating value of the waste (Btu/lb)
EA = fractional amount of excess air

The reader may refer back to Chapter 6 for additional details on this equation. Because sufficient amounts of HCl and O_2 are present, the following reaction also takes place:

$$2HCl + 0.5O_2 \leftrightarrow H_2O + Cl_2$$

In this reaction, chlorine, a hazardous pollutant, is formed. To determine the partial pressure of Cl_2 in the flue gas, the equilibrium constant (K_p) for this reaction must first be obtained. Equation (14.2.3) may be used for this purpose[1]:

$$K_P = Ae^{BX} \tag{14.2.3}$$

where $A = 0.229 \times 10^{-3}$
$B = 7340$
$X = 1/T$; T in kelvins (K)

Since T = 2784°F = 1802 K

$$K_P = 0.229 \times 10^{-3} e^{(7340/1802)}$$
$$= 0.0134\ (\text{atm})^{-1}$$

The equilibrium expression may be represented in the following form:

$$K_P = \frac{P_{H_2O} P_{Cl_2}}{P^2_{HCl} P^{0.5}_{O_2}} \tag{14.2.4}$$

The partial pressures in atmospheres of H_2O, HCl, and O_2 are given in Table 14.2.1.

$$P_{H_2O} = 0.0384 \qquad P_{HCl} = 0.0192 \qquad P_{O_2} = 0.0672$$

These are the initial partial pressures. Let Z represent the partial pressure of Cl_2 at equilibrium. The equilibrium partial pressures that appear in Eq. (14.2.4) are given here:

$$P_{H_2O} = 0.0384 + Z \qquad P_{HCl} = 0.0192 - 2Z \qquad P_{O_2} = 0.0672 - 0.5Z$$

Substituting these values into Eq. (14.2.4) and solving the equation for Z by trial and error gives

$$0.0134 = \frac{(0.0384 + Z)(Z)}{(0.0192 - 2Z)^2(0.0672 - 0.5Z)^{0.5}}$$

with

$$Z = 3.3 \times 10^{-5} \text{ atm} = 33 \text{ ppm}$$

The mole fractions and partial pressures given in Table 14.2.1 were determined without taking into account the formation of Cl_2. Since HCl and O_2 are reacting to form H_2O and Cl_2, their partial pressures will decrease and those of H_2O and Cl_2 will increase. The results of this approximate calculation for the corrected flue gas quantities are given in Table 14.2.2.

Incinerator Design

The amount of heat released by the waste during operation may be expressed as

$$q = w(\text{NHV}) \tag{14.2.5}$$

Table 14.2.2 Corrected flue gas composition (Illustrative Problem I)

Component	Mass Flow (lb/h)	Molar Flow (lbmol/h)	Mole Fraction		Partial Pressure (atm)
			Wet	Dry	
CO_2	11,792	268	0.1155	0.12	0.1155
H_2O	1,607	89.2	0.03843	0	0.03843
HCl	1,598	44.4	0.01913	0.01989	0.01913
N_2	49,336	1,762	0.7596	0.7899	0.7596
O_2	4,989	155.9	0.06718	0.06987	0.06718
Cl_2	5	0.0766	0.000033	0.00034	0.000033

where q = rate of heat released (Btu/h)
w = mass flow rate of waste (lb/h)

Since w = 5000 lb/h and the NHV = 10,769 Btu/lb, q can be calculated:

$$q = (5000 \text{ lb/h})(10{,}769 \text{ Btu/lb})$$
$$= 53.8 \times 10^6 \text{ Btu/h}$$

Allowing for an additional amount of auxiliary fuel (natural gas) firing at 10% of maximum firing rate, a 60×10^6 Btu/h incinerator size will be selected. Assuming a typical heat release rate (q_H) of 25,000 Btu/h-ft^3 for liquid injection incinerators, the volume of the incinerator may now be calculated:

$$V = q/q_H = (60 \times 10^6 \text{ Btu/h})/(25{,}000 \text{ Btu/h-ft}^3)$$
$$= 2400 \text{ ft}^3 \tag{14.2.6}$$

The diameter (D) of the incinerator can be determined using an L/D ratio of 4:

$$V = \pi D^2 L/4 = \pi D^2(4D)/4 = \pi D^3$$
$$D = (V/\pi)^{1/3} = (2400/\pi)^{1/3} = 9.139 \text{ ft} \cong 9.2 \text{ ft} \tag{14.2.7}$$

The length of the incinerator is

$$L = 4(9.2) = 36.8 \text{ ft}$$

The corrected volume is also calculated:

$$V = \pi D^2 L/4 = \pi(9.2)^2(36.8)/4 = 2446 \text{ ft}^3$$

Assuming the ideal gas law to apply, the flow rate of gas (acfm) can be determined. From Table 14.2.1, the total molar flow rate (n) is 2320 lbmol/h.

$$Q_a = \dot{V} = \dot{n}RT/P = (2320)(10.73)(3244)/(14.7) = 5.5 \times 10^6 \text{ ft}^3\text{/h}$$
$$= 91{,}667 \text{ ft}^3\text{/min (acfm on a wet basis @ 2784°F)}$$

Using Charles' law, Eq. (4.4.8), the standard cubic feet per minute of the gas can be calculated. Standard conditions will be taken as 60°F and 1 atm:

$$Q_s = Q_a\left(\frac{T_s}{T_a}\right) = 91{,}667\left(\frac{520}{3244}\right) = 14{,}694 \text{ scfm}$$

Similarly, the dry acfm may be calculated:

$$Q_{\mathrm{d}} = (2230.6 \text{ dry moles/h})(10.73)(3244)/(14.7 \times 60)$$

$$Q_{\mathrm{da}} = 88{,}031 \text{ dry acfm } (@\ 2784°\mathrm{F})$$

$$Q_{\mathrm{ds}} = 14{,}111 \text{ dscfm } (@\ 60°\mathrm{F})$$

The superficial velocity can be calculated by

$$v = \frac{Q_{\mathrm{a}}}{A} = \frac{91{,}667/60}{3.1416(9.2)(9.2)/4} = 22.98 \text{ ft/s} \tag{14.2.8}$$

where v = superficial velocity (ft/s)
A = incinerator cross sectional area (ft^2)

The residence time (θ) is calculated from

$$\theta = \frac{V}{Q_{\mathrm{a}}} = \frac{(2446)(60)}{91{,}667} = 1.6 \text{ s} \tag{14.2.9}$$

This is a typical residence time for liquid injection incinerators. The gas residence time calculated previously is an *approximate* value because it does not take into consideration flow variations that occur during chemical reaction and temperature changes in the incinerator. In actuality, a residence time distribution exists within the unit.

Since the actual volumetric flow rate is known, the particulate loading of the flue gas leaving the incinerator may now be determined. From the problem statement, the waste contains 1 wt % of particulates in the form of fly ash. The particulate emission rate can be calculated by

$$w_{\mathrm{p}} = 0.01w = 0.01\,\frac{5000}{60} = 0.833 \text{ lb/min} \tag{14.2.10}$$

where w_{p} = particulate emission rate (lb/h)
w = waste feed flow rate (lb/h)

The dry particulate loading (W_i) is calculated by

$$W_i = \frac{w_p}{Q_{\mathrm{s,d}}} \tag{14.2.11}$$

where $Q_{\mathrm{s,d}}$ = the scfm of gas on a dry basis (dscfm).
The term $Q_{\mathrm{s,d}}$ is calculated from

$$Q_{s,d} = Q_s(1 - \text{moisture content}) \tag{14.2.12}$$

The moisture content is the mole fraction of water in the flue gas. A value of 0.0384 is obtained from Table 14.2.1. Therefore,

$$Q_{s,d} = 14{,}694(1 - 0.0384) = 14{,}111 \text{ dscfm}$$

The particulate loading (gr/dscf) is calculated from Eq. (14.2.11):

$$W_i = \frac{(0.833)(7000)}{14{,}111} = 0.4132 \text{ gr/dscf}$$

Note: 1 lb = 7000 gr.

Waste Heat Boiler Design

A waste heat boiler is to be designed to cool 69,327 lb/h of flue gas from 2784 to 550°F. The unit will be designed using Ganapathy's method,[2] which was discussed in Chapter 8. Using the information given in the problem statement, the total surface area and the quantity of steam generated in the shell can be determined.

The first step is to calculate the temperature difference ratio (ϕ):

$$\phi = \frac{T_{h1} - T_c}{T_{h2} - T_c} = \frac{2784 - 312}{550 - 312} = 10.4$$

where T_{h1} = inlet temperature of flue gas
T_{h2} = exit temperature of the flue gas
T_c = temperature of saturated steam in the shell

In Fig. 8.4.1, connect $(W_h/N) = 100$ with $D_i = 1.4$ and extend the line to intersect cut line No. 1 at A. Calculate T_{ha}, the average flue gas temperature:

$$T_{ha} = 0.5(T_{h1} + T_{h2}) = 0.5(2784 + 550) = 1667°\text{F}$$

Connect T_{ha} with ϕ (10.4) and extend the resulting line to intersect cut line No. 2 at B. Join A and B to intersect the L scale at a tube length of 16 ft. Since the number of tubes (N) is 69,327/100 = 694, the total surface area can be calculated by Eq. (14.2.13).

Note: The surface area per linear foot of 1.5 in. o.d., BWG 18 tubes is 0.3925 ft^2.

$$A = (N)(L)[\text{surface per linear feet (ft}^2)] = 694(16)(0.3925)$$
$$= 4358 \text{ ft}^2 \qquad (14.2.13)$$

The heat transfer rate may now be calculated via Eq. (8.2.9):

$$q = w_h c_P (T_{h1} - T_{h2}) \qquad (8.2.9)$$

where w_h = mass flow rate of flue gas (lb/h)
c_P = average heat capacity of the flue gas (Btu/lb-°F)

Assuming an average heat capacity of approximately 0.3 Btu/lb-°F, based on calculating the heat capacity for the individual combustion products components from Table 5.2.2:

$$q = 69{,}327(0.30)(2784 - 550) = 46.4 \times 10^6 \text{ Btu/h}$$

If feed water enters at 100°F, the quantity of steam generated can be calculated from

$$w_s = \frac{q}{h_s - h_{cl}} = \frac{46.4 \times 10^6}{1183 - 68} = 4.16 \times 10^4 \text{ lb/h} \qquad (14.2.14)$$

where w_s = steam rate (lb/h)
h_s = enthalpy of saturated steam of 80 psia (Btu/lb)
H_{cl} = enthalpy of feedwater at 100°F (Btu/lb)

The enthalpy values are obtained from the steam tables.

Venturi Scrubber Calculations

The venturi is designed to capture the fly ash in the gas stream. The unit also quenches the gases, however. This quench operation is assumed to lower the temperature of the gases from 550°F to approximately the adiabatic saturation temperature of the flue gas (which is in the neighborhood of 134°F). This calculation is not shown here, but it follows the method shown in Section 9.7 for the spray quench tower method of cooling combustion gases. At 134°F, there will be an increase in water vapor flow in the combustion gases to 8325 lb/h from the original 1606 lb/h. This increases the total moles/hour from 2320 to 2691 mol/h. The saturated gases leaving the venturi scrubber are then at a flow rate of 19402 acfm (14,314 dscfm).

Using the Nukiyama and Tanasawa relationship, the mean droplet diameter (d_1) can be calculated:

$$d_l = (16{,}400/v_t) + 1.45R^{1.5}$$

$$= \frac{16{,}400}{250} + 1.45(9.5)^{1.5}$$

$$= 108\ \mu\text{m} = 3.5 \times 10^{-4}\ \text{ft} \qquad (10.2.3)$$

where v_t = throat velocity (ft/s)
R = liquid-to-gas ratio (gal/1000 ft^3)
d_l = droplet diameter (μm)

The average particle size (d_p) is

$$d_p = 1.2\ \mu\text{m} = 3.94 \times 10^{-6}\ \text{ft}$$

The temperature of the hot flue gas entering the venturi scrubber is 550°F. From Table 14.2.2, N_2 has a mole fraction of 0.7596. Since the flue gas is essentially N_2, the gas viscosity is close to that of nitrogen at 550°F (1.92 × 10^{-5} lb/ft-s).

The inertial impaction number (ψ) can be calculated using Eq. (10.2.2). The Cunningham correction factor has been neglected:

$$\psi = \frac{d_p^2 \rho_p v_t}{9\mu d_l} = \frac{(3.94 \times 10^{-6})^2(46)(250)}{(9)(1.92 \times 10^{-5})(3.54 \times 10^{-4})}$$

$$= 2.92 \qquad (10.2.2)$$

where d_p = average particle size
ρ_p = particle density

The Johnstone equation, Eq. (10.2.1), is used to calculate the overall efficiency of the venturi scrubber:

$$E_i = 1 - e - KR(\psi)^{0.5} = 1 - e^{(-0.200)(9.5)(2.92)^{0.5}} = 0.9611 = 96.11\% \qquad (10.2.1)$$

The outlet particulate concentration does not have to be corrected to 7% O_2 because the inlet particulate concentration was determined on a dry basis using 7% O_2 (see Table 14.2.2) in the combustion calculations. Therefore, the correction to 7% O_2 has been "built into" the calculation:

$$W_e = W_i - E_i W_i = 0.412 - (0.9611)0.412 = 0.0160\ \text{gr/dscf}$$

where W_e = outlet particulate loading (gr/dscf)
W_i = inlet particulate loading (gr/dscf)
E_i = overall venturi scrubber efficiency

This value is in compliance with the original RCRA regulations since the outlet particulate loading is <0.080 gr/dscf corrected to 7% O_2. However, the new MACT rules of September 30, 1999, have reduced the PM loading to 0.015 gr/dscf corrected to 7% O_2 d.v. (dry volume). This indicates that the venturi scrubber will just fail the new MACT rules. A higher ΔP will be required to ensure a safe margin below the standard. A modified design of venturi with a throat velocity of 270 ft/s, liquid-to-gas ratio of 12 gal/1000 ft^3 results in an overall efficiency of 98.226% vs. the 96.11%. This will result in an outlet particulate loading of 0.0073 gr/dscf, well below the MACT standards. The reader is left the option to perform these additional calculations.

The pressure drop across the venturi scrubber is calculated using Eq. (10.2.6)[3]:

$$\Delta P' = \frac{v_t^2 \rho_G A_t^{0.133} R^{0.78}}{1270} \tag{10.2.6}$$

In Eq. (10.2.6), A_t represents the cross-sectional area of the throat. The value of A_t is calculated as

$$A_t = \frac{Q_a}{v_t} = \frac{19{,}402}{(60)(250)} = 1.29 \text{ ft}^2$$

The gas density is calculated by the ideal gas law:

$$\rho_G = \frac{MP}{RT} = \frac{(28)(14.7)}{(10.73)(180 + 460)} = 0.060 \text{ lb/ft}^3$$

From Eq. (10.2.6),

$$\Delta P' = \frac{(250)^2(0.060)(1.29)^{0.133}(9.5)^{0.78}}{1270} = 17.7 \text{ in. } H_2O$$

As noted above for the modified venturi design (270 ft/s and 12 gal/1000 ft^3) necessary to pass the latest MACT emission standard, the pressure drop across the venturi will increase from 17.7 in. H_2O (250 ft/s and 10 gal/1000 ft^3) to 24.4 in. H_2O.

Packed Tower Absorber Design

Table 14.2.2 gives the component flow rates entering the absorber. One of the components, Cl_2, would normally be absorbed with a caustic solution since it is not readily soluble in water. However, for purposes of preliminary process design calculations, the low concentration of Cl_2 (33 ppm) will be neglected in the absorber calculation. Therefore, only the HCl will be absorbed by the water. The former RCRA regulations state that HCl emissions >4 lb/h must be controlled at an efficiency of 99%. Since equilibrium data are not available, it is assumed that m (slope of equilibrium curve) approaches zero. This is not an unreasonable assumption for most solvents that preferentially absorb the pollutant. Water and HCl fall into this category. For this condition, the number of overall gas transfer units (N_{OG}) can be calculated from

$$N_{OG} = \ln(w_1/w_2) \qquad (10.2.12a)$$

where w_1 = inlet mass flow rate of HCl
w_2 = outlet mass flow rate of HCl

If the absorber operates at a 99% removal efficiency, w_2 may be determined as follows:

$$E = (w_1 - w_2)/w_1$$

$$w_2 = 1598 - (0.99)(1598)$$

$$= 15.98 \text{ lb/h}$$

The number of gas transfer units is calculated from Eq. (10.2.12a):

$$N_{OG} = \ln(1598/15.98) = 4.605$$

Information on the height of a gas transfer unit is well documented for water systems. For the 3.5 in. diameter ceramic packing used in this absorber, the approximate H_{OG} is 5.5 ft. The height of packing (Z) is obtained from Eq. (10.2.11) with a safety factor included:

$$Z = N_{OG} H_{OG} \text{ (safety factor)} \qquad (10.2.11)$$

Experience has shown that safety factors can range from 1.25 to 1.5. Using a safety factor of 1.3, Z becomes

$$Z = (4.605)(5.5)(1.3)$$

$$= 33 \text{ ft}$$

Using a typical superficial throughput velocity of 8 ft/s, the cross-sectional area of the tower can be estimated.

$$A = \frac{Q_a}{v} = \frac{19{,}402}{8 \times 60} = 40.4 \text{ ft}^2$$

The diameter is then

$$D = \left(\frac{4A}{\pi}\right)^{0.5} = \left(\frac{4(40.4)}{\pi}\right)^{0.5} = 7.174 \text{ ft} \cong 7 \text{ ft 3 in.}$$

We want to determine the actual process gas flow rate at the absorber operating temperature of 130°F. With an HCl removal efficiency of 99%, the number of moles in the gas stream decreases from 2691 to 2574. The gas stream leaving the absorber will be essentially saturated with water vapor. Since the gas entering the absorber is already nearly saturated, this slight increase in flow rate at the entry is neglected in the following calculation. Assuming ideality, Q_a is determined:

$$Q_a = (2574)(10.73)[(460 + 130)/14.7]$$

$$= 1.108 \times 10^6 \text{ acfh} = 18{,}475 \text{ acfm}$$

Horsepower Requirements

The total pressure drop (ΔP) between the incinerator and the induced-draft fan is calculated as follows:

$$\Delta P = \Delta P_e + \Delta P_f = (3 + 5 + 6 + 24.4) + 5$$

$$= 43.4 \text{ in. } H_2O$$

where ΔP_e = pressure drop across equipment (in. H_2O)
ΔP_f = pressure drop for valves, fittings, elbows, ductwork, and expansion–contraction losses (in. H_2O)

The brake horsepower (BHP) may be calculated from

$$\text{BHP} = Q_a \Delta P(1.575 \times 10^{-4}) \text{ (fan efficiency)}$$

$$= (18{,}475 \text{ acfm})(43.4 \text{ in. } H_2O)(1.575 \times 10^{-4})/(0.65)$$

$$= 192 \tag{14.2.15}$$

The BHP represents an energy consumption rate, which is converted to kilowatts by multiplying by 0.746. The product of power in kilowatts and the number of hours of usage gives the energy consumed in kilowatt hours (kW-h). Utility companies usually charge on the basis of kilowatt-hours.

Economic Considerations

The last chapter showed that the total capital investment includes *direct costs* (e.g., cost for process equipment) and *indirect* costs (e.g., cost for contingencies). The total capital investment is recovered over the operating lifetime of the facility. The amount recovered is determined from a *present worth analysis.* However, the process of analyzing direct and indirect costs and after-tax cash flows is complicated. This was also shown in the last chapter. A present worth analysis will not be performed here. The TCC, AOC and MC will be used to calculate the TAC through the use of the CRF.

The CRF is given by Eq. (13.4.1)

$$\text{CRF} = \frac{i(1 + i)^n}{(1 + i)^n - 1} \tag{13.4.1}$$

where i = interest rate (fraction)
n = equipment life (years)

The annualized capital cost (ACC) is then

$$\text{ACC} = (\text{TCC})(\text{CRF}) = 4{,}000{,}000(1 + 0.08)^{25}(0.08)/[(1 + 0.08)^{25} - 1]$$
$$= \$374{,}715/\text{yr}$$

The TAC is given by

$$\text{TAC} = \text{ACC} + \text{AOC} + \text{MC} = 374{,}715 + 800{,}000 + 100{,}000$$
$$= \$1.275 \times 10^6/\text{yr}$$

14.3 ILLUSTRATIVE DESIGN PROBLEM II

Several chemical plants located in an industrial area are generating chlorinated wastes in solid and liquid form. The design and construction of an HWI facility has been proposed. An engineering construction firm is in the process of preparing a preliminary design of a multipurpose incinerator. Data on a somewhat similar system indicate that an operating temperature in the secondary combustion chamber (SCC) of 1800–2400°F range with 75–125% excess air (to help ensure adequate mixing and sufficient oxygen) would ensure a 99.99% DRE for this waste. Consultants recommend that a rotary kiln (L/D = 4) with an SCC (L/D = 4) be installed in order to treat both solid and liquid wastes. Preheated combustion air at 550°F is to be employed. The minimum excess air requirement calculation for both the kiln and the SCC should be based on an operating temperature of 1800°F in the kiln and 2000°F in the SCC.

The combustion gases are to be quenched in a spray tower from 2000 to 180°F, which is approximately the adiabatic saturation temperature of the flue gas produced. During the quenching operation, it is assumed that no absorption of HCl occurs.

Water, entering a packed absorption tower, is used to absorb HCl from the gas stream. The operating temperature and pressure are 120°F and 1 atm, respectively. The tower is packed with 2 in. ceramic Raschig rings.

The proposed preliminary design does not call for any particulate control equipment at this time. Only trace amounts of particulates and ash are anticipated, and it is assumed that some of these will be captured in the quencher and, to a lesser degree, in the absorber.

The pressure drop across the system including valves, elbows, fittings, and so on has been estimated to be 11 in. H_2O. According to the fan manufacturer, the overall fan-motor efficiency is 0.65.

In a preliminary cost analysis, the AOC, TCC, and MC were estimated to be \$600,000, \$3,100,000, and \$80,000 per year, respectively.

Waste characteristic data and additional system information are given here. (The notation used is that of the original EPA worksheet. See Table 6.3.2.)

Waste feed flow rate to the kiln, $m_1 + m_2 + m_3 = 5000$ lb/h
Fraction of the solid waste to the kiln, $n_1 = 0.56$ lb solid/lb feed_k
Fraction of the liquid waste to the kiln, $n_2 = 0.24$ lb liquid/lb feed_k
Fraction of the wastewater to the kiln, $n_3 = 0.20$ lb wastewater/lb feed_k
Where $_k$ = kiln waste feed
Waste feed flow rate to the SCC, $m_4 = 2500$ lb/h

Mass fractions of the waste feed stream to the kiln:

	Solids (1)	Liquid (2)	Wastewater (3)	
Carbon (C)	0.70	0.30	0.0	lb/lb waste_k
Hydrogen (H)	0.10	0.11	0.0	lb/lb waste_k
Moisture (H_2O)	0.03	0.42	1.0	lb/lb waste_k
Oxygen (O)	0.12	0.10	0.0	lb/lb waste_k
Chlorine (Cl)	0.05	0.07	0.0	lb/lb waste_k
NHV	14,262	9,067	0	Btu/lb_k

Mass fractions of the liquid waste feed to the SCC:

Carbon (C_a)	0.28	lb/lb waste_a
Hydrogen (H_a)	0.15	lb/lb waste_a
Moisture (H_2O_a)	0.40	lb/lb waste_a
Oxygen (O_a)	0.08	lb/lb waste_a
Chlorine (Cl_a)	0.09	lb/lb waste_a
NHV_a	10,602	Btu/lb_a

where a = SCC waste feed.

The HWI facility is to be designed using preliminary process design principles. Calculations should include the following:

- Air requirements
- Combustion calculations
- Incinerator design
- Coolant flow rate for quench operation
- Quench tower design
- Packed tower absorber design
- Horsepower requirements
- Total annualized costs (TAC) based on a 30-year lifetime at a 10% interest rate.

SOLUTION

As indicated in the introduction, this example will be solved in part by using the EPA worksheets presented in tabular form in Chapter 6.

Note: The net heating values of these wastes are high enough so that no auxiliary fuel is required.

Stoichiometric Air and Combustion Calculations

Calculate the composition of each component of the waste feed to the kiln (see Table 6.3.2). The mass fraction of each component is given here:

$$n_C = (n_1C_1 + n_2C_2 + n_3C_1) = (0.56)(0.70) + (0.24)(0.30) + (0.2)(0.0)$$

$$= 0.464 \text{ lb/lb feed}_k$$

$$n_H = (0.56)(0.10) + (0.24)(0.103) + (0.2)(0.0) = 0.08072 \text{ lb/lb feed}_k$$

$$n_{H_2O} = (0.56)(0.03) + (0.24)(0.42) + (0.2)(1.0) = 0.3176 \text{ lb/lb feed}_k$$

$$n_O = (0.56)(0.12) + (0.24)(0.10) + (0.2)(0.0) = 0.0912 \text{ lb/lb feed}_k$$

$$n_{Cl} = (0.56)(0.05) + (0.24)(0.07) + (0.2)(0.0) = 0.0448 \text{ lb/lb feed}_k$$

Subscripts 1, 2, and 3 refer to the solid, liquid, and wastewater, respectively.
The stoichiometric oxygen requirement for the kiln can now be calculated:

$$(O_2)_{\text{stoich}} = n_c(2.67 \text{ lb } O_2/\text{lb C}) + [n_H - (n_{Cl})/35.5](8 \text{ lb } O_2/\text{lb H}) - n_O$$

$$= (0.464)(2.67) + [0.0872 - (0.0448)/35.5](8) - 0.0912$$

$$= 1.7835 \text{ lb } O_2/\text{lb feed}_k$$

The next step is to determine the mass fraction of each component in the combustion gas per pound of kiln feed.

$$N_{CO_2} = n_C(3.67 \text{ lb } CO_2/\text{lb C}) = 0.464(3.67)$$
$$= 1.7029 \text{ lb } CO_2/\text{lb feed}_k$$
$$N_{H_2O} = n_H(9.09 \text{ lb } H_2O/\text{lb H}) + n_{H_2O} = (0.08072)(9.0) + (0.3176)$$
$$= 1.0444 \text{ lb } H_2O/\text{lb feed}_k$$
$$N_{N_2} = (O_2)_{\text{stoich}}\ (3.29 \text{ lb } N_2/\text{lb } O_2) = (1.7835)(3.29)$$
$$= 5.8677 \text{ lb } N_2/\text{lb feed}_k$$
$$N_{HCl} = n_{Cl}(1.03 \text{ lb HCl/lb Cl}) = (0.0448)(1.03)$$
$$= 0.046 \text{ lb HCl/lb feed}_k$$

The stoichiometric combustion products (CP) per pound of kiln feed are obtained by the following simple relationship:

$$CP_{kSt} = N_{CO_2} + N_{H_2O} + N_{N_2} + N_{HCl} = 8.6606 \text{ lb/lb kiln feed}$$

Rotary Kiln Excess Air Calculation

Calculate the enthalpy input to the kiln using an air preheat temperature of 550°F (see Table 6.4.6):

$$\Delta H_1' = 1.115(T_{\text{air}} - 77)(O_2)_{\text{stoich}} = 1.12(550 - 77)(1.7835)$$
$$= 940 \text{ Btu/lb feed}_k$$

The maximum heat generated in the kiln from the combustion of the waste is

$$\Delta H_2 = n_1(\text{NHV}_1) + n_2(\text{NHV}_2) + n_3(\text{NHV}_3)$$
$$= 0.56(14{,}262) + 0.24(9067) + 0.20(0) = 10{,}163 \text{ Btu/lb feed}_k$$

Assuming a 2% loss, the heat loss through the walls of the kiln is estimated as

$$q = 0.02(\Delta H_2) = 0.02(10{,}163) = 203 \text{ Btu/lb feed}_k$$

The enthalpy of the combustion products leaving the kiln is

$$
\begin{aligned}
\Delta H_3 &= 0.26(N_{CO_2} + N_{N_2} + N_{HCl})(T - 77) + 0.49(N_{H_2O})(T - 212) \\
&\quad + N_{H_2O}(1150) \\
&= 0.26(1.703 + 5.867 + 0.046)(1800 - 77) + 0.49(1.044)(1800 - 212) \\
&\quad + (1.044)(1150) \\
&= 5424.8 \text{ Btu/lb kiln feed}
\end{aligned}
$$

The enthalpy of excess air leaving the kiln is given by

$$\Delta H_4' = 1.115(T - 77)(O_2)_{\text{stoich}} = 1.115(1800 - 77)(1.7835) = 3426 \text{ Btu/lb feed}_k$$

The percentage of excess air is

$$
\begin{aligned}
\text{EA} &= 100(\Delta H_1' + \Delta H_2 - q - \Delta H_3)/(\Delta H_4' - \Delta H_1') \\
&= 100(940 + 10{,}163 - 203 - 5424.8)/(3426 - 940) \\
&= 220.24
\end{aligned}
$$

The amount of nitrogen and oxygen per pound of kiln feed added to the stoichiometric combustion products, CP_{kSt} may now be calculated:

$$
\begin{aligned}
N_{2\,\text{EA}} &= \text{EA} \times (N_{N_2}) \\
&= 2.20 \times 5.867 = 12.93 \\
O_{2\,\text{EA}} &= (N_{N_2})/3.29 \\
&= 12.93/3.29 = 3.93
\end{aligned}
$$

The total combustion products per pound kiln feed exiting the kiln are CP_T:

$$
\begin{aligned}
\text{CP}_\text{T} &= \text{CP}_{\text{kSt}} + N_{2\,\text{EA}} + O_{2\,\text{EA}} \\
&= 8.6606 + 12.93 + 3.93 \\
&= 25.52
\end{aligned}
$$

The total mass flow of kiln combustion products is

$$
\begin{aligned}
W_{\text{kT}} &= \text{CP}_\text{T} \times (m_1 + m_2 + m_3) \\
&= 25.52 \text{ lb comb. products (total)/lb kiln feed} \times 5000 \text{ lb/h (kiln feed)} \\
&= 127{,}604 \text{ lb/h}
\end{aligned}
$$

Calculate the composition of each component of the waste feed to the SCC. The mass fraction of each component is given here:

$$n_{C,a} = (C)_4 = 0.28 \text{ lb/lb feed}_a$$

$$n_{H,a} = (H)_4 = 0.15 \text{ lb/lb feed}_a$$

$$n_{H_2O,a} = (H_2O)_4 = 0.40 \text{ lb/lb feed}_a$$

$$n_{O,a} = (O)_4 = 0.08 \text{ lb/lb feed}_a$$

$$n_{Cl,a} = (Cl)_4 = 0.09 \text{ lb/lb feed}_a$$

Subscript 4 refers to the components in the liquid feed to the SCC.

The stoichiometric oxygen requirement for the SCC($_a$) can now be determined:

$$\begin{aligned}(O_2)_{stoich,a} &= n_{C,a}(2.67 \text{ lb } O_2/\text{lb C}) + [n_{H,a} - (n_{Cl,a})/35.5](8 \text{ lb } O_2/\text{lb H}) - n_{O,a} \\ &= 0.28(2.67) + [(0.15) - (0.09)/35.5](8) - 0.080 \\ &= 1.85 \text{ lb } O_2/\text{lb feed}_a\end{aligned}$$

It is now possible to calculate the mass fraction of each component in the gas per pound of SCC feed:

$$\begin{aligned}N_{CO_2,a} &= n_{C,a}(3.67 \text{ lb } Co_2/\text{lb C}) = 0.28(3.67) \\ &= 1.03 \text{ lb } CO_2/\text{lb feed}_a \\ N_{H_2O,a} &= n_{H,a}(9.0 \text{ lb } H_2O/\text{lb H}) + n_{H_2O,a} = (0.15)(9) + (0.40) \\ &= 1.75 \text{ lb } H_2O/\text{lb feed}_a \\ N_{N_2,a} &= (O_2)_{stoich,a}(3.29 \text{ lb } N_2/\text{lb } O_2) = (1.85)(3.29) \\ &= 6.087 \text{ lb } N_2/\text{lb feed}_a \\ N_{HCl,a} &= n_{Cl,a}(1.03 \text{ lb HCl/lb Cl}) = (0.09)(1.03) \\ &= 0.093 \text{ lb HCl/lb feed}_a\end{aligned}$$

The stoichiometric combustion products per pound of SCC feed (CP_a) is determined here:

$$CP_a = N_{CO_2,a} + N_{H_2O,a} + N_{N_2,a} + N_{HCl,a} = 8.96 \text{ lb/lb feed}_a$$

The ratio of total SCC feed to total kiln feed (n_{ak}) is next determined:

Solid waste to kiln: $m_1 = (0.56)(5000) = 2800$ lb/h
Liquid waste to kiln: $m_2 = (0.24)(5000) = 1200$ lb/h
Wastewater to kiln: $m_3 = (0.20)(5000) = 1000$ lb/h

Liquid waste to SCC: $m_4 = 2500$ lb/h

$$n_{ak} = m_4/(m_1 + m_2 + m_3) = 2500/(2800 + 1200 + 1000)$$
$$= 0.50 \text{ lb afterburner feed/lb kiln feed}$$

Calculate the total combustion gas flow based on the stoichiometric oxygen requirements:

$$N_{CO_2,t} = (N_{CO_2,k} + N_{CO_2,a}n_{ak})/(1 + n_{ak})$$
$$= 1.5686 \text{ lb/lb feed}_t$$
$$N_{H_2O,t} = (N_{H_2O,k} + N_{H_2O,a}n_{ak})/(1 + n_{ak})$$
$$= 1.28 \text{ lb/lb feed}_t$$
$$N_{N_2t} = (N_{N_2,k} + N_{N_2,a}n_{ak})/(1 + n_{ak})$$
$$= 5.94 \text{ lb/lb feed}_t$$
$$N_{HCl,t} = (N_{HCl,k} + N_{HCl,a}n_{ak})/(1 + n_{ak})$$
$$= 0.062 \text{ lb/lb feed}_t$$

where subscript t refers to total feed in kiln and SCC.

The total stoichiometric combustion products out of the SCC are the sum of

$$CP_t = 1.5686 + 1.28 + 5.94 + 0.062$$
$$CP_{tSt} = 8.85 \text{ lb/lb feed}_t$$

SCC Excess Air Calculation

Calculate the enthalpy input to the kiln and SCC using an air preheat temperature of 550°F (see Table 6.4.6).

$$\Delta H_1' = 1.115(T_{air} - 77)[(O_2)_{stoich,k} + (O_2)_{stoich,a}(n_{ak})]/(1 + n_{ak})$$
$$= 1.115(550 - 77)(1.7835 + (1.85)(0.5))/(1.5)$$
$$= 952 \text{ Btu/lb feed}_t$$

The heat generated in the kiln and SCC by combustion of the total waste feed is given by

$$\Delta H_2 = [\Delta H_k + n_{ak}\Delta H_{2,a}](1 + n_{ak})$$

For this equation,

$$\Delta H_k = 10{,}163 \text{ Btu/lb}_k$$

$$\Delta H_{2,a} = \text{NHV}_a = 10{,}602 \text{ Btu/lb}_a$$

$$\Delta H_2 = [10{,}163 + (0.5)(10{,}602)]/(1 + 0.5)$$

$$= 10{,}309 \text{ Btu/lb feed}_t$$

Assuming a 2% loss, the heat loss through the walls of the kiln and the SCC is

$$q = 0.02(\Delta H_2) = 206 \text{ Btu/lb feed}_t$$

The enthalpy of the combustion products leaving the SCC is given by

$$\Delta H_3 = 0.26(N_{CO_2} + N_{N_2} + N_{HClt})(T - 77) + 0.49N_{H_2O}(T - 212)$$

$$+ N_{H_2O}(1150)$$

$$= 6378 \text{ Btu/lb feed}$$

The enthalpy of the excess air leaving the SCC is

$$H_4' = 1.115(T - 77)[(O_2)_{\text{stoich}} + (O_2)_{\text{stoich,a}}(n_{ak})]/(1 + n_{ak})$$

$$= 1.115(2000 - 77)(1.7835 + (1.85)(0.5))/(1.5)$$

$$= 3870 \text{ Btu/lb feed}$$

The excess air percentage in the afterburner is

$$\text{EA} = 100(\Delta H_1 + \Delta H_2 - q - \Delta H_3)/(\Delta H_4' - \Delta H_1)$$

$$\text{EA} = 100(952 + 10{,}309 - 206 - 6378)/(3{,}870 - 952)$$

$$= 160\%$$

This represents the average excess air feed to the kiln–SCC per unit mass of feed. The calculations to follow will be based on 160% excess air. It is now possible to calculate the additional N_2 and O_2 present in the combustion gases due to excess air feed.

$$(O_2)_{EA} = EA[(O_2)_{stoich} + n_{ak}(O_2)_{stoich,a}]/(1 + n_{ak})$$
$$= 1.60[(1.7835) + (0.5)(1.85)]/(1.5)$$
$$= 2.88 \text{ lb } O_2/\text{lb feed}_t$$
$$(N_2)_{EA} = 2.88(3.29 \text{ lb } N_2/\text{lb } O_2)$$
$$= 9.49 \text{ lb } N_2/\text{lb feed}_t$$

The total combustion gas flow per pound of total feed is

$$CG_t = CP_t + (O_2)_{EA} + (N_2)_{EA} = 8.85 + 2.88 + 9.49$$
$$= 21.22 \text{ lb combustion products total per lb feed}_t$$

The total mass flow of combustion gases exiting the SCC is

$$CG_t \times m_{tf} = 21.22 \text{ lb}_t/\text{lb}_{tf} \times 7500 \text{ lb}_{tf}/\text{h}$$
$$= 159{,}194 \text{ lb/h}$$

The mass flow of each combustion gas component follows in Table 14.3.1.

Table 14.3.1 Flue gas combustion[a] (Illustrative Problem II)

Component	Mass Flow (lb/h)	Molar Flow (lbmol/h)	Mole Fraction		Partial Pressure (atm)
			Wet	Dry	
CO_2	11,764	267	0.047	0.052	0.047
H_2O	9,600	533.3	0.095		0.095
HCl	465	12.7	0.002	0.002	0.002
N_2	115,725	4,133	0.735	0.812	0.735
O_2	21,622	676	0.12	0.133	0.12
Total, wet	159,176	5,622	1.00	1.00	1.00
Total, dry	149,576	5,088			

[a]This neglects any chlorine (Cl_2) formation.

Molecular weight of wet gases = (total, wet lb/h)/(total wet lbmol/h) = 28.3
Molecular weight of dry gases = (total, dry lb/h)/(total dry lbmol/h) = 29.4
Density, ρ_w = M.W. wet/379 ft^3 per lbmol = 28.3/379 = 0.0747 lb/ft^3
Density, ρ_d = M.W. dry/379 ft^3 per lbmol = 29.40/379 = 0.0757 lb/ft^3

Mass Fraction

%CO_2: $N_{CO_2,t}/CG_t = 0.074$
%H_2O: $N_{H_2O,t}/CG_t = 0.0603$
%N_2: $[N_{N_2,t} + (N_2)_{EA}]/CG_t = 0.727$

$\%O_2$: $(O_2)_{EA}/CG_t = 0.136$
$\%HCl$: $N_{HCl,t}/CG_t = 0.003$

Table 14.3.1 gives combustion gas quantities assuming a pressure of 1 atm.

The volumes of the combustion products per pound of feed at standard conditions of 68°F and 1 atm are

CO_2: $(N_{CO_2,t})/0.114$ lb/scf = 13.7 scf/lb feed
H_2O: $(N_{H_2O,t})/0.0467$ lb/scf = 27.4 scf/lb feed
N_2: $[N_{N_2,t} + (N_2)_{EA}]/0.0727$ lb/scf = 212.24 scf/lb feed
O_2: $(O_2)_{EA}/0.083$ lb/scf = 34.7 scf/lb feed
HCl: $(N_{HCl,t})/0.0945$ lb/scf = 0.656 scf/lb feed
Total volume per lb feed = 288.75 scf/lb feed
Total volumetric flow rate (Q_s) = (7500 lb/h)(288.75 scf/lb feed) = 2.1656×10^6 scfh

The air requirement for the kiln feed (AR_k) is

$$AR_k = (O_2)_{stoich}(4.29 \text{ lb air/lb } O_2)(\text{EA ratio} + 1.0) = 1.7835(4.29)(3.20)$$
$$= 24.48 \text{ lb air/lb kiln feed}$$
$$AR_{k,t} = 24.48 \text{ (5000 lb kiln feed/h)}$$
$$= 122{,}419 \text{ lb air/h}$$

The total air requirement for both the kiln and afterburner (AR) is

$$AR = (O_2)_{stoich}(4.29 \text{ lb air/lb } O_2)(\text{EA ratio} + 1.0) = (1.805)(4.29)(2.60)$$
$$= 20.13 \text{ lb air/lb total kiln and SCC feed}$$
$$AR_t = 20.13(7500 \text{ lb/h}) = 150{,}997 \text{ lb air/h}$$

Approximately 28,578 lb air/h is added to the SCC.

Incinerator Design

The rate of heat released by the waste in the kiln is given as

$$q_k = 5000[0.56(\text{NHV}_1) + 0.24(\text{NHV}_2) + 0.20(0.0)]$$
$$= 51 \times 10^6 \text{ Btu/h} \tag{14.3.1}$$

Assuming a typical heat release rate (q_H) of 15,000 Btu/h-ft^3, the volume of the kiln is calculated.

$$V_k = q/q_H = 51 \times 10^6/15{,}000 = 3400 \text{ ft}^3 \tag{14.3.2}$$

The diameter (D_k) of the kiln can be determined using an L/D ratio of 4:

$$V_k = \pi D_k^2 L/4 = \pi D_k^3$$

$$D_k = (V_k/\pi)^{1/3} = (3400/\pi)^{1/3} = 10.24 \text{ ft} \cong 10.5 \text{ ft} \tag{14.3.3}$$

The length of the kiln is $L_k = 4(10.5) = 42$ ft. The corrected volume is

$$V_k = \pi(10.5)^2(42)/4 = 3636 \text{ ft}^3 \tag{14.3.4}$$

The heat released from the SCC feed is

$$q_a = (2500 \text{ lb/h})(10{,}602 \text{ Btu/lb}) = 26.5 \times 10^6 \text{ Btu/h}$$

Residence Time Calculation

The actual combustion gas flow rate is calculated using Charles' law:

$$Q_a = Q_s(T_a/T_s) = 2.16 \times 10^6(2000 + 460)/(60 + 460)$$

$$= 10.22 \times 10^6 \text{ acfh} = 2838 \text{ acfs} \tag{14.3.5}$$

The residence time (θ) typically required in a SCC is 2 s:

$$\theta = V_a/Q_a = \mathrm{V}_a/2838$$

$$V_a = 2 \times 2838 = 5676 \text{ ft}^3$$

Using the L/D ratio of 4, the length and diameter of the SCC are

$$D_a = [V_a/\pi]^{1/3} = [5676/\pi]^{1/3} = 12.15 \text{ ft} \cong 12 \text{ ft 3 in.} \tag{14.3.6}$$

Using 12.25 ft as the *design* diameter,

$$L_a = 4(12.25) = 49 \text{ ft}$$

$$\text{New } V_a = 5775 \text{ ft}^3$$

Quench Spray Tower Design

The quench water requirement is first calculated. For combustion gases being cooled, the quench water requirement may be approximated by[1]

$$m_{H_2O} = 1.0 m_{CG} \tag{14.3.7}$$

where m_{H_2O} and m_{CG} are the water and dry flue gas flow rates in lbmol/h (see Table 14.3.1).

The dry flue gas rate has been determined to be 5088 lbmol/h. Therefore m_{H_2O} can be obtained:

$$m_{H_2O} = (1.0)5088 = 5088 \text{ lbmol/h}$$

A simplified, reasonably accurate calculational procedure for designing a water quench spray tower is available.[1] The area can be calculated as follows:

$$A = 0.15 w_{in}/\rho \quad \text{or} \quad 0.15 Q_a \tag{14.3.8}$$

where w_{in} = total mass flow of hot gases entering quench (lb/s)
ρ = hot gas density (lb/ft^3)
A = tower cross-sectional area (ft^2)
Q_a = combustion gas flow rate in acfs from SCC

The diameter is therefore

$$D = (4A/\pi)^{0.5}$$

and the tower height is

$$Z = 2.5D \quad \text{or} \quad V = 2D^3$$

The hot gas density will be taken from Table 14.3.1 with MW = 28.3.
At 1 atm and 2000°F,

$$\rho = 28.3(14.7)/[(2000 + 460)(10.73)] = 0.01576 \text{ lb/ft}^3$$

The quench tower area, diameter, and height are calculated as follows:

$$A = 0.15(2838 \text{ acfs}) = 425.7 \text{ ft}^2$$

$$D = [(4)(425.7)/\pi]^{0.5} = 23.28 \text{ ft}$$

$$Z = 2.5(23.28) = 58.2 \text{ ft}$$

Absorber Design

Since equilibrium data are not available, it is assumed that the slope of the equilibrium curve (m) approaches zero and that Eq. (10.2.12a) is applicable:

$$N_{OG} = \ln(w_1/w_2) \tag{10.2.12a}$$

where w_1 = inlet mass flow rate of HCl
w_2 = outlet mass flow rate of HCl

From Table 14.3.1, the inlet mass flow of HCl is 465 lb/h. First determine the outlet mass flow of HCl assuming that the removal efficiency is 99%:

$$E = (w_1 - w_2)/w_1$$

$$w_2 = 465 - 0.99(465) = 4.65 \text{ lb/h of HCl}$$

The number of gas transfer units via Eq. (10.2.12a) is

$$N_{OG} = \ln(465/4.65) = 4.605$$

From data supplied by the manufacturer, the H_{OG} value for water systems is approximately 3.0 ft for 2 in. ceramic packing. Assuming a safety factor of 1.35, the packed height is calculated from Eq. (10.2.11)

$$Z = (N_{OG})(H_{OG})$$

adding a safety factor of 1.35,

$$Z = (N_{OG})(H_{OG})(\text{safety factor}) = 4.605(3.0)(1.35)$$

$$= 18.6 \text{ ft} \qquad (10.2.11)$$

Using a typical superficial throughput velocity of 6 ft/s, the cross-sectional area of the tower can be estimated. It is desired to determine the actual process gas flow rate at the operating temperature of the absorber, 120°F. With an HCl removal efficiency of 99%, the number of moles in the gas stream decreases from 5622. However, there is also a slight increase in the water content and flow rate for saturated and cooled conditions at 120°F. The flue gas water content is approximately 0.0642 lb of H_2O/lb of dry gas (see Table 14.3.1, 9600 lb/h H_2O per 149,576 lb/h wet gases) compared to the saturated humidity at 120°F of 0.081 lb of H_2O/lb of dry air. Therefore, assume the molar flow rate of flue gas at the absorber outlet to be 5761 lbmol/h. Assuming ideal gas behavior, the actual cubic feet per minute is then

$$Q_a = \frac{(5761)(10.73)(460 + 120)}{(14.7)(60)}$$

$$= 40{,}649 \text{ acfm}$$

Horsepower Requirements

The BHP may be calculated from

$$\text{BHP} = Q_a(\text{acfm})\Delta P(\text{in. } H_2O)(1.575 \times 10^{-4})/(\text{fan efficiency})$$

$$= 40{,}649(11)(1.57 \times 10^{-4})/0.65 = 107.5 \text{ hp} \qquad (14.3.9)$$

Economic Considerations

The ACC is given by

$$\text{ACC} = (\text{TCC})(\text{CRF})$$

$$\text{CRF} = \frac{(i)(1 + i)^n}{(1 + i)^n - 1}$$

$$\begin{aligned}\text{ACC} &= 3{,}100{,}000[(1 + 0.1)^{30}0.1]/[(1 + 0.1)^{30} - 1] \\ &= \$328{,}846/\text{yr} \qquad (13.4.1)\end{aligned}$$

The TAC is now determined.

$$\begin{aligned}\text{TAC} &= \text{ACC} + \text{AOC} + \text{MC} \\ &= 328{,}846 + 600{,}000 + 80{,}000 = \$1.01 \times 10^6/\text{yr} \qquad (14.3.10)\end{aligned}$$

PROBLEMS

1. Resolve Illustrative Design Problem I (Section 14.2) if the chlorobenzene is combusted with 100% excess air.

2. The construction of a hazardous waste incineration facility has been proposed. Using the approaches employed in the EPA worksheets and simple plant design principles, prepare a preliminary design of a multipurpose incineration facility fueled with natural gas that is to be operated at 2050°F with 100% excess air. Since the incinerator must be able to handle solid waste and/or liquid waste, a rotary kiln with an afterburner has been recommended. Calculations should include:

- Air requirements
- Fuel requirements
- Incinerator design
- Retention time
- Coolant flow requirement for combustion gas quench–absorber operation
- Auxiliary equipment
  - Absorption tower
  - Horsepower requirements
- Annualized costs (25 year lifetime at 8% interest rate)

Waste characteristic data and system information are given here:

- Waste feed flow rate to the kiln, $m_1 + m_2 = 4000$ lb/h

- Fraction of the solid waste to the kiln, n_1 = 0.7 lb solid/lb feed
- Fraction of the liquid waste to the kiln, n_2 = 0.3 lb liquid/lb feed
- Waste feed flow rate to the afterburner, m_3 = 1500 lb/h

Mass fractions of the waste feed stream to the kiln:

	Solids (1)	Liquid (2)	
Carbon (C)	0.70	0.30	lb/lb waste
Hydrogen (H)	0.10	0.11	lb/lb waste
Moisture (H_2O)	0.03	0.42	lb/lb waste
Oxygen (O)	0.12	0.10	lb/lb waste
Chlorine (Cl)	0.05	0.07	lb/lb waste

Mass fractions of the liquid waste feed to the afterburner:

Carbon (C_a)	0.28 lb/lb waste
Hydrogen (H_a)	0.15 lb/lb waste
Moisture (H_2O_a)	0.40 lb/lb waste
Oxygen (O_a)	0.08 lb/lb waste
Chlorine (Cl_a)	0.09 lb/lb waste

Kiln solid waste heating value, NHV_1 = 14,300 Btu/lb

Kiln liquid waste heating value, NHV_2 = 9150 Btu/lb

Afterburner waste heating value, NHV_3 = 10,600 Btu/lb

Fuel-to-waste ratio for the kiln and afterburner, n_f = 0.0 lb/lb

Excess air for the kiln and afterburner = 100%

Length-to-diameter ratio of the kiln = 4.0

Length-to-diameter ratio of the afterburner = 3.0

Recommended operating temperature during the incineration 2050°F in kiln and SCC

Minimum retention time across both incineration units = 0.75 s

Required removal efficiency of chlorine = 99.99%

Estimated pressure drop across the system = 11.0 in. H_2O

Pressure requirement of combustion air for kiln = 6.0 in. H_2O

Pressure requirement of combustion air for afterburner = 5.0 in. H_2O

Absorber description:

Packed, 2 in.-diameter ceramic Raschig rings

Operating temperature = 128°F

Enthalpy of combustion gas entering and leaving quench–absorber unit = 625 and 130 Btu/lb, respectively

Coolant water temperature entering and leaving quench–absorber unit = 106 and 177°F

Overall fan-motor efficiency = 65%
Estimated AOC $\$6 \times 10^5$/yr
Estimated TCC $\$3.1 \times 10^6$
Estimated MC = $\$8 \times 10^4$/yr

REFERENCES

1. L. Theodore, personal notes, 2000.
2. V. Ganapathy, "Size or Check Waste Heat Boilers Quickly," *Hydrocarbon Processing*, 169–170, Sept. (1984).
3. H. E. Hesketh, "Atomization and Cloud Behavior in Wet Scrubbers," US–USSR Symposium on Control of Fine Particulate Emissions, Jan. 15–18, 1974.

APPENDIX A

HWI SOFTWARE PACKAGE

A.1 INTRODUCTORY COMMENTS

An incinerator software package titled *HWI* was recently developed by two of the authors of this text under a project sponsored by the USEPA's Air Pollution Training Institute. The objectives of the project were to develop rigorous calculational procedures for the design/analysis of incinerators and to incorporate these procedures into an easy-to-use computer program. Athough written originally for hazardous waste incinerators, the package can be used for essentially any type of incinerator.

What Is the *HWI Software Package*?

The *HWI Software Package* is contained on the disk (sometimes referred to as a *floppy disk*) that accompanies this book. This package consists of three programs and two data files. The three programs are briefly described below and are discussed in detail in Sections A.3, A.4 and A.5. Two of the programs, *HWI* and *HWITRL*, serve as the basis for the USEPA Air Pollution Training Institute's self-instructional course, SI-458. The materials for this course are provided in a file named SETUP, which is also contained on the floppy disk and is provided as part of this package with permission of the Air Pollution Training Institute. The setup program also allows easier installation of the *HWI* and *HWITRL* programs to the hard drive.

The three programs that make up the HWI software package are:

1. *HWI* This program performs the hazardous waste incinerator calculations. Communication between you and the program is handled in interactive fashion; all data are inputted in response to prompts that appear

on the monitor screen during execution of the program. The data file HWI.DAT supplies required information to HWI.

2. *HWITRL* The "TRL" stands for tutorial. The purpose of HWITRL is to familiarize the user with the *HWI* program. The solutions to two problems, as they would be solved using *HWI*, are simulated in step-by-step fashion. Some explanation of the theory behind the calculations is also presented on the monitor during the execution of HWITRL.
3. *HWIF* This program is an expanded version of HWI and allows for waste blending, a secondary combustion chamber and quenching of the gas products. The data file HWIF.DAT supplies required information to HWIF.

What Type of Computer Is Required to Run *HWI*?

For the HWI software package to run on your system, your computer must be an IBM-compatible machine operating under Windows 9X or Windows 3.X. The package may be run from either the hard disk or from the HWI disk. Directions for running HWI using either method are given in Section A.2.

A.2 GETTING STARTED

Directions are given below for (A) installing HWI on a hard drive and (B) for working with the floppy disk only.

A. Program Installation on a Hard Drive

The HWI software is contained on a single 3.5-inch floppy disk. Running the SETUP program installs the HWI and HWITRL programs into a subdirectory named HWI (or some other name you may wish to employ) on the hard drive. The HWIF program and its accompanying data file, HWIF.DAT, can simply be copied from the floppy disk onto the hard drive.

To install HWI and HWITRL on any Windows 9X machine:

1. Insert the disk.
2. Select "Run" from the Start Menu.
3. Type "[floppy drive letter]:setup."

To install HWI and HWITRL on any Windows 3.X machine:

1. Insert the disk.
2. Select "Run" from the File Menu.
3. Type "[floppy drive letter]:setup."

The setup utility will begin automatically. This utility installs the HWI program files to your hard drive. It also modifies the CONFIG.sys file of your system, adding the line "device := C:\WINDOWS\SYSTEM\ANSI.sys" where "C:\WINDOWS\SYSTEM\" is the Windows System Directory. This modification will not affect any other software you may be running, and you will be given the option to back up your current CONFIG.sys file during the installation.

To use the setup utility, follow the on-screen instructions. The location of the files is optional—the default is "c:\program files\HWI" in Win9X, or c:\HWI in Win 3.X. If you wish to save the files to another location, you will need to specify another path.

After the install program is completed, you will need to restart your computer. This will be done automatically upon exit from the setup utility.

An uninstall utility is included with the HWI software. If you wish to remove the programs from your hard drive, DO NOT erase the files.

Win9X uninstall Double click on My Computer, the Control Panel, then Add/Remove Programs. Select "HWI" from the program list, and click remove.

Win3.X uninstall Double click on the HWI group, then the Uninstall icon.

To install *HWIF* on your hard drive, copy the HWIF.EXE and HWIF.DAT files from the floppy disk into the same subdirectory that you are using for the HWI and HWITRL programs.

B. Working from the Floppy Disk

1. Use the Windows Explorer to locate the file "ANSI.sys".
2. Assuming that the ANSI.sys file was found in the C:\WINDOWS\SYSTEM subdirectory, edit your system's CONFIG.sys file to add the statement: "device := C:\WINDOWS\SYSTEM\ANSI.sys". Then reboot your computer.
3. The programs can be run by accessing the floppy drive using the "My Computer" icon, Windows Explorer, or the Windows File Manager (Winfile.exe). Double-clicking on the program icon or name will start the program.

A.3 THE PROGRAM HWITRL

How to Run HWITRL

To run the program in Win 9X:

Go to the Start Menu, choose Programs, HWI, and HWITRL. (Note: HWI represents the HWI "group"; HWITRL represents the program.)

To run the program in Win 3.X:

Double click on the HWI group, then the HWITRL program icon.

If you are working from the floppy disk only, see Section A.2.

Since HWITRL is self-explanatory (i.e., instructions for its use appear on the screen while the program is running), a detailed description of it will not be given here.

What HWITRL Does

The purpose of HWITRL ("TRL" stands for "tutorial") is to familiarize you with the program HWI. The tutorial program does not actually perform any calculations the way HWI does, but it does simulate these calculations, and the results that appear on the monitor screen are the same ones that would be generated by HWI. During the execution of HWITRL, two problems will be presented and then solved. (These two problems are listed at the end of this section.) In the first, you are led in step-by-step fashion through the data input stage and then the calculation stage. During the input phase, you are requested to answer the prompts in much the same way that the input prompts are answered in the HWI program. One difference is that, in HWITRL, the answer to each prompt is checked by the program to make sure it is the correct one. If it is not, the computer responds with a rather irritating noise, and you are again requested to answer the prompt. This process continues until the prompt is answered correctly. This, of course, is not done when HWI is being run. Unlike the tutorial, the HWI program does not "know" whether or not you have inputted correct data.

During the calculational phase of the HWITRL run, the calculations that are "going on" are explained on the monitor screen and many of the intermediate results that would not be displayed by HWI are shown. Once again it should be pointed out that during the running of HWITRL, the "calculations" are not really being performed, but merely simulated.

After observing the "solving" of Problem 1, the second problem is then presented on the screen. This time, the input prompts are not as generous as before. A portion of the problem statement appears before each prompt, and you are expected to extract the necessary information from the statement to answer the prompt. Again, if an error is made, HWITRL will let you know and give you a chance to answer the prompt correctly.

Correcting Input Errors

In order to enter data to HWITRL (or, for that matter, to HWI as well), the typing of the data at the keyboard must be followed by depressing the **RETURN** or **ENTER** key. The data will not be "read" by the program until this key is depressed. This feature gives you a chance to correct the input if you have made an error. To correct or change an input that has just been typed but has not yet been "entered" by depressing the **RETURN** or **ENTER** key, simply hit the **BACKSPACE** key one or more times until the input (or the incorrect portion of it) has been erased. Then type the new input.

If the **RETURN** or **ENTER** key has been depressed before you notice the error, HWITRL lets you know immediately and gives you a chance to correct it by repeating the prompt. In fact, HWITRL will not let you proceed until you enter the data correctly. (Note: If you enter incorrect data during the running of HWI, an error message will be indicated only if it is an obvious error, for example, a letter or other invalid character appearing in what should be a numerical input.)

Fast-Forward, Reverse and Stop

HWITRL is a fairly long program and you may not want to go through the whole program at one sitting. You may also, at times, wish to go back to a previous section to review material presented earlier. The directions for terminating, moving forward, and moving backward through the program follow:

1. *Stopping the Program* If you wish to terminate the program at any time, simply depress the **CTRL** (Control) and **C** keys simultaneously and the DOS prompt (usually **C>** or **A>**) will appear.
2. *Reverse* If you wish to "back up" to some previous point in the program, input a "−1", "−2", or "−3" whenever the following prompt is displayed by HWITRL:

   *** **Depress the RETURN or ENTER key to continue** ***

   This message comes up many times during the execution of HWITRL. Inputting a "−1" will cause the program to return to the beginning of the present section. (A "section" in HWITRL generally, but not always, begins with a heading that is identified by a capital letter, e.g., "*A. Input of Waste Composition and Flow Rate*"; "*B. Calculation of Stoichiometric Air and Input of Percent Excess Air*"; etc.). Inputting a "−2" returns the program to the beginning of the previous section and "−3" to the beginning of the section before that. For example, suppose the frame with the heading: "*E. Calculation of Incinerator Temperature and Flue Gas Composition*", appears on the screen. Inputting a "−1" at this point simply brings up the same frame again, since this frame is the start of the current section. Inputting a "−2" brings up the frame with the heading: "*D. Input of Preliminary Design Parameters*", which is the start of the previous section.
3. *Fast Forward* If you wish to "jump ahead" in the program, as would be the case if you are going through the tutorial in more than one sitting and wish to skip over material already covered, input a "1", "2" or "3" in answer to the prompt:

   *** **Depress the RETURN or ENTER key to continue** ***

The "1" moves the program ahead to the beginning of the next section; "2" to the beginning of the section after that; and "3", to the section after that one. Once again, using the frame with the title: "*E. Calculation of Incinerator Temperature and Flue Gas Composition*" as an example, inputting a "1" brings up the beginning of the next section, the lead frame of which is entitled "*F. Calculation of the Flue Gas Flow Rate*".

Illustrative Problems Solved in HWITRL

The two problems that are presented and "solved" by HWITRL are listed below.

Problem 1 5000 lb/h of a hazardous waste is to be incinerated in a liquid injection unit with 100% excess air. The waste has the following composition:

Element	Mass %
Carbon	60.8
Hydrogen	4.2
Chlorine	30.0
Sulfur	5.0

a. Calculate the flue gas composition and flow rate assuming complete combustion.
b. Calculate the incinerator temperature.
c. Determine the length and diameter of the incinerator. Assume a flue gas superficial throughput velocity of 20 ft/s.

Problem 2 A hazardous waste has the following composition:

Component	Mass %
Chlorobenzene, C_6H_5Cl	58
DDT, $C_{14}H_9Cl_5$	26
Water, H_2O	16

Calculate the flue gas flow rate and composition when 6430 lb/h of the waste is incinerated in a rotary kiln incinerator with 75% excess air. Also calculate the operating temperature and dimensions of the cylindrical housing. Assume an overall *L*/*D* ratio of 4.0 for the kiln/afterburner unit.

From heat of formation data, the following values for the heat combustion (NHV) of the three components can be calculated. These are:

Chlorobenzene	714,361 cal/gmol
DDT	1,580,000 cal/gmol
Water	−10,519 cal/gmol

(The "heat of combustion" listed for water is actually the heat of vaporization. In the incinerator, the water simply vaporizes.)

A.4 THE PROGRAM HWI

What HWI Does

HWI handles all of the calculations involved in the simulation of a hazardous waste incinerator. These calculations may be divided into three main categories:

1. *Thermochemical calculations*, which relate operating incinerator temperature, excess air, and heating value of the feed (including not only waste but also possible fuel requirements)
2. *Stoichiometric calculations*, which yield the composition and flow rates of gaseous emissions from incinerators burning (hazardous) waste
3. *Preliminary incinerator design*

It should be noted that the sequence listed above is not the sequence in which these calculations are actually performed in the HWI program. The calculations of Categories 1 and 2, for example, are interdependent and must be treated in simultaneous fashion.

For What Type of Waste Is HWI Intended?

The waste (or waste–fuel mixture, when auxiliary fuel has been added to enhance the heating value) must be primarily organic and may be liquid, solid, or a combination (sludge, slurry, etc.). It may contain any or all of the following elements: carbon, hydrogen, oxygen, nitrogen, sulfur, chlorine, fluorine, bromine, and iodine. HWI will not recognize any other elements besides these. Inert solid material (designated in this section by the term "ash") may be included, but the amount should not exceed about 5% on a mass basis. It is assumed during the calculations that any ash present in the waste has negligible impact on incinerator performance and all of it leaves the incinerator as particulate matter entrained in the flue gas stream.

How to Run HWI

To run the program in Win 9X: Go to the Start Menu, choose Programs, HWI, and HWI. (Note: The first "HWI" represents the HWI "group"; the second represents the program.)

To run the program in Win 3.X: Double click on the HWI group, then the HWI program icon.

If you are working from the floppy disk only, see Section A.2.

Inputting information to HWI is accomplished at the computer terminal and is handled in question-and-answer fashion. The question or request for a particular input appears on the screen and the user types in the answer. Each typed input must be entered by depressing the **RETURN** or **ENTER** key. Until this is done the program will not "read" the information that was typed.

Most of the inputs are *numerical*; a few (e.g., chemical symbols) are *alphabetical*. Besides numerical digits, a numerical input allows only for a decimal point, a plus (+) sign, or a minus (−) sign. The plus or minus sign must appear as the first character of the number. Minus signs are mandatory for negative numbers; plus signs are optional for positive numbers. Decimal points are also optional if the number is an integer: **516.0** may simply be input as **516**, for example. Any other non-numerical character will cause an error message to come up on the monitor screen. The number **71,600**, for instance, must be typed without the comma, that is, as **71600**. Units such as % or lb/hr are never entered. If you type a number with an illegal character in it, HWI will respond by repeating the input prompt, thereby giving you a chance to type the number correctly.

The only *alphabetical* inputs to *HWI* are the chemical symbols of the elements. The symbols "recognized" by HWI are:

C for carbon
H for hydrogen
O for oxygen
Cl for chlorine
S for sulfur
N for nitrogen
F for fluorine
Br for bromine
I for iodine
WA for water
AS for ash

The last 2 are not standard element symbols; their use will be explained later. When any of these 11 symbols is typed as an input to HWI, it must be typed exactly as it appears above. For example, if **CL** is typed for chlorine instead of **Cl**, an error message will appear. In this case, the prompt will also be repeated, giving you a chance to correct the error.

Generating an "Input Data File"

As data to solve a particular problem are entered from the keyboard, that information may be stored in an "input data file" for future use. This file may be edited and used at some later date to run HWI to solve another problem, except, this time, without the necessity of reentering the data from the keyboard. This usage of the input data file is explained later in this section under the heading "Using an Input Data File for Perturbation Studies."

The first screen to appear after the title screen describes three input options and displays a prompt requesting a choice of option.

```
                 METHOD FOR INPUTTING DATA

The following options are available for data input:

  1. Data input is to be accomplished at the terminal using
     a question and answer format. The data are to be stored
     in an "input data file." (You will be asked to name this
     file.)
  2. A "data input file" that was generated during a previous
     run of HWI is to be used for data input. You will be asked
     for the name of that file.
  3. This option is used to create an "input data file" for
     future use. No calculations will be performed.

Input the number corresponding to the option to be used: ...
```

If you choose either option 1 or 3, the next screen displays a prompt asking you to type in the name (and path, if necessary) to be used for the input data file. This screen is shown below.

```
Input File:

As you input data by answering questions that appear on the
screen, an "input data file" will be generated for future use.
The name (and path, if appropriate) of that file is to be chosen
by you.

*** WARNING *** If you type in the name (and path) of an input
data file that already exists in the designated directory or
subdirectory, that input data file will be lost.

Input the name (and path, if appropriate) of the input data
file: ..........................................
```

If the file is in the same directory as the HWI package, there is no need to specify a path. However, if the file is in some other directory, the path must be typed along with the file name, so that HWI will know where to find the file. For example, if the input data file is to be named **INPUT-A.DAT** and to be in a subdirectory of the **HWI** subdirectory named **HWI-DATA**, the following would be typed in answer to the prompt:

C:\.......\HWI\HWI-DATA\INPUT-A.DAT

If you choose option 2, i.e., you wish to use a previously generated input data file, the screen shown below appears. At the prompt, the name of the input data file containing the data you wish to process is requested. If the file does not exist or if it is not in the indicated subdirectory, an error message appears and the program is terminated.

```
Input File:

The data for the calculations will be read from an input data
file created by a previous run of HWI.

Input the name (and path, if appropriate) of the input data
file: ...........................................
```

Once the input data file has been identified, no further input is required. HWI processes the information in the input data file and generates the results.

Generating an Output Data File

The HWI output consists mainly of tables of calculated results that appear on the monitor screen and also, if desired, in an "output data file." The next screen to appear is shown below.

```
                 METHOD FOR OUTPUTTING DATA

The following options are available for data output:

  1. The results are to be displayed on the monitor only.
  2. The results are to be displayed on the monitor and
     stored in an output file. You will be asked to name this
     output file.

Input the number corresponding to the option to be used: ...
```

If option 2 is chosen, the output file must be assigned a name (and path, if necessary). This process is similar to the naming of the input data file and the screen that appears is almost identical to one shown earlier.

Output File:

The results of the calculations will be stored in an output file, the name of which will be chosen by you.

*** WARNING *** If you type in the name (and path) of an output data file that already exists in the designated directory or subdirectory, that output data file will be lost.

Input the name (and path, if appropriate) of the output file:

Inputting Data to HWI from the Terminal

The required data to be inputted by the user during the running of HWI (assuming that either option 1 or option 3 for data input was chosen) include: calculational mode, waste composition, waste flow rate, air requirements, waste heating value, heat loss from the incinerator, and some design information. Each of these is described in detail below.

1. *Calculational Mode* The three key parameters in the HWI calculations are the incinerator operating temperature, the waste–fuel heating value, and the amount of excess air. The waste–fuel heating value is determined by the materials being combusted. Either one of the remaining two parameters may be treated as the independent variable. The HWI program provides two options: (1) the amount of excess air to be employed may be determined by the user, and the resulting incinerator operating temperature calculated by HWI; or (2) the desired operating temperature may be inputted by the user and the required amount of excess air calculated by the program. These two modes are described in the next frame (Main Menu frame) to appear on the monitor screen, as shown below.

MAIN MENU

The following options are available in HWI:

1. The waste (or waste-fuel mixture) composition and flow rate, plus the percent of excess air are to be specified. The incinerator operating temperature is to be calculated by the program.
2. The waste (or waste-fuel mixture) composition and flow rate, plus the incinerator operating temperature are to be specified. The percent of excess air required to achieve this temperature is to be calculated by the program.

Input the number corresponding to the option to be used: ...

If any character is entered other than a "1" or "2", or if the **RETURN** or **ENTER** key is depressed without typing anything (this is equivalent to entering zero), HWI responds with an error message and repeats the prompt.

2. *Waste Composition* In order to accommodate various methods of characterizing the chemical makeup of the waste, HWI provides the user with the following options:

1. Elemental (or ultimate analysis), atom basis
2. Elemental analysis, mass basis
3. Componential analysis, mole basis
4. Componential analysis, mass basis

In a *componential* analysis, the composition of the waste (or, when auxiliary fuel has been added, the waste–fuel mixture) is described in terms of what chemical compounds are present and the fraction or percent of each. In an elemental analysis, the composition is given in terms of what elements are present and the fraction or percent of each. Whichever of these four options is employed by the user during the input phase of the program, the composition is converted to an elemental analysis on a mole fraction basis before the stoichiometric calculations are begun.

The input menu for the composition of the waste (or waste–fuel mixture) next appears on the screen.

```
                WASTE-FUEL COMPOSITION: INPUT MENU

The composition of the waste-fuel mixture may be inputted in
any one of the following four ways:

  1. elemental composition, atom basis *
  2. elemental composition, mass basis
  3. componential composition, mole basis *
  4. componential composition, mass basis

*An atom or mole basis may NOT be used when ash is included in
the waste.
Type the number corresponding to the method to be used for
inputting the waste-fuel composition. ...................
```

Regarding the asterisked note near the bottom of the frame, the word "ash" refers to material in the waste that is presumed to be inert and has negligible effect on the incinerator performance. (This latter assumption may not necessarily be true if the waste contains a large amount of ash. An amount less than 5 mass percent of ash is considered reasonable in light of this assumption.) Since the exact composition of the ash is generally unknown, an average molecular weight for it cannot be calculated. When the waste contains ash, therefore, it is not possible to express the waste composition on either an atom (for the elemental composition) or mole (for the componential composition) basis; a mass basis must be used.

The next set of prompts to appear will vary depending on which of the four composition input options is chosen. If one of the elemental analysis options is chosen (option 1 or 2), the next prompt requests the input of the number of elements in the waste. For each element, the user is then requested to input the element's chemical symbol and its fractional or percentage contribution to the waste (atom percent or fraction for option 1, mass percent or fraction for option 2).

In the following example, the composition of a waste (chlorobenzene) consisting of 64.0% carbon (by mass), 4.4% hydrogen, and 31.6% chlorine is inputted using option 2.

Elemental Input (mass basis)

Note: This program will accept the following element symbols: C, H, O, S, N, Cl, F, Br and I. Water and ash may be treated as "elements", if desired, by using the symbols: WA and AS, respectively. AS will not be accepted when the composition is expressed on a mole or atom basis. Refer to the User's Guide for further information.

Input the number of elements in the waste-fuel mixture: 3

For Element 1

Input chemical symbol: C

Input mass percent or fraction: 64

For Element 2

Input chemical symbol: .... H

Input mass percent or fraction: 4.4

For Element 3

Input chemical symbol: Cl

Input mass percent or fraction: 31.6

Although percentages were used to input the waste composition in the example above, fractional or actual amounts may also be used in the HWI program. (In fact, any set of numbers that are in the proper ratio will work. For example, instead of 64, 4.4, and 31.6, the numbers 0.640, 0.044, and 0.316 would have the same effect; so would the numbers 32, 2.2 and 15.8.)

If the user attempts to input either a zero for any of the numerical inputs above or an element symbol unrecognizable to HWI for any of the alphabetical inputs, the input will be ignored and the prompt repeated. This will also happen if the user tries to input the element symbol "AS" when an atom percent or fraction (composition input option 1) is being used. Again, because ash has no "calculable" atomic weight, its atomic contribution to the waste mixture is indeterminable.

In most elemental analyses, the fraction or percent of any water present in the waste is given as a separate contribution, apart from any oxygen or hydrogen in the waste. In this case, the oxygen and hydrogen in the water cannot be included as part of the oxygen and hydrogen contributions to the waste. (Otherwise, those contributions would be counted twice.) To input the water fraction or percent, the symbol "WA" is used in HWI. If option 1 (atom basis) is being used and water is inputted in this manner, the fact that "WA" consists of three atoms must be taken into account when inputting the atom percent of each component. For example, if a waste consists of equal molar amounts of benzene (C_6H_6) and water, the elemental makeup of the waste may be inputted in either of the following ways:

Element	N	Atom %	Element	N	Atom %
C	6	40.00	C	6	40.00
H	8	53.33	H	6	40.00
O	1	6.67	WA	3	20.00
Total	15	100.00	Total	15	100.00

where N = number of atoms in the waste per molecule of benzene.

It should be noted here that if Dulong's approximate method is to be used to estimate the net heating value of the waste, any water present should be entered as WA rather than included in the hydrogen and oxygen in the rest of the waste. This is explained later when the inputting of the net heating value is discussed.

If one of the componential analysis options is chosen (option 3 or 4), a prompt appears for the number of compounds. For each compound, the user is requested to input the percent or fraction of that compound (mole or mass percent for option 3 or 4, respectively) and the number of elements in that compound. For each element in the compound, the user is asked for the chemical symbol and the number of atoms per molecule. Thus, in effect, the user types in the compound's chemical formula. In the next example, the composition of a chlorobenzene–water mixture (80/20 by mass) is inputted using option 4.

Componential Input (mass basis)

Note: This program will accept the following element symbols: C, H, O, S, N, Cl, F, Br, and I. Water or ash may be inputted by using the symbol WA or AS, respectively. In either case, the water or ash is treated as a single component (compound) and is considered to consist of one "element" (viz., WA or AS). AS will not be accepted when the composition is expressed on a mole basis. Refer to the User's Guide for further information.

Input the number of components (different compounds) in the waste-fuel mixture: 2

Component Number 1

Input the mass percent or fraction of Component 1: ... 80

Input the number of elements in Component 1: 3

For Element 1 of component 1:

Input chemical symbol: C

Input number of atoms: 6

For Element 2 of component 1:

Input chemical symbol: .... H

Input number of atoms: 5

For Element 3 of component 1:

Input chemical symbol: Cl

Input number of atoms: 1

Component Number 2:

Input the mass percent or fraction of Component 2: ... 20

Input the number of elements in Component 2: 2

For Element 1 of component 2:

Input chemical symbol: .... H

Input number of atoms: 2

For Element 2 of component 2:

Input chemical symbol: .... O

Input number of atoms: 1

If the waste contains only one component, the mass (or mole) percent is obviously 100, and the prompt for mass (or mole) percent or fraction does not appear.

As was the case with the elemental composition input options, a zero for any numerical input above or an element symbol unrecognizable to HWI will not be accepted by the program. Likewise, the symbol "AS" will not be accepted when a mole percent or fraction (composition input option 3) is being used. If either WA or AS is entered at a "chemical symbol" input prompt, the prompt for the number of atoms does not appear. (As noted earlier for the elemental composition input, if the net heating value is to be estimated by Dulong's method, any water present should be inputted as WA instead of as H_2O. The reason for this is given later.)

3. *Waste Flow Rate* The waste (or waste–fuel mixture) flow rate is the next input requested. With composition option 3 or 4, the user has the choice of lb/h or lbmol/h as shown in the following frame:

```
Waste-Fuel Flow Rate

The flow rate of the waste-fuel mixture may be inputted either
as lb/h or lbmol/h. Type:

"1"   if lb/h is to be used.
"2"   if lbmol/h is to be used.
```

For composition, option 1 or 2, lb/h must be used. The next prompt calls for the flow rate; in the example below, the mass flow rate in lb/h is to be inputted.

```
Input the mass flow rate of the waste-fuel mixture in
lb/h: ........................................
```

At this point in the program, a table (or two tables) showing the waste composition appears on the screen. If one of the elemental composition options (option 1 or 2) was used, a single table for the elemental makeup of the waste appears. If one of the componential composition options (option 3 or 4) was used, a table for the componential makeup appears first, followed by the elemental table.

4. *Air Requirements* If calculational mode 1 is being used (i.e., the user inputs the amount of excess air and HWI is to calculate the operating temper-

ature), the percent of excess air is inputted at this point. Since the amount of excess air chosen by the user may depend on the amount of stoichiometric or theoretical air (i.e., the exact amount of air required to convert each element in the waste to its combustion product, assuming complete combustion), this quantity is calculated by HWI and displayed on the screen prior to the prompt for the percent of excess air. This is demonstrated in the example below.

```
Air Requirements

The stoichiometric air requirement is 8.312 lb/lb of waste-
fuel mixture or 41559. lb/h for a waste input flow rate of 5000.
lb/h.

Input the percent of excess air: ........................
```

If calculational mode 2 is being used (i.e., the user inputs the desired operating temperature and HWI is to calculate the required amount of excess air), only the stoichiometric air requirement is given; the prompt for the percent of excess air does not appear.

If the air is preheated (e.g., by means of a heat exchanger), the resulting incinerator temperature will be higher for the same amount of input air. If the air is much cooler than normal ambient temperatures (around 70°F), this will also affect the incinerator temperature. Air temperatures significantly different from 70°F in either direction (hotter or colder) should be inputted here.

```
If the air is preheated, input its temperature in °F. (If not,
simply depress the RETURN or ENTER key.) ................
```

---

5. *Waste Heating Value* The net heating value (NHV) of a material is the amount of heat that is released when the material is completely combusted with oxygen. The NHV of the waste may be either inputted by the user or estimated by Dulong's method. Dulong's method is approximate and is recommended only if heat of combustion data are not known. The method is satisfactory only for wastes that contain carbon, hydrogen, oxygen, sulfur, nitrogen, and/or chlorine. If this method is used with wastes containing fluorine, bromine, and/or iodine, a warning will appear on the screen, but the calculations will continue; in this case, the effect of these three elements on the NHV is ignored. As mentioned earlier, if water is present in the waste and Dulong's method is to be used to estimate the waste's net heating value, the water contribution must be inputted as WA during the composition input step. HWI assumes that any hydrogen present in the waste (except the hydrogen indirectly inputted as part

of the WA) is combustible. This, of course, is not true if some of the hydrogen is already combined with oxygen in the water.

If NHV data are available, and if one of the componential options (option 1 or 2) was used to input the waste composition, the user has a choice of inputting the NHVs of each component or the NHV of the waste as a whole. If one of the elemental options (option 3 or 4) was used, only the overall NHV may be inputted.

```
Net Heating Value

One of the following methods may be used to supply the re-
quired net heating value of the waste-fuel mixture for the
calculations. Type:

"1"   if Dulong's method is to be used to calculate the net
      heating value. (For wastes containing C, H, O, S, N and/
      or Cl.)
"2"   if the net heating value of the total waste-fuel mix-
      ture is to be inputted.
"3"   if the net heating value of the individual components
      of the waste-fuel mixture are to be inputted.
```

If option 1 or 2 was used earlier to input the waste composition, the third choice shown in the above frame would not appear and an input of "3" would not be accepted.

If the net heating value is to be inputted by the user (either "2" or "3" above), the user is given the choice of several different units for the NHV input(s).

```
The net heating value may be inputted using any one of the
following units. Type:

"1"   for Btu/lb
"2"   for cal/g
"3"   for Btu/lbmol
"4"   for cal/gmol
```

In the following example, composition input option 4 was used to input the composition of a waste mixture consisting of three components: chlorobenzene, benzene, and water. The net heating value of each component is to be inputted individually (this is choice "3" for the method of inputting the NHV) and units of cal/gmol (choice "4" for the NHV units) are to be employed.

```
   Note: For purposes of this program, the net heating value
is defined as POSITIVE when heat is released during the incin-
eration process and NEGATIVE when heat is absorbed. The over-
all net heating value of the waste-fuel mixture must be pos-
itive, although the net heating value of a particular
component may be negative.

   Input the NHV of Component 1 in cal/gmol: .......  714361
   Input the NHV of Component 2 in cal/gmol: .......  749426
   Input the NHV of Component 3 in cal/gmol: .......  −10519
```

The "heat of combustion" or NHV for water (inputted as −10519) is actually the heat of vaporization. In the incinerator, the water simply vaporizes. The negative sign reflects the fact that this process absorbs, rather than releases, heat.

If the net heating values of the individual components are being inputted, as in the above example, and one of the components is either AS (ash) or WA (water), an input prompt for the NHV of the ash or water does not appear and HWI supplies the NHV (zero in the case of ash). In the following example, component 2 is water (inputted as WA) and component 4 is ash.

```
Input the NHV of Component 1 in cal/gmol: .........  714361
The NHV of the water (Component 2) is −10519 cal/gmol.
Input the NHV of Component 3 in cal/gmol: .........  749426
The NHV of the ash (Component 4) is zero.
```

6. *Heat Loss* A portion of the heat generated by the combustion process is always lost through the walls of the incinerator. In practice, this effect is often neglected. If the heat loss is to be taken into account, however, HWI provides for it as shown in the next prompt.

```
Input the incinerator heat loss as a percent of the net heat-
ing value of the waste-fuel mixture. (If no heat loss is to
be assumed, simply depress the RETURN or ENTER key.)  ......
```

7. *Preliminary Design Data* HWI can also be used for some preliminary incinerator design such as sizing, etc. In order to perform these design calculations, either the flue gas superficial velocity or the incinerator length-to-diameter ratio (L/D) must be specified in advance. If neither is known, inputting a "1" in response to the next prompt will cause a value of 3.0 to be assumed for the L/D ratio when HWI is being run.

Preliminary Design Data

For preliminary design purposes, either the superficial throughput velocity of the flue gas or the length-to-diameter ratio of the incinerator must be determined beforehand. (If neither is known, a length-to-diameter ratio of 3.0 may be assumed.) Type:

"1" if a length-to-diameter ratio of 3.0 is to be assumed.
"2" if the flue gas superficial velocity is to be inputted.
"3" if the length-to-diameter ratio is to be inputted.

If "2" is entered, a prompt requesting the superficial velocity appears next.

Input the superficial velocity, ft/sec: ....

If "3" is entered, the following prompt appears.

Input the length-to-diameter ratio of the incinerator: ...

Another parameter required for the design calculations is the heat release rate. (The NHV is a measure of how much heat is released by the waste; the heat release rate, which is a function of the type of incinerator and, to a lesser extent, of the nature of the waste, is a measure of how fast that heat will be released.) In actual practice, the heat release rate is often unknown and a typical (maximum) value of 25,000 Btu/h-ft^3 is often assumed. The next prompt requests a value for the heat release rate.

Input the heat release rate of the waste-fuel mixture in Btu/h-cu ft. (If the RETURN or ENTER key is depressed without entering a number, a heat release rate of 25,000 Btu/h-cu ft will be assumed.) ....

---

8. *Operating Temperature* If calculation mode 1 (calculation of the operating temperature) is being used, the input phase of HWI is now finished. If calculation mode 2 (calculation of the percent of excess air) is being employed, the desired operating temperature is next inputted.

Assuming complete combustion, when zero excess air (stoichiometric air) is used, the highest operating temperature that can theoretically be reached in the incinerator results. HWI first calculates this temperature and warns the user not to request a temperature above this maximum. In the example below, the maximum temperature (using no excess air) was calculated to be 3870°F.

```
Operating Temperature

  The maximum operating temperature (referred to as the
"theoretical adiabatic flame temperature") that can be
achieved under these conditions (i.e., the waste composi-
tion, preheated air temperature, percent heat loss that have
been inputted) is 3870°F. Do not input a temperature in excess
of this maximum.

Input the incinerator operating temperature, °F: .........
```

In this case, if the user enters a number greater than 3870, HWI will ignore the input and repeat the prompt.

Output from HWI

The HWI output consists mainly of tables of calculated results that appear on the monitor screen (and in the output data file, if that option was chosen earlier). The first output displays the composition of the waste (or waste–fuel mixture). If composition input option 3 or 4 was used, this information is contained in two tables; the first gives the composition by compound and the second, by element. If composition input option 1 or 2 was used, only the elemental composition table is generated. These tables (or this table) appear on the monitor screen right after the input prompt for the waste flow rate. The componential table includes the mass and mole percent of each component and the elemental table includes the mass and atom percent of each element. If WA (water) was entered as an element (composition input option 1 or 2), its contribution is included with oxygen and hydrogen in the elemental table. If ash is present, its contribution is included in the mass percent columns, but neglected in the atom or mole percent columns.

The remaining three output tables do not appear until after the last input. The first of these tables gives the composition of the flue gas leaving the incinerator (mass percents, mole percents, etc.). The second gives more information on the flue gas including flow rates, temperature, excess air, SO_3 and Cl_2 concentrations in ppm (parts per million by volume), and HCl mass flow rate. If ash is present, the particulate loading in gr/acf (grains per actual cubic foot) and gr/dscf (grains per dry standard cubic foot) is also given. The latter is corrected to 50% excess air, that is, the tabulated value is the particulate loading in gr/dscf that would be carried by the flue gas if 50% excess had been used.

An example of a typical flue gas data table is shown below. These results were generated from Problem 2 listed at the end of Chapter 3.

FLUE GAS DATA

Mass flow rate, lb/h	79191.
Volumetric flow rate, acfm	98255.
Volumetric flow rate, scfm (60°F)	16921.
Volumetric flow rate, dscfm (60°F)	16080.
Molar flow rate, lbmol/h	2674.
Temperature, °F	2559.
Percent excess air	75.0
Chlorine concentration, ppm	5.097E+01
Hydrogen chloride concentration, ppm	2.111E+04
HCl mass flow rate, lb/h	2057.8

The final output table to appear contains preliminary design information and is shown below. Problem 2 was also used to generate these results.

PRELIMINARY INCINERATOR DESIGN DATA

Heat generation rate, Btu/h	5.493E+07
Volume, cu ft	2197.
Diameter, ft	8.9
Length, ft	35.5
L/D ratio	4.00
Superficial velocity, ft/sec	26.5
Residence time, sec	1.34
Heat release rate, Btu/h-cu ft	25000.

Obtaining a Hard Copy of the HWI Output

A printed copy of the HWI output may be obtained by printing the output data file (assuming that this option was chosen). If the printing is performed through a word processor, a fixed-width font such as Courier (10 or 12 pitch) or Line

Printer must be used. If a variable-width font is employed, the output tables will be poorly aligned.

Using an Input Data File for Perturbation Studies

As mentioned earlier, data that have been generated by a previous run of HWI may be used as input, obviating the necessity of reentering this information from the keyboard. This feature can be particularly useful when performing perturbation studies. A perturbation study permits the user to determine the effect of changing (or "perturbing") the value of a single variable on incinerator performance. For example, suppose the user is interested in finding out how sensitive the operating temperature is to the amount of excess air being used. To accomplished this, a "base case" is first run, using a typical value for the percent of excess air (100% was used in the example below). The resulting operating temperature is found in the output and recorded. To change the excess air amount, the input data file that was generated while the "base case" was inputted is now edited. A portion of a typical input data file is shown below.

```
            HAZARDOUS WASTE INCINERATOR CALCULATIONS

PROGRAM: HWI, Version 2.0 (revised 9/96)
AUTHORS: Joseph Reynolds, Louis Theodore, Manhattan Col-
  lege, NY
PROJECT SPONSOR: Air Pollution Training Institute
                 U.S. Environmental Protection Agency
                 MD-7
                 RTP, NC 27711
DATE: 1/14/96
TIME: 11:37
INPUT FILE NAME: PR-01.DAT


Calculation Option ...........................        1
  ["1" = calculation of operating temperature]
Waste Composition Input Option ...............        4
  ["4" = componential composition, mass basis]
Number of Components .........................        1
Number of Elements in Component 1 ............        2
Chemical Symbol for Element 1 of Component 1 ....     C
Number of Atoms ..............................        6
Chemical Symbol for Element 2 of Component 1 ....     H
Number of Atoms ..............................        6
```

```
Units for Waste Flow Rate ....................          2
 ["2" = lbmol/hr]
Molar Flow Rate of Waste ..................... 1.00000E+02
Percent Excess Air ........................... 1.00000E+02
Air Temperature  ............................. 7.64000E+01
Method of Inputting Net Heating Value .......           2
 ["2" = overall net heating value]
Units for Net Heating Value ..................          3
 ["3" = Btu/lbmol]
Net Heating Value of Waste ................... 1.34900E+06
Heat Loss (Percent) .......................... 1.00000E+01
Input Method for Preliminary Design .........           2
 ["2" = input superficial velocity]
Superficial Velocity  ........................ 3.50000E+01
Heat Release Rate ............................ 2.50000E+04
```

Since the file is in ASCII format, it can easily be edited by any text editor such as that provided with Microsoft's DOS or by a word processor. When using a word processor, make sure the input data file is imported to and exported from the word processor as an ASCII or text document. The "100" (i.e., "1.0000E+02") value for the percent excess air is now changed to some other value, say, "90.0" or "9.0000E+01" (either one will work). The new input data file is now run, only this time, input option 2 is used and no data need be typed at the keyboard. The new operating temperature is found from the output and recorded. After several such runs, the operating temperature as a function of the percent excess air is known and may be conveniently represented in either graphical or tabular form.

Error and Warning Messages in HWI

Error and warning messages will appear during the running of the HWI program for any number of reasons. An error message usually indicates that some action (such as changing an input) must be taken by the user if the calculations are to continue. Most obvious input errors will simply cause HWI to repeat the prompt; in this case, since the error should be apparent to the user, no error message is displayed. Some examples of possible error messages that could appear on the screen the running of HWI are:

1. If the user enters either an invalid chemical symbol (i.e., a symbol unrecognizable to HWI), enters the same element symbol twice when using composition input option 1 or 2, or enters the same element symbol twice in the same compound when using composition input option 3 or 4, the following error message will appear:

***** ERROR. INVALID CHEMICAL SYMBOL OR ELEMENT ALREADY INPUTTED. *****

2. If the data inputted by the user result in either an operating temperature or a theoretical adiabatic flame temperature (operating temperature using zero excess air) that is unrealistically high or low, the message below, followed by a suggested remedy, will appear on the screen:

***** ERROR DURING CALCULATIONS *****

3. Occasionally, HWI is unable to converge to a solution. Sometimes this is caused by an inputted excess air percent that is too high (calculation mode 1) or an inputted net heating value that is too low. If the message: *run-time error* appears on the screen, the program has been automatically terminated by the system. The user should restart the program and either try a lower value for the excess air or check the NHV information that was inputted earlier.

The following warnings will be given during the execution of HWI if the conditions cited in the warnings apply. These warnings do not cause HWI to terminate and may be ignored by the user.

1. If the waste contains more than 5% ash (by mass), the following warning appears after the waste elemental composition table:

*****WARNING*** This waste has an ash content greater than 5%. This may yield inaccurate results.**

2. If the user indicates that Dulong's method for estimating the net heating value of the waste is to be used with a waste containing fluorine, bromine, or iodine, the following warning is given:

*****WARNING*** Using Dulong's method with a waste containing elements other than C, H, O, S, N, and/or Cl may produce inaccurate results.**

3. If the user chooses Dulong's method to estimate the net heating value of the waste after inputting water as H_2O instead of as "WA", the following warning is issued:

*****WARNING*** Apparently, one of the components in the waste is water. Dulong's method should be used ONLY WHEN the contribution of any water present has been inputted as "WA". See the User's Guide.**

4. If the calculated residence time for the waste is too short, the following warning appears after the last output table:

   *****WARNING*** A residence time less than 0.75 seconds is unusual.**

5. If the calculated superficial velocity of the flue gas is abnormally low or high, the following message is given after the last table:

   *****WARNING*** A gas superficial velocity greater than 50 or lower than 5 ft/s is unusual.**

A.5 THE PROGRAM HWIF

What HWIF Does

HWIF is an expanded version of the HWI program. The additional features are:

1. *Blending with Auxiliary Fuel for Low Btu Wastes* The HWIF program provides a third calculational option in addition to the two options provided by HWI. The third option is used when the waste's net heating value is not high enough to support combustion and an unknown quantity of auxiliary fuel must be added. In this option, the operating temperature and amount of excess air are inputted by the user, as well as information on the fuel, and are used to calculate the required waste-to-fuel ratio.
2. *Rotary Kiln and Afterburner* While the original program assumes a single combustion chamber, HWIF provides separate calculations for the rotary kiln and secondary combustion chamber (afterburner). Each unit can handle a different waste (or waste-fuel) stream. The combustion gas from the rotary kiln is assumed to pass through the secondary where it is combined with the afterburner exhaust. The compositions of both the rotary kiln combustion gas and that leaving the secondary are reported in separate tables in the output file.
3. *Air with an Oxygen Content other than 21%* The new program permits the use of oxygen-enriched or oxygen-starved air. More details on this feature can be found later in this section.
4. *Humidity* The presence of water vapor in the combustion air can have an impact on the operating temperature and quencher calculations. The original program ignores this effect.
5. *Ash* In the original program, any ash present in the waste is assumed (1) to have no effect on the stoichiometric and thermochemical calculations, and (2) to completely "fly" (i.e., all input ash ends up as par-

ticulate matter in the effluent gas stream). In HWIF, the user specifies what percentage of the ash drops out of the incinerator as bottom ash (the remainder "flies" as flyash) and also the heat capacity of the ash. More details on this feature are given in the next section entitled "Data Input to HWIF".

6. *N_2-NO and CO_2-CO equilibria* The HWI calculations accounted for the chemical reaction equilibrium between hydrogen chloride and chlorine, as well as that between sulfur dioxide and sulfur trioxide; the new program also includes nitrogen-nitric oxide and carbon dioxide-carbon monoxide equilibria as well. The amounts of all components of the flue gas, including NO, CO, Cl_2 and SO_3—which are usually overlooked in incineration calculations—are reported in tables, both on the screen and in the output data file.
7. *Quencher/Tempering Chamber* A quencher/tempering chamber module has been included in the new program. The user may opt to (a) quench the effluent gas completely to its adiabatic saturation temperature, (b) partially quench the gas to a user-specified temperature above the adiabatic saturation temperature, or (c) partially quench the gas using a user-specified water flow rate.
8. *CO-CO_2 Combustion Efficiency* Based on the results of the carbon monoxide-carbon dioxide equilibrium calculations, combustion efficiencies are calculated for both the rotary kiln and the afterburner. This combustion efficiency, E_c, is defined by

$$E_c = \frac{y_{CO_2}}{y_{CO_2} + y_{Co}} (100\%)$$

where y_{CO_2} and y_{CO} are the mole fractions of carbon dioxide and carbon monoxide, respectively. The reader should note that this calculation is based solely on chemical reaction *equilibrium* principles.

9. *Destruction and Removal Efficiency (DRE)* The user may choose to specify component DREs. Based on DREs supplied by the user for each waste component, the mass flow rate of each uncombusted waste component in the effluent gas stream is calculated. (If the waste composition is described in terms of an elemental analysis, the DRE is applied to the entire waste.)
10. *Correction to 7% Oxygen* If ash is present in the waste, the original program reports the composition of particulates in the gas stream in scfm-corrected to 50% excess air or 7% oxygen. The HWIF program also reports the compositions of nitric acid, carbon monoxide, chlorine and sulfur trioxide in ppm—corrected to 7% oxygen.

How to Run HWIF

It is assumed that the reader is already familiar with the HWI program. To run HWIF, access either the subdirectory of the hard drive into which HWIF.EXE and HWIF.DATA were copied, or the floppy disk. This can be done by using the "My Computer" icon, Windows Explorer, or the Windows File Manager (Winfile.exe). Double-clicking on the program icon or name will start the program. The required information to be supplied by the user is described below.

Data Input to HWIF

The data input list includes the following: (1) calculational mode, (2) waste composition, (3) waste flow rate, (4) information on the ash, (5) air requirements, (6) waste heating value, (7) information on the auxiliary fuel, (8) heat loss from the incinerator, (9) design information, and (10) quencher/tempering chamber data. Items (1) through (9) must be inputted separately for both the rotary kiln and the afterburner. Items (2), (3), (6), (8) and (9) are the same, or almost the same, as inputs to HWI, with which the user should already be familiar. The other input items are discussed in some detail below.

1. *Calculational Mode* The three key parameters in the HWIF calculations are the incinerator operating temperature, the waste (or waste-fuel mixture) heating value, and the amount of excess air. Any one of the three may be treated as the independent variable. The HWIF program provides three calculational options, the third of which is used when the waste's net heating value is not high enough to support combustion and an unknown quantity of auxiliary fuel must be added. The three options are: (1) for a given waste, or waste-fuel mixture, the amount of excess air to be employed is specified by the user, and the resulting incinerator operating temperature calculated by HWIF; (2) for a given waste, or waste-fuel mixture, the desired operating temperature is inputted by the user, and the amount of excess air required to achieve that temperature is calculated by the program; and (3) for a low heating value waste to which an unspecified amount of auxiliary fuel must be added, the operating temperature and amount of excess air are inputted by the user and the required waste-to-fuel ratio is calculated. In the third option, the particular fuel to be used must also be inputted. This input is discussed later (see subsection 7 entitled "Information on the auxiliary fuel").
2. *Waste Composition* See Section A.4 on HWI. The list of elements accepted by HWIF also includes fluorine, bromine and iodine.
3. *Waste Flow Rate* See Section A.4 on HWI.
4. *Information on the Ash* The presence of ash in the waste has no direct effect on the stoichiometric calculations since the ash is assumed to be chemically inert. However, ash does have the capability of absorbing

heat and should be accounted for in the energy balance used to determine the temperature. The fact that, in practice, some of the ash (fly ash) ends up in the flue gas stream, and the remainder (bottom ash) drops out of the incinerator, is also included in the calculations. In this section, three inputs are requested: (a) the percent of the total ash content that "flies" with the gas stream, (b) the heat capacity of the fly ash, and (c) the difference in temperature between the flue gas exiting the incinerator and the bottom ash left behind. The default value for (b), the heat capacity of the ash, is 0.2 Btu/lb-°F.

5. *Air Requirements* Inputs in this section include the percent excess air, temperature of preheated (or cooler-than-normal) air, oxygen content and humidity. If either calculational mode 1 or 3 is being used, the amount of excess air is inputted at this point. Since the amount of excess air may depend on the amount of stoichiometric or theoretical air (i.e., the exact amount of air required to convert each element in the waste to its combustion product, assuming complete combustion), this quantity is calculated and displayed on the screen prior to the prompt for the percent of excess air.

   If the air is preheated (for example, by means of a heat exchanger), the resulting incinerator temperature will be higher for the same amount of input air. The next prompt asks for the temperature of the preheated air. The volume or mole percent of oxygen in the air, if oxygen-enriched or oxygen-starved air is being used, is requested next. The default value is 21% oxygen. The humidity of the air, which has some impact on both the temperature and quencher calculations, is the next input. The units requested are the usual ones for absolute humidity, lb water per lb of BDA (bone dry air).
6. *Waste Heating Value* See Section A.4 on HWI.
7. *Information on the Auxiliary Fuel* This input section is employed only when calculational mode 3 (i.e., the required waste-to-fuel ratio is to be determined by the program) is chosen. The auxiliary fuel to be used must be chosen by the user. The choices provided by HWIF are: methane, propane, No. 2 fuel oil, or a "user-defined fuel." If a user-defined fuel is employed, the elemental analysis (see 2, "Waste composition") of the fuel using either mass or mole percentages must also be inputted.
8. *Heat Loss* See Section A.4 on HWI.
9. *Design Information* See Section A.4 on HWI.
10. *Quencher/Tempering Chamber Data* For the quencher/tempering chamber calculations, the following options are provided: (a) the gas is completely quenched to saturation; HWIF calculates the adiabatic saturation temperature and the required minimum water flow rate; (b) the gas is partially quenched to a user-specified final temperature and HWIF provides the required water flow rate; (c) the gas is partially quenched

using a user-specified water flow rate and the program determines the final gas temperature, and (d) these calculations are omitted. For the first three options, the quench water temperature must be inputted; for option (b), the desired final gas temperature, and for option (a), the quench water flow rate must also be supplied by the user.

APPENDIX B
METRIC AND SI SYSTEMS

B.1 METRIC SYSTEM

The need for a single worldwide coordinated measurement system was recognized over 300 years ago. Gabriel Mouton, Vicar of St. Paul in Lyons, proposed in 1670 a comprehensive decimal measurement system based on the length of 1 min of arc of a great circle of Earth. In 1671, Jean Picard, a French astronomer, proposed the length of a pendulum beating seconds as the unit of length. (Such a pendulum would have been fairly easily reproducible, thus facilitating the widespread distribution of uniform standards.) Other proposals were made, but over a century elapsed before any action was taken.

In 1790, in the midst of the French Revolution, the National Assembly of France requested the French Academy of Sciences to "deduce an invariable standard for all the measures and all the weights." The commission appointed by the academy created a system that was, at once, simple and scientific. The unit of length was to be a portion of Earth's circumference. Measures for capacity (volume) and mass (weight) were to be derived from the unit of length, thus relating the basic units of the system to each other and to nature. Furthermore, the larger and smaller versions of each unit were to be created by multiplying or dividing the basic units by 10 and its multiples. This feature provided a great convenience to users of the system by eliminating the need for such calculations as dividing by 16 (to convert ounces to pounds) or by 12 (to convert inches to feet). Similar calculations in the metric system could be performed simply by shifting the decimal point. Thus, the metric system is a "base 10" or "decimal" system.

The commission assigned the name "metre" (which we now spell *meter*) to the unit of length. This name was derived from the Greek word *metron*, meaning "a measure." The physical standard representing the meter was to be constructed so that it would equal one ten-millionth of the distance from the North Pole to the equator along Earth's meridian running near Dunkirk in France and Barcelona in Spain.

The metric unit of mass, called the "gram," was defined as the mass of one cubic centimeter (a cube that is 1/100 of a meter on each side) of water at its temperature of maximum density. The cubic decimeter (a cube 1/10 of a meter on each side) was chosen as the unit of fluid capacity. This measure was given the name "liter."

Although the metric system was not accepted with enthusiasm at first, adoption by other nations occurred steadily after France made its use compulsory in 1840. The standardized character and decimal features of the metric system made it well suited to scientific and engineering work. Consequently, it is not surprising that the rapid spread of the system coincided with an age of rapid technological development. In the United States, by act of Congress in 1866, it was made "lawful throughout the United States of America to employ the weights and measures of the metric system in all contracts, dealings, or court proceedings."

By the late 1860s, even better metric standards were needed to keep pace with scientific advances. In 1875, an international treaty, the "Treaty of the Meter," set up well-defined metric standards for length and mass and established permanent machinery to recommend and adopt further refinements in the metric system. This treaty, known as the Metric Convention, was signed by 17 countries, including the United States.

As a result of the treaty, metric standards were constructed and distributed to each nation that ratified the convention. Since 1893, the internationally agreed-to metric standards have served as the fundamental weights and measures standards of the United States.

By 1900 a total of 35 nations, including the major nations of continental Europe and most of South America, had officially accepted the metric system. Today, with the exception of the United States and a few small countries, the entire world is using the metric system predominantly or is committed to such use. In 1971, the secretary of commerce, in transmitting to Congress the results of a 3-year study authorized by the Metric Study Act of 1968, recommended that the United States change to predominant use of the metric system through a coordinated national program.

The International Bureau of Weights and Measures located at Sevres, France, serves as a permanent secretariat for the Metric Convention, coordinating the exchange of information about the use and refinement of the metric system. As measurement science develops more precise and easily reproducible ways of defining the measurement units, the General Conference of Weights and Measures—the diplomatic organization made up of adherents to the convention—meets periodically to ratify improvements in the system and the standards.

B.2 SI SYSTEM

In 1960, the General Conference adopted an extensive revision and simplification of the system. The name *Le Système Internationale d'Unites* (International System of Units), with the international abbreviation SI, was adopted for this modernized metric system. Further improvements in and additions to SI were made by the General Conference in 1964, 1968, and 1971.

The basic units in the SI system are the kilogram (mass), meter (length), second (time), kelvin (temperature), ampere (electric current), candela (the unit of luminous intensity), and radian (angular measure). All are commonly used by the engineer. The Celsius scale of temperature (0°C = 273.15 K) is commonly used with the absolute Kelvin scale. The important derived units are the newton (SI unit of force), the joule (SI unit of energy), the watt (SI unit of power), the pascal (SI unit of pressure), and the hertz (unit of frequency). There are a number of electrical units: coulomb (charge), farad (capacitance), henry (inductance), volt (potential), and weber (magnetic flux). One of the major advantages of the metric system is that larger and smaller units are given in powers of 10. In the SI system a further simplification is introduced by recommending only those units with multipliers of 10^3. Thus, for lengths in engineering, the micrometer (previously micron), millimeter, and kilometer are recommended, and the centimeter is generally avoided. A further simplification is that the decimal point may be replaced by a comma (as in France, Germany, and South Africa), while the other numbers, before and after the comma, will be separated by spaces between groups of three; for example, one million dollars will be "$1 000 000.00." More details are provided in the following section.

Seven Base Units

1. *Length—meter (m).* The meter (common international spelling, *metre*) is defined as 1,650,763.73 wavelengths in vacuum of the orange-red line of the spectrum of krypton-86. The SI unit of area is the square meter (m^2). The SI unit of volume is the cubic meter (m^3). The liter (0.001 cubic meter), although not an SI unit, is commonly used to measure fluid volume.
2. *Mass—kilogram (kg).* The standard for the unit of mass, the kilogram, is a cylinder of platinum–iridium alloy kept by the International Bureau of Weights and Measures at Sevres. A duplicate in the custody of the National Bureau of Standards serves as the mass standard for the United States. This is the only base unit still defined by an artifact. The SI unit of force is the newton (N). One newton is the force which, when applied to a 1-k mass, will give the kilogram mass an acceleration of 1 meter per second per second. $1 \text{ N} = 1 \text{ kg} \cdot \text{m/s}^2$. The SI unit for pressure is the pascal (Pa). $1 \text{ Pa} = 1 \text{ N/m}^2$. The Sl unit for work and energy of any kind is the joule (J). $1 \text{ J} = 1 \text{ N} \cdot \text{m}$. The SI unit for power of any kind is the watt (W). $1 \text{ W} = 1 \text{ J/s}$.
3. *Time—second (s).* The second is defined as the duration of 9,192,631,770 cycles of radiation associated with a specified transition of the cesium-

133 atom. It is realized by tuning an oscillator to the resonance frequency of cesium-133 atoms as they pass through a system of magnets, and a number of periods or cycles per second is called frequency. The SI unit for frequency is the hertz (Hz). One hertz equals one cycle per second (1 Hz = 1 cps). The SI unit for speed is the meter per second (m/s). The SI unit for acceleration is the (meter per second) per second (m/s^2).

4. *Electric current—ampere (A).* The ampere is defined as that current which, if maintained in each of two long parallel wires separated by one meter in free space, would produce a force between the two wires (due to their magnetic fields) of 2×10^{-7} newton for each meter of length. The SI unit of voltage is the volt (V). 1 V = 1 W/A. The SI unit of electrical resistance is the ohm (Ω). 1 Ω = 1 V/A.
5. *Temperature—kelvin (K).* The kelvin is defined as the fraction 1/273.16 of the thermodynamic temperature of the triple point of water. The temperature 0 K is called "absolute zero." On the commonly used Celsius temperature scale, water freezes at about 0°C and boils at about 100°C. The °C is defined as an interval of 1 K, and the Celsius temperature 0°C is defined as 273.15 K; 1.8 Fahrenheit degrees are equal to 1.0°C or 1.0 K; the Fahrenheit scale uses 32°F as a temperature corresponding to 0°C.
6. *Amount of substance—mole (mol).* The mole is the amount of substance of a system that contains as many elementary entities as there are atoms in 0.012 kg of carbon-12. When the mole is used, the elementary entities must be specified and may be atoms, molecules, ions, electrons, other particles, or specified groups of such particles. The SI unit of concentration (of amount of substance) is the mole per cubic meter (mol/m^3).
7. *Luminous intensity—candela (cd).* The candela is defined as the luminous intensity of 1/600,000 of a square meter of a blackbody at the temperature of freezing platinum (2045 K). The SI unit of light flux is the lumen (lm). A source having an intensity of 1 candela in all directions radiates a light flux of 4 lumens.

Two Supplementary Units

1. *Plane angle—radian (rad).* The radian is the plane angle with its vertex at the center of a circle that is subtended by an arc equal in length to the radius.
2. *Solid angle—steradian (sr).* The steradian is the solid angle with its vertex at the center of a sphere that is subtended by an area of the spherical surface equal to that of a square with sides equal in length to the radius.

Multiples and prefixes assigned to SI units are given in Table 4.2.4. Common abbreviations for SI units are given in Table 4.2.3.

INDEX